Foundations of Engineering Mechanics

V.A. Svetlitsky, Statics of Rods

Springer
Berlin
Heidelberg
New York
Barcelona
Hong Kong
London
Milano
Paris
Singapore
Tokyo

V.A. Svetlitsky

Statics of Rods

Translated by E. Evseev and K. Ramodanova

With 191 Figures

Series Editors:
V. I. Babitsky
Department of Mechanical Engineering
Loughborough University
LE11 3TU Loughborough, Leicestershire
Great Britain

J. Wittenburg
Institut für Technische Mechanik
Universität Karlsruhe (TH)
Kaiserstrasse 12
76128 Karlsruhe / Germany

Author:
Valery A. Svetlitsky
Chusovskaya str. 11-7-12
107207 Moscow / Russia

Translators:
Evgeny Evseev, Kseniya Ramodanova
Ap. 34, 50 Skhodnenskaya Street
123363 Moscow / Russia
E-mail: evseev@ami.rospac.ru

ISBN 978-3-642-53646-5 ISBN 978-3-540-45593-6 (eBook)
DOI 10.1007/978-3-540-45593-6

Cataloging-in-Publication Data applied for
Die Deutsche Bibliothek - CIP-Einheitsaufnahme
Svetlitsky, Valerie A.:
Statics of rods / V. A. Svetlitsky. Transl. by E. Evseev and K. Ramodanova. - Berlin ; Heidelberg ; New York ; Barcelona ; Hong Kong ; London ; Milan ; Paris ; Singapore; Tokyo : Springer, 2000
(Foundations of engineering mechanics)

Softcover reprint of the hardcover 1st edition 2000

Typesetting: Camera-ready copy from translators
Cover-Design: de'blik, Berlin
Printed on acid-free paper SPIN: 10728545 62/3020 - 5 4 3 2 1 0

Preface

The volume is devoted to mechanics of rods, which is a branch of mechanics of deformable bodies. The main goal of the book is to present systematically theoretical fundamentals of mechanics of rods as well as numerical methods used for practical purposes.

The monograph is concerned with the most general statements of the problems in mechanics of rods. Various types of external loads that a rod may be subject to are discussed. Advanced technique that includes vector analysis, linear algebra, and distributions is used in the derivation of linear and nonlinear equilibrium equations. The use of this technique helps us to make transformations and rearrangement of equations more transparent and compact.

Theoretical basics of rods interacting with external and internal flows of fluid and the derivation of the formulas for the hydrodynamic and aerodynamic forces are presented.

The book consists of six chapters and appendices and may be conventionally divided into two parts. That is, Chapters 1 to 3 contain, in the main, theoretical material, whereas Chapters 4 to 6 illustrate the application of the theoretical results to problems of practical interest. Problems for self-study are found in Chapters 1, 3, 4, and 5. The solutions to most of the problems are given in Appendix B.

The monograph is addressed to scientists, institutional and industrial researchers, lecturers, and graduate students.

Moscow, March 2000 *Valery Svetlitsky*

Table of Contents

Introduction

Various types of rods are widely used in engineering. Rods are elements of machines, building constructions, and a great number of gauges. Rods find their application as key components of measurement instruments and accumulators of mechanical energy. In electrical engineering, rods are used as sensing elements of accelerometers, frequency detectors, mechanical high-cut filters.

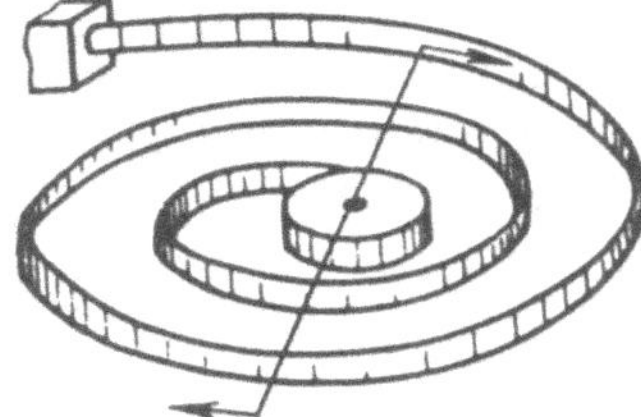

Fig. 0.1.

Most timing devices have rods of complex shape as their key elements, e.g. a spiral balance spring (Fig. 0.1) and tuning forks (Fig. 0.2).

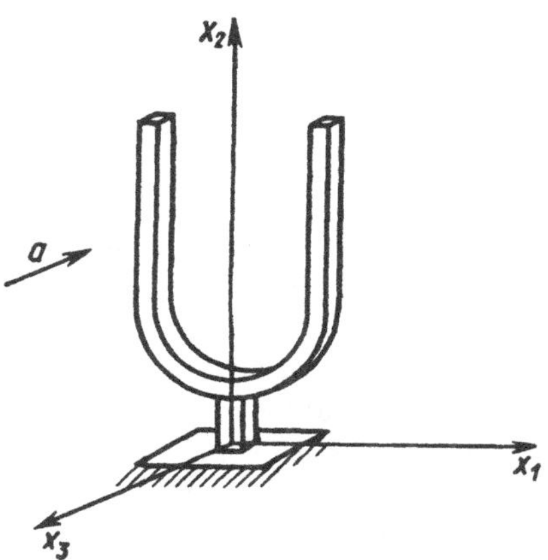

Fig. 0.2.

Timing devices which have rods as parts of their construction are not only timers but also transducers of steady signals. To control efficiently space vehicles, one must be able to define the current time and measure time slices with high accuracy. The more accurate characteristics of the elastic element, the more precise the readings of the timing device.

Fig. 0.3.

The usage conditions of an element are of great importance. In practice, some unwanted forces acting upon a sensitive element of a gauge may appear (e.g. an object, on which the gauge is mounted, may move with acceleration or be under vibration). The acceleration may dramatically change the elastic characteristics of the element and lead to the loss of stability of the element. Therefore, the problem of determination of the values of acceleration at which the loss of stability occurs is of practical interest (e.g. loss of stability of the plane configuration of the spring illustrated in Fig. 0.1).

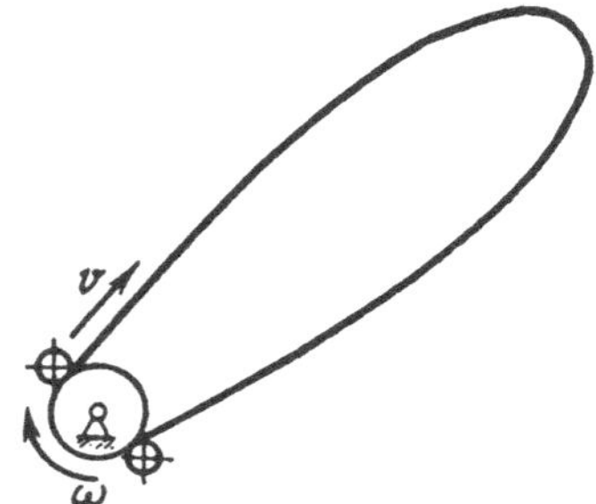

Fig. 0.4.

The loss of stability may take place when a rod is significantly deformed but still preserves its plane shape (e.g. out-of-plane deformation of the spring). To examine this phenomenon, one must use nonlinear equilibrium equations.

The recent advances in engineering have brought forth new problems related to statics and dynamics of rods. Examples are as follows: an analysis of mechanical strength of a flexible conductor that is used to control a moving object (Fig. 0.3), examination of stability of stationary motions of a band-like radiator (Fig. 0.4) or a ballistic aerial, coiling and uncoiling of wires, threads or bands of rolled metal. Nowadays, such a band (Fig. 0.5) can move with

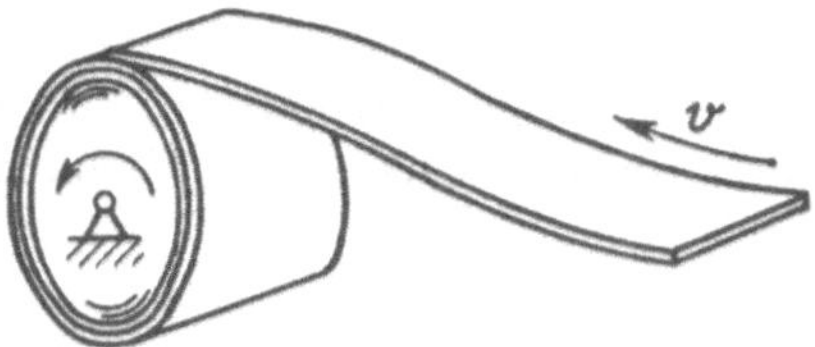

Fig. 0.5.

speed varying from thirty to forty meters per second, hence, dynamic effects cannot be neglected.

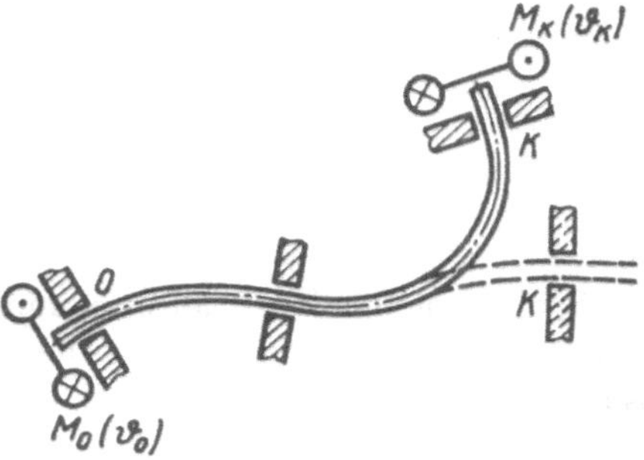

Fig. 0.6.

Space-curved elastic elements (which can be treated as rods) are structural members of most engineering constructions. For example, a flexible shaft can be used to transmit forces and moments (Fig. 0.6). A driving force is applied at a point O. The computer control must be organized in such a way that the cross sections K and O rotate simultaneously by the same angle while the position of K may change substantially (the dashed line in Fig. 0.6 shows a possible location of the cross section K). The change of the shape of the shaft results in the change of the stress state of the shaft, therefore, misalignment of the angles ϑ_O and ϑ_K may appear.

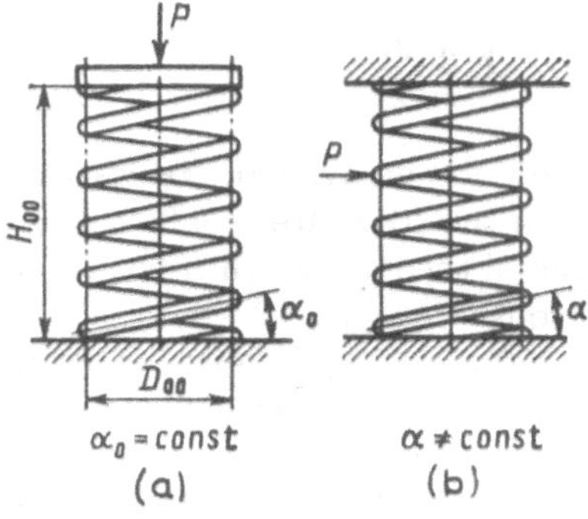

Fig. 0.7.

Some shock-absorption systems are based on springs that can be treated as spatially curvilinear rods (see Figs. 0.7 and 0.8). A cylindrical spring may have a fixed angle of helix α_0 (Fig. 0.7*a*) or a varying angle of helix α (Fig. 0.7*b*).

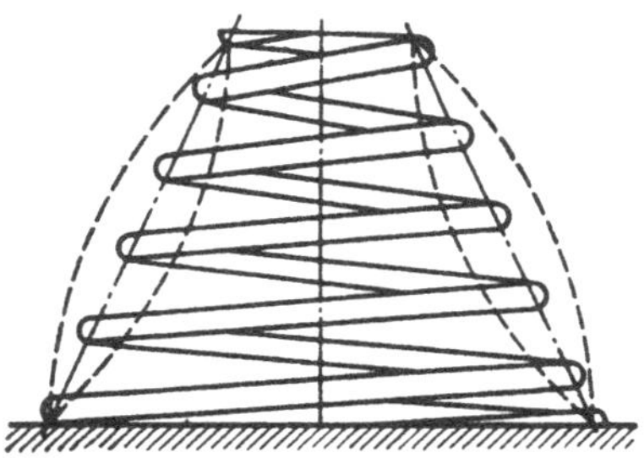

Fig. 0.8.

A conical spring is shown in Fig. 0.8. Springs of this kind can be manufactured by winding rods about various surfaces. Those surfaces are usually surfaces of revolution of negative or positive Gaussian curvature.

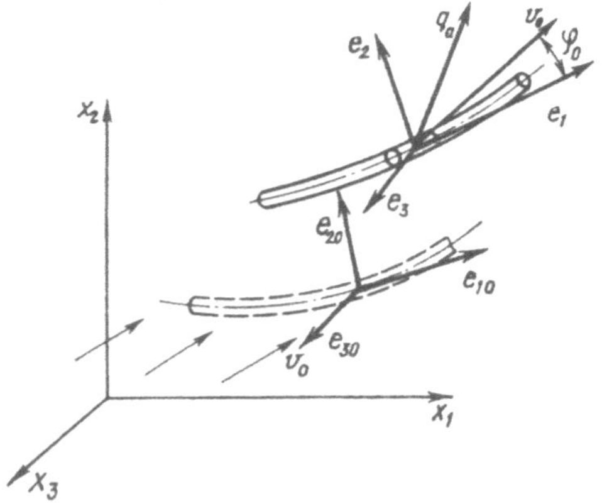

Fig. 0.9.

The theory of space-curved rods is suitable for analysis of statics and static stability of rods when the axis of a rod in the stress-free configuration is a spatial curve (Figs. 0.7 and 0.8). This general theory may turn out to be useful in statics and static stability of plane rods.

Investigation of interaction between rods and air or liquid flows bring up many problems related to statics and dynamics of rods. These problems are of great practical and theoretical interest. It should be noted that difficulty in determination of the interaction forces may arise. This difficulty can be explained by the fact that a rod (a cable or a wire) immersed into a flow can take a shape that differs significantly from its original shape (see dashed lines in Fig. 0.9). An aerodynamic force $\mathbf{q}_\alpha$ depends on the angle φ_α between the tangent to the rod axis (the vector $\mathbf{e}_1$) and the vector of the flow velocity $\mathbf{v}_0$.

Figures 0.10 to 0.18 illustrate interaction of rod-like elements with air and liquid flows.

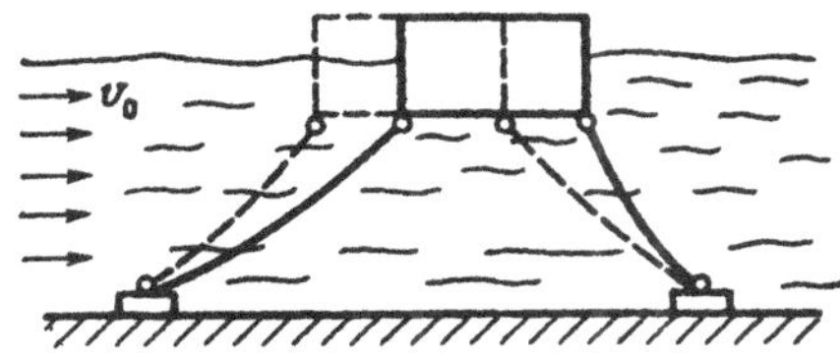

Fig. 0.10.

Figure 0.10 shows an anchor-type system designed to retain floating objects. The anchor cables are not always absolutely flexible rods since their twisting and bending stiffnesses are not small.

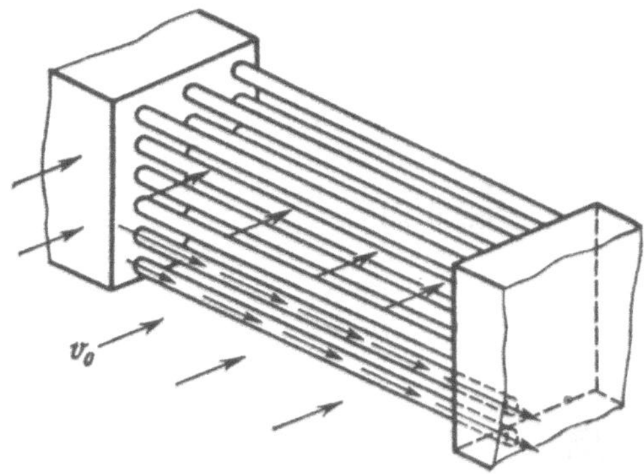

Fig. 0.11.

A cooling system is shown in Fig. 0.11. The thin-walled tubes containing a running liquid are placed into the flow. The aerodynamic forces that are functions of velocity of the external flow $\mathbf{v}_0$ may cause considerable static stresses in the tubes. Besides, Kármán vortices may lead to vibrations of the tubes.

Figure 0.12 illustrates a rod (a model of an airplane wing) in an air flow. For a certain value of the flow velocity, the loss of stability of the wing may occur. (Static loss of stability is called *divergence*, dynamic loss of stability is called *flutter*.)

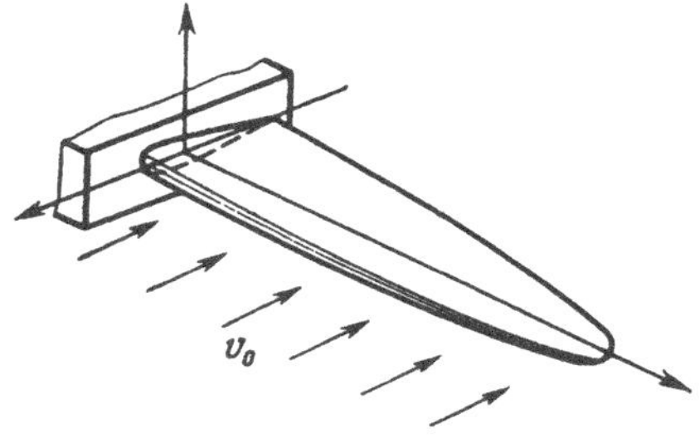

Fig. 0.12.

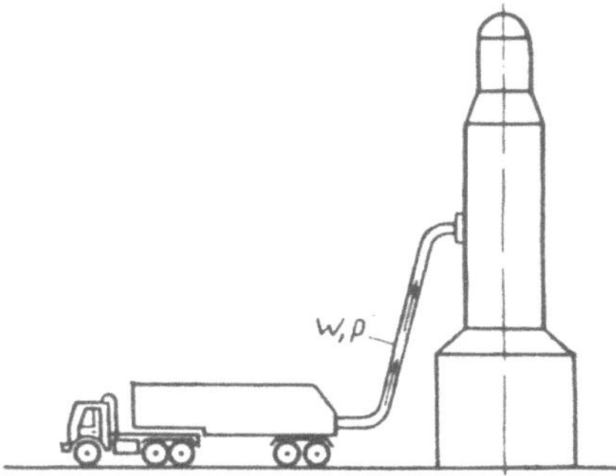

Fig. 0.13.

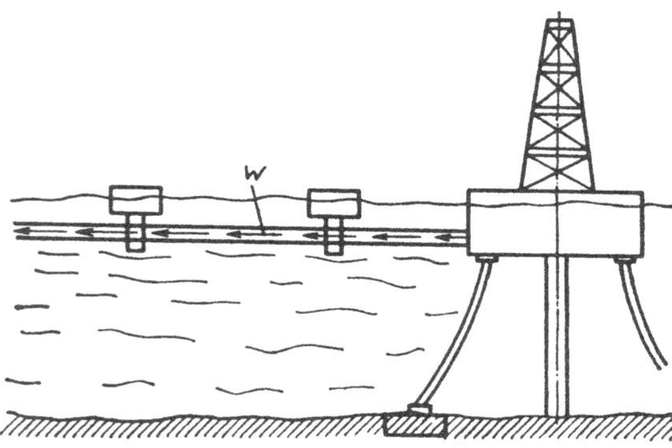

Fig. 0.14.

Tubes containing liquid flows (e.g. pipelines) are widespread structural members. An internal steady flow results in a static load applied to the tube while a nonsteady flow results in a dynamic load. Figures 0.13 to 0.15 illustrate tubes used for liquid or fuel transfer.

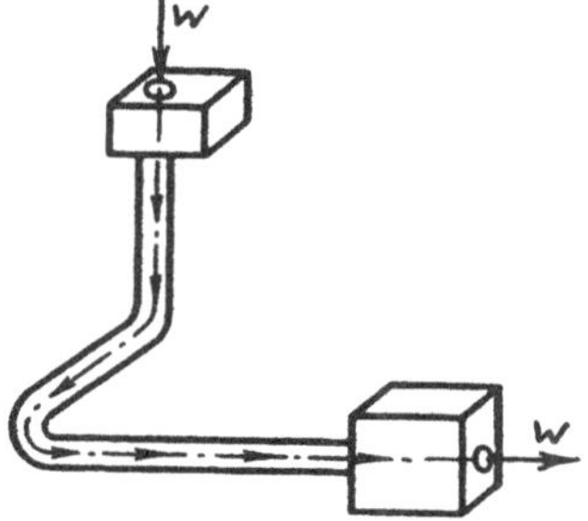

Fig. 0.15.

Cutting tools are cooled with liquid flows of high pressure. Such a flow running inside a drilling bit is shown in Fig. 0.16.

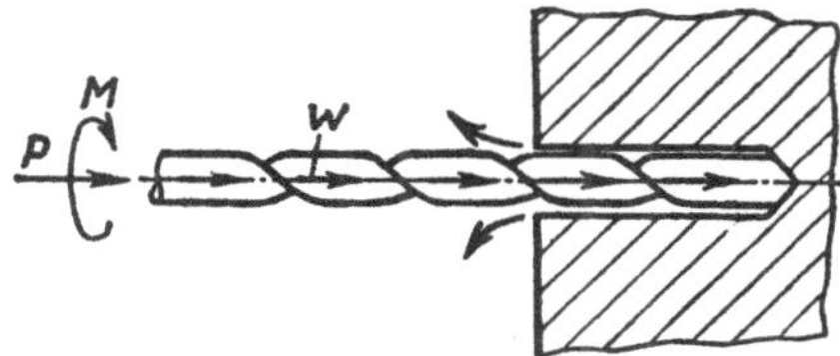

Fig. 0.16.

A paddle (see Fig. 0.17) has a special passage for a flow of a cooling liquid.

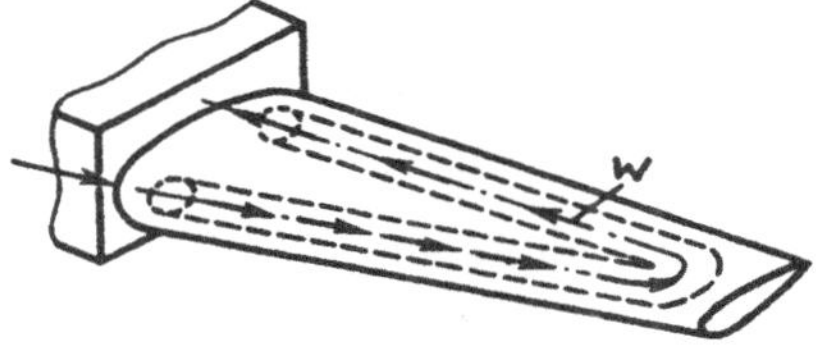

Fig. 0.17.

An airborne fueling is shown in Fig. 0.18. Here, the hosepipe interacts with the external flow of air as well as with the internal flow of fuel.

Straight rods, being a special case of curvilinear rods, hold a central position in statics and dynamics of rods. Various structural elements can be mathematically treated as straight rods (Figs. 0.19– 0.23).

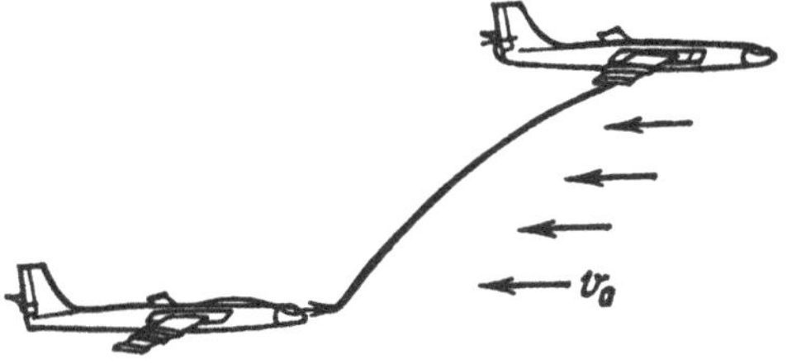

Fig. 0.18.

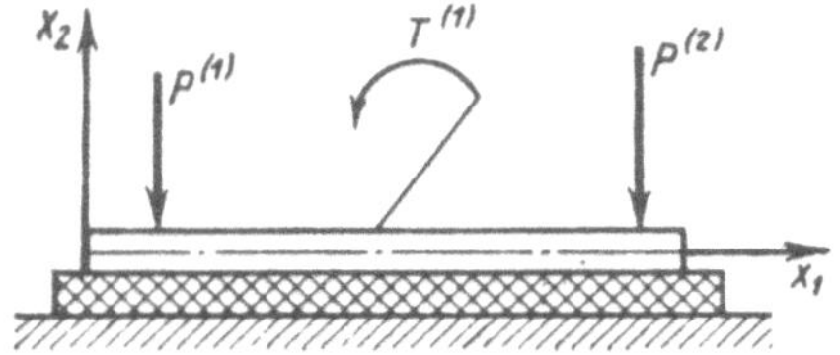

Fig. 0.19.

A rod on an elastic foundation is shown in Fig. 0.19. This foundation is not necessarily soil, it may be an elastic separator used for shock absorption.

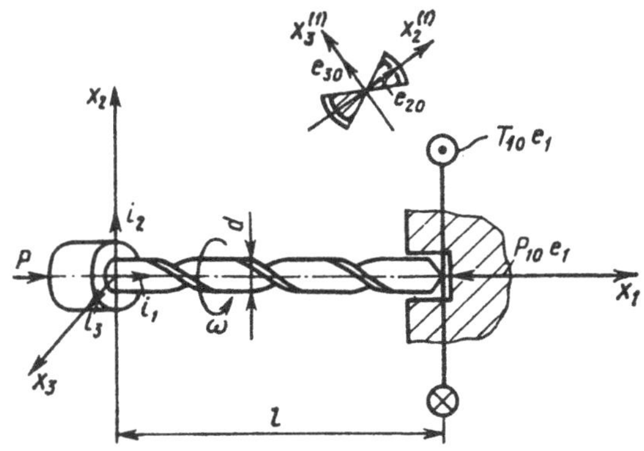

Fig. 0.20.

A drilling bit subjected to a concentrated compressing force **P** and a concentrated torque **T** is shown in Fig. 0.20. The regime of drilling (especially for deep-hole drilling when the ratio l/d is large, e.g. $l/d > 150$) must be so selected as to prevent the loss of stability of the bit. The key feature of this problem is that the location of the principal axes $(x_2^{(1)}, x_3^{(1)})$ of a cross section of the bit with respect to the Cartesian coordinate system (x_2, x_3) depends on the coordinate x_1.

A straight rod immersed into a liquid or air flow is shown in Fig. 0.21. This external flow produces distributed aerodynamic forces $\mathbf{q}_a$ and a distributed

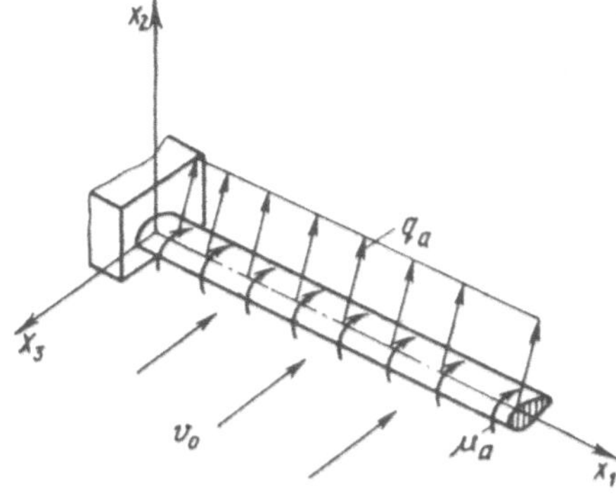

Fig. 0.21.

aerodynamic moment μ_a. Under certain conditions, the loss of stability may occur.

Problems of interaction between rods and external liquid or air flows are discussed in greater detail in Sect. 6.2.

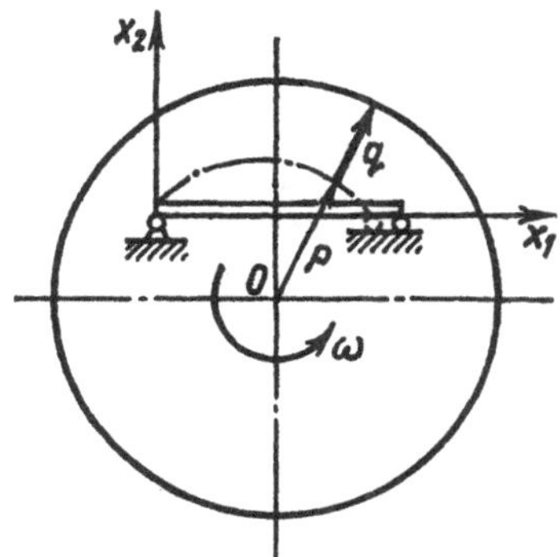

Fig. 0.22.

A straight rod mounted on a rotating disk is shown in Fig. 0.22. A distributed load of magnitude

$$|\mathbf{q}| = m_0 \rho \omega^2$$

is applied to the rod; here m_0 is the mass of the rod per unit length. For the gauge to work properly, deflections of the rod must be small. The difficulty is that for some values of the angular velocity such that $\omega > \omega_*$ (here ω_* is the critical value of the angular velocity) deflections of the rod may be considerably large. The deflections corresponding to $\omega > \omega_*$ can be found from the nonlinear equations governing the equilibrium of the rod.

Although the examples considered above are taken from different fields of engineering, they fall within the domain of mechanics of rods and, consequently, can be examined from a unified point of view.

1. Equilibrium Equations

This chapter is concerned with methods of derivation of equilibrium equations for three-dimensional curvilinear rods. The vector form of the equations used below simplifies rearrangement and makes the representation of the equations more compact. Two coordinate systems are introduced. One of them is an immovable Cartesian frame, the other is a frame rigidly connected with a rod axis. Under certain conditions, the appropriate choice of a coordinate system may help advance further in the analysis of the problem.

Special cases of equations for small and large displacements of points of a rod axis (linear and nonlinear equations, respectively) are examined in detail.

1.1 Vector Equilibrium Equations

1.1.1 Basic Definitions and Hypothesis

A body is called a *rod* if the dimensions of its cross sections are small compared with its length and the radius of curvature of the rod axis. A *rod axis* or an *axial line* of a rod is a line through the geometrical centers of the cross sections. Two general types of axial lines are distinguished. The shape of the axial line of a load-free rod is a characteristic of its natural configuration. The *axial line of a rod* subjected to a load is called an *elastic* axial line. The key feature of flexible rods is that the shape of an elastic axial line may essentially differ from the shape of the axial line in the natural (stress-free) configuration, however, deformations of the rod are assumed to obey Hooke's law.

Problems discussed below are nonlinear in geometrical sense but linear in physical sense. It is a common knowledge that the principles of superposition and invariability of initial sizes are valid for problems linear in both geometrical sense and physical sense. The principle of superposition states that forces applied to a rod act upon the rod independently. The principle of invariability of initial dimensions states that displacements that appear as a result of deformation of a rod are negligibly small and consequently the geometrical parameters of the rod remain practically unchanged during deformation. For geometrically nonlinear problems these principles are no longer valid. This fact complicates the analysis of nonlinear problems compared to the analysis of linear ones.

The equilibrium equations of a rod will be derived below on the basis of the following assumptions:

(1) Plane cross sections of a rod orthogonal to the axial line remain plane after deformation (shear strains are neglected). This assumption is known as the Bernoulli hypothesis.

(2) Dimensions of a cross section are small compared to the rod length and the radius of curvature of the axial line.

(3) The axial line of a rod is inextensible.

(4) Different but statically equivalent local loads result in the same stress state in a rod except the immediate vicinity of the points of application of the loads. This fact is known as Saint-Venant's principle.

As noted above, the magnitude of stress is ruled by Hooke's law. Hence, all further conclusions are valid providing the maximum of magnitude of normal stresses lies within certain limits of proportionality. These limits are predetermined by the elastic properties of material.

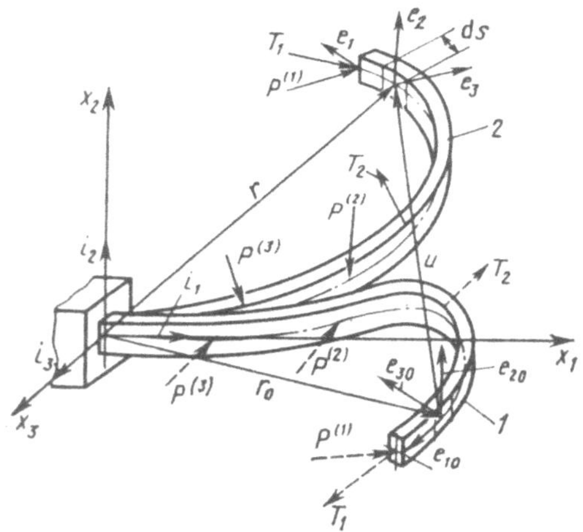

Fig. 1.1.

Let us introduce into consideration two rectangular coordinate systems. One is an immovable Cartesian system with unit axis vectors $\mathbf{i}_j$, the other, whose unit vectors are denoted by $\mathbf{e}_i$, is rigidly attached to the rod axial line (Fig. 1.1). The axial coordinate s is measured from a fixed point; s is the length of an element of the axial line. For simplicity and brevity of notation of equilibrium equations and equations of motion, let us choose the attached coordinate system in the following way. Let the origin of coordinates be at the geometrical center of a cross section. Let the axis corresponding to the unit vector $\mathbf{e}_1$ be directed along the tangent to the rod axial line such that the coordinate s increases in the direction of $\mathbf{e}_1$. Let the other axes be directed along the principal axes of the cross section.

Consider two configurations of a rod shown in Fig. 1.1. Configuration *1* corresponds to the stress-free natural state of the rod; the rod in configuration *2* is deformed. Under the action of slowly increasing forces $\mathbf{P}_i$ and moments $\mathbf{T}_i$, the rod undergoes deformation and passes from configuration *1* to configuration *2*. Figure 1.1 clearly shows us that elastic displacements may be so large that the shape of the rod axis in the natural state differs substantially from its shape in a deformed state. During deformation, external moments and forces may also change their directions. Dashed lines in Fig. 1.1 show the direction of the vectors $\mathbf{P}_i$ and $\mathbf{T}_i$ at the very instant they were applied to the rod.

Fig. 1.2.

To solve a nonlinear problem of statics of flexible rods, it is essential to know the behavior of external loads during deformation. Moreover, changes in boundary conditions, say, a moving hinge (Fig. 1.2), must be also taken into consideration. Forces applied to a rod can be dead or follower. A *dead* force is of the same direction during deformation while a *follower* one keeps its direction with respect to a rod, that is, during deformation it makes the constant angles with the axes of the coordinate frame attached the rod axis. In the general case, there is a number of concentrated (dead or follower) forces and moments; distributed forces and moments could be also applied to a rod.

1.1.2 Vector Equilibrium Equations

A rod element of length $\mathrm{d}s$ is shown in Fig. 1.3. Let $\mathbf{Q}$ be the vector of an internal force such that $\mathbf{Q} = Q_1\mathbf{e}_1 + Q_2\mathbf{e}_2 + Q_3\mathbf{e}_3$, where Q_1 is an axial force, Q_2 and Q_3 are shear forces; let $\mathbf{M}$ be the vector of an internal moment such that $\mathbf{M} = M_1\mathbf{e}_1 + M_2\mathbf{e}_2 + M_3\mathbf{e}_3$, where M_1 is a twisting moment, M_2 and M_3 are bending moments. Let q_1, q_2, and q_3 be the components of the vector of an external load $\mathbf{q}$ in the attached coordinate system; let μ_1, μ_2, and μ_3 be the components of the vector of an external distributed moment $\boldsymbol{\mu}$ in the same coordinate system. The directions of the axes of the natural trihedron defined by the unit vectors $\mathbf{e}_2$ and $\mathbf{e}_3$ coincide with the directions of the principal axes of a cross section.

Since the element is in equilibrium, the sum of the forces and the sum of the moments equal zero. Hence, we have two vector equations

$$\mathrm{d}\mathbf{Q} + \mathbf{q}\,\mathrm{d}s = 0\,; \tag{1.1}$$

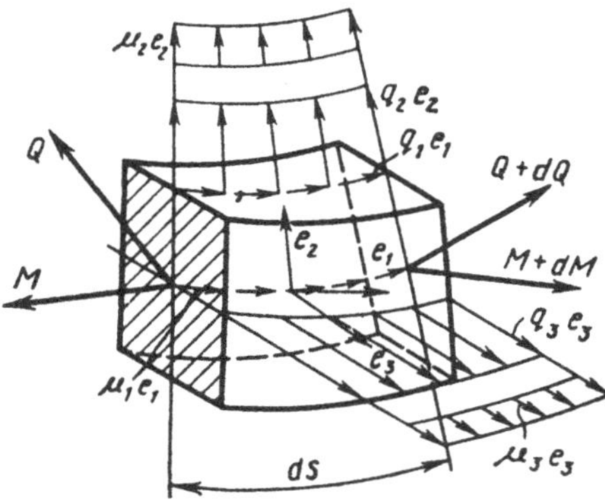

Fig. 1.3.

$$\mathrm{d}\mathbf{M} + \mathbf{e}_1 \times \mathbf{Q}\,\mathrm{d}s + \boldsymbol{\mu}\,\mathrm{d}s = 0 \tag{1.2}$$

or

$$\frac{\mathrm{d}\mathbf{Q}}{\mathrm{d}s} + \mathbf{q} = 0\,; \tag{1.3}$$

$$\frac{\mathrm{d}\mathbf{M}}{\mathrm{d}s} + \mathbf{e}_1 \times \mathbf{Q} + \boldsymbol{\mu} = 0\,. \tag{1.4}$$

These equations are valid in any coordinate system. The vectors $\mathbf{Q}$, $\mathbf{M}$, and $\mathbf{e}_1$ in (1.3) and (1.4) are unknown. The distributed loads $\mathbf{q}$ and $\boldsymbol{\mu}$, the concentrated moments and forces applied to the rod (Fig. 1.3) as well as the boundary conditions (e.g. clamped ends, simply supported ends, etc.) are assumed to be known.

Concentrated forces and moments can be introduced into (1.3) and (1.4) by means of the δ-function (see Appendix 4). Then, we obtain the following equilibrium equations:

$$\frac{\mathrm{d}\mathbf{Q}}{\mathrm{d}s} + \mathbf{q} + \sum_{i=1}^{n} \mathbf{P}^{(i)}\,\delta\,(s - s_i) = 0\,; \tag{1.5}$$

$$\frac{\mathrm{d}\mathbf{M}}{\mathrm{d}s} + \mathbf{e}_1 \times \mathbf{Q} + \boldsymbol{\mu} + \sum_{\nu=1}^{\rho} \mathbf{T}^{(\nu)}\,\delta\,(s - s_\nu) = 0\,; \tag{1.6}$$

here s_i and s_ν are the coordinates of the points of application of the concentrated forces and moments. The distributed loads $\mathbf{q}$ and $\boldsymbol{\mu}$ are applied to a part of the rod. We can write

$$\mathbf{q} = \mathbf{q}(s)\,[\,H(s) - H(s - s_i)\,]\,;$$
$$\boldsymbol{\mu} = \boldsymbol{\mu}(s)\,[\,H(s) - H(s - s_\nu)\,]\,;$$

here H is the Heaviside function (see Appendix 4). The expressions for $\mathbf{q}$ and $\boldsymbol{\mu}$ can be rewritten as follows:

$$\mathbf{q} = \begin{cases} \mathbf{q}(s) \neq 0\,, & 0 \leq s \leq s_i\,; \\ \mathbf{q}(s) = 0\,, & s > s_i\,; \end{cases}$$

$$\boldsymbol{\mu} = \begin{cases} \boldsymbol{\mu}(s) \neq 0\,, & 0 \leq s \leq s_\nu\,; \\ \boldsymbol{\mu}(s) = 0\,, & s > s_\nu\,. \end{cases}$$

Equations (1.5) and (1.6) are valid for initially unloaded rods. Generally, **Q** and **M** cannot be determined from the system of equations (1.5)–(1.6) because the number of unknowns exceeds the number of equations. The fact is that the components of the vector $\mathbf{e}_1$ in (1.6) are unknown since they depend on a deformed configuration of the rod.

1.1.3 Relationship Between the Vectors M and æ

Consider a rod element in a deformed configuration. The element is attributed to the attached coordinate system (Fig. 1.4). Let $æ_2$ and $æ_3$ be curvatures of the projections of the rod axis on the coordinate planes containing the vectors $(\mathbf{e}_1,\ \mathbf{e}_3)$ and $(\mathbf{e}_1,\ \mathbf{e}_2)$, respectively. Because the radius vector $\boldsymbol{\rho}$ is directed along the binormal axis rotated by an angle ϑ_{10} with respect to the principal axes, we obtain (see (A.2.4))

$$æ_2 = \frac{\sin\vartheta_{10}}{\rho} = \Omega_3 \sin\vartheta_{10}\,; \qquad æ_3 = \frac{\cos\vartheta_{10}}{\rho} = \Omega_3 \cos\vartheta_{10}\,. \tag{1.7}$$

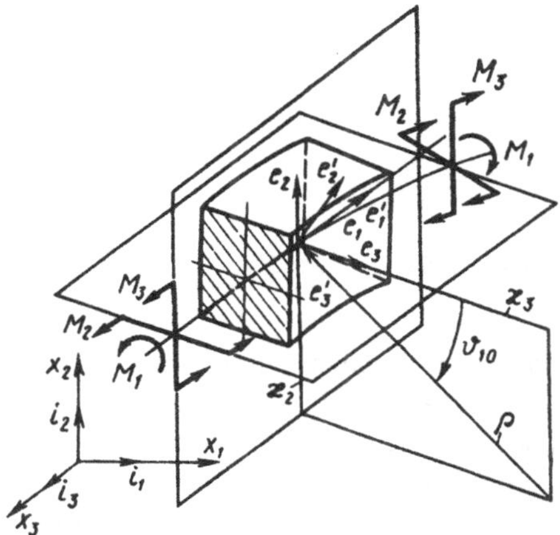

Fig. 1.4.

The moments M_2 and M_3 bend the rod element in two mutually orthogonal planes. The moment M_1 twists the rod element. Let the measure of this twist be denoted by $æ_1$. Assuming the quantities M_1, M_2, and M_3 be proportional to the changes of the curvatures of the rod axis and twist, we get

$$M_1 = A_{11}\left(æ_1 - æ_{10}\right), \qquad æ_1 = \Omega_1 + \frac{d\vartheta_{10}}{ds};$$
$$M_2 = A_{22}\left(æ_2 - æ_{20}\right);$$
$$M_3 = A_{33}\left(æ_3 - æ_{30}\right). \tag{1.8}$$

Here $æ_{i0}$ are the curvatures and the twist before deformation, Ω_1 is the twist (the second curvature) of the axial line, and A_{ii} are the bending and twisting stiffnesses; in the case of a rod of variable cross section, these stiffnesses are functions of s.

Recall that the quantities A_{ii} satisfy the relations

$$A_{11} = GJ_{\mathrm{t}}; \qquad A_{22} = EJ_{x_{20}}; \qquad A_{33} = EJ_{x_{30}},$$

where E and G are the moduli of elasticity of the first and second kind, respectively, J_{x_2} and J_{x_3} are the moments of inertia of the cross section of the rod with respect to the principal axes x_{30} and x_{20} (for example, for a rectangular cross section, we have $J_{x_{20}} = b^3h/12$ and $J_{x_{30}} = h^3b/12$), J_{t} is a geometrical characteristic of a cross section under twisting (for a bar of rectangular cross section, we have $J_{\mathrm{t}} = \beta b^3 h$, where $h > b$ and β is a Saint-Venant coefficient (Feodosiev (1974)).

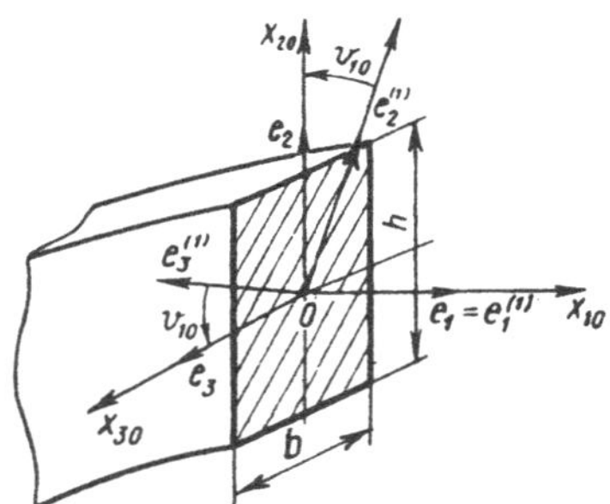

Fig. 1.5.

As it was mentioned above, a cross section can be attributed to the two coordinate systems: the natural coordinate system with the basis vectors $\{\mathbf{e}_i^{(1)}\}$ and the other one with the basis vectors $\{\mathbf{e}_i\}$ directed along the principal axes x_{20} and x_{30} (Fig. 1.5). The use of the coordinate system associated with the principal axes is extremely advantageous because in this system, the relationship between the internal moments M_i and the increments of $æ_i$ are given by the linear relations (1.8). In the basis $\{\mathbf{e}_i\}$, (1.8) can be presented as a vector equation

$$\mathbf{M} = \mathrm{A}\left(æ - æ_0^{(1)}\right), \tag{1.9}$$

where

$$A = \begin{bmatrix} A_{11} & 0 & 0 \\ 0 & A_{22} & 0 \\ 0 & 0 & A_{33} \end{bmatrix} . \tag{1.10}$$

It should be noted that the vector $æ_0^{(1)}$ differs from the vector $æ_0$. The vector $æ_0$ is a characteristic of the initial state of the rod (see Appendix 3). In the basis $\{\mathbf{e}_{i0}\}$, we have

$$æ_0 = æ_{10}\mathbf{e}_{10} + æ_{20}\mathbf{e}_{20} + æ_{30}\mathbf{e}_{30} . \tag{1.11}$$

To determine the increments of the curvatures $\Delta æ_i$, let us assume that the components of the vector $æ_0^{(1)}$ in the basis $\{\mathbf{e}_i\}$ remain unaltered during deformation. Thus, we obtain

$$æ_0^{(1)} = æ_{i0}\mathbf{e}_i = æ_{10}\mathbf{e}_1 + æ_{20}\mathbf{e}_2 + æ_{30}\mathbf{e}_3 . \tag{1.12}$$

Let us rewrite (1.9) using the Cartesian basis $\{\mathbf{i}_j\}$. We have

$$\mathbf{M} = \sum_{j=1}^{3} M_{x_j}\mathbf{i}_j . \tag{1.13}$$

In view of (A.58), we obtain $\mathbf{i}_j = \mathrm{L}^{(1)}\mathbf{e}_j$ or

$$\mathbf{i}_j = \sum_{\nu=1}^{3} l_{\nu j}^{(1)}\mathbf{e}_\nu . \tag{1.14}$$

Substituting (1.14) into (1.13), we arrive at

$$\mathbf{M} = \mathrm{L}^{(1)}\mathbf{M}_x . \tag{1.15}$$

Finally, from (1.19) it follows that

$$\mathbf{M}_x = (\mathrm{L}^{(1)})^{\mathrm{T}} \mathrm{A} \left(æ - æ_0^{(1)}\right) , \qquad \mathbf{M}_x = \sum_{j=1}^{3} M_{x_j}\mathbf{i}_j . \tag{1.16}$$

Equation (1.16) relates the Cartesian components of the vector $\mathbf{M}$ to the increments of the curvatures $(æ_i - æ_{i0})$ represented in the attached system.

1.1.4 Relationship Between the Vectors æ and ϑ

The relationship between the vector æ and the vector $\boldsymbol{\vartheta}$ is derived in Sect. A.2.5 of the Appendix. It can be written as

$$æ = \mathrm{L}_1 \frac{\tilde{\mathrm{d}}\boldsymbol{\vartheta}}{\mathrm{d}s} + \mathrm{L}æ_0^{(1)}, \tag{1.17}$$

where L and L_1 are the matrices (A.44) and (A.122), respectively.

For a further analysis, it is convenient to subtract the vector $æ_0^{(1)}$ from both sides of (1.17). As a result, we have

$$æ - æ_0^{(1)} = \mathrm{L}_1 \frac{\tilde{\mathrm{d}}\boldsymbol{\vartheta}}{\mathrm{d}s} + \mathrm{L}_2 æ_0^{(1)}, \qquad \mathrm{L}_2 = \mathrm{L} - \mathrm{E}; \tag{1.18}$$

here E is a unit matrix. Substituting (1.9) into (1.18), we get

$$\mathrm{L}_1 \frac{\tilde{\mathrm{d}}\boldsymbol{\vartheta}}{\mathrm{d}s} + \mathrm{L}_2 æ_0^{(1)} - \mathrm{A}^{-1}\mathbf{M} = 0. \tag{1.19}$$

1.1.5 Displacement of an Axial Line

Let us derive the equation in the displacement vector $\mathbf{u}$. It can be seen from Fig. 1.6 that

$$\mathbf{u} = \sum_{i=1}^{3} u_i \mathbf{e}_i = \mathbf{r} - \mathbf{r}_0. \tag{1.20}$$

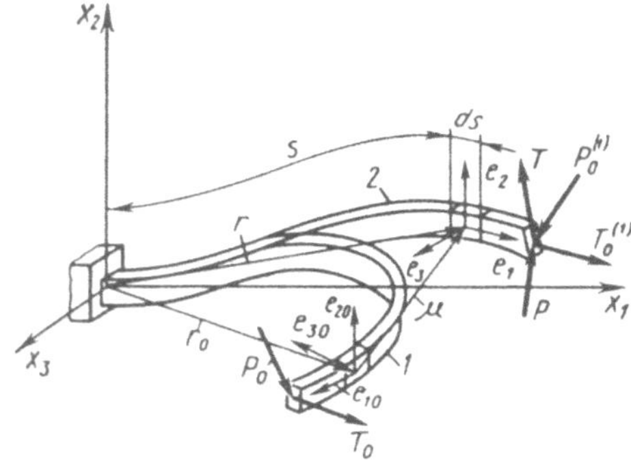

Fig. 1.6.

Differentiating (1.20) with respect to s, we obtain

$$\frac{\mathrm{d}\mathbf{u}}{\mathrm{d}s} = \frac{\mathrm{d}\mathbf{r}}{\mathrm{d}s} - \frac{\mathrm{d}\mathbf{r}_0}{\mathrm{d}s} = \mathbf{e}_1 - \mathbf{e}_{10}, \tag{1.21}$$

where $\mathbf{e}_{10}$, $\mathbf{e}_{20}$, and $\mathbf{e}_{30}$ are the unit base vectors related to the undeformed configuration of the rod. Let us assume that the following quantities related to the undeformed configuration of the rod are known: the vector æ_0, the matrix of direction cosines, and the matrix L^0 (see (A.55)) whose elements l_{ij}^0 enter into the following relation:

$$\mathbf{e}_{i_0} = \sum_{j=1}^{3} l_{ij}^0 \mathbf{i}_j \,, \tag{1.22}$$

The unit vectors $\mathbf{e}_i$ and $\mathbf{e}_{i0}$ satisfy the following relations:

$$\mathbf{e}_k = \sum_{\rho=1}^{3} l_{k\rho} \mathbf{e}_{\rho 0} \,; \qquad \mathbf{e}_{\rho 0} = \sum_{\nu=1}^{3} l_{\nu\rho} \mathbf{e}_\nu \,.$$

Here $l_{k\rho}$ are the elements of the matrix L (see (A.44)); these elements are functions of the unknown angles ϑ_1, ϑ_2, and ϑ_3, that is,

$$l_{ij} = l_{ij}(\vartheta_1\,, \vartheta_2\,, \vartheta_3)\,.$$

Eliminating $\mathbf{e}_{10}$ from (1.21), we obtain the following equation in the basis $\{\mathbf{e}_i\}$:

$$\frac{\mathrm{d}\mathbf{u}}{\mathrm{d}s} = (1 - l_{11})\,\mathbf{e}_1 - l_{21}\mathbf{e}_2 - l_{31}\mathbf{e}_3 \,, \qquad \mathbf{u} = \sum_{i=1}^{3} u_i \mathbf{e}_i \,. \tag{1.23}$$

In the basis $\{\mathbf{e}_{i0}\}$, (1.23) can be rewritten as follows:

$$\frac{\mathrm{d}\mathbf{u}}{\mathrm{d}s} = (l_{11} - 1)\,\mathbf{e}_{10} + l_{12}\mathbf{e}_{20} + l_{13}\mathbf{e}_{30} \,, \qquad \mathbf{u} = \sum_{i=1}^{3} u_{i0} \mathbf{e}_{i0} \,. \tag{1.24}$$

Relative displacements of the points of the rod axis can be found from either (1.23) or (1.24).

Let us derive the equations for the displacement $\mathbf{u}$ in the Cartesian basis $\{\mathbf{i}_j\}$. Since L^0 is the matrix of transformation from the basis $\{\mathbf{i}_j\}$ to the basis $\{\mathbf{e}_{j0}\}$, we get (see (A.58))

$$\mathbf{e}_1 = (\mathrm{L}^0)^{\mathrm{T}}\mathrm{L}^{\mathrm{T}}\mathbf{i}_1 \,; \qquad \mathbf{e}_{10} = (\mathrm{L}^0)^{\mathrm{T}}\,\mathbf{i}_1 \,. \tag{1.25}$$

In view of (1.21), we obtain

$$\frac{\mathrm{d}\mathbf{u}}{\mathrm{d}s} - (\mathrm{L}^0)^{\mathrm{T}}(\mathrm{L}^T - \mathrm{E})\,\mathbf{i}_1 = 0 \,, \qquad \mathbf{u} = \sum_{j=1}^{3} u_{x_j} \mathbf{i}_j \,. \tag{1.26}$$

Equation (1.26) can be solved for the Cartesian components u_{x_j} of the vector $\mathbf{u}$. These components are the absolute displacements of the axial points.

For rods of straight natural configuration, we have $L^0 = E$, thus, (1.26) can be rewritten as follows:

$$\frac{d\mathbf{u}}{ds} - (L^T - E)\,\mathbf{i}_1 = 0 \tag{1.27}$$

or

$$\frac{d\mathbf{u}}{ds} - (l_{11} - 1)\,\mathbf{i}_1 - l_{12}\mathbf{i}_2 - l_{13}\mathbf{i}_3 = 0\,. \tag{1.28}$$

1.1.6 Nondimensional Form of Equations

Let us introduce some nondimensional variables (they are marked by a tilde) by the following relations:

$$\begin{aligned}
&\eta = \frac{s}{l}\,; \qquad \tilde{æ} = æ l\,; \qquad \tilde{A}_{ii}(\eta) = \frac{A_{ii}(s)}{A_{33}(0)}\,; \qquad \tilde{\mathbf{Q}} = \frac{\mathbf{Q}l^2}{A_{33}(0)}\,;\\
&\tilde{\mathbf{q}} = \frac{\mathbf{q}l^3}{A_{33}(0)}\,; \qquad \tilde{\mathbf{M}} = \frac{\mathbf{M}l}{A_{33}(0)}\,; \qquad \tilde{\boldsymbol{\mu}} = \frac{\boldsymbol{\mu}l^2}{A_{33}(0)}\,;\\
&\tilde{\mathbf{P}}^{(i)} = \frac{\mathbf{P}^{(i)}l^2}{A_{33}(0)}\,; \qquad \tilde{\mathbf{T}}^{(\nu)} = \frac{\mathbf{T}^{(\nu)}l}{A_{33}(0)}\,;
\end{aligned} \tag{1.29}$$

here $A_{33}(0)$ is the magnitude of the bending stiffness at the origin of the coordinate system. The δ-function in (1.5) and (1.6) has the dimension of $(\text{length})^{-1}$. By use of the nondimensional coordinate η (see Appendix 4), we can write

$$\delta\,[\,l\,(\eta - \eta_i)\,] = \frac{1}{l}\,\tilde{\delta}\,(\eta - \eta_i)\,, \tag{1.30}$$

where $\tilde{\delta}$ is a nondimensional function.

Substituting (1.29) into (1.5), (1.6), (1.9), (1.19), (1.23), and (1.26), we obtain the following system of nondimensional nonlinear equilibrium equations[1]:

$$\frac{d\mathbf{Q}}{d\eta} + \mathbf{q} + \sum_{i=1}^{n} \mathbf{P}^{(i)}\,\delta\,(\eta - \eta_i) = 0\,; \tag{1.31}$$

$$\frac{d\mathbf{M}}{d\eta} + \mathbf{e}_1 \times \mathbf{Q} + \boldsymbol{\mu} + \sum_{\nu=1}^{\rho} \mathbf{T}^{(\nu)}\,\delta\,(\eta - \eta_\nu) = 0\,; \tag{1.32}$$

$$\mathbf{M} = \mathrm{A}\left(æ - æ_0^{(1)}\right)\,; \tag{1.33}$$

[1] From here on, all the equations are written in nondimensional form.

$$\mathrm{L}_1 \frac{\tilde{\mathrm{d}}\boldsymbol{\vartheta}}{\mathrm{d}\eta} + \mathrm{L}_2 \text{æ}_0^{(1)} - \mathrm{A}^{-1}\mathbf{M} = 0\,, \qquad \mathrm{L}_2 = \mathrm{L} - \mathrm{E}\,; \tag{1.34}$$

$$\frac{\mathrm{d}\mathbf{u}}{\mathrm{d}\eta} + (l_{11} - 1)\,\mathbf{e}_1 + l_{21}\mathbf{e}_2 + l_{31}\mathbf{e}_3 = 0\,; \tag{1.35}$$

$$\frac{\mathrm{d}\mathbf{u}}{\mathrm{d}\eta} - (\mathrm{L}^0)^{\mathrm{T}}(\mathrm{L}^{\mathrm{T}} - \mathrm{E})\,\mathbf{i}_1 = 0\,. \tag{1.36}$$

In (1.35), the vector $\mathbf{u}$ is referred to the attached basis,

$$\mathbf{u} = \sum_{j=1}^{3} u_j \mathbf{e}_j\,;$$

in (1.36), this vector is referred to the Cartesian basis,

$$\mathbf{u} = \sum_{j=1}^{3} u_{xj} \mathbf{i}_j\,.$$

Equations (1.31)–(1.34) and (1.35) or (1.36) are seen to be a system of equations in the five unknown vectors $\mathbf{Q}$, $\mathbf{M}$, $\boldsymbol{\vartheta}$, æ, and $\mathbf{u}$. Let us examine this system in greater detail.

Equations (1.31) and (1.32) are valid in any coordinate system, that is, they remain unchanged under any coordinate transformation. Equations (1.33)–(1.35) are valid only in the attached coordinate system $\{\mathbf{e}_j\}$. The vectors in (1.31) and (1.32) can be presented either in the immovable coordinate system,

$$\mathbf{Q} = \sum_{j=1}^{3} Q_{x_j} \mathbf{i}_j\,; \qquad \mathbf{q} = \sum_{j=1}^{3} q_{x_j} \mathbf{i}_j\,; \qquad \mathbf{P}^{(i)} = \sum_{j=1}^{3} P_{x_j}^{(i)} \mathbf{i}_j\,,$$

or in the attached one,

$$\mathbf{Q} = \sum_{j=1}^{3} Q_j \mathbf{e}_j\,; \qquad \mathbf{q} = \sum_{j=1}^{3} q_j \mathbf{e}_j\,; \qquad \mathbf{P}^{(i)} = \sum_{j=1}^{3} P_j^{(i)} \mathbf{e}_j\,.$$

In (1.33), (1.34), and (1.35), the vectors $\mathbf{M}$, $\boldsymbol{\vartheta}$, æ, $\text{æ}_0^{(1)}$ and $\mathbf{u}$ are assumed to be referred to the basis $\{\mathbf{e}_j\}$, that is,

$$\mathbf{M} = \sum_{j=1}^{3} M_j \mathbf{e}_j\,; \qquad \text{æ} = \sum_{j=1}^{3} \text{æ}_j \mathbf{e}_j\,; \qquad \text{æ}_0^{(1)} = \sum_{j=1}^{3} \text{æ}_{j0} \mathbf{e}_j\,;$$

$$\boldsymbol{\vartheta} = \sum_{j=1}^{3} \vartheta_j \mathbf{e}_j\,; \qquad \mathbf{u} = \sum_{j=1}^{3} u_j \mathbf{e}_j\,.$$

Equation (1.35) is valid only in the attached coordinate system while (1.36) is valid only in the Cartesian coordinate system. The transformation of equations related to the basis $\{\mathbf{e}_j\}$ into equations related to the basis $\{\mathbf{i}_j\}$ has been already discussed (see (1.16)). According to this procedure, in the basis $\{\mathbf{i}_j\}$, the relation (1.34) takes the form

$$(\mathrm{L}^{(1)})^{-1}\mathrm{AL}_1\,\frac{\tilde{\mathrm{d}}\boldsymbol{\vartheta}}{\mathrm{d}\eta}+(\mathrm{L}^{(1)})^{-1}\mathrm{AL}_2\,\text{æ}_0^{(1)}-\mathbf{M}_x=0\,. \tag{1.37}$$

A more detailed discussion of equilibrium equations referred to the coordinate systems $\{\mathbf{e}_j\}$ and $\{\mathbf{i}_j\}$ is presented in Sect. 1.3.

1.1.7 Boundary Conditions

There are two distinct types of boundary conditions: homogeneous and inhomogeneous. The total set of boundary conditions for a space-curved rod consists of 12 relations: 6 at the left-hand end (at $\eta=0$) and 6 at the right-hand end (at $\eta=1$).

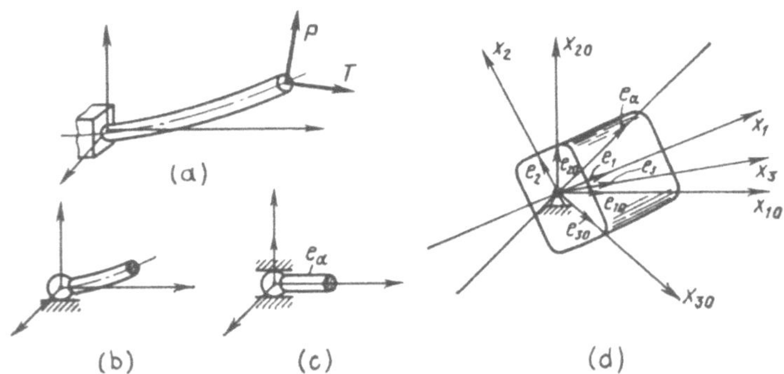

Fig. 1.7.

In the case of a cantilever rod (Fig. 1.7*a*), the boundary conditions are as follows:

(1) at $\eta=0\,,\quad \mathbf{u}=0\quad$ and $\quad\boldsymbol{\vartheta}=0\,;$

(2) at $\eta=1\,,\quad \mathbf{Q}=\mathbf{P}^{(3)}\quad$ and $\quad\mathbf{M}=\mathbf{T}^{(2)}\,.$

If no concentrated forces and moments are applied to the rod at its right-hand end, that is, $\mathbf{P}^{(3)}=\mathbf{T}^{(2)}=0$, then, at $\eta=1$, we have $\mathbf{Q}=0$ and $\mathbf{M}=0$.

If the ends of the rod are simply supported, the boundary conditions depend on the number of degrees of freedom that the support provides. For example, if the support permits rotation of the rod with respect to any of the three axes (Fig. 1.7*b*), then the corresponding boundary conditions are $\mathbf{u} = 0$ and $\mathbf{M} = 0$.

If one of the ends is fixed in such a way that no displacements in the direction defined by the unit vector $\mathbf{e}_\alpha$ are possible (Fig. 1.7*c*), then the following equality must hold: $\mathbf{u} \cdot \mathbf{e}_\alpha = 0$. In the case of two such vectors, we get $\mathbf{u} \cdot \mathbf{e}_{\alpha_1} = 0$ and $\mathbf{u} \cdot \mathbf{e}_{\alpha_2} = 0$.

The similar restrictions can be imposed on the angles of rotation of the end cross sections (e.g. $\vartheta_1 \neq 0$ while $\vartheta_2 = \vartheta_3 = 0$). Boundary conditions of more special form also exist. For example, a cross section is allowed to rotate only about an axis defined by a unit vector $\mathbf{e}_\alpha$ (Fig. 1.7*d*).

1.2 External Loads

1.2.1 Types of External Loads

The general vector equilibrium equations of a rod loaded by external forces and moments (1.31)–(1.35) are derived in Sect. 1.1. We assume that all the necessary information on the behavior of the external forces and moments for large deflections of the rod axis is given.

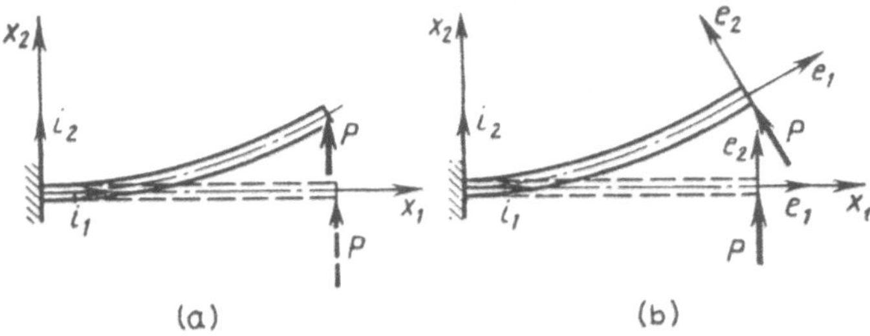

Fig. 1.8.

Let us consider in detail types of behavior of external loads which enter into (1.31) and (1.32). Since (1.31)–(1.35) are valid for large displacements, it is important to know the behavior of the external forces during deformation. We distinguish two basic types of forces, dead and follower. A force is termed *dead* if its magnitude and direction do not change during deformation (Fig. 1.8*a*). A *follower* force, in contrast, keeps its direction with respect to a certain coordinate system rigidly attached to the rod (Fig. 1.8*b*). For example, the gravity force is a dead force.

The vector equilibrium equations are valid for all types of loads (e.g. dead or follower). The type of loading plays important role when we write the vector equations in a particular coordinate system (e.g. in the basis $\{\mathbf{e}_i\}$

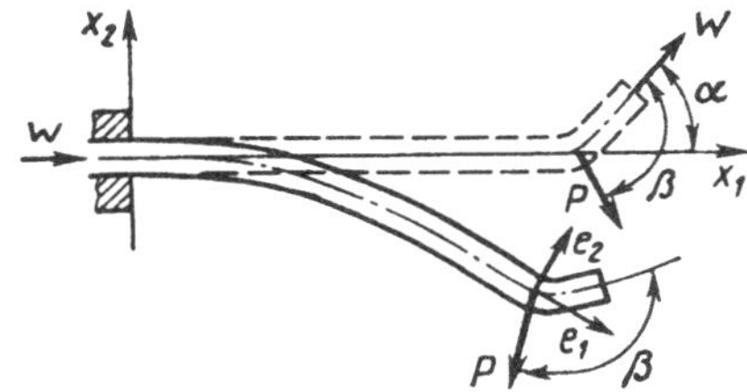

Fig. 1.9.

or in the basis $\{\mathbf{i}_j\}$), or when we transform these equations to the scalar form suitable for application of numerical methods.

If an external load is *dead*, it is convenient to rewrite the equations with respect to the immovable Cartesian basis $\{\mathbf{i}_j\}$. In this case, the magnitudes of the projections of forces and moments $P_{x_j}^{(i)}$, $T_{x_j}^{(\nu)}$, q_{x_j}, and μ_{x_j} do not depend on a deformed configuration of the rod. More precisely, $P_{x_j}^{(i)}$ and $T_{x_j}^{(\nu)}$ are constants, while q_{x_j} and μ_{x_j} are given functions. Meanwhile, the projections of a follower load on immovable axes depend on a deformed configuration of the rod.

An example of a follower force $\mathbf{P}$ is shown in Fig. 1.9. There is a flow of liquid of velocity $\mathbf{w}$ inside a hollow cantilever rod. The rod has a bend near its right-hand end that makes the angle α with the axial line of the rod. Due to this bend, the concentrated force $\mathbf{P}$ occurs. This force is of constant direction with respect to the basis $\{\mathbf{e}_j\}$ and $|\mathbf{P}|$ depends on the flow velocity $\mathbf{w}$ at $\eta = 1$.

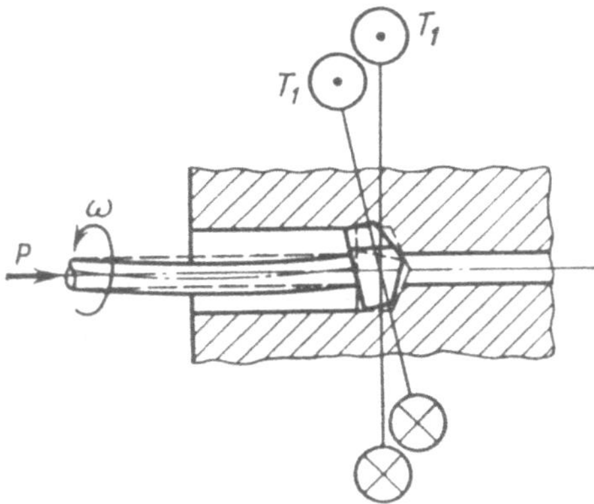

Fig. 1.10.

A process of deep-hole drilling is schematically illustrated in Fig. 1.10. Here $\boldsymbol{\omega}$ is the velocity of rotation of the drilling bit. At loss of static stability of the drilling bit or in the case of small lateral oscillations of the bit, the dominant part of the applied moment $\mathbf{T}_1$ can be treated as a follower moment.

An example of a distributed follower load $\mathbf{q}$ is given in Fig. 1.11. An ideal liquid runs with velocity $\mathbf{w}$ ($|\mathbf{w}|$ = const) inside a hollow curvilinear pipeline (a hollow rod of circular cross section). The distributed load acts in the direction of the principal normal to the axial line, hence,

$$\mathbf{q} = -\frac{m\omega^2}{\rho(\eta)}\,\mathbf{e}_2\,; \tag{1.38}$$

here m is the mass of the liquid per unit length of the rod and $\rho(\eta)$ is the radius of curvature of the rod axis.

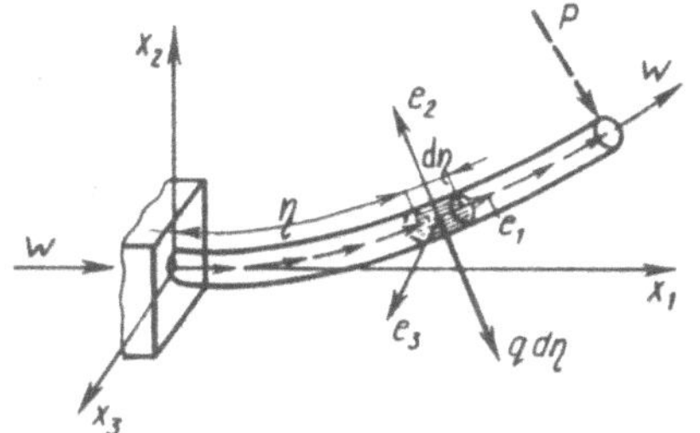

Fig. 1.11.

Under the action of an external force $\mathbf{P}$ shown as a dashed line in Fig. 1.11, the curvature of the axial line may change, hence, the distributed load $\mathbf{q}$ also changes (see (1.38)). For any deformed configuration of the rod, we have $(\mathbf{q}||\mathbf{e}_2)$.

Thus, we have an example of a distributed follower load whose magnitude depends on the deflection of the axial line, whereas the direction remains unchanged with respect to the axial line.

Consider this problem in detail. Let the force $\mathbf{P}$ be a follower one. Using the attached coordinate system, we can write

$$\mathbf{P} = \sum_{i=1}^{3} P_i\mathbf{e}_i\,,$$

where P_i are constants (they are constants only in the basis $\{\mathbf{e}_i\}$). In the immovable coordinate frame, we get

$$\mathbf{P} = \sum_{i=1}^{3} P_i\mathbf{e}_i = \sum_{j=1}^{3} P_{x_j}\mathbf{i}_j\,,$$

where

$$P_{x_j} = \sum_{i=1}^{3} P_i l_{ij}^{(1)}(\eta)\,. \tag{1.39}$$

Here $l_{ij}^{(1)}$ are the elements of the matrix $L^{(1)}$ (see (A.57)).

In the vector notation, (1.39) can be rewritten as follows (see (A.60)):

$$\mathbf{P}_x = (L^{(1)})^{\mathrm{T}}\mathbf{P} . \tag{1.40}$$

The projections P_{x_j} are functions of $l_{ij}^{(1)}(\eta)$ and, consequently, their magnitudes are not constant.

The distributed loads can be written in the same way:

$$\mathbf{q} = \sum_{j=1}^{3} q_{x_j}\mathbf{i}_j ; \qquad \boldsymbol{\mu} = \sum_{j=1}^{3} \mu_{x_j}\mathbf{i}_j$$

or

$$\mathbf{q}_x = (L^{(1)})^{\mathrm{T}}\mathbf{q} ; \qquad \boldsymbol{\mu}_x = (L^{(1)})^{\mathrm{T}}\boldsymbol{\mu} ,$$

where

$$q_{x_j} = \sum_{i=1}^{3} q_i l_{ij}^{(1)} ; \qquad \mu_{x_j} = \sum_{i=1}^{3} \mu_i l_{ij}^{(1)} .$$

For a dead force $\mathbf{P}$, we get

$$\mathbf{P} = \sum_{j=1}^{3} P_{x_j}\mathbf{i}_j ,$$

where P_{x_j} are specified constants (in the basis $\{\mathbf{i}_j\}$). In view of the relation

$$\mathbf{i}_j = \sum_{i=1}^{3} l_{ij}^{(1)}\mathbf{e}_i ,$$

we have

$$\mathbf{P} = \sum_{i=1}^{3} P_i\mathbf{e}_i , \qquad P_i = \sum_{j=1}^{3} P_{x_j} l_{ij}^{(1)} .$$

We see that the projections P_i depend on a deformed configuration of the rod.

If distributed loads are dead, the following expressions are valid in the attached coordinate system:

$$\mathbf{q}_i = \sum_{j=1}^{3} q_{x_j} l_{ij}^{(1)} ; \qquad \boldsymbol{\mu}_i = \sum_{j=1}^{3} \mu_{x_j} l_{ij}^{(1)} ;$$

here q_{x_j} and μ_{x_j} are given functions. In vector notation, the dead forces are

$$\mathbf{q} = L^{(1)}\mathbf{q}_x ; \quad \boldsymbol{\mu} = L^{(1)}\boldsymbol{\mu}_x ; \quad \mathbf{P} = L^{(1)}\mathbf{P}_x ; \quad \mathbf{T} = L^{(1)}\mathbf{T}_x . \tag{1.41}$$

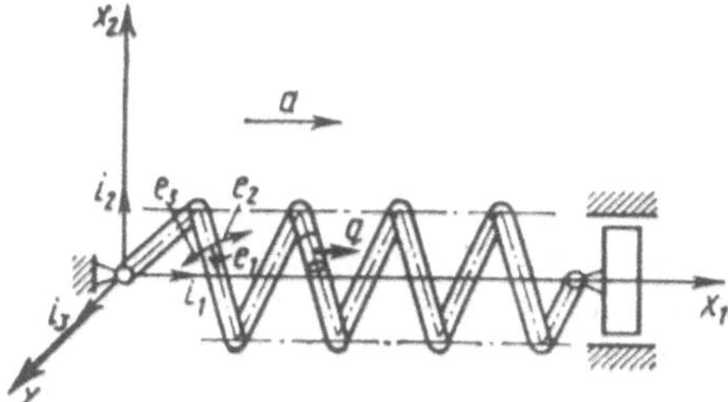

Fig. 1.12.

The following examples illustrate forces of constant direction with respect to an immovable Cartesian frame. It was mentioned above that the gravity force provides an example of such a force. However, dead forces are not necessarily of gravity nature. Figure 1.12 illustrates a cylindrical spring (say, an elastic element of an accelerometer) that is mounted on a moving object. Denote the acceleration of the object by $\mathbf{a}$. The rod is subjected to a distributed load

$$\mathbf{q} = \mathbf{a}m_0 = a\mathbf{i}_1 m_0 \,, \quad a = |\mathbf{a}|$$

whose magnitude and direction with respect to the basis $\{\mathbf{i}_j\}$ remain the same during deformation. In the attached frame, we get

$$\mathbf{q} = al_{11}^{(1)}\mathbf{e}_1 + al_{21}^{(1)}\mathbf{e}_2 + al_{31}^{(1)}\mathbf{e}_3 \,.$$

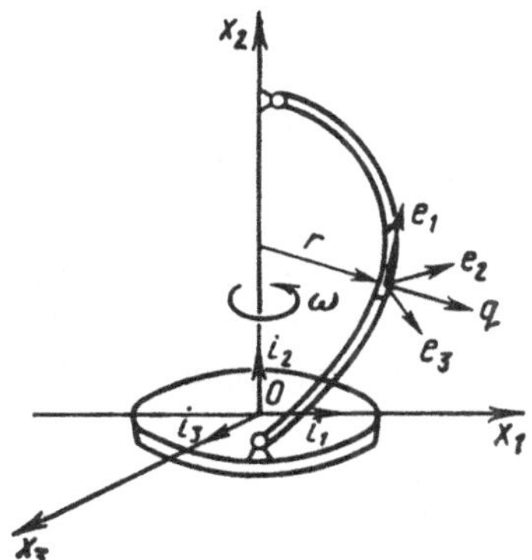

Fig. 1.13.

As another example, let us consider a rod mounted on a rotating disk (Fig. 1.13). The disk is referred to a Cartesian coordinate system such that the axis x_3 keeps its direction in space while the axes x_1 and x_2 are rigidly attached to the disk. The distributed load applied to the rod is

$$\mathbf{q} = \mathbf{r}\,\omega^2 m_0\,,$$

where $\mathbf{r} = r_{x_1}\mathbf{i}_1 + r_{x_3}\mathbf{i}_3$ and m_0 is a mass per unit length. In the rotating Cartesian frame, the vector $\mathbf{q}$ can be written as follows:

$$\mathbf{q} = x_1\omega^2 m_0\mathbf{i}_1 + x_3\omega^2 m_0\mathbf{i}_3\,.$$

The absolute value of $\mathbf{q}$ depends on a deformed configuration of the rod but its direction remains the same with respect to the rotating frame (in the plane x_1Ox_3). Therefore, $\mathbf{q}$ is a dead load with respect to just one axis of the Cartesian coordinate system.

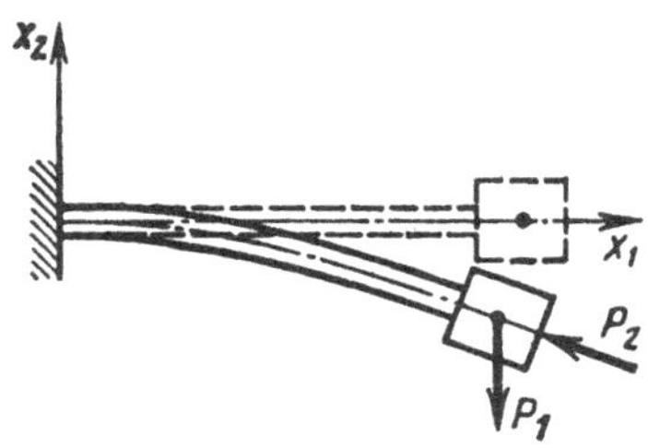

Fig. 1.14.

The vector $\mathbf{q}$ may be also written in the basis $\{\mathbf{e}_i\}$. By use of the transformation matrix $\mathrm{L}^{(1)}$, we get

$$\mathbf{q} = x_1\omega^2 m_0(l_{11}^{(1)}\mathbf{e}_1 + l_{21}^{(1)}\mathbf{e}_2 + l_{31}^{(1)}\mathbf{e}_3) + x_3\omega^2 m_0(l_{13}^{(1)}\mathbf{e}_1 + l_{23}^{(1)}\mathbf{e}_2 + l_{33}^{(1)}\mathbf{e}_3)$$

or, regrouping,

$$\mathbf{q} = (x_1 l_{11}^{(1)} + x_3 l_{13}^{(1)})m_0\omega^2\mathbf{e}_1 + (x_1 l_{21}^{(1)} + x_3 l_{23}^{(1)})m_0\omega^2\mathbf{e}_2 + (x_1 l_{31}^{(1)} + x_3 l_{33}^{(1)})m_0\omega^2\mathbf{e}_3\,.$$

The behavior of external loads is essential for representation of the rod equilibrium equations in scalar form both in immovable and attached coordinate systems.

In applications, loads of more complicated nature may occur. For example, it may happen that some of the applied loads are follower while the others are dead. It may happen that some of the projections of applied loads are dead or follower.

A cantilever rod with a jet engine joined to one of its ends is shown in Fig. 1.14. There are two forces applied to the rod. The first one is the gravity force $\mathbf{P}_1$, which is a dead force, and the second force is a reactive follower force $\mathbf{P}_2$.

Another example is shown in Fig. 1.15. During deformation, the applied load is directed at a fixed point A. In this case, the load projections both on attached and immovable coordinate axes depend on a deformed configuration of the rod. Here, the absolute value of the applied load is a constant: $mg = \text{const}$.

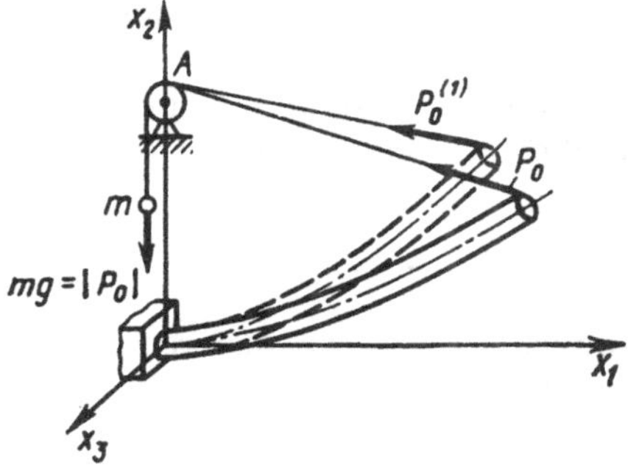

Fig. 1.15.

1.2.2 Increments of External Loads

To solve numerically the nonlinear equilibrium equations of a rod, one should know the explicit formulas for increments of vectors of external loads $\mathbf{q}$, $\boldsymbol{\mu}$, $\mathbf{P}^{(i)}$, and $\mathbf{T}^{(\nu)}$. These formulas can be also used for determination of critical loads in the analysis of static stability. From here on, forces applied to a rod as well as geometrical parameters are presented in nondimensional form. Examples of determination of increments of a vector are given in Appendix 3. One of the examples deals with the case when angles of rotation of the attached frame are small (see (A.143)).

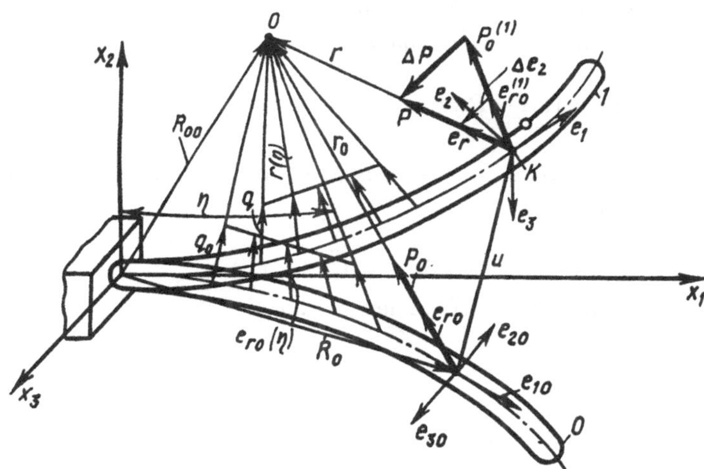

Fig. 1.16.

Consider now a more complicated case. Let a force applied to a rod be permanently directed at a fixed point (Fig. 1.16). The further analysis is restricted to the case of forces and moments of constant magnitude. Two configurations of the rod (*0* and *1*) are shown in Fig. 1.16. The subscript '0' corresponds to the unloaded configuration of the rod.

In the general case, the components of the vector $\mathbf{u}$ are not small. Then, the equilibrium equations are nonlinear differential equations and the vectors $\mathbf{P}_0$ and $\mathbf{P}$ can be written as follows:

$$\mathbf{P}_0 = |\mathbf{P}_0|\mathbf{e}_{r_0}\,; \qquad \mathbf{P} = |\mathbf{P}_0|\mathbf{e}_r\,;$$

here $\mathbf{e}_{r_0}$ and $\mathbf{e}_r$ are the unit vectors directed along the vectors $\mathbf{r}_0$ and $\mathbf{r}$, respectively. It gives $\mathbf{r}_0 = |\mathbf{r}_0|\mathbf{e}_{r_0}$ and $\mathbf{r} = |\mathbf{r}|\mathbf{e}_r$.

The increment of the vector $\mathbf{P}$ is

$$\Delta\mathbf{P} = \mathbf{P} - \mathbf{P}_0^{(1)}$$

or

$$\Delta\mathbf{P} = |\mathbf{P}_0|(\mathbf{e}_r - \mathbf{e}_{r_0}^{(1)}) = |\mathbf{P}_0|\,\Delta\mathbf{e}_r\,.$$

The vectors $\mathbf{e}_{r_0}^{(1)}$ and $\mathbf{e}_{r_0}$ have identical components in the bases $\{\mathbf{e}_j\}$ and $\{\mathbf{e}_{j_0}\}$, respectively. Hence, we write

$$\mathbf{e}_{r_0}^{(1)} = \sum_{i=1}^{3} \beta_i \mathbf{e}_i\,,$$

where $\beta_i = (\mathbf{e}_{r_0}\,,\,\mathbf{e}_{i_0})$ are the components of $\mathbf{e}_{r_0}$ in the basis $\{\mathbf{e}_{i_0}\}$.

To obtain the expressions for $\Delta\mathbf{P}$ in the basis $\{\mathbf{e}_j\}$, we must represent the vectors $\mathbf{e}_{r_0}^{(1)}$ and $\mathbf{e}_r$ in this basis. The vector $\mathbf{e}_{r_0}$ satisfies the relation

$$\mathbf{e}_{r_0} = \frac{\mathbf{r}_0}{|\mathbf{r}_0|}\,, \qquad \mathbf{r}_0 = \mathbf{R}_{00} - \mathbf{R}_0\,.$$

The vector $\mathbf{r}_0$ can be written in the basis $\{\mathbf{e}_{j0}\}$ in terms of its projections in the basis $\{\mathbf{i}_j\}$ as follows:

$$\mathbf{r}_{0e} = \mathrm{L}^0\mathbf{r}_0\,;$$

here L^0 is the transformation matrix. Then, we have

$$\mathbf{e}_{r_0} = \frac{\mathrm{L}^0\mathbf{r}_0}{|\mathbf{r}_0|}\,.$$

The expression for the vector $\mathbf{e}_r$ is

$$\mathbf{e}_r = \frac{\mathbf{r}}{|\mathbf{r}|}\,, \qquad \mathbf{r} = \mathbf{r}_0 - \mathbf{u}\,.$$

Solving the equilibrium equations for $\mathbf{u}$, we find its components in the basis $\{\mathbf{e}_j\}$. Consequently, it is more convenient to represent the vector $\mathbf{r}_0$ in the same basis. We obtain $\mathbf{r}_0^{(1)} = \mathrm{L}\mathrm{L}^0\mathbf{r}_0$, where L is the matrix of transformation from the basis $\{\mathbf{e}_{i0}\}$ to the basis $\{\mathbf{e}_i\}$. On rearrangement, we arrive at

$$\mathbf{e}_r = \frac{\mathrm{LL}^0\mathbf{r}_0 - \mathbf{u}}{|\mathrm{LL}^0\mathbf{r}_0 - \mathbf{u}|}\,.$$

Thus, the increment of the vector $\mathbf{P}$ may be written as follows:

$$\Delta\mathbf{P} = |\mathbf{P}_0| = \left(\frac{\mathrm{LL}^0\mathbf{r}_0 - \mathbf{u}}{|\mathrm{LL}^0\mathbf{r}_0 - \mathbf{u}|} - \frac{\mathrm{L}^0\mathbf{r}_0}{|\mathbf{r}_0|}\right)\,. \tag{1.42}$$

Similarly, we get

$$\Delta\mathbf{q} = |\mathbf{q}_0| = \left(\frac{\mathrm{LL}^0\mathbf{r}_0 - \mathbf{u}}{|\mathrm{LL}^0\mathbf{r}_0 - \mathbf{u}|} - \frac{\mathrm{L}^0\mathbf{r}_0}{|\mathbf{r}_0|}\right)\,; \tag{1.43}$$

$$\Delta\boldsymbol{\mu} = |\boldsymbol{\mu}_0| = \left(\frac{\mathrm{LL}^0\mathbf{r}_0 - \mathbf{u}}{|\mathrm{LL}^0\mathbf{r}_0 - \mathbf{u}|} - \frac{\mathrm{L}^0\mathbf{r}_0}{|\mathbf{r}_0|}\right)\,; \tag{1.44}$$

$$\Delta\mathbf{T} = |\mathbf{T}_0| = \left(\frac{\mathrm{LL}^0\mathbf{r}_0 - \mathbf{u}}{|\mathrm{LL}^0\mathbf{r}_0 - \mathbf{u}|} - \frac{\mathrm{L}^0\mathbf{r}_0}{|\mathbf{r}_0|}\right)\,. \tag{1.45}$$

The elements of the matrix L in (1.42)–(1.45) can be determined from the solution of the equilibrium equations. The elements of L^0 are assumed to be known as they are related to the stress-free natural state of the rod. The elements l_{ij} of L depend on the angles of rotation ϑ_i of the attached frame (see Sect. A.1.6 of the Appendix). Since the applied forces and moments are concentrated ones, the elements l_{ij} depend on the angles of rotation $\vartheta_i(\eta_K)$ of the axes related to the point of application of these forces and moments. In the case of distributed loads, the elements of L and L^0 are functions of the coordinate η. The increments of forces and moments are used in numerical analysis of the nonlinear equilibrium equations.

Now let us assume that the displacements u_j of the axial points and the angles of rotation ϑ_j are small. Then, we can write (see (A.47))

$$\mathrm{L} = \mathrm{E} + \Delta\mathrm{L}_1\,. \tag{1.46}$$

Therefore, in view of (1.46), the magnitude of the vector $\mathbf{r} = \mathrm{LL}^0\mathbf{r}_0 - \mathbf{u}$ can be expressed as follows (only terms linear in $\Delta\mathrm{L}_1$ and $\mathbf{u}$ are retained):

$$J = |\mathrm{LL}^0\mathbf{r}_0 - \mathbf{u}| = \sqrt{|\mathrm{L}^0\mathbf{r}_0|^2 + 2\,(\mathrm{L}^0\mathbf{r}_0\cdot\Delta\mathbf{a})}\,, \qquad \Delta\mathbf{a} = \Delta\mathrm{L}_1\mathrm{L}^0\mathbf{r}_0 - \mathbf{u}\,. \tag{1.47}$$

Expanding the right-hand side of (1.47) in a series and retaining only the linear terms, we get

$$J = A\left[1 + \frac{1}{A^2}(\mathrm{L}^0\mathbf{r}_0\cdot\Delta\mathbf{a})\right]\,, \qquad A = |\mathrm{L}^0\mathbf{r}_0| = |\mathbf{r}_0|\,. \tag{1.48}$$

By use of (1.46) and (1.48), (1.42) can be rewritten as follows:

$$\Delta\mathbf{P} = |\mathbf{P}_0| \left[\frac{L^0\mathbf{r}_0}{A^3} \left(L^0\mathbf{r}_0 \cdot \mathbf{u}\right) - \frac{\mathbf{u}}{A} + \frac{1}{A}\Delta L_1 L^0\mathbf{r}_0 - \frac{L^0\mathbf{r}_0}{A^3}\left(L^0\mathbf{r}_0 \cdot \Delta L_1 L^0\mathbf{r}_0\right) \right];$$

in view of the equation $L^0\mathbf{r}_0 \cdot \Delta L_1 L^0\mathbf{r}_0 = 0$, we get

$$\Delta\mathbf{P} = |\mathbf{P}_0| \left[\frac{\mathbf{e}_{r_0}}{A^3} \left(\mathbf{e}_{r_0} \cdot \mathbf{u}\right) - \frac{\mathbf{u}}{A} + \Delta L_1 \mathbf{e}_{r_0} \right], \tag{1.49}$$

where

$$\mathbf{e}_{r_0} = \frac{L^0\mathbf{r}_0}{|L^0\mathbf{r}_0|} = \frac{L^0\mathbf{r}_0}{|\mathbf{r}_0|}. \tag{1.50}$$

From (1.50) it follows that the vector $\mathbf{e}_{r_0}$ is represented in the basis $\{\mathbf{e}_{j0}\}$. As noted above, the matrix L^0 is the matrix of transformation from the basis $\{\mathbf{i}_j\}$ to the basis $\{\mathbf{e}_{i0}\}$. The origin of the frame $\{\mathbf{e}_{i0}\}$ is placed at a fixed point (e.g. the point K in Fig. 1.16). Hence, the elements of L^0 depend on η_K. The unit vector $\mathbf{e}_{r_0}$ in (1.49) is a function of the coordinate η_K of the point of application of the force $\mathbf{P}_0$.

Let the rod be subjected to a follower distributed force $\mathbf{q}_0$ directed at a fixed point O. The matrix L^0 and the unit vectors $\mathbf{e}_{r_0}(\eta) = L^0(\eta)\,\mathbf{r}(\eta)/|\mathbf{r}(\eta)|$ are functions of the axial coordinate η. Omitting derivation, we obtain

$$\Delta\mathbf{q} = |\mathbf{q}_0| \left[\frac{\mathbf{e}_{r_0}}{A} \left(\mathbf{e}_{r_0} \cdot \mathbf{u}\right) - \frac{\mathbf{u}}{A} + \Delta L_1 \mathbf{e}_{r_0} \right]; \tag{1.51}$$

$$\Delta\boldsymbol{\mu} = |\boldsymbol{\mu}_0| \left[\frac{\mathbf{e}_{r_0}}{A} \left(\mathbf{e}_{r_0} \cdot \mathbf{u}\right) - \frac{\mathbf{u}}{A} + \Delta L_1 \mathbf{e}_{r_0} \right]; \tag{1.52}$$

$$\Delta\mathbf{T} = |\mathbf{T}_0| \left[\frac{\mathbf{e}_{r_0}}{A} \left(\mathbf{e}_{r_0} \cdot \mathbf{u}\right) - \frac{\mathbf{u}}{A} + \Delta L_1 \mathbf{e}_{r_0} \right]. \tag{1.53}$$

The unit vector $\mathbf{e}_{r_0}$ in (1.51) and (1.52) is a function of the coordinate η. The vector $\mathbf{e}_{r_0}$ in (1.49) and (1.53) depends on the coordinate η_K of the point of application of the force $\mathbf{P}_0$ and the moment $\mathbf{T}_0$.

Consider an example. Let us determine the projections of $\Delta\mathbf{P}$ on the attached axes. Denote the projections of $\mathbf{e}_{r_0}$ by $\mathbf{e}_{r_{0j}}$. We have

$$\Delta P_1 = \frac{e_{r01}}{A}\left(\sum_{j=1}^{3} e_{r_{0j}} u_j\right) - \frac{u_1}{A} + \vartheta_3 e_{r02} - \vartheta_2 e_{r03};$$

$$\Delta P_2 = \frac{e_{r02}}{A}\left(\sum_{j=1}^{3} e_{r_{0j}} u_j\right) - \frac{u_2}{A} + \vartheta_3 e_{r01} + \vartheta_1 e_{r03};$$

$$\Delta P_3 = \frac{e_{r03}}{A}\left(\sum_{j=1}^{3} e_{r_{0j}} u_j\right) - \frac{u_3}{A} + \vartheta_2 e_{r01} + \vartheta_1 e_{r02}. \tag{1.54}$$

For numerical analysis of the equilibrium equations, it is more convenient to represent the increments as linear combinations of the vectors $\mathbf{u}$ and $\boldsymbol{\vartheta}$. For example, (1.54) can be rewritten as follows:

$$\Delta\mathbf{P} = B^{(1)}\boldsymbol{\vartheta} + B^{(2)}\mathbf{u}, \tag{1.55}$$

where

$$B^{(1)} = \begin{bmatrix} 0 & -e_{ro3} & e_{ro2} \\ e_{ro3} & 0 & -e_{ro1} \\ -e_{ro2} & e_{ro1} & 0 \end{bmatrix};$$

$$B^{(2)} = \begin{bmatrix} \dfrac{e_{ro1}^2 - 1}{A} & \dfrac{e_{ro1}e_{ro2}}{A} & \dfrac{e_{ro1}e_{ro3}}{A} \\ \dfrac{e_{ro1}e_{ro2}}{A} & \dfrac{e_{ro2}^2 - 1}{A} & \dfrac{e_{ro2}e_{ro3}}{A} \\ \dfrac{e_{ro1}e_{ro2}}{A} & \dfrac{e_{ro3}e_{ro2}}{A} & \dfrac{e_{ro3}^2 - 1}{A} \end{bmatrix}.$$

If a rod interacts with an air flow (see Chap. 6), force increments are not functions of $\mathbf{u}$ alone but they also depend on the first derivative of $\mathbf{u}$. The expression for the increments may contain a term $A_P^{(2)}\mathbf{u}'$.[2]

Similarly to (1.55), we get

$$\Delta\mathbf{q} = C^{(1)}\boldsymbol{\vartheta} + C^{(2)}\mathbf{u};$$
$$\Delta\boldsymbol{\mu} = C^{(3)}\boldsymbol{\vartheta} + C^{(4)}\mathbf{u};$$
$$\Delta\mathbf{T} = B^{(3)}\boldsymbol{\vartheta} + B^{(4)}\mathbf{u}. \tag{1.56}$$

The increments of forces and moments (1.55) and (1.56) will be used in the analysis of the equilibrium equations.

1.3 Equilibrium Equations in the Attached and Cartesian Coordinate Systems

1.3.1 Vector Equilibrium Equations in the Attached Coordinate System

To derive the equilibrium equations in a certain basis, we should represent all the vectors in this basis, say, $\{\mathbf{e}_i\}$, which is the basis related to the principal axes of a cross section. Note that the unit base vectors themselves are functions of η. This fact is emphasized by the notation $\mathbf{e}_i(\eta)$. Using (A.129), we can rewrite (1.31)–(1.35) in terms of local derivatives[3]:

[2] The prime is used to designate the differentiation with respect to the axial coordinate.

[3] The local derivatives are marked by a tilde.

$$\frac{\tilde{d}\mathbf{Q}}{d\eta} + \text{æ} \times \mathbf{Q} + \mathbf{P} = 0\,; \tag{1.57}$$

$$\frac{\tilde{d}\mathbf{M}}{d\eta} + \text{æ} \times \mathbf{M} + \mathbf{e}_1 \times \mathbf{Q} + \mathbf{T} = 0\,; \tag{1.58}$$

$$\mathbf{M} = \mathrm{A}\left(\text{æ} - \text{æ}_0^{(1)}\right)\,; \tag{1.59}$$

$$\mathrm{L}_1\,\frac{\tilde{d}\boldsymbol{\vartheta}}{d\eta} + \mathrm{L}_2\text{æ}_0^{(1)} - \mathrm{A}^{-1}\mathbf{M} = 0 \qquad \text{or} \qquad \mathrm{L}_1\,\frac{\tilde{d}\boldsymbol{\vartheta}}{d\eta} + \mathrm{L}\text{æ}_0^{(1)} - \text{æ} = 0\,; \tag{1.60}$$

$$\frac{\tilde{d}\mathbf{u}}{d\eta} + \text{æ} \times \mathbf{u} + (l_{11} - 1)\,\mathbf{e}_1 + l_{21}\mathbf{e}_2 + l_{31}\mathbf{e}_3 = 0\,; \tag{1.61}$$

here

$$\mathbf{P} = \mathbf{q} + \sum_{i=1}^{n} \mathbf{P}^{(i)}\,\delta\,(\eta - \eta_i)\,, \tag{1.62}$$

$$\mathbf{T} = \boldsymbol{\mu} + \sum_{\nu=1}^{\rho} \mathbf{T}^{(\nu)}\,\delta\,(\eta - \eta_\nu) \tag{1.63}$$

and

$$\mathrm{L}_1 = \begin{bmatrix} \cos\vartheta_2\cos\vartheta_3 & 0 & -\sin\vartheta_2 \\ -\sin\vartheta_3 & 1 & 0 \\ \sin\vartheta_2\cos\vartheta_3 & 0 & \cos\vartheta_2 \end{bmatrix},$$

$$\mathrm{L}_2 = \begin{bmatrix} \cos\vartheta_2\cos\vartheta_3 - 1 & \begin{matrix}\cos\vartheta_2\sin\vartheta_3\cos\vartheta_1 \\ +\sin\vartheta_2\sin\vartheta_1\end{matrix} & \begin{matrix}\cos\vartheta_2\sin\vartheta_3\sin\vartheta_1 \\ -\sin\vartheta_2\cos\vartheta_1\end{matrix} \\ -\sin\vartheta_3 & \cos\vartheta_1\cos\vartheta_3 - 1 & \cos\vartheta_3\sin\vartheta_1 \\ \sin\vartheta_2\cos\vartheta_3 & \begin{matrix}\sin\vartheta_2\sin\vartheta_3\cos\vartheta_1 \\ -\cos\vartheta_2\sin\vartheta_1\end{matrix} & \begin{matrix}\sin\vartheta_2\sin\vartheta_3\sin\vartheta_1 \\ +\cos\vartheta_2\cos\vartheta_1 - 1\end{matrix} \end{bmatrix}.$$

In general, the forces and moments $\mathbf{q}$, $\mathbf{P}^{(i)}$, $\boldsymbol{\mu}$, and $\mathbf{T}^{(\nu)}$ in (1.57) and (1.58) depend on the displacements u_j of the axial points and the angles of rotation ϑ_j of the attached frame. We assume that these relations are given a priori. Different types of external loads are fully considered in Sect. 1.2.

If the applied loads are dead and we deal with the equations in the attached frame (1.57)–(1.61), than the transformation formulas (1.41) must be used. From here on, the notation for forces and moments introduced in Sect. 1.1 will be used, that is, $\mathbf{q}$, $\mathbf{P}^{(i)}$, $\boldsymbol{\mu}$, and $\mathbf{T}^{(\nu)}$. In what follows, the explicit relations governing the behavior of external loads during deformation will be obtained.

1.3.2 Equilibrium Equations in the Attached Coordinate Frame

The use of the attached coordinate frame in the equilibrium analysis may happen to be profitable. Moreover, in this frame, the physical meaning of the components Q_i and M_i of the vectors $\mathbf{Q}$ and $\mathbf{M}$ is obvious: Q_1 is a compressive (axial) force, Q_2 and Q_3 are transverse shearing forces, M_1 is a twisting moment; M_2 and M_3 are bending moments. Using (1.62) and (1.63) in (1.57)–(1.61), we obtain (a tilde in the notation of the local derivatives is dropped)

$$\frac{dQ_1}{d\eta} + Q_3æ_2 - Q_2æ_3 + q_1 + \sum_{i=1}^{n} P_1^{(i)}\,\delta(\eta - \eta_i) = 0\,;$$
$$\frac{dQ_2}{d\eta} + Q_1æ_3 - Q_3æ_1 + q_2 + \sum_{i=1}^{n} P_2^{(i)}\,\delta(\eta - \eta_i) = 0\,;$$
$$\frac{dQ_3}{d\eta} + Q_2æ_1 - Q_1æ_2 + q_3 + \sum_{i=1}^{n} P_3^{(i)}\,\delta(\eta - \eta_i) = 0\,; \tag{1.64}$$
$$\frac{dM_1}{d\eta} + M_3æ_2 - M_2æ_3 + \mu_1 + \sum_{\nu=1}^{\rho} T_1^{(\nu)}\,\delta(\eta - \eta_\nu) = 0\,;$$
$$\frac{dM_2}{d\eta} + M_1æ_3 - M_3æ_1 - Q_3 + \mu_2 + \sum_{\nu=1}^{\rho} T_2^{(\nu)}\,\delta(\eta - \eta_\nu) = 0\,;$$
$$\frac{dM_3}{d\eta} + M_2æ_1 - M_1æ_2 + Q_2 + \mu_3 + \sum_{\nu=1}^{\rho} T_3^{(\nu)}\,\delta(\eta - \eta_\nu) = 0\,; \tag{1.65}$$
$$M_1 = A_{11}(æ_1 - æ_{10})\,;$$
$$M_2 = A_{22}(æ_2 - æ_{20})\,;$$
$$M_3 = A_{33}(æ_3 - æ_{30})\,; \tag{1.66}$$
$$\begin{aligned} æ_1 = {} & \left(\frac{d\vartheta_1}{d\eta} + æ_0\right)\cos\vartheta_2\cos\vartheta_3 - \frac{d\vartheta_3}{d\eta}\sin\vartheta_2 \\ & + (\sin\vartheta_2\sin\vartheta_1 + \cos\vartheta_1\cos\vartheta_2\sin\vartheta_3)\,æ_{20} \\ & + (\cos\vartheta_2\sin\vartheta_3\sin\vartheta_1 - \sin\vartheta_2\sin\vartheta_1)\,æ_{30}\,; \end{aligned}$$
$$\begin{aligned} æ_2 = {} & \frac{d\vartheta_2}{d\eta} - \left(\frac{d\vartheta_1}{d\eta} + æ_{10}\right)\sin\vartheta_3 \\ & + \cos\vartheta_3\cos\vartheta_1\,æ_{20} + \cos\vartheta_3\sin\vartheta_1\,æ_{30}\,; \end{aligned}$$
$$\begin{aligned} æ_3 = {} & \frac{d\vartheta_3}{d\eta}\cos\vartheta_2 + \left(\frac{d\vartheta_1}{d\eta} + æ_{10}\right)\sin\vartheta_2\cos\vartheta_3 \\ & + (\sin\vartheta_2\sin\vartheta_3\cos\vartheta_1 - \cos\vartheta_2\sin\vartheta_1)\,æ_{20} \\ & + (\cos\vartheta_2\cos\vartheta_1 + \sin\vartheta_1\sin\vartheta_2\sin\vartheta_3)\,æ_{30}\,; \end{aligned} \tag{1.67}$$
$$\frac{du_1}{d\eta} + u_3æ_2 - u_2æ_3 + l_{11} - 1 = 0\,;$$

$$\frac{du_2}{d\eta} + u_1 æ_3 - u_3 æ_1 + l_{21} = 0 ;$$
$$\frac{du_3}{d\eta} + u_2 æ_1 - u_1 æ_2 + l_{31} = 0 . \tag{1.68}$$

The parameters $æ_{i0}$ in (1.66) and (1.67) are assumed to be known.

For the components of the forces and the moments in (1.64) and (1.65), the following notation will be used:

$$P_j = q_j + \sum_{i=1}^{n} P_j^{(i)} \delta(\eta - \eta_i) ; \tag{1.69}$$

$$T_j = \mu_j + \sum_{\nu=1}^{\rho} \mathbf{T}_j^{(\nu)} \delta(\eta - \eta_\nu) , \qquad j = 1, 2, 3 . \tag{1.70}$$

Eliminating M_i from (1.65), we can rewrite the system (1.65) as follows:

$$\frac{d}{d\eta}[A_{11}(æ_1 - æ_{10})]$$
$$+ A_{33}(æ_3 - æ_{30}) æ_2 - A_{22}(æ_2 - æ_{20}) æ_3 + T_1 = 0 ;$$
$$\frac{d}{d\eta}[A_{22}(æ_2 - æ_{20})]$$
$$+ A_{11}(æ_1 - æ_{10}) æ_3 - A_{33}(æ_3 - æ_{30}) æ_1 - Q_3 + T_2 = 0 ;$$
$$\frac{d}{d\eta}[A_{33}(æ_3 - æ_{30})]$$
$$+ A_{22}(æ_2 - æ_{20}) æ_1 - A_{11}(æ_1 - æ_{10}) æ_2 + Q_2 + T_3 = 0 . \tag{1.71}$$

1.3.3 Special Cases of Equilibrium Equations in the Attached Coordinate Frame

Let us consider some nonlinear problems concerned with an initially deformed rod of constant cross section. A follower force and a moment are applied at the end $\eta = 1$ as indicated in Fig. 1.17. The effect that the force and the moment produce can be obtained by the appropriate choice of the boundary conditions. Consequently, the force and the moment do not appear in the equilibrium equations and the systems of equations (1.64) and (1.71) take the following form:

$$\frac{dQ_1}{d\eta} + Q_3 æ_2 - Q_2 æ_3 = 0 ;$$
$$\frac{dQ_2}{d\eta} + Q_1 æ_3 - Q_3 æ_1 = 0 ;$$
$$\frac{dQ_3}{d\eta} + Q_2 æ_1 - Q_1 æ_2 = 0 ; \tag{1.72}$$

$$\begin{aligned}
&A_{11}\frac{\mathrm{d}}{\mathrm{d}\eta}(æ_1 - æ_{10}) \\
&\qquad + A_{33}(æ_3 - æ_{30})\,æ_2 - A_{22}(æ_2 - æ_{20})\,æ_3 = 0\,; \\
&A_{22}\frac{\mathrm{d}}{\mathrm{d}\eta}(æ_2 - æ_{20}) \\
&\qquad + A_{11}(æ_1 - æ_{10})\,æ_3 - A_{33}(æ_3 - æ_{30})\,æ_1 - Q_3 = 0\,; \\
&A_{33}\frac{\mathrm{d}}{\mathrm{d}\eta}(æ_3 - æ_{30}) \\
&\qquad + A_{22}(æ_2 - æ_{20})\,æ_1 - A_{11}(æ_1 - æ_{10})\,æ_2 + Q_2 = 0\,.
\end{aligned} \tag{1.73}$$

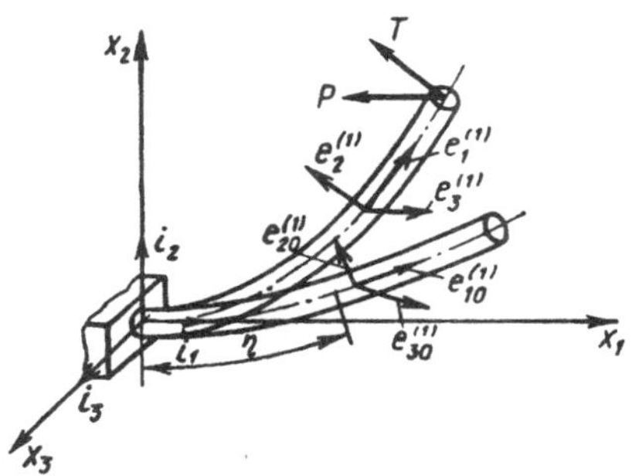

Fig. 1.17.

In the case under consideration, the systems (1.67) and (1.68) remain valid. If the rod axis is initially a plane curve lying in the x_1Ox_2 plane, then $æ_{10} = æ_{20} = 0$, and, from (1.73), we get

$$\begin{aligned}
&A_{11}\frac{\mathrm{d}æ_1}{\mathrm{d}\eta} + (A_{33} - A_{22})\,æ_2æ_3 - A_{33}æ_{30}æ_2 = 0\,; \\
&A_{22}\frac{\mathrm{d}æ_2}{\mathrm{d}\eta} + (A_{11} - A_{33})\,æ_1æ_3 + A_{33}æ_{30}æ_1 - Q_3 = 0\,; \\
&A_{33}\frac{\mathrm{d}æ_3}{\mathrm{d}\eta} + (A_{22} - A_{11})\,æ_1æ_2 - A_{33}\frac{\mathrm{d}æ_{30}}{\mathrm{d}\eta} + Q_2 = 0\,.
\end{aligned} \tag{1.74}$$

If the rod in its natural configuration is straight (Fig. 1.18), then $æ_{i0} = 0$ and, in view of (1.67) and (1.74), we get

$$\begin{aligned}
æ_1 &= \frac{\mathrm{d}\vartheta_1}{\mathrm{d}\eta}\cos\vartheta_2\cos\vartheta_3 - \frac{\mathrm{d}\vartheta_3}{\mathrm{d}\eta}\sin\vartheta_2\,; \\
æ_2 &= \frac{\mathrm{d}\vartheta_2}{\mathrm{d}\eta} - \frac{\mathrm{d}\vartheta_1}{\mathrm{d}\eta}\sin\vartheta_3\,; \\
æ_3 &= \frac{\mathrm{d}\vartheta_3}{\mathrm{d}\eta}\cos\vartheta_2 + \frac{\mathrm{d}\vartheta_1}{\mathrm{d}\eta}\sin\vartheta_2\cos\vartheta_3\,;
\end{aligned} \tag{1.75}$$

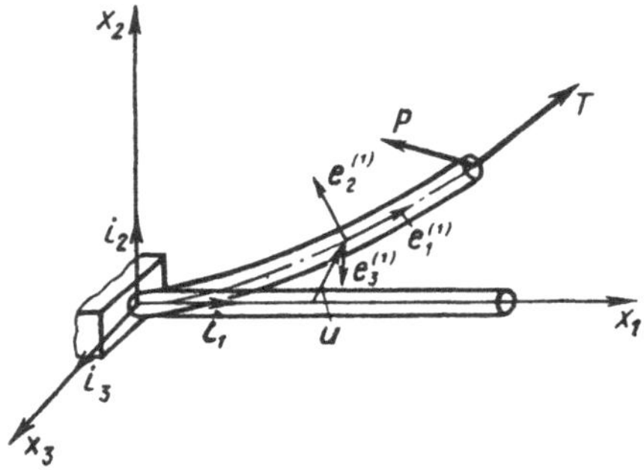

Fig. 1.18.

$$A_{11}\frac{\mathrm{d}æ_1}{\mathrm{d}\eta}+(A_{33}-A_{22})\,æ_2æ_3=0\,;$$
$$A_{22}\frac{\mathrm{d}æ_2}{\mathrm{d}\eta}+(A_{11}-A_{33})\,æ_1æ_3-Q_3=0\,;$$
$$A_{33}\frac{\mathrm{d}æ_3}{\mathrm{d}\eta}+(A_{22}-A_{11})\,æ_1æ_2+Q_2=0\,. \tag{1.76}$$

For initially straight rods, the angles ϑ_i in (1.75) are the angles of rotation of the basis $\{\mathbf{e}_i\}$ with respect to the basis $\{\mathbf{i}_j\}$.

As an example, consider a straight rod such that $A_{22}=A_{33}=\text{const}$. This rod may be of square or circular cross section. A force **P** and a follower moment **T** are applied to the rod. Equations (1.76) become

$$A_{11}\frac{\mathrm{d}æ_1}{\mathrm{d}\eta}=0\,;$$
$$A_{22}\frac{\mathrm{d}æ_2}{\mathrm{d}\eta}+(A_{11}-A_{33})\,æ_1æ_3-Q_3=0\,;$$
$$A_{33}\frac{\mathrm{d}æ_3}{\mathrm{d}\eta}-(A_{11}-A_{22})\,æ_1æ_2+Q_2=0\,. \tag{1.77}$$

Integration of the first equation of (1.77) yields

$$æ_1=c_1=\text{const}\,. \tag{1.78}$$

For $\eta=1$, we have $æ_1=T_1/A_{11}$ and, hence, $c_1=T_1/A_{11}$.

In view of (1.78), the internal twisting moment is of the same magnitude at any point of the rod axis irrespective of a deflected configuration of the rod.

Consider a straight rod (Fig. 1.18) subjected to a concentrated follower moment **T**, whereas $\mathbf{P}=0$. The direction of **T** is of no significance. In this case, $Q_i=0$. During the deformation, the components of **T** are constants in the basis $\{\mathbf{e}_i\}$. The first equation of (1.77) gives $æ_1=T_1/A_{11}=\text{const}$. The other two equations yield

$$
\begin{aligned}
æ_2 &= c_1 \cos \alpha\eta + c_2 \sin \alpha\eta \,; \\
æ_3 &= -c_1 \sin \alpha\eta + c_2 \cos \alpha\eta \,,
\end{aligned}
\tag{1.79}
$$

where $\alpha = (A_{11} - A_{22})T_1/(A_{11}A_{22})$.

The constants c_1 and c_2 can be determined by introduction of the boundary-condition equations at $\eta = 1$ into (1.79). Since $æ_2 = T_2/A_{22}$ and $æ_3 = T_3/A_{22}$, the system of equations (1.79) can be easily solved in c_1 and c_2:

$$
\begin{aligned}
c_1 &= \frac{T_2}{A_{22}} \cos \alpha - \frac{T_3}{A_{22}} \sin \alpha \,; \\
c_2 &= \frac{T_2}{A_{22}} \sin \alpha + \frac{T_3}{A_{22}} \cos \alpha \,.
\end{aligned}
$$

Once the components $æ_1$, $æ_2$, and $æ_3$ of the vector $\boldsymbol{æ}$ are determined, the following equations in the unknown angles ϑ_i can be obtained:

$$
\begin{aligned}
&\frac{d\vartheta_1}{d\eta} \cos\vartheta_2 \cos\vartheta_3 - \frac{d\vartheta_3}{d\eta} \sin\vartheta_2 - æ_1 = 0 \,; \\
&\frac{d\vartheta_2}{d\eta} - \frac{d\vartheta_1}{d\eta} \sin\vartheta_3 - æ_2 = 0 \,; \\
&\frac{d\vartheta_3}{d\eta} \cos\vartheta_2 + \frac{d\vartheta_1}{d\eta} \sin\vartheta_2 \cos\vartheta_3 - æ_3 = 0
\end{aligned}
\tag{1.80}
$$

or

$$
\frac{d\boldsymbol{\vartheta}}{d\eta} + \mathrm{L}_1^{-1} \boldsymbol{æ} = 0 \,;
\tag{1.81}
$$

here L_1^{-1} is the inverse matrix of L_1 (A.122). The functions $\vartheta_i(\eta)$ may be obtained numerically from (1.80) under the zero boundary condition $\vartheta_i(0) = 0$.

Now let us determine the displacements of the axial points. As the rod in the unloaded configuration is straight, (1.21) in nondimensional notation becomes

$$
\frac{d\mathbf{u}}{d\eta} = \mathbf{e}_1 - \mathbf{e}_{10} \,.
\tag{1.82}
$$

In the case under consideration, the bases $\{\mathbf{e}_{i0}\}$ and $\{\mathbf{i}_j\}$ coincide; hence, we get

$$
\mathbf{e}_{10} = \mathbf{i}_1 \,; \qquad \mathbf{e}_1 = l_{11}\mathbf{i}_1 + l_{12}\mathbf{i}_2 + l_{13}\mathbf{i}_3 \,,
$$

where l_{1i} are the elements of L:

$$
\begin{aligned}
l_{11} &= \cos\vartheta_2 \cos\vartheta_3 \,; \\
l_{12} &= \cos\vartheta_2 \sin\vartheta_3 \cos\vartheta_1 + \sin\vartheta_2 \sin\vartheta_1 , ; \\
l_{13} &= \cos\vartheta_2 \sin\vartheta_3 \sin\vartheta_1 - \sin\vartheta_2 \cos\vartheta_1 \,.
\end{aligned}
$$

In view of (1.82), we obtain

$$\frac{\mathrm{d}u_{x_1}}{\mathrm{d}\eta} = \cos\vartheta_2\cos\vartheta_3 - 1\,;$$
$$\frac{\mathrm{d}u_{x_2}}{\mathrm{d}\eta} = \cos\vartheta_2\sin\vartheta_3\cos\vartheta_1 + \sin\vartheta_2\sin\vartheta_1\,;$$
$$\frac{\mathrm{d}u_{x_3}}{\mathrm{d}\eta} = \cos\vartheta_2\sin\vartheta_3\sin\vartheta_1 - \sin\vartheta_2\cos\vartheta_1\,. \tag{1.83}$$

In this particular case, $\vartheta_{x_1} = \vartheta_1$, $\vartheta_{x_2} = \vartheta_2$, and $\vartheta_{x_3} = \vartheta_3$. The absolute displacements of the axial points (the displacements measured in an immovable coordinate system) can be determined by integration of (1.83) under the zero boundary conditions $u_{x_i}(0) = 0$.

In general, for external loads of more complicated nature, the explicit solution of (1.83) cannot be obtained; thus, numerical methods are to be applied.

1.3.4 Vector Equilibrium Equations in the Cartesian Coordinate System

As noted above, the use of the attached axes is extremely advantageous when external loads are follower and their components in the attached basis are known. If the components of an external load are given in the Cartesian basis, it would be more convenient to refer the equilibrium equations to this frame. Using the matrix of transformation, it is possible to derive the load components in any coordinate system. The use of the attached axes is effective in the analysis of rod equilibrium because the formula (1.9), which relates the internal moment to the increment of the elastic deformations æ, is of comparatively simple form if referred to the frame $\{\mathbf{e}_i\}$. In general, the equations related to the Cartesian basis are of more complicated form as compared to those related to the attached basis. This is particularly true for a rod whose axial line is a spatial curve.

Let the vectors referred to the basis $\{\mathbf{i}_j\}$ be denoted by the subscript 'x', that is, $\mathbf{Q}_x = \sum Q_{x_j}\mathbf{i}_j$, $\mathbf{M}_x = \sum M_{x_j}\mathbf{i}_j$, etc. The formulas of coordinate transformation (see (A.59) and (A.60)) are

$$\mathbf{a} = \mathrm{L}^{(1)}\mathbf{a}_x$$

for the transformation from the basis $\{\mathbf{i}_j\}$ to the basis $\{\mathbf{e}_i\}$ and

$$\mathbf{a}_x = (\mathrm{L}^{(1)})^{\mathrm{T}}\mathbf{a}$$

for the transformation from the basis $\{\mathbf{e}_i\}$ to the basis $\{\mathbf{i}_j\}$.

Consider the vector equations (1.31)–(1.35). In a Cartesian coordinate system, the material derivative coincide with the local one, hence, (1.31) and (1.32) preserve their form, whereas, the vectors in these equations are related to the basis $\{\mathbf{i}_j\}$, that is,

$$\frac{d\mathbf{Q}_x}{d\eta} + \mathbf{P}_x = 0\,; \tag{1.84}$$

$$\frac{d\mathbf{M}_x}{d\eta} + (\mathrm{L}^{(1)})^{\mathrm{T}}\mathbf{i}_1 \times \mathbf{Q}_x + \mathbf{T}_x = 0\,, \tag{1.85}$$

where

$$\mathbf{P}_x = \mathbf{q}_x + \sum_{i=1}^{n} \mathbf{P}_x^{(i)}\,\delta\,(\eta - \eta_i)\,;$$

$$\mathbf{T}_x = \boldsymbol{\mu}_x + \sum_{\nu=1}^{\rho} \mathbf{T}_x^{(\nu)}\,\delta\,(\eta - \eta_\nu)\,.$$

Equation (1.33) is referred to the basis $\{\mathbf{e}_i\}$. In the basis $\{\mathbf{i}_j\}$, it takes the form (see (1.16))

$$\mathbf{M}_x = (\mathrm{L}^{(1)})^{\mathrm{T}}\mathrm{A}\,(\text{æ} - \text{æ}_0^{(1)})\,. \tag{1.86}$$

Equations (1.34) and (1.35) related to the basis $\{\mathbf{i}_j\}$ were obtained above (see (1.37) and (1.36))

$$\frac{d\boldsymbol{\vartheta}}{d\eta} + \mathrm{L}_1^{-1}\mathrm{L}_2\text{æ}_0^{(1)} - \mathrm{L}_1^{-1}\mathrm{A}^{-1}\mathrm{L}^{(1)}\mathbf{M}_x = 0\,; \tag{1.87}$$

$$\frac{d\mathbf{u}}{d\eta} - \left[(\mathrm{L}^{(1)})^{\mathrm{T}} - (\mathrm{L}^{0})^{\mathrm{T}}\right]\mathbf{i}_1 = 0\,, \tag{1.88}$$

where

$$L_1^{-1} = \frac{1}{\cos\vartheta_3}\begin{bmatrix} \cos\vartheta_2 & 0 & \sin\vartheta_2 \\ \cos\vartheta_2\,\sin\vartheta_3 & \cos\vartheta_3 & \sin\vartheta_2\,\sin\vartheta_3 \\ -\sin\vartheta_2\,\cos\vartheta_3 & 0 & \cos\vartheta_2\,\cos\vartheta_3 \end{bmatrix}.$$

1.3.5 Equilibrium Equations in the Cartesian Coordinate System

In a Cartesian coordinate system, the systems of equations (1.84)–(1.88) are

$$\frac{dQ_{x_1}}{d\eta} + q_{x_1} + \sum_{i=1}^{n} P_{x_1}^{(i)}\,\delta\,(\eta - \eta_i) = 0\,;$$

$$\frac{dQ_{x_2}}{d\eta} + q_{x_2} + \sum_{i=1}^{n} P_{x_2}^{(i)}\,\delta\,(\eta - \eta_i) = 0\,;$$

$$\frac{dQ_{x_3}}{d\eta} + q_{x_3} + \sum_{i=1}^{n} P_{x_3}^{(i)}\,\delta\,(\eta - \eta_i) = 0\,; \tag{1.89}$$

$$\frac{dM_{x_1}}{d\eta} + l_{12}^{(1)}Q_{x_3} - l_{13}{}^{(1)}Q_{x_2} + \mu_{x_1} + \sum_{\nu=1}^{\rho} T_{x_1}^{(\nu)}\,\delta\,(\eta - \eta_\nu) = 0\,;$$

$$\frac{dM_{x_2}}{d\eta} + l_{13}^{(1)} Q_{x_1} - l_{11}^{(1)} Q_{x_3} + \mu_{x_2} + \sum_{\nu=1}^{\rho} T_{x_2}^{(\nu)} \delta(\eta - \eta_\nu) = 0 ;$$

$$\frac{dM_{x_3}}{d\eta} + l_{11}^{(1)} Q_{x_2} - l_{12}^{(1)} Q_{x_1} + \mu_{x_3} + \sum_{\nu=1}^{\rho} T_{x_3}^{(\nu)} \delta(\eta - \eta_\nu) = 0 ; \tag{1.90}$$

$$\frac{d\vartheta_j}{d\eta} + \sum_{\nu=1}^{3} (l_{j\nu}^{(2)} æ_{\nu 0} - l_{j\nu}^{(3)} M_{x_\nu}) = 0 ; \tag{1.91}$$

$$M_{x_j} = \sum_{\nu=1}^{3} l_{\nu j}^{(1)} A_{(\nu\nu)} (æ_\nu - æ_{\nu 0}) ; \tag{1.92}$$

$$\frac{du_{x_j}}{d\eta} - (l_{1j}^{(1)} - l_{1j}^{(0)}) = 0 , \tag{1.93}$$

where

$$P_{x_j} = q_{x_j} + \sum_{i=1}^{n} P_{x_j}^{(i)} \delta(\eta - \eta_i) ;$$

$$T_{x_j} = \mu_{x_j} + \sum_{\nu=1}^{\rho} T_{x_j}^{(\nu)} \delta(\eta - \eta_\nu) ;$$

here, $l_{j\nu}^{(2)}$ and $l_{j\nu}^{(3)}$ are the elements of the matrices $\mathrm{L}^{(2)} = \mathrm{L}_1^{-1}\mathrm{L}_2$ and $\mathrm{L}^{(3)} = \mathrm{L}_1^{-1}\mathrm{A}^{-1}\mathrm{L}^{(1)}$, respectively.

Thus, the choice of a coordinate system (immovable or attached) is to be done with regard to the peculiarities of a particular problem.

1.4 Equilibrium Equations for Small Displacements and Angles of Rotation

1.4.1 Equilibrium Equations in the Attached Coordinate System

Let us examine the equilibrium equations of a space-curved rod assuming the components of the vectors $\mathbf{u}$, $\Delta æ = æ - æ^{(0)}$, and $\boldsymbol{\vartheta}$ to be small. To clarify some subtle points, it is convenient to deal with the equilibrium equations in the attached basis.

Consider the nonlinear equilibrium equations (1.57)–(1.61):

$$\frac{d\mathbf{Q}}{d\eta} + æ \times \mathbf{Q} + \mathbf{P} = 0 ; \tag{1.94}$$

$$\frac{d\mathbf{M}}{d\eta} + æ \times \mathbf{M} + \mathbf{e}_1 \times \mathbf{Q} + \mathbf{T} = 0 ; \tag{1.95}$$

$$\mathbf{M} = \mathrm{A}\,(æ - æ_0^{(1)}) ; \tag{1.96}$$

$$\mathrm{L}_1 \frac{\mathrm{d}\boldsymbol{\vartheta}}{\mathrm{d}\eta} + \mathrm{L}_2 æ_0^{(1)} - \mathrm{A}^{-1}\mathbf{M} = 0\,, \qquad \mathrm{L}_2 = \mathrm{L} - \mathrm{E}\,; \tag{1.97}$$

$$\frac{\mathrm{d}\mathbf{u}}{\mathrm{d}\eta} + æ \times \mathbf{u} + (l_{11} - 1)\,\mathbf{e}_1 + l_{21}\mathbf{e}_2 + l_{31}\mathbf{e}_3 = 0\,. \tag{1.98}$$

We assume that the components of $\boldsymbol{\vartheta}$ and $\mathbf{u}$ and the increments of æ ($\Delta æ = æ - æ_0^{(1)}$) are small, whereas, in general, the components of $\mathbf{Q}$ and $\mathbf{M}$ are not small. No restrictions are applied on the behavior of external loads and their magnitudes. The vectors $\mathbf{P}$ and $\mathbf{T}$ are functions of the vectors $\mathbf{q}$, $\mathbf{P}^{(\nu)}$, $\boldsymbol{\mu}$, and $\mathbf{T}^{(\nu)}$. In the general case, the directions and the magnitudes of these vectors may change during deformation, i.e. the following relations take place: $\mathbf{q} = \mathbf{q}_0 + \Delta\mathbf{q}$, $\mathbf{P}^{(i)} = \mathbf{P}_0^{(i)} + \Delta\mathbf{P}^{(i)}$ $\boldsymbol{\mu} = \boldsymbol{\mu}_0 + \Delta\boldsymbol{\mu}$, and $\mathbf{T}^{(\nu)} = \mathbf{T}_0^{(\nu)} + \Delta\mathbf{T}^{(\nu)}$. The subscript '0' is used to indicate parameters that correspond to the case of an absolutely rigid rod (the vectors $\mathbf{q}_0$, $\mathbf{P}_0^{(i)}$, $\boldsymbol{\mu}_0$, and $\mathbf{T}_0^{(\nu)}$ are known). The vectors $\Delta\mathbf{q}$, $\Delta\mathbf{P}^{(i)}$, $\Delta\boldsymbol{\mu}$, and $\Delta\mathbf{T}^{(\nu)}$ are increments of the external loads. These increments depend on the displacements of the axial points and the angles of rotation of the attached coordinate system. The increments can be presented by relations which generalize equations (1.55) and (1.56):

$$\begin{aligned}
&\Delta\mathbf{q} = \mathrm{C}^{(2)}\mathbf{u} + \mathrm{C}^{(3)}\mathbf{u}' + \mathrm{C}^{(1)}\boldsymbol{\vartheta}\,;\\
&\Delta\mathbf{P}^{(i)} = \mathrm{B}_i^{(2)}\mathbf{u} + \mathrm{B}_i^{(3)}\mathbf{u}' + \mathrm{B}_i^{(1)}\boldsymbol{\vartheta}\,;\\
&\Delta\boldsymbol{\mu} = \mathrm{C}^{(5)}\mathbf{u} + \mathrm{C}_{\mathbf{1}}^{(6)}\mathbf{u}' + \mathrm{C}^{(4)}\boldsymbol{\vartheta}\,;\\
&\Delta\mathbf{T}^{(\nu)} = \mathrm{B}_\nu^{(5)}\mathbf{u} + \mathrm{B}_\nu^{(6)}\mathbf{u}' + \mathrm{B}_\nu^{(4)}\boldsymbol{\vartheta}\,.
\end{aligned} \tag{1.99}$$

The relationship between the increments of the coordinates of a vector and small angles of rotation of the attached axes is considered in Appendix 3 (see (A.143)). In view of (1.94) and (1.95), we obtain

$$\frac{\mathrm{d}\mathbf{Q}}{\mathrm{d}\eta} + æ_0^{(1)} \times \mathbf{Q} + \Delta æ \times \mathbf{Q} + \mathbf{P}_0 + \Delta\mathbf{P} = 0\,; \tag{1.100}$$

$$\frac{\mathrm{d}\mathbf{M}}{\mathrm{d}\eta} + æ_0^{(1)} \times \mathbf{M} + \Delta æ \times \mathbf{M} + \mathbf{e}_1 \times \mathbf{Q} + \mathbf{T}_0 + \Delta\mathbf{T} = 0\,, \tag{1.101}$$

where

$$\mathbf{P}_0 = \mathbf{q}_0 + \sum_{i=1}^{n} \mathbf{P}_0^{(i)} \delta(\eta - \eta_i)\,;$$

$$\Delta\mathbf{P} = \Delta\mathbf{q} + \sum_{i=1}^{n} \Delta\mathbf{P}^{(i)} \delta(\eta - \eta_i)\,;$$

$$\mathbf{T}_0 = \boldsymbol{\mu}_0 + \sum_{\nu=1}^{\rho} \mathbf{T}_0^{(\nu)} \delta(\eta - \eta_\nu)\,;$$

$$\Delta\mathbf{T} = \Delta\boldsymbol{\mu} + \sum_{\nu=1}^{\rho} \Delta\mathbf{T}^{(\nu)} \delta(\eta - \eta_\nu)\,.$$

The determination of the direction cosine of $\mathbf{e}_{10}$ and the components of $æ_0$ is presented in Appendix 5.

Equations (1.100) and (1.101) are exact nonlinear equations because they contain the products $\Delta æ \times \mathbf{Q}$ and $\Delta æ \times \mathbf{M}$. As noted above, the components of $\Delta æ$ are small while the components of $\mathbf{Q}$ and $\mathbf{M}$ are not necessarily small. If the components of $\Delta æ$ are small compared to the components of $\Delta æ_0^{(1)}$, then the products $\Delta æ \times \mathbf{Q}$ and $\Delta æ \times \mathbf{M}$ can be ignored. In this case, the approximate values $\tilde{\mathbf{Q}}$ and $\tilde{\mathbf{M}}$ can be determined from (1.100) and (1.101). Note that the products $\Delta æ \times \mathbf{Q}$ and $\Delta æ \times \mathbf{M}$ cannot be neglected unless $æ_{j0} \neq 0$. If $æ_{j0} = 0$, the terms $\Delta æ \times \mathbf{Q}$ and $\Delta æ \times \mathbf{M}$ are to be linearized in the following way. The components of $\mathbf{Q}$ and $\mathbf{M}$ must be replaced by those of the components Q_{j0} and M_{j0} that do not alter the initially straight configuration of the rod. For example,

$$\Delta æ \times \mathbf{Q} = \Delta æ \times Q_{10}\mathbf{e}_1 \,; \qquad \Delta æ \times \mathbf{M} = \Delta æ \times \mathbf{M}_{10}\mathbf{e}_1 \,,$$

where Q_{10} is a compressive force and M_0 is a twisting moment.

In the basis $\{\mathbf{e}_j\}$, the vector equations (1.94) and (1.95) are as follows:

$$\frac{\mathrm{d}\tilde{\mathbf{Q}}}{\mathrm{d}\eta} + æ_0 \times \tilde{\mathbf{Q}} + \mathbf{P}_0 + \Delta \mathbf{P} = 0\,; \qquad (1.102)$$

$$\frac{\mathrm{d}\tilde{\mathbf{M}}}{\mathrm{d}\eta} + æ_0 \times \tilde{\mathbf{M}} + \mathbf{e}_1 \times \tilde{\mathbf{Q}} + \mathbf{T}_0 + \Delta \mathbf{T} = 0\,. \qquad (1.103)$$

For small increments of the curvatures $\Delta æ_j$, (1.96) becomes

$$\mathbf{M} = \mathrm{A}\Delta æ\,. \qquad (1.104)$$

If the angles of rotation of the attached axes are small, the matrices L_1, L, and L_2 can be written as follows:

$$\mathrm{L}_1 = \begin{bmatrix} 1 & 0 & -\vartheta_2 \\ \vartheta_3 & 1 & 0 \\ 0 & 0 & 1 \end{bmatrix}; \qquad \mathrm{L} = \begin{bmatrix} 1 & \vartheta_3 & -\vartheta_2 \\ -\vartheta_3 & 1 & \vartheta_1 \\ \vartheta_2 & -\vartheta_1 & 1 \end{bmatrix};$$

$$\mathrm{L}_2 = \begin{bmatrix} 0 & \vartheta_3 & -\vartheta_2 \\ -\vartheta_3 & 0 & \vartheta_1 \\ \vartheta_2 & -\vartheta_1 & 0 \end{bmatrix}.$$

In view of the smallness of $\boldsymbol{\vartheta}$, we put $\mathrm{L}_1 = \tilde{\mathrm{d}}\boldsymbol{\vartheta}/\mathrm{d}\eta \approx \mathrm{d}\boldsymbol{\vartheta}/\mathrm{d}\eta$ (it means that $\mathrm{L}_1 = \mathrm{E}$) in (1.97). Therefore, for small angles ϑ_i, (1.97) takes the form

$$\frac{\mathrm{d}\boldsymbol{\vartheta}}{\mathrm{d}\eta} + æ_0 \times \boldsymbol{\vartheta} - \mathrm{A}^{-1}\mathbf{M} = 0\,. \qquad (1.105)$$

If ϑ_i is small, we have $l_{11} - 1 \approx 0$, $l_{21} = -\vartheta_3$, and $l_{31} \approx \vartheta_2$. For small $\mathbf{u}_j$, we put $æ \times \mathbf{u} \approx æ_0 \times \mathbf{u}$ and, consequently, (1.98) can be rewritten as follows:

$$\frac{d\mathbf{u}}{d\eta} + æ_0 \times \mathbf{u} - \vartheta_3 \mathbf{e}_2 + \vartheta_2 \mathbf{e}_3 = 0 . \tag{1.106}$$

The vectors $\boldsymbol{\vartheta}$, $\Delta æ$, and $\mathbf{u}$ in (1.104)–(1.106) are just approximations to their actual values, i.e. $\boldsymbol{\vartheta} \approx \tilde{\boldsymbol{\vartheta}}$, $\Delta æ \approx \Delta \tilde{æ}$, and $\mathbf{u} \approx \tilde{\mathbf{u}}$.

1.4.2 Equilibrium Equations of the Zeroth Approximation in the Attached Basis

Let us use the superscript '(0)' in the notation of the vectors that satisfy (1.102)–(1.106): $\tilde{\mathbf{Q}} = \mathbf{Q}^{(0)}$, $\tilde{\mathbf{M}} = \mathbf{M}^{(0)}$, etc. The following system of equilibrium equations

$$\frac{d\mathbf{Q}^{(0)}}{d\eta} + æ_0 \times \mathbf{Q}^{(0)} + \mathbf{P}_0 + \Delta\mathbf{P}^{(0)} = 0 ; \tag{1.107}$$

$$\frac{d\mathbf{M}^{(0)}}{d\eta} + æ_0 \times \mathbf{M}^{(0)} + \bar{\mathbf{e}}_{10} \times \bar{\mathbf{Q}}^{(0)} + \mathbf{T}_0 + \Delta\mathbf{T}^{(0)} = 0 ; \tag{1.108}$$

$$\mathbf{M}^{(0)} = \mathrm{A}\, \Delta æ^{(0)} ; \tag{1.109}$$

$$\frac{d\boldsymbol{\vartheta}^{(0)}}{d\eta} + æ_0 \times \boldsymbol{\vartheta}^{(0)} - \mathrm{A}^{-1}\mathbf{M}^{(0)} = 0 ; \tag{1.110}$$

$$\frac{d\mathbf{u}^{(0)}}{d\eta} + æ_0 \times \mathbf{u}^{(0)} - \vartheta_3^{(0)} \mathbf{e}_2 + \vartheta_2^{(0)} \mathbf{e}_3 = 0 \tag{1.111}$$

will be termed the *equilibrium equations of the zeroth approximation.* The increments of the forces (see (1.99)) in (1.107) and (1.108) are functions of $\mathbf{u}^{(0)}$, $\mathbf{u}'^{(0)}$, and $\boldsymbol{\vartheta}^{(0)}$. If the applied loads are follower, then, in the attached basis, their increments equal zero: $\Delta\mathbf{q}^{(0)} = \Delta\mathbf{P}^{(i)(0)} = \Delta\boldsymbol{\mu}^{(0)} = \Delta\mathbf{T}^{(\nu)(0)} = 0$, hence, $\Delta\mathbf{P}^{(0)} = \Delta\mathbf{T}^{(0)} = 0$.

The components of $\boldsymbol{\vartheta}^{(0)}$ are small angles of rotation of the attached axes with respect to their initial directions. The components of $\mathbf{u}^{(0)}$ are displacements of the axial points measured from their original location. The vector products $æ_0 \times \mathbf{Q}^{(0)}$, $æ_0 \times \mathbf{M}^{(0)}$, $æ_0 \times \boldsymbol{\vartheta}^{(0)}$, and $æ_0 \times \mathbf{u}^{(0)}$ in (1.107)–(1.111) can be rewritten as follows (see Sect. A.1.3 of the Appendix):

$$æ_0 \times \mathbf{Q}^{(0)} = \mathrm{A}_{æ} \mathbf{Q}^{(0)} ;$$

here

$$\mathrm{A}_{æ} = \begin{bmatrix} 0 & -æ_{30} & æ_{20} \\ æ_{30} & 0 & -æ_{10} \\ -æ_{20} & æ_{10} & 0 \end{bmatrix} .$$

(For brevity, only the formula for $æ_0 \times \mathbf{Q}^{(0)}$ is presented here.)

Thus, the following system of the four vector differential equations in the four unknowns $\mathbf{Q}^{(0)}$, $\mathbf{M}^{(0)}$, $\mathbf{u}^{(0)}$, and $\boldsymbol{\vartheta}^{(0)}$ is obtained:

$$\frac{d\mathbf{Q}^{(0)}}{d\eta} + A_{æ}\mathbf{Q}^{(0)} + \mathbf{P}_0 + \Delta\mathbf{P}^{(0)} = 0; \tag{1.112}$$

$$\frac{d\mathbf{M}^{(0)}}{d\eta} + A_{æ}\mathbf{M}^{(0)} + A_1\mathbf{Q}^{(0)} + \mathbf{T}_0 + \Delta\mathbf{T}^{(0)} = 0; \tag{1.113}$$

$$\frac{d\boldsymbol{\vartheta}^{(0)}}{d\eta} + A_{æ}\boldsymbol{\vartheta}^{(0)} - A^{-1}\mathbf{M}^{(0)} = 0; \tag{1.114}$$

$$\frac{d\mathbf{u}^{(0)}}{d\eta} + A_{æ}\mathbf{u}^{(0)} + A_1\boldsymbol{\vartheta}^{(0)} = 0; \tag{1.115}$$

here

$$A_1 = \begin{bmatrix} 0 & 0 & 0 \\ 0 & 0 & -1 \\ 0 & 1 & 0 \end{bmatrix}.$$

Once $\mathbf{M}^{(0)}$ is found, $\Delta æ^{(0)}$ can be determined from (1.109).

The equilibrium equations of the zeroth approximation in terms of projections on the attached axes can be written as follows:

$$\begin{aligned} &\frac{dQ_1^{(0)}}{d\eta} + æ_{20}Q_3^{(0)} - æ_{30}Q_2^{(0)} + P_{10} + \Delta P_1^{(0)} = 0; \\ &\frac{dQ_2^{(0)}}{d\eta} + æ_{30}Q_1^{(0)} - æ_{10}Q_3^{(0)} + P_{20} + \Delta P_2^{(0)} = 0; \\ &\frac{dQ_3^{(0)}}{d\eta} + æ_{10}Q_2^{(0)} - æ_{20}Q_1^{(0)} + P_{30} + \Delta P_3^{(0)} = 0; \end{aligned} \tag{1.116}$$

$$\begin{aligned} &\frac{dM_1^{(0)}}{d\eta} + æ_{20}M_3^{(0)} - æ_{30}M_2^{(0)} + T_{10} + \Delta T_1^{(0)} = 0; \\ &\frac{dM_2^{(0)}}{d\eta} + æ_{30}M_1^{(0)} - æ_{10}M_3^{(0)} - Q_3^{(0)} + T_{20} + \Delta T_2^{(0)} = 0; \\ &\frac{dM_3^{(0)}}{d\eta} + æ_{10}M_2^{(0)} - æ_{20}M_1^{(0)} + Q_2^{(0)} + T_{30} + \Delta T_3^{(0)} = 0; \end{aligned} \tag{1.117}$$

$$\begin{aligned} &\frac{d\vartheta_1^{(0)}}{d\eta} + æ_{20}\vartheta_3^{(0)} - æ_{30}\vartheta_2^{(0)} - \frac{1}{A_{11}}M_1^{(0)} = 0; \\ &\frac{d\vartheta_2^{(0)}}{d\eta} + æ_{30}\vartheta_1^{(0)} - æ_{10}\vartheta_3^{(0)} - \frac{1}{A_{22}}M_2^{(0)} = 0; \\ &\frac{d\vartheta_3^{(0)}}{d\eta} + æ_{10}\vartheta_2^{(0)} - æ_{20}\vartheta_1^{(0)} - \frac{1}{A_{33}}M_3^{(0)} = 0; \end{aligned} \tag{1.118}$$

$$\frac{du_1^{(0)}}{d\eta} + æ_{20}u_3^{(0)} - æ_{30}u_2^{(0)} = 0;$$

$$\frac{du_2^{(0)}}{d\eta} + æ_{30}u_1^{(0)} - æ_{10}u_3^{(0)} - \vartheta_3^{(0)} = 0;$$

$$\frac{du_3^{(0)}}{d\eta} + æ_{10}u_2^{(0)} - æ_{20}u_1^{(0)} + \vartheta_2^{(0)} = 0\,; \tag{1.119}$$

here

$$P_{j0} + \Delta P_j^{(0)} = q_{j0} + \Delta q_j^{(0)} + \sum_{i=1}^{n}(P_{j0}^{(i)} + \Delta P_j^{(i)(0)})\,\delta\,(\eta - \eta_i)\,; \tag{1.120}$$

$$T_j + \Delta T_j^{(0)} = \mu_{j0} + \Delta\mu_j^{(0)} + \sum_{\nu=1}^{\rho}(T_{j0}^{(\nu)} + \Delta T_j^{(\nu)(0)})\,\delta\,(\eta - \eta_\nu)\,. \tag{1.121}$$

1.4.3 Equilibrium Equations of the Zeroth Approximation in the Cartesian Coordinate System

Let us derive the equations governing the equilibrium of a rod on the assumption that the generalized displacements u_j and ϑ_j are small. In a Cartesian basis, the relations (1.84) and (1.85) become

$$\frac{d\mathbf{Q}_x^{(0)}}{d\eta} + \mathbf{P}_{x0} + \Delta\mathbf{P}_x^{(0)} = 0\,; \tag{1.122}$$

$$\frac{d\mathbf{M}_x^{(0)}}{d\eta} + \mathrm{A}_L\mathbf{Q}_x^{(0)} + \mathbf{T}_{x0} + \Delta\mathbf{T}_x^{(0)} = 0\,, \tag{1.123}$$

where

$$(\mathrm{L}^1)^{\mathrm{T}}\mathbf{i}_1 \times \mathbf{Q}_x^{(0)} \approx (\mathrm{L}^0)^{\mathrm{T}}\mathbf{i}_1 \times \mathbf{Q}_x^{(0)} = \mathrm{A}_L\mathbf{Q}_x^{(0)}\,;$$

$$\mathrm{A}_L = \begin{bmatrix} 0 & -l_{13}^0 & l_{12}^0 \\ l_{13}^0 & 0 & -l_{11}^0 \\ -l_{12}^0 & l_{11}^0 & 0 \end{bmatrix}.$$

Here l_{ij}^0 are the elements of the matrix L^0, which is the matrix of the transformation from the basis $\{\mathbf{i}_j\}$ to the basis $\{\mathbf{e}_{i0}\}$.

If the angles of rotation of the attached axes are small, we can put $\mathbf{M}^{(0)} = \mathrm{L}^0\mathbf{M}_x^{(0)}$, hence, (1.86) can be written as

$$\mathbf{M}_x^{(0)} = (\mathrm{L}^0)^{\mathrm{T}}\mathrm{A}\Delta æ^{(0)}\,. \tag{1.124}$$

In view of $\mathbf{M} = \mathrm{L}^0\mathbf{M}_x$, (1.87) referred to the Cartesian basis is as follows:

$$\frac{d\boldsymbol{\vartheta}^{(0)}}{d\eta} + \mathrm{A}_æ\boldsymbol{\vartheta}^{(0)} - \mathrm{A}^{(1)}\mathbf{M}_x^{(0)} = 0\,, \qquad \mathrm{A}^{(1)} = \mathrm{A}^{-1}\mathrm{L}^0\,. \tag{1.125}$$

Finally, we arrive at the following system of the vector equations in the basis $\{\mathbf{i}_j\}$:

$$\frac{d\mathbf{Q}_x^{(0)}}{d\eta} + \mathbf{P}_{x0} + \Delta\mathbf{P}_x^{(0)} = 0\,; \tag{1.126}$$

$$\frac{d\mathbf{M}_x^{(0)}}{d\eta} + \mathrm{A}_L\mathbf{Q}_x^{(0)} + \mathbf{T}_{x0} + \Delta\mathbf{T}_x^{(0)} = 0\,; \tag{1.127}$$

$$\frac{d\boldsymbol{\vartheta}^{(0)}}{d\eta} + \mathrm{A}_{æ}\boldsymbol{\vartheta}^{(0)} - \mathrm{A}^{(1)}\mathbf{M}_x^{(0)} = 0\,; \tag{1.128}$$

$$\frac{d\mathbf{u}_x^{(0)}}{d\eta} + \mathrm{A}_L^{(1)}\boldsymbol{\vartheta}^{(0)} = 0\,; \tag{1.129}$$

here

$$\mathrm{A}_L^{(1)} = \begin{bmatrix} 0 & l_{31}^0 & -l_{21}^0 \\ 0 & l_{32}^0 & -l_{22}^0 \\ 0 & l_{33}^0 & -l_{23}^0 \end{bmatrix}.$$

In terms of projections on the Cartesian coordinate axes, the system of equations (1.126)–(1.129) is as follows:

$$\frac{dQ_{x_1}^{(0)}}{d\eta} + P_{x_10} + \Delta P_{x_1} = 0\,;$$

$$\frac{dQ_{x_2}^{(0)}}{d\eta} + P_{x_20} + \Delta P_{x_2} = 0\,;$$

$$\frac{dQ_{x_3}^{(0)}}{d\eta} + P_{x_30} + \Delta P_{x_3} = 0\,; \tag{1.130}$$

$$\frac{dM_{x_1}^{(0)}}{d\eta} + l_{12}^0 Q_{x_3}^{(0)} - l_{13}^0 Q_{x_2}^{(0)} + T_{x_10} + \Delta T_{x_1} = 0\,;$$

$$\frac{dM_{x_2}^{(0)}}{d\eta} + l_{13}^0 Q_{x_1}^{(0)} - l_{11}^0 Q_{x_3}^{(0)} + T_{x_20} + \Delta T_{x_2} = 0\,;$$

$$\frac{dM_{x_3}^{(0)}}{d\eta} + l_{11}^0 Q_{x_2}^{(0)} - l_{12}^0 Q_{x_1}^{(0)} + T_{x_30} + \Delta T_{x_3} = 0\,; \tag{1.131}$$

$$\frac{d\vartheta_1^{(0)}}{d\eta} - æ_{30}\vartheta_2^{(0)} + æ_{20}\vartheta_3^{(0)} - \sum_{j=1}^{3}\frac{l_{1j}^0}{A_{11}}M_{xj}^{(0)} = 0\,;$$

$$\frac{d\vartheta_2^{(0)}}{d\eta} + æ_{30}\vartheta_1^{(0)} - æ_{10}\vartheta_3^{(0)} - \sum_{j=1}^{3}\frac{l_{2j}^0}{A_{22}}M_{xj}^{(0)} = 0\,;$$

$$\frac{d\vartheta_3^{(0)}}{d\eta} - æ_{20}\vartheta_1^{(0)} + æ_{10}\vartheta_2^{(0)} - \sum_{j=1}^{3}\frac{l_{3j}^0}{A_{33}}M_{xj}^{(0)} = 0\,; \tag{1.132}$$

$$\frac{du_{x_1}^{(0)}}{d\eta} + l_{31}^0\vartheta_2^{(0)} - l_{21}^0\vartheta_3^{(0)} = 0\,;$$

$$\frac{du_{x2}^{(0)}}{d\eta} + l_{32}^0 \vartheta_2^{(0)} - l_{22}^0 \vartheta_3^{(0)} = 0 ;$$
$$\frac{du_{x3}^{(0)}}{d\eta} + l_{33}^0 \vartheta_2^{(0)} - l_{23}^0 \vartheta_3^{(0)} = 0 . \quad (1.133)$$

1.4.4 Increments of External Loads

Let us examine in detail the expressions for increments of the external loads $\Delta \mathbf{q}$, $\Delta \mathbf{P}^{(i)}$, $\Delta \boldsymbol{\mu}$, and $\Delta \mathbf{T}^{(\nu)}$. These increments appear in $\Delta \mathbf{P}^{(0)}$ and $\Delta \mathbf{T}^{(0)}$. If the displacements u_j and the angles of rotation of the attached axes ϑ_j are small, then the change of external loads can also be treated as small. Hence, we have $\mathbf{P}^{(i)} = \mathbf{P}_0^{(i)} + \Delta \mathbf{P}^{(i)(0)}$, $\mathbf{T}^{(\nu)} = \mathbf{T}_0^{(\nu)} + \Delta \mathbf{T}^{(\nu)(0)}$, $\mathbf{q} = \mathbf{q}_0 + \Delta \mathbf{q}^{(0)}$, and $\boldsymbol{\mu} = \boldsymbol{\mu}_0 + \Delta \boldsymbol{\mu}^{(0)}$, where the components of $\mathbf{P}_0^{(i)}$, $\mathbf{T}_0^{(\nu)}$, $\mathbf{q}_0$, and $\boldsymbol{\mu}_0$ are known while $\Delta \mathbf{P}^{(i)(0)}$, $\Delta \mathbf{T}^{(\nu)(0)}$, $\Delta \mathbf{q}^{(0)}$, and $\Delta \boldsymbol{\mu}^{(0)}$ are vector increments whose components are small. In the general case, the components of $\Delta \mathbf{P}_0^{(i)(0)}$, $\Delta \mathbf{T}_0^{(\nu)(0)}$, $\Delta \mathbf{q}^{(0)}$, and $\Delta \boldsymbol{\mu}^{(0)}$ are functions of the known components of $\mathbf{P}_0^{(i)}$, $\mathbf{T}_0^{(\nu)}$, $\mathbf{q}_0^{(0)}$, and $\boldsymbol{\mu}_0^{(0)}$, respectively, and the components of $\mathbf{u}$ and $\boldsymbol{\vartheta}$ (see (1.155) and (1.156)).

Suppose that the applied loads are dead. Their components are constants in the Cartesian basis. In the attached basis $\{\mathbf{e}_{j0}\}$, we have $\mathbf{P}_0^{(i)} = \mathrm{L}^0 \mathbf{P}_x^{(i)}$, where L^0 is the matrix of transformation from the basis $\{\mathbf{i}_j\}$ to the basis $\{\mathbf{e}_{j0}\}$. In the basis $\{\mathbf{e}_j\}$, we have

$$\mathbf{P}^{(i)} = \mathrm{L}\mathrm{L}^0 \mathbf{P}_x^{(i)} , \quad (1.134)$$

where L is the matrix of transformation from the basis $\{\mathbf{e}_{j0}\}$ to the basis $\{\mathbf{e}_i\}$. The matrix L can be presented as a sum $\mathrm{L} = \mathrm{E} + \Delta \mathrm{L}_1$; for small values of ϑ_j, we have

$$\Delta \mathrm{L}_1 = \begin{bmatrix} 0 & \vartheta_3 & -\vartheta_2 \\ -\vartheta_3 & 0 & \vartheta_1 \\ \vartheta_2 & -\vartheta_1 & 0 \end{bmatrix} . \quad (1.135)$$

Consequently, (1.134) can be rewritten as follows:

$$\mathbf{P}^{(i)} = \mathrm{L}^0 \mathbf{P}_x^{(i)} + \Delta \mathrm{L}_1 \mathrm{L}^0 \mathbf{P}_x^{(i)} \quad (1.136)$$

or

$$\mathbf{P}^{(i)} = \mathbf{P}_0^{(i)} + \Delta \mathbf{P}^{(i)} = \mathbf{P}_0^{(i)} + \mathrm{B}_i^{(1)} \boldsymbol{\vartheta} ; \quad (1.137)$$

here

$$\mathrm{B}_i^{(1)} = \begin{bmatrix} 0 & -P_{30}^{(i)} & P_{20}^{(i)} \\ P_{30}^{(i)} & 0 & -P_{10}^{(i)} \\ -P_{20}^{(i)} & P_{10}^{(i)} & 0 \end{bmatrix} . \quad (1.138)$$

Similarly to (1.137), we can write

$$\mathbf{q} = \mathbf{q}_0 + \Delta\mathbf{q}; \qquad \boldsymbol{\mu} = \boldsymbol{\mu}_0 + \Delta\boldsymbol{\mu}; \qquad \mathbf{T}^{(\nu)} = \mathbf{T}_0^{(\nu)} + \Delta\mathbf{T}^{(\nu)};$$

here

$$\Delta\mathbf{q} = \mathrm{C}^{(1)}\boldsymbol{\vartheta}; \qquad \Delta\boldsymbol{\mu} = \mathrm{C}^{(3)}\boldsymbol{\vartheta}; \qquad \Delta\mathbf{T}^{(\nu)} = \mathrm{B}_{\vartheta}^{(3)}\boldsymbol{\vartheta}, \tag{1.139}$$

where

$$\mathrm{C}^{(1)} = \begin{bmatrix} 0 & -q_{30} & q_{20} \\ q_{30} & 0 & -q_{10} \\ -q_{20} & q_{10} & 0 \end{bmatrix};$$
$$\mathrm{C}^{(3)} = \begin{bmatrix} 0 & -\mu_{30} & \mu_{20} \\ \mu_{30} & 0 & -\mu_{10} \\ -\mu_{20} & \mu_{10} & 0 \end{bmatrix};$$
$$\mathrm{B}_{\nu}^{(3)} = \begin{bmatrix} 0 & -T_{30}^{(\nu)} & T_{20}^{(\nu)} \\ T_{30}^{(\nu)} & 0 & -T_{10}^{(\nu)} \\ -T_{20}^{(\nu)} & T_{10}^{(\nu)} & 0 \end{bmatrix}. \tag{1.140}$$

For example, in the attached basis, the vector $\mathbf{q}$ has the following components:

$$\begin{aligned} q_1 &= q_{10} + q_{20}\vartheta_3 - q_{30}\vartheta_2; \\ q_2 &= q_{20} + q_{30}\vartheta_1 - q_{10}\vartheta_3; \\ q_3 &= q_{30} + q_{10}\vartheta_2 - q_{20}\vartheta_1. \end{aligned} \tag{1.141}$$

In the equilibrium equations of the zeroth approximation (1.112) and (1.113), the increments $\Delta\mathbf{q}^{(0)}$, $\Delta\mathbf{P}^{(i)(0)}$, $\Delta\boldsymbol{\mu}^{(0)}$, and $\Delta\mathbf{T}^{(\nu)(0)}$, which appear in $\Delta\mathbf{P}^{(0)}$ and $\Delta\mathbf{T}^{(0)}$, are functions of $\boldsymbol{\vartheta}^{(0)}$:

$$\begin{aligned} \Delta\mathbf{q}^{(0)} &= \mathrm{C}^{(1)}\boldsymbol{\vartheta}^{(0)}; \qquad & \Delta\mathbf{P}^{(i)(0)} &= \mathrm{B}_i^{(1)}\boldsymbol{\vartheta}^{(0)}; \\ \Delta\boldsymbol{\mu}^{(0)} &= \mathrm{C}^{(3)}\boldsymbol{\vartheta}^{(0)}; \qquad & \Delta\mathbf{T}^{(\nu)(0)} &= \mathrm{B}_{\nu}^{(3)}\boldsymbol{\vartheta}^{(0)}. \end{aligned} \tag{1.142}$$

Similar relations for the increments can be obtained when external loads depend on the displacement vector $\mathbf{u}$.

External loads of more complicated nature (all the forces are directed at a fixed point, see Fig. 1.16) were discussed in Sect. 1.2. The vectors of the zeroth approximation $\mathbf{Q}^{(0)}$, $\mathbf{M}^{(0)}$, $\boldsymbol{\vartheta}^{(0)}$, $\Delta\boldsymbol{æ}^{(0)}$, $\mathbf{u}^{(0)}$ can be used to evaluate a more accurate approximation. Let $\mathbf{Q} = \mathbf{Q}^{(0)} + \mathbf{Q}^{(1)}$, $\mathbf{M} = \mathbf{M}^{(0)} + \mathbf{M}^{(1)}$, $\mathbf{u} = \mathbf{u}^{(0)} + \mathbf{u}^{(1)}$, $\boldsymbol{\vartheta} = \boldsymbol{\vartheta}^{(0)} + \boldsymbol{\vartheta}^{(1)}$, and $\mathbf{M}^{(1)} = \mathrm{A}\Delta\boldsymbol{æ}^{(1)}$. The vectors with the superscript '(1)' will be called vectors of the first approximation. Unlike the components of $\mathbf{Q}^{(0)}$ and $\mathbf{M}^{(0)}$, the components of $\mathbf{Q}^{(1)}$ and $\mathbf{M}^{(1)}$ are small.

1.4.5 Equilibrium Equations of the First Approximation in the Attached Coordinate System

Let $\mathbf{Q} = \mathbf{Q}^{(0)} + \mathbf{Q}^{(1)}$, $æ = æ_0 + \Delta æ^{(0)} + \Delta æ^{(1)}$, and $\mathbf{P} = \mathbf{P}_0 + \Delta\mathbf{P}^{(0)} + \Delta\mathbf{P}^{(1)}$ in (1.94), and let the left-hand side of (1.94) be denoted by $\mathbf{b}$. As a result, we have

$$\mathbf{b} = \frac{\mathrm{d}\mathbf{Q}}{\mathrm{d}\eta} + æ \times \mathbf{Q} + \mathbf{P} = 0\,. \tag{1.143}$$

The equality (1.143) must hold for any changes of the vectors $\mathbf{Q}$, $æ$, and $\mathbf{P}$. Hence, the linearized equation (1.143)

$$\mathbf{b} = \mathbf{b}_0 + \Delta\mathbf{b} = 0$$

yields two equations

$$\mathbf{b}_0 = 0 \tag{1.144}$$

and

$$\Delta\mathbf{b} = 0\,. \tag{1.145}$$

On rearrangement, (1.144) and (1.145) are as follows:

$$\mathbf{b}_0 = \frac{\mathrm{d}\mathbf{Q}^{(0)}}{\mathrm{d}\eta} + æ_0 \times \mathbf{Q}^{(0)} + \mathbf{P}_0 + \Delta\mathbf{P}^{(0)} = 0\,; \tag{1.146}$$

$$\Delta\mathbf{b} = \frac{\mathrm{d}\mathbf{Q}^{(1)}}{\mathrm{d}\eta} + æ_0^{(0)} \times \mathbf{Q}^{(1)} + \Delta æ^{(1)} \times \mathbf{Q}^{(0)} + \Delta æ^{(0)} \times \mathbf{Q}^{(0)} + \Delta\mathbf{P}^{(1)} = 0\,; \tag{1.147}$$

here $æ_0^{(0)} = æ_0 + \Delta æ(0)$.

The vectors $\mathbf{Q}^{(1)}$ and $\Delta æ^{(1)}$ in (1.147) are unknown quantities, while the vectors $\mathbf{Q}^{(0)}$ and $\Delta æ^{(0)}$ can be determined from the equations of the zeroth approximation.

Applying this method to the system of equations (1.95)–(1.98), we readily obtain the equations of the zeroth and the first approximation.

(1) The equations of the zeroth approximation coincide with the system of equations (1.107)–(1.111):

$$\frac{\mathrm{d}\mathbf{Q}^{(0)}}{\mathrm{d}\eta} + æ_0 \times \mathbf{Q}^{(0)} + \mathbf{P}_0 + \Delta\mathbf{P}^{(0)} = 0\,; \tag{1.148}$$

$$\frac{\mathrm{d}\mathbf{M}^{(0)}}{\mathrm{d}\eta} + æ_0 \times \mathbf{M}^{(0)} + \mathbf{e}_1 \times \mathbf{Q}^{(0)} + \mathbf{T}_0 + \Delta\mathbf{T}^{(0)} = 0\,; \tag{1.149}$$

$$\frac{\mathrm{d}\boldsymbol{\vartheta}^{(0)}}{\mathrm{d}\eta} + æ_0 \times \boldsymbol{\vartheta}^{(0)} - \mathrm{A}^{-1}\mathbf{M}^{(0)} = 0\,, \qquad \Delta æ^{(0)} = \mathrm{A}^{-1}\mathbf{M}^{(0)}\,; \tag{1.150}$$

$$\frac{\mathrm{d}\mathbf{u}^{(0)}}{\mathrm{d}\eta} + æ_0 \times \mathbf{u}^{(0)} + \mathrm{A}_1\boldsymbol{\vartheta}^{(0)} = 0\,, \tag{1.151}$$

For example, in the case of dead loads, the increments of the vectors entering into $\Delta\mathbf{P}^{(0)}$ and $\Delta\mathbf{T}^{(0)}$ (for simplicity, let the increments be functions of $\boldsymbol{\vartheta}$ alone) are

$$\begin{aligned}
&\Delta\mathbf{q}^{(0)} = \mathrm{C}^{(1)}\boldsymbol{\vartheta}^{(0)}; \qquad \Delta\mathbf{P}^{(i)(0)} = \mathrm{B}_i^{(1)}\boldsymbol{\vartheta}^{(0)}\delta(\eta-\eta_i);\\
&\Delta\boldsymbol{\mu}^{(0)} = \mathrm{C}^{(3)}\boldsymbol{\vartheta}^{(0)}; \qquad \Delta\mathbf{T}^{(\nu)(0)} = \mathrm{B}_\nu^{(3)}\boldsymbol{\vartheta}^{(0)}\delta(\eta-\eta_\nu).
\end{aligned} \tag{1.152}$$

Hence, we have

$$\begin{aligned}
&\Delta\mathbf{P}^{(0)} = \mathrm{C}^{(1)}\boldsymbol{\vartheta}^{(0)} + \sum_{i=1}^{n}\mathrm{B}_i^{(1)}\boldsymbol{\vartheta}^{(0)}\delta(\eta-\eta_i);\\
&\Delta\mathbf{T}^{(0)} = \mathrm{C}^{(3)}\boldsymbol{\vartheta}^{(0)} + \sum_{\nu=1}^{\rho}\mathrm{B}_\nu^{(3)}\boldsymbol{\vartheta}^{(0)}\delta(\eta-\eta_\nu).
\end{aligned} \tag{1.153}$$

In the case of follower loads, we have

$$\Delta\mathbf{q}^{(0)} = \Delta\mathbf{P}^{(i)(0)} = \Delta\boldsymbol{\mu}^{(0)} = \Delta\mathbf{T}^{(\nu)(0)} = 0.$$

(2) The equations of the first approximation are as follows:

$$\frac{\mathrm{d}\mathbf{Q}^{(1)}}{\mathrm{d}\eta} + \text{æ}_0^{(0)}\times\mathbf{Q}^{(1)} + \Delta\text{æ}^{(1)}\times\mathbf{Q}^{(0)}$$

$$= -\Delta\mathbf{q}^{(1)} - \Delta\text{æ}^{(0)}\times\mathbf{Q}^{(0)} - \sum_{i=1}^{n}\Delta\mathbf{P}^{(i)(1)}\delta_i; \tag{1.154}$$

$$\frac{\mathrm{d}\mathbf{M}^{(1)}}{\mathrm{d}\eta} + \text{æ}_0^{(0)}\times\mathbf{M}^{(1)} + \Delta\text{æ}^{(1)}\times\mathbf{M}^{(0)} + \mathrm{A}_1\mathbf{Q}^{(1)}$$

$$= -\Delta\boldsymbol{\mu}^{(1)} - \Delta\text{æ}^{(0)}\times\mathbf{M}^{(0)} - \sum_{\nu=1}^{\rho}\Delta\mathbf{T}^{(\nu)(1)}\delta_\nu; \tag{1.155}$$

$$\frac{\mathrm{d}\boldsymbol{\vartheta}^{(1)}}{\mathrm{d}\eta} + \text{æ}_0^{(0)}\times\boldsymbol{\vartheta}^{(1)} - \mathrm{A}^{-1}\mathbf{M}^{(1)} = 0, \qquad \Delta\text{æ}^{(1)} = \mathrm{A}^{-1}\mathbf{M}^{(1)}; \tag{1.156}$$

$$\frac{\mathrm{d}\mathbf{u}^{(1)}}{\mathrm{d}\eta} + \text{æ}_0^{(0)}\times\mathbf{u}^{(1)} + \mathrm{A}_1\boldsymbol{\vartheta}^{(1)} = 0, \tag{1.157}$$

For example, in the case of dead loads, the increments of the vectors in the expressions for $\Delta\mathbf{P}^{(1)}$ and $\Delta\mathbf{T}^{(1)}$ are

$$\begin{aligned}
&\Delta\mathbf{q}^{(1)} = \mathrm{A}_q\boldsymbol{\vartheta}^{(1)}; \qquad \Delta\mathbf{P}^{(i)(1)} = \mathrm{A}_P^{(i)}\boldsymbol{\vartheta}^{(1)}\delta(\eta-\eta_i);\\
&\Delta\boldsymbol{\mu}^{(1)} = \mathrm{A}_\mu\boldsymbol{\vartheta}^{(1)}; \qquad \Delta\mathbf{T}^{(\nu)(1)} = \mathrm{A}_T^{(\nu)}\boldsymbol{\vartheta}^{(1)}\delta(\eta-\eta_\nu).
\end{aligned} \tag{1.158}$$

Hence, we have

$$\Delta \mathbf{P}^{(1)} = \mathrm{C}^{(1)}\boldsymbol{\vartheta}^{(1)} + \sum_{i=1}^{n} \mathrm{B}_i^{(1)}\boldsymbol{\vartheta}^{(1)}\delta(\eta - \eta_i);$$

$$\Delta \mathbf{T}^{(1)} = \mathrm{C}^{(3)}\boldsymbol{\vartheta}^{(1)} + \sum_{\nu=1}^{\rho} \mathrm{B}_\nu^{(3)}\boldsymbol{\vartheta}^{(1)}\delta(\eta - \eta_\nu). \tag{1.159}$$

(For follower loads, we have $\Delta \mathbf{q}^{(1)} = \Delta \boldsymbol{\mu}^{(1)} = \Delta \mathbf{P}^{(i)(1)} = \Delta \mathbf{T}^{(\nu)(1)} = 0$.)

The equations of the first approximation can be written as follows:

$$\frac{\mathrm{d}\mathbf{Q}^{(1)}}{\mathrm{d}\eta} + \mathrm{A}_{æ}^{(0)}\mathbf{Q}^{(1)} + \mathrm{A}_Q^{(0)}\Delta æ^{(1)} + \mathrm{C}^{(1)}\boldsymbol{\vartheta}^{(1)}$$

$$= -\Delta æ^{(0)} \times \mathbf{Q}^{(0)} - \sum_{i=1}^{n} \mathrm{B}_i^{(1)}\boldsymbol{\vartheta}^{(1)}\delta(\eta - \eta_i); \tag{1.160}$$

$$\frac{\mathrm{d}\mathbf{M}^{(1)}}{\mathrm{d}\eta} + \mathrm{A}_{æ}^{(0)}\mathbf{M}^{(1)} + \mathrm{A}_M^{(0)}\Delta æ^{(1)} + \mathrm{A}_1\mathbf{Q}^{(1)} + \mathrm{C}^{(2)}\boldsymbol{\vartheta}^{(1)}$$

$$= -\Delta æ^{(0)} \times \mathbf{M}^{(0)} - \sum_{\nu=1}^{\rho} \mathrm{B}_\nu^{(3)}\boldsymbol{\vartheta}^{(1)}\delta(\eta - \eta_\nu); \tag{1.161}$$

$$\frac{\mathrm{d}\boldsymbol{\vartheta}^{(1)}}{\mathrm{d}\eta} + \mathrm{A}_{æ}^{(0)}\boldsymbol{\vartheta}^{(1)} - \mathrm{A}^{-1}\mathbf{M}^{(1)} = 0; \tag{1.162}$$

$$\frac{\mathrm{d}\mathbf{u}^{(1)}}{\mathrm{d}\eta} + \mathrm{A}_{æ}^{(0)}\mathbf{u}^{(1)} + \mathrm{A}_1\boldsymbol{\vartheta}^{(1)} = 0; \tag{1.163}$$

here

$$\mathrm{A}_{æ}^{(0)} = \begin{bmatrix} 0 & -æ_{30}^{(0)} & æ_{20}^{(0)} \\ æ_{30}^{(0)} & 0 & -æ_{10}^{(0)} \\ -æ_{20}^{(0)} & æ_{10}^{(0)} & 0 \end{bmatrix};$$

$$\mathrm{A}_Q^{(0)} = \begin{bmatrix} 0 & Q_3^{(0)} & -Q_2^{(0)} \\ -Q_3^{(0)} & 0 & Q_1^{(0)} \\ Q_2^{(0)} & -Q_1^{(0)} & 0 \end{bmatrix};$$

$$\mathrm{A}_M^{(0)} = \begin{bmatrix} 0 & M_3^{(0)} & -M_2^{(0)} \\ -M_3^{(0)} & 0 & M_1^{(0)} \\ M_2^{(0)} & -M_1^{(0)} & 0 \end{bmatrix};$$

$$\Delta æ^{(1)} = \mathrm{A}^{-1}\mathbf{M}^{(1)}; \qquad æ_{j0}^{(0)} = æ_{j0} + \Delta æ_j^{(0)}.$$

By means of the approach presented above, the equations of the second or higher approximation can be obtained (see Appendix 3).

The application of numerical methods to the equations of the first and higher approximation is discussed in Chap. 2. In most applications and theoretical courses, only equations of the zeroth approximation (1.107)–(1.111) are considered. The validity of the assumption of smallness of displacements of the axial points and the angles of rotation, as well as the assumption of smallness of the components of $\mathbf{Q}^{(1)}$ and $\mathbf{M}^{(1)}$ is usually out of discussion. Using (1.154)–(1.157) (or (1.164)–(1.168)), one can estimate the accuracy of the solutions of the equations of the zeroth approximation. For example, if $Q_j^{(1)}$, $M_j^{(1)}$, $u_j^{(1)}$, $\vartheta_j^{(1)}$ and $\Delta æ_j^{(1)}$ are found, we can determine the absolute values of the maximum deviations on the interval of integration $0 \leq \eta \leq \eta_k$:

$$\max \Delta_{Q_j} = \max \left| \frac{Q_j^{(0)} + Q_j^{(1)}}{Q_j^{(0)}} \right| ; \qquad \max \Delta_{M_j} = \left| \frac{M_j^{(0)} + M_j^{(1)}}{M_j^{(0)}} \right| ; \quad \ldots$$

$$\max \Delta_{\Delta æ_j} = \left| \frac{\Delta æ_j^{(0)} + \Delta æ_j^{(1)}}{\Delta æ_j^{(0)}} \right| .$$

If the following inequalities are satisfied

$$\max \Delta_{Q_j} \leq \Delta ; \quad \max \Delta_{M_j} \leq \Delta ; \quad \ldots ; \quad \max \Delta_{\Delta æ_j} \leq \Delta , \qquad (1.164)$$

where $j = 1, 2, 3$ and Δ is the admissible error, we conclude that the solution of the equations of the zeroth approximation yields the required accuracy. If at least one of the inequalities (1.64) fails, the solution is to be found as a sum of the zeroth and the first approximations: $Q_j = Q_j^{(0)} + Q_j^{(1)}$, $M_j = M_j^{(0)} + M_j^{(1)}$, etc. In such a situation, to estimate the accuracy of this solution, we are to use the equations of the second approximation. Thus, we get the following relations: $Q_j = Q_j^{(0)} + Q_j^{(1)} + Q_j^{(2)}$, $M_j = M_j^{(0)} + M_j^{(1)} + M_j^{(2)}$, etc. The corresponding deviations are $\max \Delta_{Q_j}$, $\max \Delta_{M_j}$, etc. If the inequalities

$$\max \Delta_{Q_j} \leq \Delta ; \qquad \max \Delta_{M_j} \leq \Delta ; \qquad \ldots \qquad (1.165)$$

are valid, we conclude that the solution is of the required accuracy.

1.5 Problems

•*1.1.* [4] An axis of a rod in both deformed and undeformed configurations is a plane curve (Fig. 1.19). Derive the equilibrium equations from the general equilibrium equations (1.64)–(1.66). Determine the conditions under which equilibrium of the rod may take place, i.e. what are the necessary and sufficient conditions for the rod axis to undergo an in-plane deformation?

•*1.2.* A cantilever beam of circular cross section is subjected to a dead force $\mathbf{P}^{(1)}$ and a distributed follower load $\mathbf{q}$ (Fig. 1.20). During the deformation, the loads $\mathbf{P}^{(1)}$ and $\mathbf{q}$ lie in the coordinate plane $x_1 O x_2$. Assume that the beam axis remains a plane curve during the deformation. Derive the equilibrium equations of the zeroth approximation referred to the attached axes (assume that the displacements of the axial points are small).

[4] The solutions of the problems labeled with a bullet can be found in the Appendix.

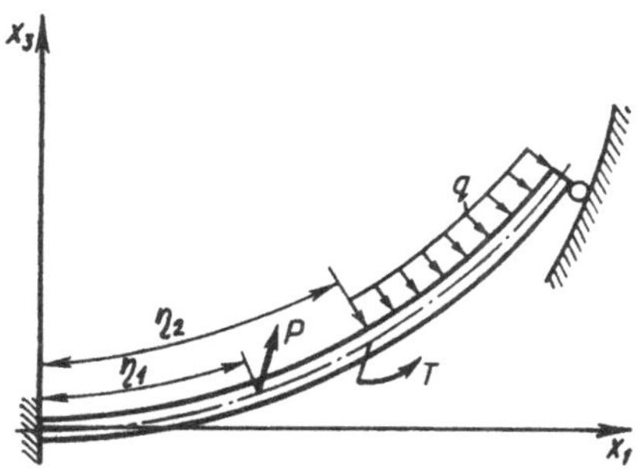

Fig. 1.19.

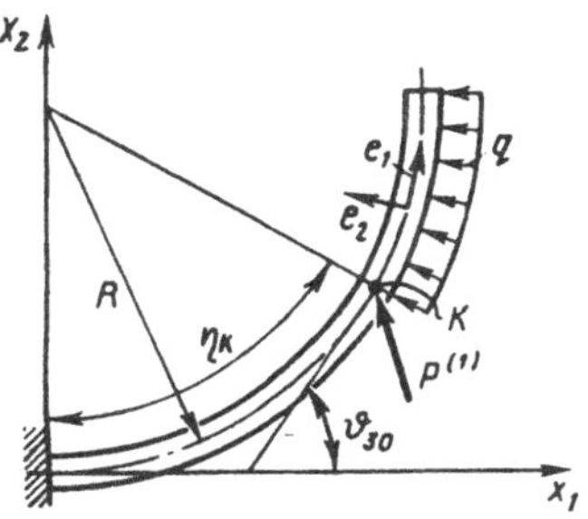

Fig. 1.20.

•*1.3.* A cantilever beam of circular cross section is rigidly mounted on an object moving with an acceleration **a**. The vector of acceleration is parallel to the plane x_1Ox_2 (Fig. 1.21). Assume that displacements of the axial points are small. Let m_0 be the mass of the beam per unit length. A mass point m is joined to the beam. In the natural configuration, the beam axis is a plane curve lying in the plane x_1Ox_2. Derive the equilibrium equations of the zeroth approximation for the beam.

•*1.4.* Consider a rod of circular cross section rigidly mounted on an object rotating with an angular velocity ω (Fig. 1.22). In the natural configuration, the rod axis is straight. The axis of rotation is perpendicular to the plane x_1Ox_2 at the origin. Examine the forces acting on the rod. Find their components in the attached basis. Assume that displacements of the axial points are large.

•*1.5.* Determine the components of the follower force (Fig. 1.23) in an immovable (Cartesian) and attached coordinate systems.

•*1.6.* A ring of circular cross section is subjected to a distributed twisting moment μ_1 of constant magnitude (Fig. 1.24). Determine the stress field in the ring.

•*1.7.* Consider a ring of asymmetrical cross section ($A_{33} \neq A_{22}$) (see Fig. 1.24). Derive the relationship between the angle of rotation of a cross section and μ_1.

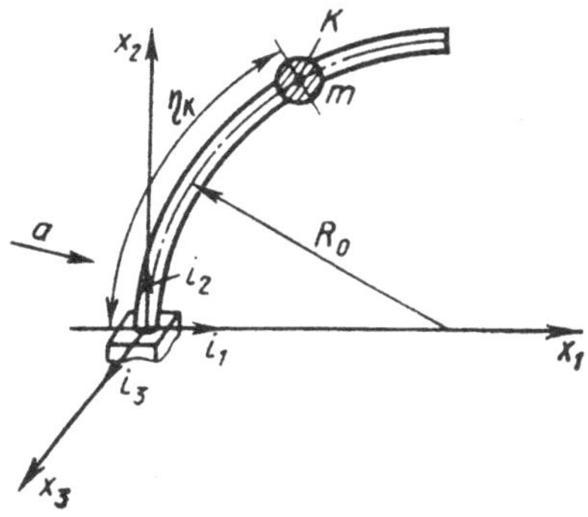

Fig. 1.21.

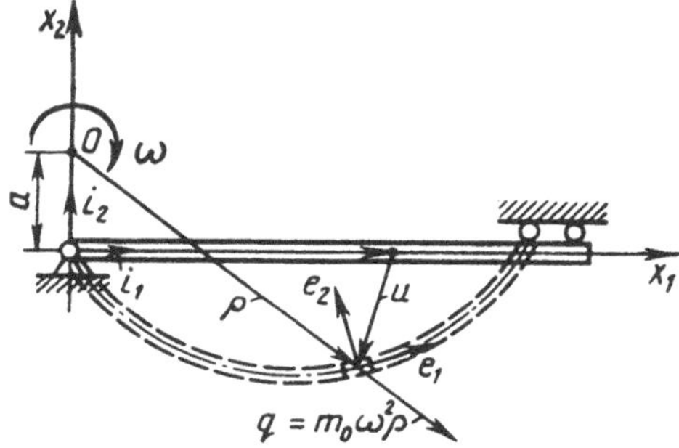

Fig. 1.22.

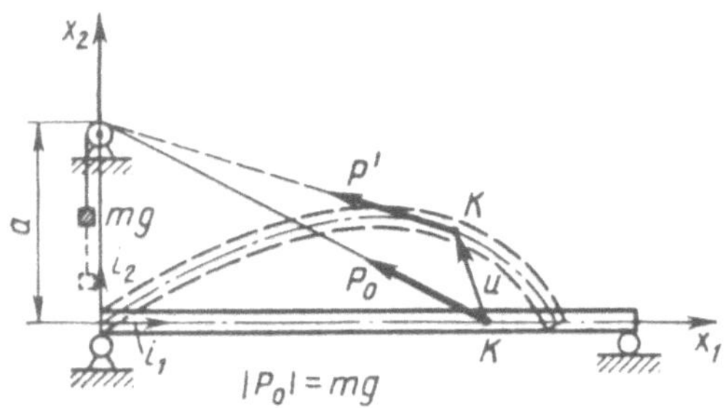

Fig. 1.23.

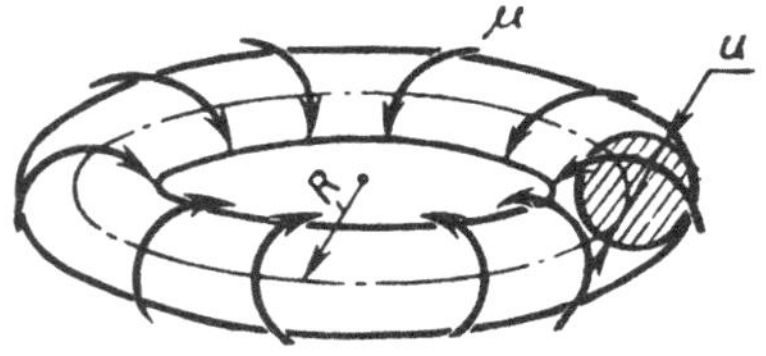

Fig. 1.24.

2. Integration of Equilibrium Equations

In this chapter, some computer-oriented methods for the analysis of linear and nonlinear equilibrium equations are discussed. For nonlinear equations, the method of step-by-step loading is presented. The key feature of this method is that we solve linear equations at each step of loading. Of course, there are other methods used in analyses of applied problems.

It is well known that the solution of the Cauchy problem for an ordinary differential equation must satisfy some conditions at one end of the interval of integration. Unlike the Cauchy problem, a problem of statics of rods is a two-point boundary-value problem, i.e. its solution must satisfy some conditions at both ends of the interval of integration. In the general case, the methods derived for the Cauchy problem cannot be applied to problems of statics of rods. At present, there exist widely used methods: the method of initial parameters, the sweep method (Balabukh et al. (1984)), the finite element method (Smirnov et al. (1981)), etc. Generally, solutions of linear equilibrium equations, say, (1.112)–(1.115), cannot be found explicitly unless the elements of the matrix $A_{æ}$ are constants. In more detail this situation is examined in Sect. 5.2 by the example of helical rods (cylindrical springs). In the general case, linear differential equations with variable coefficients can be solved only approximately.

2.1 Integration of Linear Equilibrium Equations

2.1.1 Equilibrium Equations of the Zeroth Approximation

The general form of the equilibrium equations of the zeroth approximation related to the attached coordinate system, (1.112)–(1.115), and those related to the Cartesian coordinate system, (1.130)–(1.133), were obtained in Sect. 1.4. These equations are valid for arbitrary external loads. Let us consider particular types of external loads.

Follower Forces In the case of follower forces, the equations referred to the attached frame, (1.112)–(1.115), are best suited for numerical analysis. We have

$$\frac{d\mathbf{Q}^{(0)}}{d\eta} + A_{æ}\mathbf{Q}^{(0)} = -\mathbf{P}_0 ; \tag{2.1}$$

$$\frac{d\mathbf{M}^{(0)}}{d\eta} + A_{æ}\mathbf{M}^{(0)} + A_1\mathbf{Q}^{(0)} = -\mathbf{T}_0 ; \tag{2.2}$$

$$\frac{d\boldsymbol{\vartheta}^{(0)}}{d\eta} + A_{æ}\boldsymbol{\vartheta}^{(0)} - A^{-1}\mathbf{M}^{(0)} = 0 ; \tag{2.3}$$

$$\frac{d\mathbf{u}^{(0)}}{d\eta} + A_{æ}\mathbf{u}^{(0)} + A_1\boldsymbol{\vartheta}^{(0)} = 0 . \tag{2.4}$$

The system of equations (2.1)–(2.4) can be written as a vector equation

$$\mathbf{Y}^{(0)\prime} + A^{(0)}\mathbf{Y}^{(0)} = \mathbf{f}^{(0)} , \tag{2.5}$$

where

$$\mathbf{Y}^{(0)} = \begin{bmatrix} \mathbf{Q}^{(0)} \\ \mathbf{M}^{(0)} \\ \boldsymbol{\vartheta}^{(0)} \\ \mathbf{u}^{(0)} \end{bmatrix} ; \qquad \mathbf{f}^{(0)} = \begin{bmatrix} -\mathbf{P}_0 \\ -\mathbf{T}_0 \\ \mathbf{0} \\ \mathbf{0} \end{bmatrix} ;$$

$$A^{(0)} = \begin{bmatrix} A_{æ} & 0 & 0 & 0 \\ A_1 & A_{æ} & 0 & 0 \\ 0 & -A^{-1} & A_{æ} & 0 \\ 0 & 0 & A_1 & A_{æ} \end{bmatrix} .$$

The elements of the matrix $A^{(0)}$ are functions of η, hence, numerical methods are to be applied. The general solution of the linear inhomogeneous equation may be written as

$$\mathbf{Y}^{(0)} = K(\eta)\,\mathbf{C} + \mathbf{Y}_1^{(0)}(\eta) , \tag{2.6}$$

where $K(\eta)$ is the fundamental matrix, $\mathbf{C}$ is an initial-data vector, and $\mathbf{Y}_1^{(0)}$ is a partial solution of the inhomogeneous equation. The components of $\mathbf{C}$ are to be found from the boundary conditions at $\eta = 0$ and $\eta = 1$.

Consider now the method of numerical determination of the matrix $K(\eta)$ and the partial solution $\mathbf{Y}_1^{(0)}$. The matrix $K(\eta)$ can be derived from the homogeneous equation

$$\mathbf{Y}^{(0)\prime} + A^{(0)}\mathbf{Y}^{(0)} = 0 , \tag{2.7}$$

which is to be solved 12 times for the following initial-data vectors:

$$(1)\begin{bmatrix} 1 \\ 0 \\ 0 \\ \vdots \\ 0 \\ 0 \end{bmatrix} ; \quad (2)\begin{bmatrix} 0 \\ 1 \\ 0 \\ \vdots \\ 0 \\ 0 \end{bmatrix} ; \quad (3)\begin{bmatrix} 0 \\ 0 \\ 1 \\ \vdots \\ 0 \\ 0 \end{bmatrix} ; \quad \ldots ; \quad (12)\begin{bmatrix} 0 \\ 0 \\ 0 \\ \vdots \\ 0 \\ 1 \end{bmatrix} . \tag{2.8}$$

Hence, the 12 solutions are the columns of the matrix $\mathrm{K}(\eta)$. The matrix $\mathrm{K}(\eta)$ is determined in such a way that $\mathrm{K}(0) = \mathrm{E}$. To find $\mathrm{K}(\eta)$ more accurately, the Picard method presented in the next subsection may be employed.

A partial solution of (2.5) can be derived either by means of numerical methods under the zero initial data or by use of the Green matrix. In the latter case, we can present a partial solution of (2.5) in the following manner:

$$\mathbf{Y}_1^{(0)}(\eta) = \int_0^\eta \mathrm{K}(\eta)\,\mathrm{K}^{-1}(\zeta)\,\mathbf{f}^{(0)}(\zeta)\,\mathrm{d}\zeta\,; \tag{2.9}$$

here the matrix $\mathrm{K}(\eta)\,K^{-1}(\zeta) = \mathrm{G}(\eta\,,\,\zeta)$ is known as a *Green matrix* and $\mathrm{K}^{-1}(\zeta)$ is the inverse of the fundamental matrix $\mathrm{K}(\zeta)$. The expanded form of (2.9) follows:

$$\begin{aligned}\mathbf{Y}_1^{(0)}(\eta) &= -\int_0^\eta \mathrm{K}(\eta)\,\mathrm{K}^{-1}(\zeta) \times \begin{bmatrix} \mathbf{q}_0(\zeta) \\ \boldsymbol{\mu}_0(\zeta) \\ \mathbf{0} \\ \mathbf{0} \end{bmatrix} d\zeta \\ &\quad - \int_0^\eta \mathrm{K}(\eta)\,\mathrm{K}^{-1}(\zeta) \times \begin{bmatrix} \sum\limits_{i=1}^{n} \mathbf{P}_0^{(i)}\,\delta(\zeta - \eta_i) \\ \sum\limits_{\nu=1}^{\rho} \mathbf{T}_0^{(\nu)}\,\delta(\zeta - \eta_\nu) \\ \mathbf{0} \\ \mathbf{0} \end{bmatrix} \mathrm{d}\zeta \\ &= -\boldsymbol{\Phi}_0 - \boldsymbol{\Phi}\,. \end{aligned} \tag{2.10}$$

Consider now the terms containing the δ-function. In view of the properties of the δ-function (see Appendix 4), the vector $\boldsymbol{\Phi}$ can be written as follows:

$$\begin{aligned}\boldsymbol{\Phi} &= \int_0^\eta \mathrm{K}(\eta)\,\mathrm{K}^{-1}(\zeta) \times \begin{bmatrix} \sum\limits_{i=1}^{n} \mathbf{P}_0^{(i)}\,\delta(\zeta - \eta_i) \\ \sum\limits_{\nu=1}^{\rho} \mathbf{T}_0^{(\nu)}\,\delta(\zeta - \eta_\nu) \\ \mathbf{0} \\ \mathbf{0} \end{bmatrix} \mathrm{d}\zeta \\ &= \sum_{i=1}^{n} \mathrm{K}(\eta)\,\mathrm{K}^{-1}(\eta_i) \begin{bmatrix} \mathbf{P}_0^{(i)}\,H(\eta - \eta_i) \\ \mathbf{0} \\ \mathbf{0} \\ \mathbf{0} \end{bmatrix} \\ &\quad + \sum_{\nu=1}^{\rho} \mathrm{K}(\eta)\,\mathrm{K}^{-1}(\eta_\nu) \begin{bmatrix} \mathbf{0} \\ \mathbf{T}_0^{(\nu)}\,H(\eta - \eta_\nu) \\ \mathbf{0} \\ \mathbf{0} \end{bmatrix}; \end{aligned} \tag{2.11}$$

here $H(\eta - \eta_i)$ and $H(\eta - \eta_\nu)$ are the Heaviside functions.

The vector $\boldsymbol{\Phi}$ can be presented as a sum of the vectors $\boldsymbol{\Phi}^{(i)}$ and $\boldsymbol{\Phi}^{(\nu)}$, which are functions of $\mathbf{P}_0^{(i)}$ and $\mathbf{T}_0^{(\nu)}$, respectively, i.e.

$$\boldsymbol{\Phi}(\eta) = \sum_{i=1}^{n} \boldsymbol{\Phi}^{(i)}(\eta)\, H(\eta - \eta_i) + \sum_{\nu=1}^{\rho} \boldsymbol{\Phi}^{(\nu)}(\eta)\, H(\eta - \eta_\nu) . \tag{2.12}$$

The vectors $\boldsymbol{\Phi}^{(i)}$ and $\boldsymbol{\Phi}^{(\nu)}$ are related to η in the following way:

$$\boldsymbol{\Phi}^{(i)}(\eta) = \begin{cases} \mathbf{0}, & \text{for} \quad \eta < \eta_i ; \\ \mathrm{K}(\eta)\,\mathrm{K}^{-1}(\eta_i)\,\mathbf{P}_0^{(i)}, & \text{for} \quad \eta > \eta_i ; \end{cases}$$

$$\boldsymbol{\Phi}^{(\nu)}(\eta) = \begin{cases} \mathbf{0}, & \text{for} \quad \eta < \eta_\nu ; \\ \mathrm{K}(\eta)\,\mathrm{K}^{-1}(\eta_\nu)\,\mathbf{T}_0^{(\nu)}, & \text{for} \quad \eta > \eta_\nu . \end{cases}$$

It is more convenient to obtain the partial solution as a function of $\mathbf{q}_0$ and $\boldsymbol{\mu}_0$ through the solution of the equation

$$\mathbf{Y}' + \mathrm{B}\mathbf{Y}^{(0)} = \mathbf{f}_0^{(0)}, \qquad \mathbf{f}_0^{(0)} = (-\mathbf{q}_0\,, -\boldsymbol{\mu}_0\,, \mathbf{0}\,, \mathbf{0})^{\mathrm{T}}\,, \tag{2.13}$$

under the zero initial condition $\mathbf{Y}(0) = 0$. As a result, we get

$$\mathbf{Y}_1^{(0)}(\eta) = -\boldsymbol{\Phi}_0 - \sum_{i=1}^{n} \boldsymbol{\Phi}^{(i)} - \sum_{\nu=1}^{\rho} \boldsymbol{\Phi}^{(\nu)}\,, \tag{2.14}$$

where $\boldsymbol{\Phi}_0$ is a partial solution of (2.13). Note that if it is required to determine a partial solution containing the δ-function, it is sufficient to obtain the values of $\mathrm{K}^{-1}(\zeta)$ only for a finite set of arguments. These values can be easily calculated during the determination of the fundamental matrix $\mathrm{K}(\eta)$.

The components of the vector $\mathbf{C}$ must satisfy the following relations:

$$\mathbf{Y}^{(0)}(0) = \mathbf{C}\,; \tag{2.15}$$

$$\mathbf{Y}^{(0)}(1) = \mathrm{K}(1)\,\mathbf{C} + \mathbf{Y}_1^{(0)}(1)\,. \tag{2.16}$$

Generally, the set of boundary conditions for a rod consists of 12 relations, six for each end. As an example, consider a rod with clamped ends. In this case, the boundary conditions are homogeneous, that is,

(1) at $\eta = 0\,, \quad \mathbf{u}^{(0)}(0) = \boldsymbol{\vartheta}^{(0)}(0) = 0\,;$

(2) at $\eta = 1\,, \quad \mathbf{u}^{(0)}(1) = \boldsymbol{\vartheta}^{(0)}(1) = 0\,.$

In view of (2.15), we get

$$\mathbf{Y}^{(0)}(0) = \begin{bmatrix} \mathbf{Q}^{(0)}(0) \\ \mathbf{M}^{(0)}(0) \\ \mathbf{0} \\ \mathbf{0} \end{bmatrix} = \begin{bmatrix} c_1 \\ c_2 \\ \vdots \\ c_{12} \end{bmatrix} . \tag{2.17}$$

It follows that $c_7 = c_8 = \ldots = c_{12} = 0$. The other six unknowns can be determined from the inhomogeneous algebraic equation (2.16). This equation can be rewritten in the expanded form as follows:

$$k_{7.1}c_1 + k_{7.2}c_2 + k_{7.3}c_3 + k_{7.4}c_4 + k_{7.5}c_5 + k_{7.6}c_6 + Y_{1.7}^{(0)} = 0\,;$$

$$\ldots$$

$$k_{12.1}c_1 + k_{12.2}c_2 + k_{12.3}c_3 + k_{12.4}c_4 + k_{12.5}c_5 + k_{12.6}c_6 + Y_{1.12}^{(0)} = 0\,; \qquad (2.18)$$

here k_{ij} are the elements of the matrix $\mathrm{K}\,(1)$ and $Y_{1.i}^{(0)}$ are the components of the vector $\mathbf{Y}_1^{(0)}(1)$. Having determined c_j from (2.18), we can obtain the solution of the equilibrium equation of the zeroth approximation in the form (2.6). This procedure of numerical integration of linear equations (two-point boundary-value problems) may be used for any set of boundary conditions.

Dead Forces Suppose that a rod is subjected to dead forces. If we need to apply numerical methods, it is convenient to present the equilibrium equations in the immovable Cartesian frame because in this frame the components of the applied forces are constants. Hence, the equilibrium equations of the zeroth approximation referred to the Cartesian frame, (1.30)–(1.33), take the following form:

$$\frac{d\mathbf{Q}_x^{(0)}}{d\eta} + \mathbf{P}_{x_0} = 0\,; \qquad (2.19)$$

$$\frac{d\mathbf{M}_x^{(0)}}{d\eta} + \mathrm{A}_L\mathbf{Q}_x^{(0)} + \mathbf{T}_{x_0} = 0\,; \qquad (2.20)$$

$$\frac{d\boldsymbol{\vartheta}^{(0)}}{d\eta} + \mathrm{A}_{\text{æ}}\boldsymbol{\vartheta}^{(0)} - \mathrm{A}^{-1}\mathbf{M}_x^{(0)} = 0\,; \qquad (2.21)$$

$$\frac{d\mathbf{u}_x^{(0)}}{d\eta} + \mathrm{A}_L\boldsymbol{\vartheta}^{(0)} = 0\,. \qquad (2.22)$$

The system of equations (2.19)–(2.22) can be written as a vector equation

$$\mathbf{Y}_x^{(0)\prime} + \mathrm{A}_x^{(0)}\mathbf{Y}_x^{(0)} = \mathbf{f}_x^{(0)}\,, \qquad (2.23)$$

where

$$\mathbf{Y}_x^{(0)} = \begin{bmatrix} \mathbf{Q}_x^{(0)} \\ \mathbf{M}_x^{(0)} \\ \boldsymbol{\vartheta}^{(0)} \\ \mathbf{u}_x^{(0)} \end{bmatrix}; \qquad \mathbf{f}_x^{(0)} = \begin{bmatrix} -\mathbf{P}_{x0} \\ -\mathbf{T}_{x0} \\ \mathbf{0} \\ \mathbf{0} \end{bmatrix};$$

$$\mathrm{A}_x^{(0)} = \begin{bmatrix} 0 & 0 & 0 & 0 \\ \mathrm{A}_L & 0 & 0 & 0 \\ 0 & -\mathrm{A}^{-1} & \mathrm{A}_x & 0 \\ 0 & 0 & \mathrm{A}_L & 0 \end{bmatrix}.$$

Equation (2.23) can be solved by means of the method that we applied to (2.5). However, there exists an alternative method. It is clear that (2.19) is uncoupled from the other equations. Integration of (2.19) yields

$$\mathbf{Q}_x^{(0)} = -\int_0^\eta \mathbf{P}_{x0}\,\mathrm{d}\zeta + \mathbf{C}_Q\,, \tag{2.24}$$

where the vector $\mathbf{C}_Q$ is equal to the vector $\mathbf{Q}_x^{(0)}$ at $\eta = 0$. Equation (2.24) can be written in the expanded form,

$$\mathbf{Q}_x^{(0)}(\eta) = -\int_0^\eta \mathbf{q}_{x0}\,\mathrm{d}\zeta - \sum_{i=1}^{n} \mathbf{P}_{x0}^{(i)} H(\eta - \eta_i) + \mathbf{C}_Q$$

or

$$\mathbf{Q}_x^{(0)}(\eta) = \mathbf{Q}_{x0}^{(0)}(\eta) + \mathbf{C}_Q\,. \tag{2.25}$$

Equation (2.20) can be solved for $\mathbf{M}_x^{(0)}$ as follows:

$$\mathbf{M}_x^{(0)}(\eta) = \int_0^\eta \mathrm{A}_L[\mathbf{f}_q + \sum \mathbf{P}_{x0}^{(i)} H(\zeta - \eta_i) + \mathbf{C}_Q]\,\mathrm{d}\zeta + \mathbf{C}_M$$

or

$$\mathbf{M}_x^{(0)}(\eta) = \mathbf{M}_{x0}^{(0)}(\eta) + \mathbf{C}_Q\eta + \mathbf{C}_M\,. \tag{2.26}$$

Then, in view of (2.2), we obtain

$$\boldsymbol{\vartheta}^{(0)}(\eta) = \mathrm{K}^{(\vartheta)}(\eta)\mathbf{C}_\vartheta + \int_0^\eta \mathrm{G}^{(\nu)}(\eta\,,\,\zeta)\mathrm{A}^{-1}\mathbf{M}_x^{(0)}(\zeta)\,\mathrm{d}\zeta$$

or

$$\boldsymbol{\vartheta}^{(0)}(\eta) = \mathrm{K}^{(\vartheta)}(\eta)\mathbf{C}_\vartheta + \mathrm{D}_1\mathbf{C}_Q + \mathrm{D}_2\mathbf{C}_M + \mathbf{b}_1\,, \tag{2.27}$$

where

$$\mathrm{D}_1 = \int_0^\eta \mathrm{G}(\eta\,,\,\zeta)\,\mathrm{A}^{-1}\zeta\,\mathrm{d}\zeta\,;$$

$$\mathrm{D}_2 = \int_0^\eta \mathrm{G}(\eta\,,\,\zeta)\,\mathrm{A}^{-1}\,\mathrm{d}\zeta\,;$$

$$\mathbf{b}_1 = \int_0^\eta \mathrm{G}(\eta\,,\,\zeta)\,\mathrm{A}^{-1}\mathbf{M}_{x0}^{(0)}(\zeta)\,\mathrm{d}\zeta\,.$$

Once $\boldsymbol{\vartheta}^{(0)}$ is found, we have

$$\mathbf{u}_x^{(0)}(\eta) = \mathrm{K}^{(u)}(\eta)\mathbf{C}_u + \int_0^\eta \mathrm{G}^{(u)}(\eta\,,\,\zeta)\mathrm{A}_L(\zeta)\,\boldsymbol{\vartheta}^{(0)}(\zeta)\,\mathrm{d}\zeta$$

or

$$\mathbf{u}_x^{(0)}(\eta) = \mathrm{K}^{(u)}(\eta)\mathbf{C}_u + \mathrm{D}_3\mathbf{C}_\vartheta + \mathrm{D}_4\mathbf{C}_Q + \mathrm{D}_5\mathbf{C}_M + \mathbf{b}_2\,, \tag{2.28}$$

where

$$\mathrm{D}_3 = \int_0^\eta \mathrm{GA}_L\mathrm{K}^{(\vartheta)}\,\mathrm{d}\zeta\,; \qquad \mathrm{D}_4 = \int_0^\eta \mathrm{GA}_L\mathrm{D}_1\,\mathrm{d}\zeta\,;$$
$$\mathrm{D}_5 = \int_0^\eta \mathrm{GA}_L\mathrm{D}_2\,\mathrm{d}\zeta\,; \qquad \mathbf{b}_2 = \int_0^\eta \mathrm{GA}_L\mathbf{b}_1\,\mathrm{d}\zeta\,.$$

The vectors $\mathbf{C}_Q$, $\mathbf{C}_M$, $\mathbf{C}_\vartheta$, and $\mathbf{C}_u$ are to be determined from the boundary conditions.

Arbitrary Forces Assume that the increments of the applied forces depend linearly on the vectors of generalized displacements (see (1.99)). Let us consider the equations in the attached coordinate system. The equations of the zeroth approximation (1.112)–(1.115) take the form

$$\frac{\mathrm{d}\mathbf{Q}^{(0)}}{\mathrm{d}\eta} + \mathrm{A}_{æ}\mathbf{Q}^{(0)} + \mathbf{P}^{(0)} + \Delta\mathbf{P}^{(0)} = 0\,; \tag{2.29}$$

$$\frac{\mathrm{d}\mathbf{M}^{(0)}}{\mathrm{d}\eta} + \mathrm{A}_{æ}\mathbf{M}^{(0)} + \mathrm{A}_1\mathbf{Q}^{(0)} + \mathbf{T}^{(0)} + \Delta\mathbf{T}^{(0)} = 0\,; \tag{2.30}$$

$$\frac{\mathrm{d}\boldsymbol{\vartheta}^{(0)}}{\mathrm{d}\eta} + \mathrm{A}_{æ}\boldsymbol{\vartheta}^{(0)} - \mathrm{A}^{-1}\mathbf{M}^{(0)} = 0\,; \tag{2.31}$$

$$\frac{\mathrm{d}\mathbf{u}^{(0)}}{\mathrm{d}\eta} + \mathrm{A}_{æ}\mathbf{u}^{(0)} + \mathrm{A}_1\boldsymbol{\vartheta}^{(0)} = 0\,. \tag{2.32}$$

For simplicity of notation, let the rod be subjected to distributed loads $\mathbf{q}$ and $\boldsymbol{\mu}$, a concentrated force $\mathbf{P}$, and a concentrated moment $\mathbf{T}$. In the general case (see (1.99)), we get

$$\begin{aligned}\Delta\mathbf{P}^{(0)} &= \mathrm{C}^{(1)}\boldsymbol{\vartheta}^{(0)} + \mathrm{C}^{(2)}\mathbf{u}^{(0)} + \mathrm{C}^{(3)}\mathbf{u}^{(0)\prime} \\ &\quad + (\mathrm{B}^{(1)}\boldsymbol{\vartheta}^{(0)} + \mathrm{B}^{(2)}\mathbf{u}^{(0)} + \mathrm{B}^{(3)}\mathbf{u}^{(0)\prime})\,\delta(\eta-\eta_P)\,;\end{aligned} \tag{2.33}$$

$$\begin{aligned}\Delta\mathbf{T}^{(0)} &= \mathrm{C}^{(4)}\boldsymbol{\vartheta}^{(0)} + \mathrm{C}^{(5)}\mathbf{u}^{(0)} + \mathrm{C}^{(6)}\mathbf{u}^{(0)\prime} \\ &\quad + (\mathrm{B}^{(4)}\boldsymbol{\vartheta}^{(0)} + \mathrm{B}^{(5)}\mathbf{u}^{(0)} + \mathrm{B}^{(6)}\mathbf{u}^{(0)\prime})\,\delta(\eta-\eta_T)\,.\end{aligned} \tag{2.34}$$

By use of (2.32), we can eliminate $\mathbf{u}^{(0)\prime}$ from (2.33) and (2.34) and, therefore, we obtain formulas for $\Delta\mathbf{P}^{(0)}$ and $\Delta\mathbf{T}^{(0)}$:

$$\begin{aligned}\Delta\mathbf{P}^{(0)} &= \mathrm{C}_0^{(1)}\boldsymbol{\vartheta}^{(0)} + \mathrm{C}_0^{(2)}\mathbf{u}^{(0)} \\ &\quad + \mathrm{B}_0^{(1)}\boldsymbol{\vartheta}^{(0)}\,\delta(\eta-\eta_P) + \mathrm{B}_0^{(2)}\mathbf{u}^{(0)}\,\delta(\eta-\eta_P)\,;\end{aligned} \tag{2.35}$$

$$\begin{aligned}\Delta\mathbf{T}^{(0)} &= \mathrm{C}_0^{(3)}\boldsymbol{\vartheta}^{(0)} + \mathrm{C}_0^{(4)}\mathbf{u}^{(0)} \\ &\quad + \mathrm{B}_0^{(3)}\boldsymbol{\vartheta}^{(0)}\,\delta(\eta-\eta_T) + \mathrm{B}_0^{(4)}\mathbf{u}^{(0)}\,\delta(\eta-\eta_T)\,;\end{aligned} \tag{2.36}$$

here

$$\mathrm{C}_0^{(1)} = \mathrm{C}^{(1)} - \mathrm{C}^{(3)}\mathrm{A}_1\,; \qquad \mathrm{C}_0^{(2)} = \mathrm{C}^{(2)} - \mathrm{C}^{(3)}\mathrm{A}_{æ}\,;$$
$$\mathrm{C}_0^{(3)} = \mathrm{C}^{(3)} - \mathrm{C}^{(6)}\mathrm{A}_1\,; \qquad \mathrm{C}_0^{(4)} = \mathrm{C}^{(4)} - \mathrm{C}^{(6)}\mathrm{A}_{æ}\,;$$
$$\mathrm{B}_0^{(1)} = \mathrm{B}^{(1)} - \mathrm{B}^{(3)}\mathrm{A}_1\,; \qquad \mathrm{B}_0^{(2)} = \mathrm{B}^{(2)} - \mathrm{B}^{(3)}\mathrm{A}_{æ}\,;$$
$$\mathrm{B}_0^{(3)} = \mathrm{B}^{(3)} - \mathrm{B}^{(6)}\mathrm{A}_1\,; \qquad \mathrm{B}_0^{(4)} = \mathrm{B}^{(4)} - \mathrm{B}^{(6)}\mathrm{A}_{æ}\,.$$

Thus, the system of equations (2.29)–(2.36) can be written as a vector equation

$$\mathbf{Y}^{(0)\prime} + \mathrm{A}^{(0)}\mathbf{Y}^{(0)} = \mathbf{f}^{(0)}\,, \tag{2.37}$$

where

$$\mathrm{A}^{(0)} = \begin{bmatrix} \mathrm{A}_{æ} & 0 & \mathrm{C}_0^{(1)} & \mathrm{C}_0^{(2)} \\ \mathrm{A}_1 & \mathrm{A}_{æ} & \mathrm{C}_0^{(3)} & \mathrm{C}_0^{(4)} \\ 0 & -\mathrm{A}^{-1} & \mathrm{A}_{æ} & 0 \\ 0 & 0 & \mathrm{A}_1 & \mathrm{A}_{æ} \end{bmatrix};$$

$$\mathbf{f}^{(0)} = \begin{bmatrix} -\mathbf{P}^{(0)} - (\mathrm{B}_0^{(1)}\boldsymbol{\vartheta}^{(0)} + \mathrm{B}_0^{(2)}\mathbf{u}^{(0)})\,\delta(\eta - \eta_i) \\ -\mathbf{T}^{(0)} - (\mathrm{B}_0^{(3)}\boldsymbol{\vartheta}^{(0)} + \mathrm{B}_0^{(4)}\mathbf{u}^{(0)})\,\delta(\eta - \eta_\nu) \\ \mathbf{0} \\ \mathbf{0} \end{bmatrix}.$$

The vector $\mathbf{f}^{(0)}$ can be represented as follows:

$$\mathbf{f}^{(0)} = \mathbf{f}_0 + \mathrm{B}_1\mathbf{Y}^{(0)}\,\delta\,(\eta - \eta_P) + \mathrm{B}_2\mathbf{Y}^{(0)}\,\delta\,(\eta - \eta_T)\,; \tag{2.38}$$

here

$$\mathbf{f}_0 = \begin{bmatrix} -\mathbf{P}^{(0)} \\ -\mathbf{T}^{(0)} \\ \mathbf{0} \\ \mathbf{0} \end{bmatrix}; \qquad \mathrm{B}_1 = \begin{bmatrix} 0 & 0 & \mathrm{B}_0^{(1)} & 0 \\ 0 & 0 & \mathrm{B}_0^{(3)} & 0 \\ 0 & 0 & 0 & 0 \\ 0 & 0 & 0 & 0 \end{bmatrix};$$

$$\mathrm{B}_2 = \begin{bmatrix} 0 & 0 & 0 & \mathrm{B}_0^{(2)} \\ 0 & 0 & 0 & \mathrm{B}_0^{(4)} \\ 0 & 0 & 0 & 0 \\ 0 & 0 & 0 & 0 \end{bmatrix}.$$

The solution of (2.37) is

$$\mathbf{Y}^{(0)} = \mathrm{K}\,(\eta)\,\mathbf{C} + \int_0^{\eta} \mathrm{G}\,(\eta\,,\,\zeta)\,[\,\mathbf{f}_0 + \mathrm{B}_1\mathbf{Y}^{(0)}\,\delta\,(\zeta - \eta_P) + \mathrm{B}_2\mathbf{Y}^{(0)}\,\delta\,(\zeta - \eta_T)\,]\,\mathrm{d}\zeta$$

or

$$\begin{aligned}\mathbf{Y}^{(0)} &= \mathrm{K}(\eta)\,\mathbf{C} + \int_0^{\eta} \mathrm{G}(\eta\,,\,\zeta)\,\mathrm{d}\zeta \cdot \mathbf{f}_0 \\ &+ \mathrm{G}(\eta\,,\,\eta_P)\,\mathrm{B}_1(\eta_P)\,\mathbf{Y}^{(0)}(\eta_P)\,H(\eta-\eta_P) \\ &+ \mathrm{G}(\eta\,,\,\eta_T)\,\mathrm{B}_2(\eta_T)\,\mathbf{Y}^{(0)}(\eta_T)\,H(\eta-\eta_T)\,,\end{aligned} \tag{2.39}$$

where

$$\mathrm{G}(\eta\,,\,\eta_P) = \mathrm{K}(\eta)\,\mathrm{K}^{-1}(\eta_P)\,; \qquad \mathrm{G}(\eta\,,\,\eta_T) = \mathrm{K}(\eta)\,\mathrm{K}^{-1}(\eta_T)\,.$$

The partial solution corresponding to the vector $\mathbf{f}^{(0)}$ can be obtained without use of the Green matrix G. To do this, it is sufficient to solve the inhomogeneous equation (2.37) for the zero initial data.

If $\eta = 0$, the matrix $\mathrm{K}(\eta)$ is the unit matrix. Hence, in the case of homogeneous boundary conditions, six components of the vector $\mathbf{C}$ equal zero. The other six components may be found from the boundary conditions at $\eta = 1$.

The equations of the first and higher-order approximation can be solved in a similar way.

2.1.2 Picard Iteration Method for Determination of the Fundamental Matrix K (η)

The general solution of a system of inhomogeneous linear equations is $\mathbf{Y}(\eta) = \mathrm{K}(\eta)\,\mathbf{C} + \mathbf{Y}_1(\eta)$ (see (2.6)); here the matrix $\mathrm{K}(\eta)$ satisfies the following homogeneous equation:

$$\frac{\mathrm{dK}}{\mathrm{d}\eta} + \mathrm{A}(\eta)\,\mathrm{K} = 0\,, \qquad \mathrm{K}(0) = \mathrm{E}\,. \tag{2.40}$$

Integrating this equation, we get

$$\mathrm{K}(\eta) = -\int_0^{\eta} \mathrm{A}(\zeta)\,\mathrm{K}(\zeta)\,\mathrm{d}\zeta + \mathrm{E}\,. \tag{2.41}$$

The relation (2.41) enables us to use the Picard iteration method for determination of $\mathrm{K}(\eta)$. Let us use the superscript '(1)' to denote the first approximation of the matrix $\mathrm{K}(\eta)$. Substituting $\mathrm{K}^{(1)}(\eta)$ into the right-hand side of (2.41), we readily get the second approximation of $\mathrm{K}(\eta)$,

$$\mathrm{K}^{(2)}(\eta) = -\int_0^{\eta} \mathrm{A}(\zeta)\,\mathrm{K}^{(1)}(\zeta)\,\mathrm{d}\zeta + \mathrm{E}\,. \tag{2.42}$$

Similarly, a kth approximation is as follows:

$$\mathrm{K}^{(k)}(\eta) = -\int_0^{\eta} \mathrm{A}(\zeta)\,\mathrm{K}^{(k-1)}(\zeta)\,\mathrm{d}\zeta + \mathrm{E}\,. \tag{2.43}$$

If the elements of $\mathrm{K}^{(k)}$ are of the required accuracy, that is, the inequalities

$$\left| \frac{K_{ij}^{(k)} - K_{ij}^{(k-1)}}{K_{ij}^{(k)}} \right| 100\% \leq \Delta \,, \tag{2.44}$$

are valid, then the matrix $\mathrm{K}^{(k)}(\eta)$ is a sought solution of (2.40).

This variant of the method of iterations enables us to find the matrix K (η) for the whole interval $0 \leq \eta \leq 1$. Due to some peculiarities of the equations, the convergence of the iteration process for the whole interval $0 \leq \eta \leq 1$ may be poor. Hence, it is more convenient to divide this interval into n subintervals. Then, the matrix K (η) can be determined for each subinterval. Using (2.43), we get

$$\mathrm{K}^{(k_1)}(\eta) = -\int_0^{\eta_1} \mathrm{A}\,(\zeta)\, \mathrm{K}^{(k_1-1)}(\zeta)\, \mathrm{d}\zeta + \mathrm{E} \,, \tag{2.45}$$

where η_1 is the endpoint of the first subinterval.

For the second subinterval, we have

$$\mathrm{K}^{(k_2)}(\eta) = -\int_{\eta_1}^{\eta_2} \mathrm{A}\,(\zeta)\, \mathrm{K}^{(k_2-1)}(\zeta)\, \mathrm{d}\zeta + \mathrm{K}^{(k_1)} \,, \tag{2.46}$$

where $\mathrm{K}^{(k_1)}$ is the matrix corresponding to the subinterval $0 \leq \eta \leq \eta_1$ (see (2.45)) and k_1 is the number of iterations we have performed to get the required accuracy.

Similarly, for an mth subinterval, we have

$$\mathrm{K}^{(k_m)} = -\int_{\eta_{m-1}}^{\eta_m} \mathrm{A}\,(\zeta)\, \mathrm{K}^{(k_m-1)}(\zeta)\, \mathrm{d}\zeta + \mathrm{K}^{(k_m-1)} \,. \tag{2.47}$$

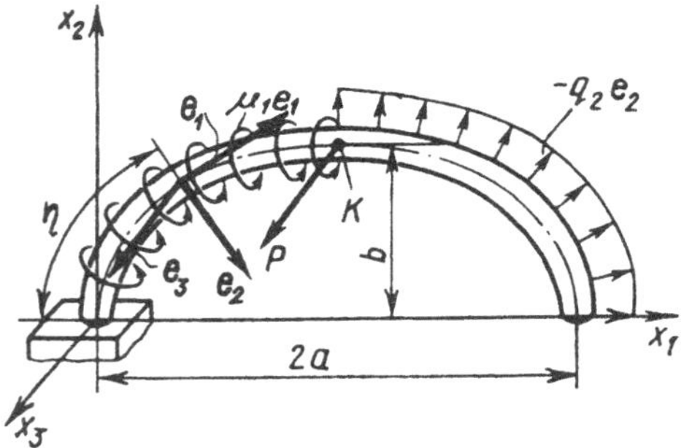

Fig. 2.1.

Let us consider some examples. Let an axial line of a curvilinear rod be an ellipse (Fig. 2.1) defined by

$$\left(\frac{x_1 - a}{a}\right)^2 + \left(\frac{x_2}{b}\right)^2 = 1 . \tag{2.48}$$

The quantities in (2.48) are assumed to be nondimensional. The rod is subjected to a distributed twisting moment $\mu_1 \mathbf{e}_1$, a distributed load $q_2 \mathbf{e}_2$, and a concentrated force $P_3 \mathbf{e}_3$. Let us determine the strain-stress state of the rod. Assuming that the loads are follower, we can write the equations of the zeroth approximation as follows:

$$\begin{aligned}
&Q_1^{(0)\prime} - æ_{30} Q_2^{(0)} = 0 ;\\
&Q_2^{(0)\prime} + æ_{30} Q_1^{(0)} - q_2 \, H(\eta - 0.5) = 0 ;\\
&Q_3^{(0)\prime} = P_3 \, \delta(\eta - 0.5) ;
\end{aligned} \tag{2.49}$$

$$\begin{aligned}
&M_1^{(0)\prime} - æ_{30} M_2^{(0)} + \mu_1 \left[H(\eta) - H(\eta - 0.5) \right] = 0 ;\\
&M_2^{(0)\prime} + æ_{30} M_1^{(0)} - Q_3^{(0)} = 0 ;\\
&M_3^{(0)\prime} + Q_2^{(0)} = 0 ;
\end{aligned} \tag{2.50}$$

$$\begin{aligned}
&\vartheta_1^{(0)\prime} - æ_{30} \vartheta_2^{(0)} - \frac{M_1^{(0)}}{A_{11}} = 0 ;\\
&\vartheta_2^{(0)\prime} + æ_{30} \vartheta_1^{(0)} - \frac{M_2^{(0)}}{A_{22}} = 0 ;\\
&\vartheta_3^{(0)\prime} - \frac{M_3^{(0)}}{A_{33}} = 0 ;
\end{aligned} \tag{2.51}$$

$$\begin{aligned}
&u_1^{(0)\prime} - æ_{30} u_2^{(0)} = 0 ;\\
&u_2^{(0)\prime} + æ_{30} u_1^{(0)} - \vartheta_3^{(0)} = 0 ;\\
&u_3^{(0)\prime} + \vartheta_2^{(0)} = 0 .
\end{aligned} \tag{2.52}$$

The curvature of the rod axis $æ_{30}(\eta)$ in (2.49)–(2.52) is unknown. Let us evaluate $æ_{30}$ in accordance with the method described in Appendix 5. Differentiating (2.48), we get

$$x_1' \, \frac{x_1 - a}{a^2} + x_2' \, \frac{x_2}{b^2} = 0 . \tag{2.53}$$

It is evident that x_1' and x_2' satisfy the relation

$$x_1'^{\,2} + x_2'^{\,2} = 1 . \tag{2.54}$$

Thus, the derivatives x_1' and x_2' can be found from (2.53) and (2.54). Solving for x_2' from (2.53), we get

$$x_2' = -x_1' \, \frac{(x_1 - a)\, b^2}{a^2 x_2} . \tag{2.55}$$

In view of (2.48), we arrive at

$$x_2 = b\sqrt{1 - \frac{(x_1 - a)^2}{a^2}} \,. \tag{2.56}$$

Elimination of x_2 and x_2' from (2.53) and (2.54), respectively, and rearrangement yield the following equation in x_1:

$$x_1' - a\sqrt{\frac{a^2 - (x_1 - a)^2}{a^4 - (a^2 - b^2)(x_1 - a)^2}} = 0 \,. \tag{2.57}$$

Since $x_1 = 0$ at $\eta = 0$, we can find x_1 from (2.57). Once x_1 is found, we determine x_1' and x_1''; then, x_2 can be obtained from (2.56) and x_2' and x_2'' can be found from (2.55).

Thus, the curvature æ$_{30}$ is

$$æ_{30} = \sqrt{x_1''^2 + x_2''^2} \,.$$

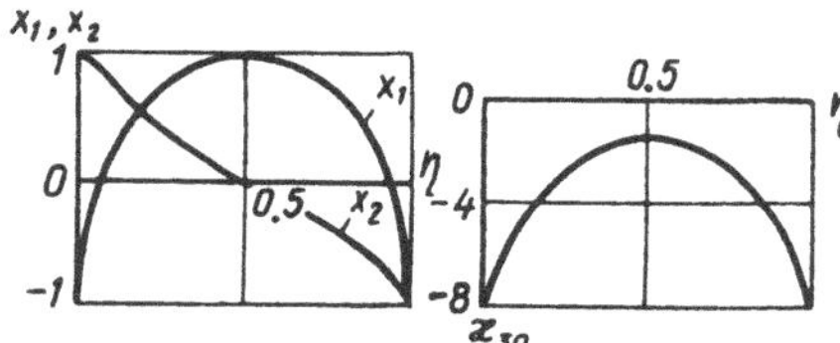

Fig. 2.2.

The plots of x_1', x_2', and æ$_{30}$ versus the axial nondimensional coordinate η are shown in Fig. 2.2. These plots correspond to $a = 0.4$ and $b = 0.224$.

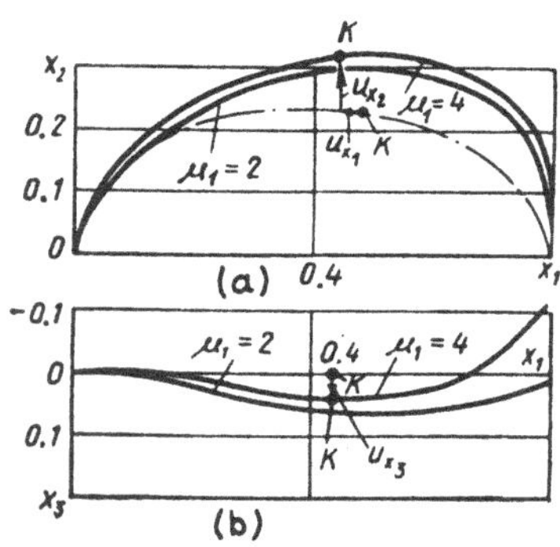

Fig. 2.3.

The equations were solved numerically for $q_2 = 2$; $P_3 = 1.5$; $\mu_1 = 2$, and $\mu_1 = 4$. When dealing with (2.49) and (2.50), it is more convenient to introduce a new independent variable $\eta_1 = 1 - \eta$. At $\eta_1 = 0$, the magnitudes of $Q_j^{(0)}$ and $M_j^{(0)}$ are known (they equal zero). Since at $\eta = 0$ the equalities $\vartheta_j = 0$ and $u_j = 0$ hold, we can analyze (2.51) and (2.52) without introduction of the new variable.

The projections of the rod axis on the coordinate planes x_1Ox_2 and x_1Ox_3 are shown in Fig. (2.3)*a,b*. Constructing the plots, we used the numerical values of the displacements $u_j(\eta)$ found from (2.49)–(2.52), hence, the displacements u_{x_j} related to the Cartesian frame are as follows:

$$u_{x_j} = \sum_{i=1}^{3} u_i l_{ij}^{(0)} ;$$

here $l_{ij}^{(0)}$ are the elements of the matrix $\mathrm{L}^{(0)}$. The displacements u_{x_j} of an arbitrary point K are shown in Fig. 2.3*a,b*.

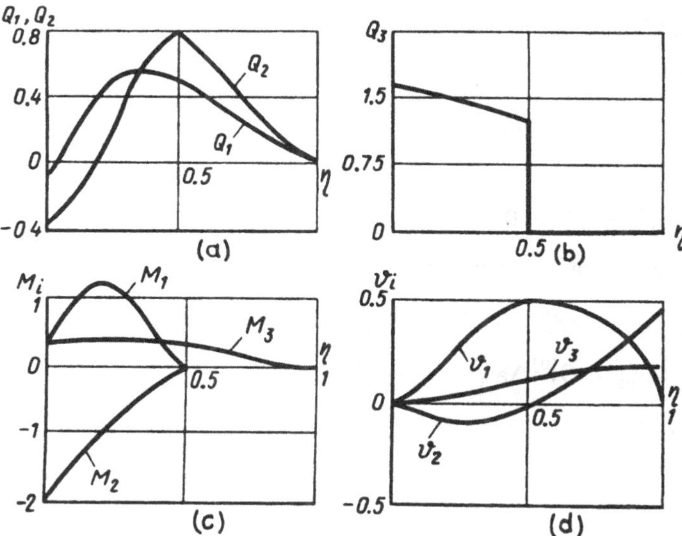

Fig. 2.4.

The plots of $Q_j(\eta)$, $M_j(\eta)$, and $\vartheta_j(\eta)$ versus η are given in Fig. 2.4.

As another example, let us consider a rod of circular cross section shown in Fig. 2.5*a*. Owing to the symmetry of the cross section, external loads that lie in the plane x_1Ox_2 may cause only in-plane deformation of the rod. The distributed load applied to the rod is $q_2 = q_{20}\eta$.

Let us consider the equations of the zeroth, the first, and the second approximation. The loads applied to the rod are follower.

The equations of the zeroth approximation are as follows:

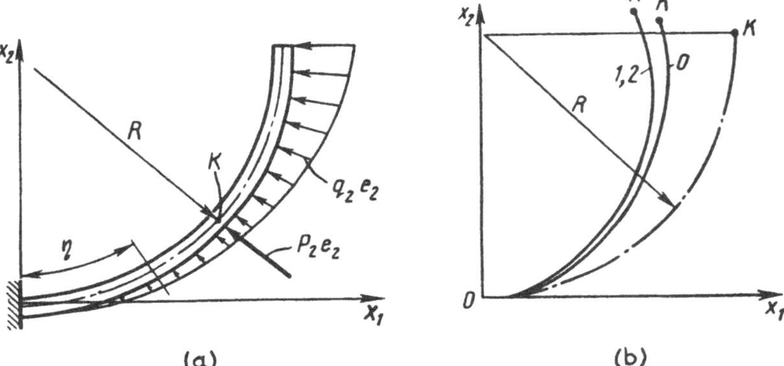

Fig. 2.5.

$$
\begin{aligned}
& Q_1^{(0)\prime} - æ_{30} Q_2^{(0)} = 0\,, \qquad æ_{30} = \frac{1}{R^0} = \frac{\pi}{2}\,; \\
& Q_2^{(0)\prime} + æ_{30} Q_1^{(0)} = -q_2 - P_2\,\delta\,(\eta - 0.5)\,; \\
& M_3^{(0)\prime} + Q_2^{(0)} = 0\,; \\
& \vartheta_3^{(0)\prime} - M_3^{(0)} = 0\,, \qquad M_3^{(0)} = \Delta æ_3^{(0)}\,; \\
& u_1^{(0)\prime} - æ_{30} u_2^{(0)} = 0\,; \\
& u_2^{(0)\prime} + æ_{30} u_1^{(0)} - \vartheta_3^{(0)} = 0\,.
\end{aligned}
$$

The equations of the first approximation are as follows:

$$
\begin{aligned}
& Q_1^{(1)\prime} - Q_2^{(1)} æ_{30}^{(0)} - Q_2^{(0)} \Delta æ_3^{(1)} - Q_2^{(0)} \Delta æ_3^{(0)}\,; \\
& Q_2^{(1)\prime} + Q_1^{(1)} æ_{30}^{(0)} + Q_1^{(0)} \Delta æ_3^{(1)} = -Q_1^{(0)} \Delta æ_3^{(0)}\,; \\
& M_3^{(1)\prime} + Q_2^{(1)} = 0\,; \\
& \vartheta_3^{\prime\,(1)} - M_3^{(1)} = 0\,; \\
& u_1^{(1)\prime} - æ_{30}^{(0)} u_2^{(1)} = 0\,; \\
& u_2^{(1)\prime} + æ_{30}^{(0)} u_1^{(1)} - \vartheta_3^{(1)} = 0\,;
\end{aligned}
$$

here $æ_{30}^{(0)} = æ_{30} + \Delta æ_3^{(0)}$.

The method presented in the previous example may be applied to the equations of the zeroth and the first approximation. The solutions of the equation of the zeroth, the first, and the second approximation must satisfy the same set of boundary conditions. Combining these solutions, we introduce new variables

$$
\begin{aligned}
& \tilde{Q}_j^{(1)} = Q_j^{(0)} + Q_j^{(1)}\,; \qquad \tilde{M}_3^{(1)} = M_3^{(0)} + M_3^{(1)}\,; \\
& \tilde{\vartheta}_3^{(1)} = \vartheta_3^{(0)} + \vartheta_3^{(1)}\,; \qquad \tilde{u}_i^{(1)} = u_i^{(0)} + u_i^{(1)}\,, \qquad i = 1\,,2\,.
\end{aligned}
$$

Using these variables in the equations of the second approximation, we get

$$
\begin{aligned}
&Q_1^{(2)\prime} - Q_2^{(2)} æ_{30}^{(1)} - \tilde{Q}_2^{(1)} \Delta æ_3^{(2)} = \tilde{Q}_2^{(1)} \Delta \tilde{æ}_3^{(1)}\,; \\
&Q_2^{(2)\prime} + Q_1^{(2)} æ_{30}^{(1)} + \tilde{Q}_1^{(1)} \Delta æ_3^{(2)} = -\tilde{Q}_1^{(1)} \Delta æ_3^{(1)}\,; \\
&M_3^{(2)\prime} + Q_2^{(2)} = 0\,, \qquad M_3^{(2)} = \Delta æ_3^{(2)}\,; \\
&\vartheta_3'^{(2)} - M_3^{(2)} = 0\,; \\
&u_1^{(2)\prime} - æ_{30}^{(1)} u_2^{(2)} = 0\,; \\
&u_2^{(2)\prime} + æ_{30}^{(1)} u_1^{(2)} - \vartheta_3^{(2)} = 0\,,
\end{aligned}
$$

where $æ_{30}^{(1)} = æ_{30} + \Delta æ_3^{(0)} + \Delta æ_3^{(1)}$ and $\Delta \tilde{æ}_3^{(1)} = \Delta æ_3^{(0)} + \Delta æ_3^{(1)}$.

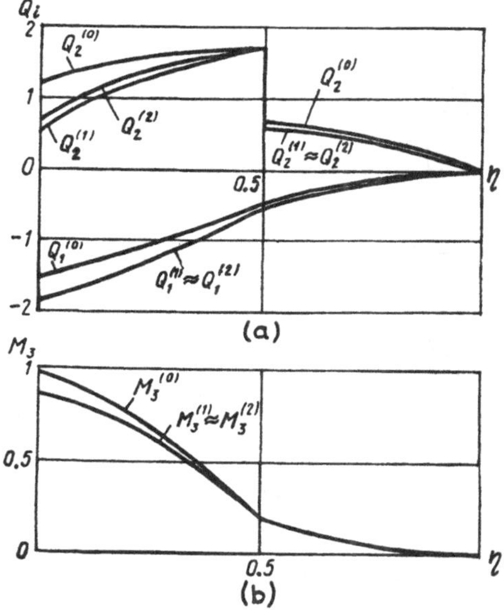

Fig. 2.6.

The values of the second approximation follow:

$$
\tilde{Q}_i^{(2)} = \sum_{k=0}^{2} Q_i^{(k)}\,; \qquad \tilde{u}_i^{(2)} = \sum_{k=0}^{2} u_i^{(k)}\,;
$$

$$
\tilde{M}_3^{(2)} = \sum_{k=0}^{2} M_3^{(k)}\,; \qquad \tilde{\vartheta}_3^{(2)} = \sum_{k=0}^{2} \vartheta_3^{(k)}\,.
$$

The shapes of the axial line for $q_{20} = 2$ and $P_2 = 1$ are shown in Fig. 2.5*b*. Configurations *0*, *1*, and *2* correspond to the zeroth, the first, and the second approximations, respectively.

It is clearly seen from the plot that axial displacements corresponding to the zeroth approximation (curve *0*) differ from the corrected values (curve *1*). The first approximation of the absolute displacement of the point K differs from the zeroth approximation of this displacement by 25% (see Fig. 2.5). The second approximation of the displacement of the point K is very close to the first approximation and, therefore, this approximation may be taken as the exact solution provided the required accuracy is 1%.

The plots of $Q_1(\eta)$, $Q_2(\eta)$, and $M_3(\eta)$ versus η are given in Fig. 2.6*a,b*. It can be seen from the plots that the maximum error of the zeroth approximation is

$$\delta_{\text{max}} = \left| \frac{1.20 - 0.75}{0.75} \right| 100\% = 60\% .$$

This estimate corresponds to Q_2 at $\eta = 0$ for the set of the nondimensional magnitudes of the applied forces.

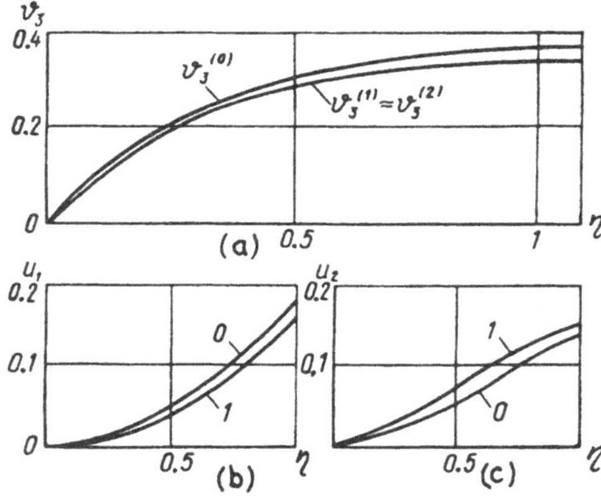

Fig. 2.7.

The plots of the zeroth and the first approximation of the displacements u_1 and u_2 and the angles of rotation ϑ_3 versus η are presented in Fig. 2.7*a-c*. The difference between the first and the second approximations is insignificant.

It is clearly seen from this example that a zeroth approximation may be rough. To estimate the accuracy of the zeroth approximation, one should solve the equations of the first approximation at least.

2.2 Equilibrium Equations for Rods with Lateral Supports

2.2.1 Rods with Lateral Hinge Supports

Let us consider a rod with a lateral hinge support. Loads that are applied to the rod cause a concentrated reaction force $\mathbf{R}$ applied to the hinge (a dashed line in Fig. 2.8*a*). This force may be treated as an external force applied to the rod. The direction and the magnitude of the vector $\mathbf{R}$ are unknown a priori. The displacement of the axial point with coordinate η_{10} equals zero, hence, we get

$$\mathbf{u}^{(0)}(\eta_{10}) = 0. \tag{2.58}$$

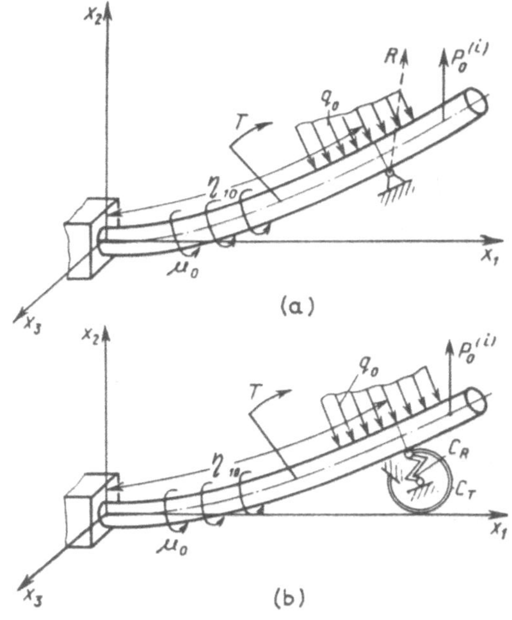

Fig. 2.8.

Because of the presence of the concentrated force $\mathbf{R}$, a partial solution takes the form

$$\mathbf{Y}_{11}^{(0)} = \mathbf{Y}_1^{(0)} + \int_0^{\eta} \mathrm{K}(\eta)\,\mathrm{K}^{-1}(\zeta) \times \begin{bmatrix} \mathbf{R}\,\delta(\zeta - \eta_{10}) \\ \mathbf{0} \\ \mathbf{0} \\ \mathbf{0} \end{bmatrix} \mathrm{d}\zeta, \tag{2.59}$$

where the vector $\mathbf{Y}_1^{(0)}$ coincides with the vector (2.9). According to the properties of the δ-function, (2.59) takes the form

$$\mathbf{Y}_{11}^{(0)} = \mathbf{Y}_1^{(0)} + \mathrm{K}^{(1)}\mathbf{R}^{(1)} H(\eta - \eta_{10}), \qquad \mathrm{K}^{(1)} = \mathrm{K}(\eta)\,\mathrm{K}^{-1}(\eta_{10}). \tag{2.60}$$

In the presence of the reaction $\mathbf{R}$, the general solution (2.6) becomes

$$\mathbf{Y}^{(0)} = \mathrm{K}(\eta)\,\mathbf{C} + \mathbf{Y}_1^{(0)} + \mathrm{K}^{(1)}\mathbf{R}^{(1)} H(\eta - \eta_{10}), \tag{2.61}$$

where $\mathbf{R}^{(1)} = (\mathbf{R}\,,\,\mathbf{0}\,,\,\mathbf{0}\,,\,\mathbf{0})^{\mathrm{T}}$.

The solution (2.60) must satisfy the 12 boundary conditions and the three additional relations (2.58), $u_j^{(0)}(\eta_{10}) = 0$, $j = 1\,,2\,,3$. As a result, we have the following 15 equations in the 15 unknowns:

six equations for $\eta = 0$,

$$\mathbf{Y}^{(0)}(0) = \mathrm{E}\mathbf{C}\,; \tag{2.62}$$

six equations for $\eta = 1$,

$$\mathbf{Y}^{(0)}(1) = \mathrm{K}\,(1)\,\mathbf{C} + \mathbf{Y}_1^{(0)}(1) + \mathrm{K}\,(1)\,\mathbf{R}\,; \tag{2.63}$$

three equations for $\eta = \eta_{10}$,

$$\mathbf{Y}^{(0)}(\eta_{10}) = \mathrm{K}\,(\eta_{10})\,\mathbf{C} + \mathbf{Y}_1^{(0)}(\eta_{10})\,. \tag{2.64}$$

If the boundary conditions at $\eta = 0$ are homogeneous, then the six components of the vector $\mathbf{C}$ equal zero. Having found the arbitrary constants and the components of $\mathbf{R}$, we obtain the general solution of the equilibrium equations for a rod with a lateral hinge support. The similar method can be applied in the case of an arbitrary number of supports. This method is a version of the generalized Krylov method for space-curved rods.

2.2.2 Rods with Lateral Elastic Supports

A curvilinear rod with a lateral elastic support is shown in Fig. 2.8*b*. The linear elastic stiffness of the support is c_R, the angular stiffness is c_T. If the rod is loaded, then concentrated forces and moments occur. These forces and moments are applied to the cross sections at which the supports are fixed. By use of the δ-function, these forces and moments may be introduced into the equilibrium equations. First, let us consider the most simple case when no restriction is imposed on the generalized displacements of the axial points at which the elastic supports are attached. We have

$$\mathbf{R}^{(i)} = -\mathrm{c}_R^{(i)}\mathbf{u}\,(\eta_i)\,; \qquad \mathbf{T}^{(\nu)} = -\mathrm{c}_T^{(\nu)}\boldsymbol{\vartheta}\,(\eta_\nu)\,;$$

here $\mathrm{c}_R^{(i)}$ and $\mathrm{c}_T^{(\nu)}$ are 3×3 matrices, whose elements are the stiffnesses of the elastic supports.The elastic reactions may depend on the components of $\mathbf{u}$ and $\boldsymbol{\vartheta}$ in more complicated manner. For example, suppose the reaction depends only on the displacements in a certain direction defined by a unit vector $\mathbf{e}_\alpha$. In this case, we get

$$\mathbf{R}^{(i)} = -\mathrm{c}_R^{(i)}\mathbf{u}_\alpha^{(0)}\,, \qquad \mathbf{u}_\alpha = (\mathbf{u}^{(0)}\,\mathbf{e}_\alpha)\,\mathbf{e}_\alpha\,. \tag{2.65}$$

In view of the relation $\mathbf{e}_\alpha = \sum_{i=1}^{3} k_{i\alpha}\mathbf{e}_{i0}$, (2.65) can be rewritten as follows:

$$\mathbf{R}^{(i)} = -c_R^{(i)} \mathrm{K}_\alpha \mathbf{u}^{(0)}, \tag{2.66}$$

where

$$\mathrm{K}_\alpha = \begin{bmatrix} k_{1\alpha}^2 & k_{1\alpha}k_{2\alpha} & k_{1\alpha}k_{3\alpha} \\ k_{2\alpha}k_{1\alpha} & k_{2\alpha}^2 & k_{2\alpha}k_{3\alpha} \\ k_{3\alpha}k_{1\alpha} & k_{3\alpha}k_{2\alpha} & k_{3\alpha}^2 \end{bmatrix}. \tag{2.67}$$

Since

$$\mathbf{f}^{(0)} = \begin{bmatrix} -\mathbf{q}_0 - \sum\limits_{i=1}^{n} c_P^{(i)} \mathbf{u}^{(0)}\, \delta(\eta - \eta_i) \\ -\boldsymbol{\mu}_0 - \sum\limits_{\nu=1}^{\rho} c_T^{(\nu)} \boldsymbol{\vartheta}^{(0)}\, \delta(\eta - \eta_\nu) \\ \mathbf{0} \\ \mathbf{0} \end{bmatrix}, \tag{2.68}$$

we see that generalized displacements appear in the equation of the zeroth approximation (2.5). We assume here that only distributed loads are applied to the rod.

Solving (2.5) and taking into account (2.38) (this formula was used in the analysis of (2.37)), we get

$$\mathbf{f}^{(0)} = \boldsymbol{\Phi}_0^{(0)} + \sum_{i=1}^{n} \Phi_P^{(i)}\, \delta(\eta - \eta_i) \mathbf{Y}^{(0)} + \sum_{\nu=1}^{\rho} \Phi_T^{(\nu)}\, \delta(\eta - \eta_\nu) \mathbf{Y}^{(0)}, \tag{2.69}$$

where the components of the vector $\boldsymbol{\Phi}_0^{(0)}$ are functions of $\mathbf{q}_0$ and $\boldsymbol{\mu}_0$. The matrices $\Phi_P^{(i)}$ and $\Phi_T^{(\nu)}$ are

$$\Phi_P^{(i)} = \begin{bmatrix} 0 & \dots & 0 & (c_P^{(i)})_{1.10} & (c_P^{(i)})_{1.11} & (c_P^{(i)})_{1.12} \\ 0 & \dots & 0 & (c_P^{(i)})_{2.10} & (c_P^{(i)})_{2.11} & (c_P^{(i)})_{2.12} \\ 0 & \dots & 0 & (c_P^{(i)})_{3.10} & (c_P^{(i)})_{3.11} & (c_P^{(i)})_{3.12} \\ 0 & \dots & 0 & 0 & 0 & 0 \\ \dots & \dots & \dots & \dots & \dots & \dots \end{bmatrix}; \tag{2.70}$$

$$\Phi_T^{(\nu)} = \begin{bmatrix} 0 & \dots & 0 & 0 & 0 & 0 & 0 & 0 & 0 \\ 0 & \dots & 0 & 0 & 0 & 0 & 0 & 0 & 0 \\ 0 & \dots & 0 & (c_T^{(\nu)})_{77} & (c_T^{(\nu)})_{78} & (c_T^{(\nu)})_{79} & 0 & 0 & 0 \\ 0 & \dots & 0 & (c_T^{(\nu)})_{87} & (c_T^{(\nu)})_{88} & (c_T^{(\nu)})_{89} & 0 & 0 & 0 \\ 0 & \dots & 0 & (c_T^{(\nu)})_{97} & (c_T^{(\nu)})_{98} & (c_T^{(\nu)})_{99} & 0 & 0 & 0 \\ 0 & \dots & 0 & 0 & 0 & 0 & 0 & 0 & 0 \\ \dots & \dots & \dots & \dots & \dots & \dots & \dots & \dots & \dots \end{bmatrix}. \tag{2.71}$$

The general solution of (2.61) can be written as follows:

$$\mathbf{Y}^{(0)} = \mathrm{K}\mathbf{C} + \mathbf{Y}_{10}^{(0)} - \sum_{i=1}^{n} \mathrm{G}_P^{(i)}(\eta)\, \mathbf{Y}^{(0)}(\eta_i)\, H(\eta - \eta_i)$$

$$- \sum_{\nu=1}^{\rho} \mathrm{G}_T^{(\nu)}(\eta)\, \mathbf{Y}^{(0)}(\eta_\nu)\, H(\eta - \eta_\nu)\,; \tag{2.72}$$

here

$$\begin{aligned} \mathrm{G}_P^{(i)}(\eta) &= \mathrm{K}(\eta)\, \mathrm{K}^{-1}(\eta_i)\, \varPhi_P^{(i)}\,; \\ \mathrm{G}_T^{(\nu)}(\eta) &= \mathrm{K}(\eta)\, \mathrm{K}^{-1}(\eta_\nu)\, \varPhi_T^{(\nu)}\,. \end{aligned} \tag{2.73}$$

It should be noted that the vector $\mathbf{C}$ and a part of the components of the vectors $\mathbf{Y}^{(0)}(\eta_i)\,(u_j^{(0)}(\eta_i))$ and $\mathbf{Y}^{(0)}(\eta_\nu)\,(u_j^{(0)}(\eta_\nu))$ can be found in accordance with the method presented at the beginning of this section.

2.2.3 Rods with Predetermined Displacement of Some Cross Sections

During assembly of elastic constructions, some structural elements must be deformed before installation for better adjustment. Usually, this is a result of errors in their manufacturing. Thus, it is important to elaborate the method for determination of the stress state resulting from predetermined linear and angular displacements of some cross sections of a rod.

A rod in its natural configuration is shown in Fig. 2.9 as a dashed line. During the process of assembly, a cross section K is displaced by a force and the rod takes the form shown in Fig. 2.9 as a full line. The displacement of the cross section K is defined by the vector $\mathbf{u}_K$. The vectors $\mathbf{Q}$ and $\mathbf{M}$ are to be determined. The components of the vector $\mathbf{u}_K$ are assumed to be small. Then, in view of the equations of the zeroth approximation (1.112)–(1.115) (see also (2.5) with $\mathbf{f}^{(0)} = 0$), we can rewrite (2.6) as follows:

$$\mathbf{Y}^{(0)} = \mathrm{K}(\eta)\, \mathbf{C}\,. \tag{2.74}$$

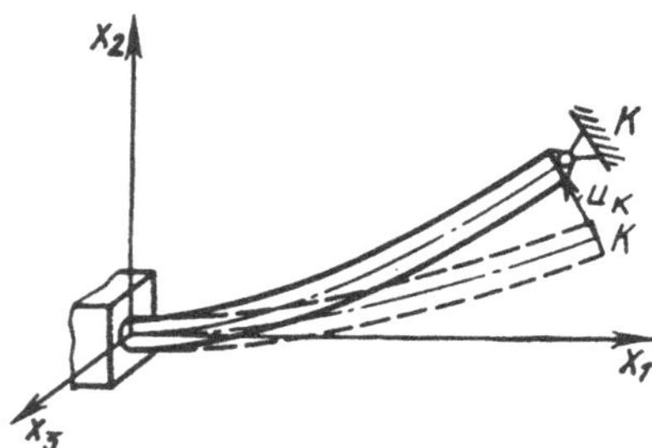

Fig. 2.9.

The components of the vector $\mathbf{C}$ can be found from the following boundary conditions:

$$\begin{aligned}
&(1) \quad \text{at} \quad \eta = 0\,, \quad \boldsymbol{\vartheta}^{(0)} = \mathbf{u}^{(0)} = 0 \quad (\text{i.e.} \quad Y_7^{(0)} = \ldots = Y_{12}^{(0)} = 0)\,;\\
&(2) \quad \text{at} \quad \eta = 1\,, \quad \mathbf{M}^{(0)}(1) = 0 \quad (\text{i.e.} \quad Y_4^{(0)} = Y_5^{(0)} = Y_6^{(0)} = 0)\\
&\qquad\qquad\qquad\quad \text{and} \quad \mathbf{u}^{(0)}(1) = \mathbf{u}_k\,.
\end{aligned}$$

From these relations, it follows that $c_7 = c_8 = \ldots = c_{12} = 0$ at $\eta = 0$. The boundary conditions at $\eta = 1$ yield a system of six linear inhomogeneous equations in the six unknowns $\mathbf{C}\,(c_1\,,\, c_2\,, \ldots\,,\, c_6)$.

In practice, more complicated cases may appear, that is, more than one cross section of an element are to be displaced during assembly.

2.3 Method of Step-by-Step Loading

2.3.1 Equilibrium Equations for One Step of Loading

Methods of integration of the equilibrium equations for small displacement of a rod axis were presented in Sect. 2.2. However, these methods can be used for numerical solution of equilibrium equations for arbitrary finite displacements of a rod axis. Consider stepwise loading of a rod. During each small step of loading, the increments of internal forces Q_j and M_j are assumed to be small. Thus, at each step of loading, we get a system of linear equations, which is equivalent to the equations of the first approximation. For the mth step, we have

$$\begin{aligned}
&\boldsymbol{\mu}^{(m)} = \beta m \boldsymbol{\mu}\,; \qquad \mathbf{q}_n^{(m)} = \beta m \mathbf{q}\,;\\
&\mathbf{P}^{(i)(m)} = \beta m \mathbf{P}^{(i)}\,; \qquad \mathbf{T}^{(\nu)(m)} = \beta m \mathbf{T}^{(\nu)}\,,
\end{aligned} \tag{2.75}$$

where $\beta < 1$ and m is the step number; the physical meaning of the constant β is a fraction of the magnitude of the total load per one step. For example, $\beta = 0.1$ means that the magnitude of the load added at each step makes $1/10$ of the magnitude of the total load.

Now, let us consider the equilibrium equations for particular types of external loads.

Follower Forces Suppose that the equilibrium equations are obtained in the attached coordinate system. In this case, the components of the forces $\beta\mathbf{q}$, $\beta\mathbf{P}^{(i)}$, $\beta\boldsymbol{\mu}$, and $\beta\mathbf{T}^{(\nu)}$ applied to the rod at each step do not depend on the generalized displacements u_j and ϑ_j. Thus, the right-hand sides of the equilibrium equations are known. The left-hand sides are similar to those of the equations of the first approximation. Then, for the mth step, we get

$$\frac{\mathrm{d}\mathbf{Q}^{(m)}}{\mathrm{d}\eta} + \mathrm{A}_{æ}^{(m-1)}\mathbf{Q}^{(m)} + \mathrm{A}_{Q}^{(m-1)}\,\Delta æ^{(m)}$$

$$= -\beta\mathbf{q} - \sum_{i=1}^{n} \beta\mathbf{P}^{(i)}\,\delta(\eta-\eta_i) - \Delta æ^{(m-1)} \times \mathbf{Q}^{(m-1)}\,; \tag{2.76}$$

$$\frac{\mathrm{d}\mathbf{M}^{(m)}}{\mathrm{d}\eta} + \mathrm{A}_{æ}^{(m-1)}\mathbf{M}^{(m)} + \mathrm{A}_{M}^{(m-1)}\,\Delta æ^{(m)} + \mathrm{A}_1\mathbf{Q}^{(m)}$$

$$= -\beta\boldsymbol{\mu} - \sum_{\nu=1}^{\rho} \beta\mathbf{T}^{(\nu)}\,\delta(\eta-\eta_\nu) - \Delta æ^{(m-1)} \times \mathbf{M}^{(m-1)}\,; \tag{2.77}$$

$$\mathbf{M}^{(m)} = \mathrm{A}\,\Delta æ^{(m)}\,; \tag{2.78}$$

$$\frac{\mathrm{d}\boldsymbol{\vartheta}^{(m)}}{\mathrm{d}\eta} + \mathrm{A}_{æ}^{(m-1)}\boldsymbol{\vartheta}^{(m)} - \Delta æ^{(m)} = 0\,; \tag{2.79}$$

$$\frac{\mathrm{d}\mathbf{u}^{(m)}}{\mathrm{d}\eta} + \mathrm{A}_{æ}^{(m-1)}\mathbf{u}^{(m)} + \mathrm{A}_1\boldsymbol{\vartheta}^{(m)} = 0\,, \tag{2.80}$$

where

$$\mathrm{A}_{æ}^{(m-1)} = \begin{bmatrix} 0 & -æ_3^{(m-1)} & æ_2^{(m-1)} \\ æ_3^{(m-1)} & 0 & -æ_1^{(m-1)} \\ -æ_2^{(m-1)} & æ_1^{(m-1)} & 0 \end{bmatrix};$$

$$\mathrm{A}_{Q}^{(m-1)} = \begin{bmatrix} 0 & \tilde{Q}_3^{(m-1)} & -\tilde{Q}_2^{(m-1)} \\ -\tilde{Q}_3^{(m-1)} & 0 & \tilde{Q}_1^{(m-1)} \\ \tilde{Q}_2^{(m-1)} & -\tilde{Q}_1^{(m-1)} & 0 \end{bmatrix};$$

$$\mathrm{A}_{M}^{(m-1)} = \begin{bmatrix} 0 & \tilde{M}_3^{(m-1)} & -\tilde{M}_2^{(m-1)} \\ -\tilde{M}_3^{(m-1)} & 0 & \tilde{M}_1^{(m-1)} \\ \tilde{M}_2^{(m-1)} & -\tilde{M}_1^{(m-1)} & 0 \end{bmatrix}.$$

Here

$$æ_j^{(m-1)} = æ_{j0} + \sum_{k=1}^{m-1} \Delta æ_j^{(k)}\,;$$

$$\tilde{Q}_j^{(m-1)} = \sum_{k=1}^{m-1} Q_j^{(k)}\,; \qquad \tilde{M}_j^{(m-1)} = \sum_{k=1}^{m-1} M_j^{(k)}\,.$$

The elements of the matrices $\mathrm{A}_{æ}^{(m-1)}$, $\mathrm{A}_{Q}^{(m-1)}$, and $\mathrm{A}_{M}^{(m-1)}$ are to be found from the equilibrium equations at the previous steps of loading.

Dead Forces The components of the vectors $\mathbf{q}_x$, $\mathbf{P}_x^{(i)}$, $\boldsymbol{\mu}_x$, and $\mathbf{T}_x^{(\nu)}$ in the basis $\{\mathbf{i}_j\}$ are known. In the basis $\{\mathbf{e}_j\}$, we have $\mathbf{q} = \mathrm{L}^{(1)}\mathbf{q}_x$, $\mathbf{P}^{(i)} = \mathrm{L}^{(1)}\mathbf{P}_x^{(i)}$, $\boldsymbol{\mu} = \mathrm{L}^{(1)}\boldsymbol{\mu}_x$, and $\mathbf{T}^{(\nu)} = \mathrm{L}^{(1)}\mathbf{T}_x^{(\nu)}$, where $\mathrm{L}^{(1)}$ is the matrix of transformation (A.57) from the basis $\{\mathbf{i}_j\}$ to the basis $\{\mathbf{e}_j\}$.

During deformation, a dead load may change its direction with respect to the attached basis. Therefore, the components of the vectors $\mathbf{q}$, $\mathbf{P}^{(i)}$, $\boldsymbol{\mu}$, and $\mathbf{T}^{(\nu)}$ in the attached basis are functions of increments of the angles ϑ_j. These angles are characteristics of the relative position of the basis $\mathbf{e}_{j0}$ with respect to the basis $\mathbf{e}_{kj}$. The matrix of transformation from the basis $\{\mathbf{i}_j\}$ to the basis $\{\mathbf{e}_j\}$ is $\mathrm{L}^{(1)} = \mathrm{L}\mathrm{L}^0$, where L^0 is the matrix of transformation from the basis $\{\mathbf{i}_j\}$ to the basis $\{\mathbf{e}_{j0}\}$. The basis $\{\mathbf{e}_{j0}\}$ is attributed to the natural configuration of the rod. The matrix L is the matrix of transformation from the basis $\{\mathbf{e}_{i0}\}$ to the basis $\{\mathbf{e}_i\}$. The basis $\{\mathbf{e}_i\}$ is connected with the rod at the mth step of loading. The elements of the matrix $\mathrm{L}\,(l_{ij})$ depend on the angles as follows:

$$\vartheta_{i\,,m} = \sum_{\rho=1}^{m} \vartheta_j^{(\rho)}\,;$$

here $\vartheta_i^{(\rho)}$ are small angles of rotation of the basis $\{\mathbf{e}_j^{(\rho)}\}$ $(\rho = 1\,,2\,,\ldots\,,m)$ at each step of loading. At an arbitrary mth step, the matrix L satisfies the relation

$$\mathrm{L} = \mathrm{L}_m = \Delta\mathrm{L}_m\mathrm{L}^{(m-1)}\,, \tag{2.81}$$

where

$$\Delta\mathbf{L}_m = \begin{bmatrix} 1 & \vartheta_3^{(m)} & -\vartheta_2^{(m)} \\ -\vartheta_3^{(m)} & 1 & \vartheta_1^{(m)} \\ \vartheta_2^{(m)} & -\vartheta_1^{(m)} & 1 \end{bmatrix} = \mathrm{E} + \Delta\mathbf{L}^{(m)} \tag{2.82}$$

is the matrix of transformation from the basis $\{\mathbf{e}_j^{(m-1)}\}$ to the basis $\{\mathbf{e}_j^{(m)}\}$, The elements of $\mathrm{L}^{(m-1)}$ depend on the angles in the following way:

$$\vartheta_{j\,,m-1} = \sum_{\rho=1}^{m-1} \vartheta_j^{(\rho)}\,.$$

The angles $\vartheta_j^{(\rho)}$ $(\rho = 1\,,2\,,\ldots\,,m-1)$ can be determined from the equations obtained at the previous $(m-1)$ steps. At the mth step, we have

$$\mathrm{L}_m^{(1)} = (\mathrm{E} + \Delta\mathrm{L}^{(m)})\,\mathrm{L}^{(m-1)}\mathrm{L}^0 = \mathrm{L}^{(m-1)}\mathrm{L}^0 + \Delta\mathrm{L}^{(m)}\mathrm{L}^{(m-1)}\mathrm{L}^0\,. \tag{2.83}$$

Therefore, in the case of dead loads, the load applied to the rod at the mth step can be written as follows:

$$\begin{aligned} \Delta\mathbf{q}^{(m)} &= \mathrm{L}^{(m-1)}\mathrm{L}^0\beta\mathbf{q}_x + \Delta\mathrm{L}^{(m)}\mathrm{L}^{(m-1)}\mathrm{L}^0\beta\mathbf{q}_x\,; \\ \Delta\mathbf{P}^{(i)(m)} &= \mathrm{L}^{(m-1)}\mathrm{L}^0\beta\mathbf{P}_x^{(i)} + \Delta\mathrm{L}^{(m)}\mathrm{L}^{(m-1)}\mathrm{L}^0\beta\mathbf{P}_x^{(i)}\,; \\ \Delta\boldsymbol{\mu}^{(m)} &= \mathrm{L}^{(m-1)}\mathrm{L}^0\beta\boldsymbol{\mu}_x + \Delta\mathrm{L}^{(m)}\mathrm{L}^{(m-1)}\mathrm{L}^0\beta\boldsymbol{\mu}_x\,; \\ \Delta\mathbf{T}^{(\nu)(m)} &= \mathrm{L}^{(m-1)}\mathrm{L}^0\beta\mathbf{T}_x^{(\nu)} + \Delta\mathrm{L}^{(m)}\mathrm{L}^{(m-1)}\mathrm{L}^0\beta\mathbf{T}_x^{(\nu)}\,. \end{aligned} \tag{2.84}$$

The terms containing $\Delta\mathrm{L}^{(m)}$ are

$$\beta\,(\Delta\mathrm{L}^{(m)}\mathrm{L}^{(m-1)}\mathrm{L}^{0}\mathbf{q}_x) = \beta\mathrm{A}_{qk}^{(m-1)}\boldsymbol{\vartheta}^{(m)}\,;$$
$$\beta\,(\Delta\mathrm{L}^{(m)}\mathrm{L}^{(m-1)}\mathrm{L}^{0}\mathbf{P}_x^{(i)}) = \beta\mathrm{A}_{Pk}^{(i)(m-1)}\boldsymbol{\vartheta}^{(m)}\,;$$
$$\beta\,(\Delta\mathrm{L}^{(m)}\mathrm{L}^{(m-1)}\mathrm{L}^{0}\boldsymbol{\mu}_x) = \beta\mathrm{A}_{\mu k}^{(m-1)}\boldsymbol{\vartheta}^{(m)}\,;$$
$$\beta\,(\Delta\mathrm{L}^{(m)}\mathrm{L}^{(m-1)}\mathrm{L}^{0}\mathbf{T}_x^{(\nu)}) = \beta\mathrm{A}_{Tk}^{(\nu)(m-1)}\boldsymbol{\vartheta}^{(m)}\,,$$

where

$$\mathrm{A}_{qk}^{(m-1)} = \begin{bmatrix} 0 & -\sum_{j=1}^{3} c_{3j}^{(m-1)} q_{xj} & \sum_{j=1}^{3} c_{2j}^{(m-1)} q_{xj} \\ \sum_{j=1}^{3} c_{3j}^{(m-1)} q_{xj} & 0 & -\sum_{j=1}^{3} c_{1j}^{(m-1)} q_{xj} \\ -\sum_{j=1}^{3} c_{2j}^{(m-1)} q_{xj} & \sum_{j=1}^{3} c_{1j}^{(m-1)} q_{xj} & 0 \end{bmatrix};$$

$$\mathrm{A}_{Pk}^{(i)(m-1)} = \begin{bmatrix} 0 & -\sum_{j=1}^{3} c_{3j}^{(m-1)} P_{xj}^{(i)} & \sum_{j=1}^{3} c_{2j}^{(m-1)} P_{xj}^{(i)} \\ \sum_{j=1}^{3} c_{3j}^{(m-1)} P_{xj}^{(i)} & 0 & -\sum_{j=1}^{3} c_{1j}^{(m-1)} P_{xj}^{(i)} \\ -\sum_{j=1}^{3} c_{2j}^{(m-1)} P_{xj}^{(i)} & \sum_{j=1}^{3} c_{1j}^{(m-1)} P_{xj}^{(i)} & 0 \end{bmatrix};$$

$$\mathrm{A}_{\mu k}^{(m-1)} = \begin{bmatrix} 0 & -\sum_{j=1}^{3} c_{3j}^{(m-1)} \mu_{xj} & \sum_{j=1}^{3} c_{2j}^{(m-1)} \mu_{xj} \\ \sum_{j=1}^{3} c_{3j}^{(m-1)} \mu_{xj} & 0 & -\sum_{j=1}^{3} c_{1j}^{(m-1)} \mu_{xj} \\ -\sum_{j=1}^{3} c_{2j}^{(m-1)} \mu_{xj} & \sum_{j=1}^{3} c_{1j}^{(m-1)} \mu_{xj} & 0 \end{bmatrix};$$

$$\mathrm{A}_{Tk}^{(\nu)(m-1)} = \begin{bmatrix} 0 & -\sum_{j=1}^{3} c_{3j}^{(m-1)} T_{xj}^{(\nu)} & \sum_{j=1}^{3} c_{2j}^{(m-1)} T_{xj}^{(\nu)} \\ \sum_{j=1}^{3} c_{3j}^{(m-1)} T_{xj}^{(\nu)} & 0 & -\sum_{j=1}^{3} c_{1j}^{(m-1)} T_{xj}^{(\nu)} \\ -\sum_{j=1}^{3} c_{2j}^{(m-1)} T_{xj}^{(\nu)} & \sum_{j=1}^{3} c_{1j}^{(m-1)} T_{xj}^{(\nu)} & 0 \end{bmatrix};$$

here $c_{ij}^{(m-1)}$ are the elements of the matrix $\mathrm{C}^{(m-1)} = \mathrm{L}^{(m-1)}\mathrm{L}^{(0)}$.

Finally, in the case of dead loads, the equilibrium equations at the mth step are as follows:

$$\frac{\mathrm{d}\mathbf{Q}^{(m)}}{\mathrm{d}\eta} + \mathrm{A}_{æ}^{(m-1)}\mathbf{Q}^{(m)} + \mathrm{A}_{Q}^{(m-1)}\,\Delta æ^{(m)} + \beta\mathrm{A}_{qk}^{(m-1)}\boldsymbol{\vartheta}^{(m)}$$

$$= -\beta \sum_{i=1}^{n} A_{Pk}^{(i)(m-1)} \boldsymbol{\vartheta}^{(m)} \, \delta(\eta - \eta_i)$$

$$- \beta C^{(m-1)} \mathbf{q}_x - \beta \sum_{i=1}^{n} C^{(m-1)} \mathbf{P}_x^{(i)} \, \delta(\eta - \eta_i) \, ; \tag{2.85}$$

$$\frac{d\mathbf{M}^{(m)}}{d\eta} + A_{æ}^{(m-1)} \mathbf{M}^{(m)} + \beta A_{\mu k}^{(m-1)} \boldsymbol{\vartheta}^{(m)}$$

$$= -\beta \sum_{\nu=1}^{\rho} A_{Tk}^{(\nu)(m-1)} \boldsymbol{\vartheta}^{(m)} \, \delta(\eta - \eta_\nu)$$

$$- \beta C^{(m-1)} \boldsymbol{\mu}_x - \beta \sum_{\nu=1}^{\rho} C^{(m-1)} \mathbf{T}_x^{(\nu)} \, \delta(\eta - \eta_\nu) \, ; \tag{2.86}$$

$$\mathbf{M}^{(m)} = A \, \Delta æ^{(m)} \, ; \tag{2.87}$$

$$\frac{d\boldsymbol{\vartheta}^{(m)}}{d\eta} + A_{æ}^{(m-1)} \boldsymbol{\vartheta}^{(m)} - \Delta æ^{(m)} = 0 \, ; \tag{2.88}$$

$$\frac{d\mathbf{u}^{(m)}}{d\eta} + A_{æ}^{(m-1)} \mathbf{u}^{(m)} + A_1 \boldsymbol{\vartheta}^{(m)} = 0 \, . \tag{2.89}$$

2.3.2 Integration of the Equilibrium Equations

At any step of loading, the systems of equations (2.76)–(2.80) and (2.85)–(2.89) consist of linear differential equations. Eliminating $\Delta æ^{(m)}$, we arrive at

$$\mathbf{Y}^{(m)} + A^{(m-1)} \mathbf{Y}^{(m)} = \beta \mathbf{f}^{(m-1)} \, , \tag{2.90}$$

where

$$\mathbf{Y}^{(m)} = \begin{bmatrix} \mathbf{Q}^{(m)} \\ \mathbf{M}^{(m)} \\ \boldsymbol{\vartheta}^{(m)} \\ \mathbf{u}^{(m)} \end{bmatrix} ;$$

$$A^{(m-1)} = \begin{bmatrix} A_{æ}^{(m-1)} & A_{Q}^{(m-1)} A^{-1} & 0 & 0 \\ A_1 & A_{æ}^{(m-1)} + A_{M}^{(m-1)} A^{-1} & 0 & 0 \\ 0 & -A^{-1} & A_{æ}^{(m-1)} & 0 \\ 0 & 0 & A_1 & A_{æ}^{(m-1)} \end{bmatrix} ;$$

$$\mathbf{f}^{(m-1)} = \begin{bmatrix} -\sum\limits_{i=1}^{n} (\mathrm{A}_{Pk}^{(i)(m-1)} \boldsymbol{\vartheta}^{(m)} + \mathrm{C}^{(m-1)} \mathbf{P}_x^{(i)})\, \delta(\eta - \eta_i) \\ - \mathrm{C}^{(m-1)} \mathbf{q}_x \\ -\sum\limits_{\nu=1}^{\rho} (\mathrm{A}_{Tk}^{(\nu)(m-1)} \boldsymbol{\vartheta}^{(m)} + \mathrm{C}^{(m-1)} \mathbf{T}_x^{(\nu)})\, \delta(\eta - \eta_\nu) \\ - \mathrm{C}^{(m-1)} \boldsymbol{\mu}_x \\ \mathbf{0} \\ \mathbf{0} \end{bmatrix} .$$

The solution of (2.90) can be written as

$$\mathbf{Y}^{(m)} = \mathrm{K}^{(m)} \mathbf{C}^{(m)} + \mathbf{Y}_1^{(m)} . \tag{2.91}$$

The method of numerical determination of the fundamental matrix $\mathrm{K}^{(m)}$ is given in Sect. 2.1. The components of the vector $\mathbf{C}^{(m)}$ are to be determined from the boundary conditions at $\eta = 1$. It should be noted that if some elements of $\mathrm{K}^{(m)}(1)$ are rapidly growing functions, then $\det \mathrm{K}$ is ill-conditioned and, therefore, the components of the vector $\mathbf{C}^{(m)}$ may be evaluated with large errors.

Consider a method that allows us to determine the fundamental matrix more accurately. This method is based on the Picard iteration method (see Sect. 2.1). The fundamental matrix $\mathrm{K}^{(m)}$ satisfies the relation

$$\mathrm{K}^{(\mathrm{m})'} + \mathrm{A}^{(m-1)} \mathrm{K}^{(m)} = 0 ,$$

which yields

$$\mathrm{K}^{(m)} = - \int_0^\eta \mathrm{A}^{(m-1)} \mathrm{K}^{(m)} \, \mathrm{d}\zeta + \mathrm{E} . \tag{2.92}$$

The fundamental matrix $\mathrm{K}^{(m)}$ determined by the method of initial parameters can be taken as the first approximation. Therefore, the corrected matrix $\mathrm{K}^{(m)}$ (i.e. the second approximation) can be obtained from (2.92) as follows:

$$\mathrm{K}^{(m)(2)} = - \int_0^\eta \mathrm{A}^{(m-1)} \mathrm{K}^{(m)(1)} \, \mathrm{d}\zeta + \mathrm{E} . \tag{2.93}$$

Let us consider an equation

$$\delta_{ij} = \max \left| \frac{K_{ij}^{(\rho)}(\eta) - K_{ij}^{(\rho-1)}(\eta)}{K_{ij}^{(\rho)}(\eta)} \right| 100\% ,$$

where ρ is the iteration index. If the inequality $\delta_{ij} \leq \Delta$ holds for $\max \delta_{ij}$, then the required accuracy is obtained.

At each step, the values of the elements of the fundamental matrix $\mathrm{K}(\eta)$ may be found more accurately in the following way. Let us divide the interval of integration into a few subintervals (this procedure was applied in Sect. 2.2). For each subinterval, we get

$$\mathrm{K}^{(m)(n)} = -\int_{\eta_{\nu-1}}^{\eta_\nu} \mathrm{A}^{(m-1)}\mathrm{K}^{(m)(n-1)}\,\mathrm{d}\zeta + \mathrm{K}^{(m)}(\eta_{\nu-1})\,. \tag{2.94}$$

Once the corrected fundamental matrix $\mathrm{K}_\nu^{(m)}(\eta)$ is found, we readily determine the corrected fundamental matrix $\mathrm{K}^{(m)}(\eta)$ for the whole interval of integration at the mth step of loading. Then, using $\mathrm{K}^{(m)}(\eta)$, we find $(\mathrm{K}^{(m)}(\eta_i))^{-1}$ and $(\mathrm{K}^{(m)}(\eta_\nu))^{-1}$. These terms enter into the partial solution.

2.3.3 Method of Successive Approximations

Equilibrium of a spatial rod is governed by the following set of five nonlinear equations (see (1.57)–(1.61)):

$$\begin{aligned}
&\frac{\mathrm{d}\mathbf{Q}}{\mathrm{d}\eta} + \mathbf{F}_1(\mathbf{Q}\,,\,\mathbf{M}\,,\,æ\,,\,\boldsymbol{\vartheta}\,,\,\mathbf{u}) = \mathbf{b}_P\,;\\
&\frac{\mathrm{d}\mathbf{M}}{\mathrm{d}\eta} + \mathbf{F}_2(\mathbf{Q}\,,\,\mathbf{M}\,,\,æ\,,\,\boldsymbol{\vartheta}\,,\,\mathbf{u}) = \mathbf{b}_T\,;\\
&\mathbf{L}_1\frac{\mathrm{d}\boldsymbol{\vartheta}}{\mathrm{d}\eta} + \mathbf{F}_3(\boldsymbol{\vartheta}\,,\,\mathbf{M}\,,\,æ_0^{(1)}) = 0\,;\\
&\frac{\mathrm{d}\mathbf{u}}{\mathrm{d}\eta} + \mathbf{F}_4(æ\,,\,\boldsymbol{\vartheta}) = 0\,;
\end{aligned} \tag{2.95}$$

here

$$\mathbf{M} = \mathrm{A}\,(æ - æ_0^{(1)})\,. \tag{2.96}$$

Suppose that the variables æ and ϑ_j are known and the external loads are linear in the vectors $\boldsymbol{\vartheta}$, $\mathbf{u}'$, and $\mathbf{u}$. Under these conditions, (2.95) become linear differential equations. Hence, the following numerical method can be applied. We assume that the zeroth approximations of $æ = æ_0$ and ϑ_{j0} are known. Consequently, (2.95) take the form

$$\begin{aligned}
&\frac{\mathrm{d}\mathbf{Q}^{(1)}}{\mathrm{d}\eta} + æ_0 \times \mathbf{Q}^{(1)} + \mathrm{P}\,(\boldsymbol{\vartheta}^{(1)}\,,\,\mathbf{u}^{(1)\prime}\,,\,\mathbf{u}^{(1)}) = \mathbf{b}_P\,;\\
&\frac{\mathrm{d}\mathbf{M}^{(1)}}{\mathrm{d}\eta} + æ_0 \times \mathbf{M}^{(1)} + \mathbf{e}_1 \times \mathbf{Q}^{(1)} + \mathrm{T}\,(\boldsymbol{\vartheta}^{(1)}\,,\,\mathbf{u}^{(1)\prime}\,,\,\mathbf{u}^{(1)}) = \mathbf{b}_T\,;\\
&\frac{\mathrm{d}\boldsymbol{\vartheta}^{(1)}}{\mathrm{d}\eta} + \mathbf{L}_{10}^{-1}(\mathbf{L}_{20}æ_0^{(1)} - \mathrm{A}^{-1}\mathbf{M}^{(1)}) = 0\,;\\
&\frac{\mathrm{d}\mathbf{u}^{(1)}}{\mathrm{d}\eta} + æ_0 \times \mathbf{u}^{(1)} + (l_{11}^{(0)} - 1)\,\mathbf{e}_1 + l_{21}^{(0)}\mathbf{e}_2 + l_{31}^{(0)}\mathbf{e}_3 = 0\,,
\end{aligned} \tag{2.97}$$

where the elements of the matrices L_{10}^{-1} and L_{20} are functions of the known angles ϑ_{j0} and the elements $l_{j1}^{(0)}$ of the matrix $L^{(0)}$ are functions of the angles ϑ_{j0}.

Having determined the unknown quantities (including the unknown vector $\mathbf{M}^{(1)}$) from the linear equations (2.97) and using them in (2.96), we readily obtain the next approximation

$$æ^{(1)} = A^{-1}\mathbf{M}^{(1)} + æ_0^{(1)} . \tag{2.98}$$

Once $æ^{(1)}$ is found, the next approximations $\mathbf{Q}^{(2)}$, $\mathbf{M}^{(2)}$, $\boldsymbol{\vartheta}^{(2)}$, and $\mathbf{u}^{(2)}$ can be obtained from the system of equations

$$\begin{aligned}
&\frac{d\mathbf{Q}^{(2)}}{d\eta} + æ^{(1)} \times \mathbf{Q}^{(2)} + \mathbf{P}\,(\boldsymbol{\vartheta}^{(2)}, \mathbf{u}^{(2)\prime}, \mathbf{u}^{(2)}) = \mathbf{b}_P ;\\
&\frac{d\mathbf{M}^{(2)}}{d\eta} + æ^{(1)} \times \mathbf{M}^{(2)} + \mathbf{e}_1 \times \mathbf{Q}^{(2)} + \mathbf{T}\,(\boldsymbol{\vartheta}^{(2)}, \mathbf{u}^{(2)\prime}, \mathbf{u}^{(2)}) = \mathbf{b}_T ;\\
&\frac{d\boldsymbol{\vartheta}^{(2)}}{d\eta} + L_{11}^{-1}(L_{21}æ_0^{(1)} - A^{-1}\mathbf{M}^{(2)}) = 0 ;\\
&\frac{d\mathbf{u}^{(2)}}{d\eta} + æ^{(1)} \times \mathbf{u}^{(2)} + (l_{11}^{(0)(1)} - 1)\mathbf{e}_1 + l_{21}^{(0)(1)}\mathbf{e}_2 + l_{31}^{(0)(1)}\mathbf{e}_3 = 0 ,
\end{aligned} \tag{2.99}$$

where the elements of the matrices L_{11}^{-1} and L_{21} are functions of the angles $\vartheta_j^{(1)}$ and the elements $l_{j1}^{(0)(1)}$ of the matrix $L^{(0)}$ are functions of the angles $\vartheta_j^{(1)}$. The approximate values of the variables defined from (2.97) and (2.99) as well as the values of a higher approximation must satisfy the same boundary conditions. The number of iterations depends on the desired accuracy.

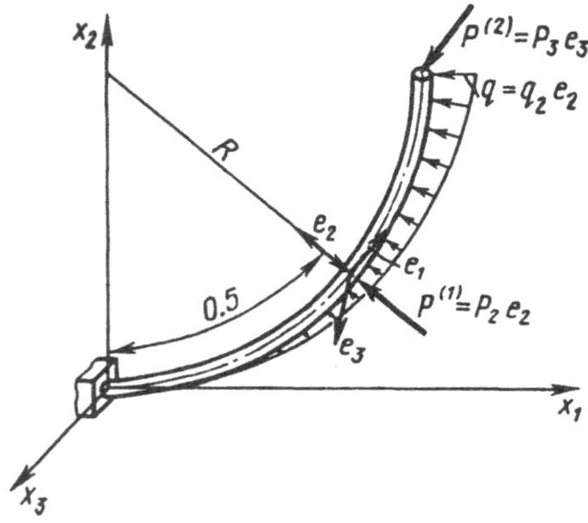

Fig. 2.10.

Consider a numerical example. A curvilinear rod subjected to follower forces is shown in Fig. 2.10. Unlike the problem considered in Sect. 2.1, the force $\mathbf{P}^{(2)} = P_3\mathbf{e}_3$ is perpendicular to the plane x_1Ox_2.

Let us apply the method of step-by-step loading. At any step, the expanded form of (2.76)–(2.80) reads

$$\frac{\mathrm{d}\mathbf{Q}^{(1)}}{\mathrm{d}\eta} + \mathrm{A}_{æ}^{(0)}\mathbf{Q}^{(1)} + \mathrm{A}_{Q}^{(0)}\,\Delta æ^{(1)}$$

$$= -\beta\mathbf{q} - \beta\mathbf{P}^{(1)}\,\delta(\eta - 0.5) - \beta\mathbf{P}^{(2)}\,\delta(\eta - 1)\,; \tag{2.100}$$

$$\frac{\mathrm{d}\mathbf{M}^{(1)}}{\mathrm{d}\eta} + \mathrm{A}_{æ}^{(0)}\mathbf{M}^{(1)} + \mathrm{A}_{M}^{(0)}\,\Delta æ^{(1)} + \mathrm{A}_1\mathbf{Q}^{(0)} = 0\,; \tag{2.101}$$

$$\mathbf{M}^{(1)} = \mathrm{A}\,\Delta æ^{(1)}\,; \tag{2.102}$$

$$\frac{\mathrm{d}\boldsymbol{\vartheta}^{(1)}}{\mathrm{d}\eta} + \mathrm{A}_{æ}^{(0)}\boldsymbol{\vartheta}^{(1)} - \Delta æ^{(1)} = 0\,; \tag{2.103}$$

$$\frac{\mathrm{d}\mathbf{u}^{(1)}}{\mathrm{d}\eta} + \mathrm{A}_{æ}^{(0)}\mathbf{u}^{(1)} + \mathrm{A}_1\boldsymbol{\vartheta}^{(1)} = 0\,. \tag{2.104}$$

The magnitudes of the nondimensional loads are $P_2 = 2$, $P_3 = 2$, and $q_0 = 3$. Suppose that the process of loading is divided into three steps, i.e. $\beta \approx 0.33$.

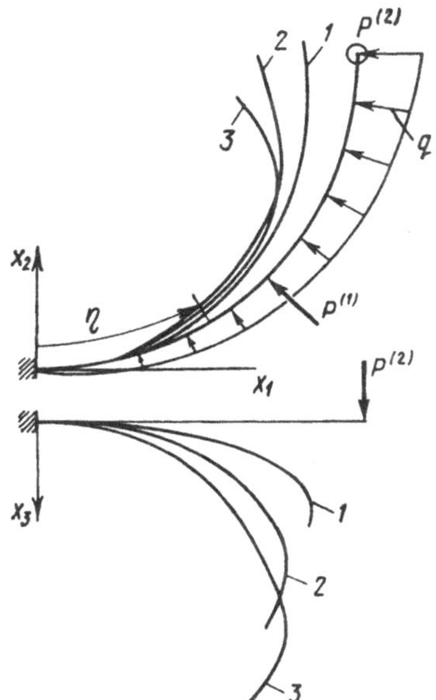

Fig. 2.11.

At the first step, all the elements of the matrices $\mathrm{A}_Q^{(0)}$ and $\mathrm{A}_M^{(0)}$ equal zero. The elements of $\mathrm{A}_{æ}^{(0)}$ depend on the curvatures $æ_{i0}$, which are characteristics of the rod axis in the natural configuration. In the problem under consideration, $æ_{10} = æ_{20} = 0$ and $æ_{30} = 1/R$.

We integrate (2.100) and (2.101) starting at the free end of the rod, where the boundary conditions for $\mathbf{Q}^{(1)}$ and $\mathbf{M}^{(1)}$ are stated. Equations (2.103) and (2.104) should be integrated starting at the clamped end, where the components of the vectors $\boldsymbol{\vartheta}^{(1)}$ and $\mathbf{u}^{(1)}$ are known. Having found the first approximation of the vectors $\mathbf{Q}^{(1)(1)}$, $\mathbf{M}^{(1)(1)}$, etc., we can obtain the corrected values of the elements of the matrix $\mathrm{A}_{æ}^{(0)}$ in the following manner:

$$\begin{aligned} æ_1^{(0)(1)} &= \Delta æ_1^{(1)}\,; \\ æ_2^{(0)(1)} &= \Delta æ_2^{(1)}\,; \\ æ_3^{(0)(1)} &= æ_{30} + \Delta æ_3^{(1)}\,. \end{aligned}$$

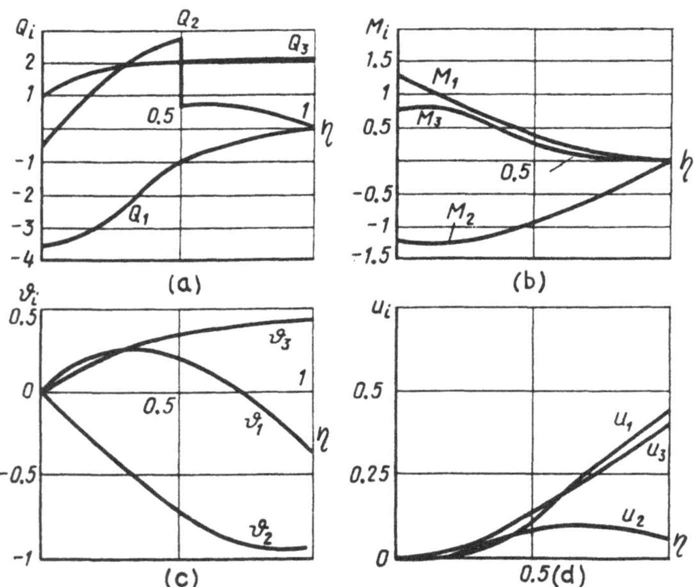

Fig. 2.12.

Substituting the matrix $\mathrm{A}_{æ}^{(0)(1)}$ into (2.100)–(2.104) and solving them, we obtain the vectors $\mathbf{Q}^{(1)(2)}$, $\mathbf{M}^{(1)(2)}$, etc. The elements of the matrix $\mathrm{A}_{æ}^{(0)(2)}$ can also be corrected as follows:

$$æ_1^{(0)(2)} = \Delta æ_1^{(2)}\,;$$

$$æ_2^{(0)(2)} = \Delta æ_2^{(2)} ;$$
$$æ_3^{(0)(2)} = æ_{30} + \Delta æ_3^{(2)} .$$

To achieve the accuracy of 0.05%, four iterations at each step need be performed. The projections of the rod axis on the planes x_1Ox_2 and x_1Ox_3 at each of the three steps are shown in Fig. 2.11.

The plots of the shear forces Q_2 and Q_3 and the axial force Q_1 (they are the components of the vector $\mathbf{Q}$) versus η are shown in Fig. 2.12*a*. The plots of the components of the vector $\mathbf{M}$ versus η are shown in Fig. 2.12*b*. The components of the vectors $\boldsymbol{\vartheta}$ and $\mathbf{u}$ plotted against η are illustrated in Fig. 2.12*c*,*d*.

3. Static Stability of Rods

In this chapter, the static stability of rods is examined. The theory of static stability of curvilinear rods is developed for the case when a loss of stability may occur with respect to a new equilibrium state of the rod. In this state, the shape of the rod axis may differ significantly from its shape in the natural configuration. Much attention is given to the analysis of deformation of a rod subjected to loads of different types (dead forces, follower forces or their combinations).

3.1 Basic Concepts

3.1.1 State of Equilibrium

In general, for a rod subjected to an external load, there exists a set of possible equilibrium states. An equilibrium state can be either stable or unstable. If an equilibrium configuration is unstable, then, due to small disturbances, a rod jumps into a new, stable equilibrium configuration. The jump from an unstable equilibrium configuration to a stable one is called a *loss of static stability*. If the stable equilibrium configuration lies in the immediate vicinity of the unstable one, we say that an *instability in the small* (a local instability) takes place, otherwise we say that an *instability in large* (a total instability) takes place (Alfutov (2000)).

3.1.2 Examples

First, recall some basic problems related to the static stability of rods. A rod with simply supported ends subjected to a compressive dead force $\mathbf{P}$ is shown in Fig. 3.1*a*. As soon as $|\mathbf{P}|$ reaches its critical value $|\mathbf{P}_*|$, the straight-line configuration becomes unstable. Being subjected to some small stochastic disturbances, the rod jumps into another state of equilibrium shown in Fig. 3.1*a* as a dashed line. It can be seen that the new equilibrium configuration differs slightly from the initial one. If we adopt the assumption of closeness of the stable and of the corresponding unstable configurations, then we can use *linear equations* for postbuckling analysis. If the critical loads determined from these equations are real, then this assumption is valid, otherwise (when either the critical loads cannot be found from the linear equations or they are complex) the assumption fails. If the latter is the case, then two approaches to the problem are possible.

(1) We can use nonlinear equations assuming the finiteness of displacement of rod points and angles of rotation of cross sections after loss of stability (i.e. we consider *instability in large*).

(2) We can study the stability of motion of a rod in the immediate vicinity of the initial state of equilibrium.

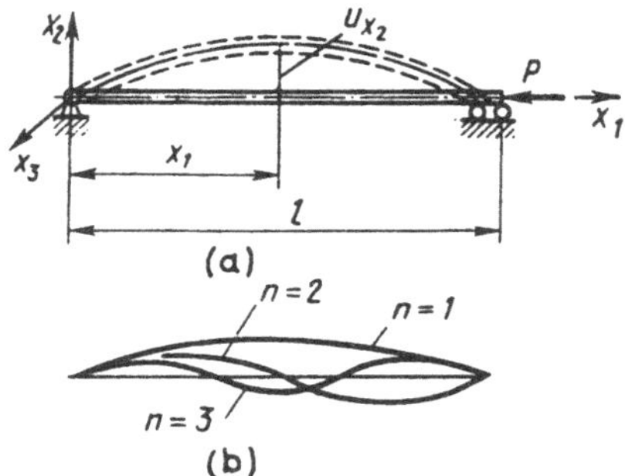

Fig. 3.1.

Let us assume that displacements of axial points are small and a loss of stability occurs in the plane of the drawing (Fig. 3.1*a*). Then, the following linear equilibrium equation can be obtained:

$$x_2'' + \frac{P}{A_{33}} x_2 = 0\,, \qquad A_{33} = EJ_{x_3}\,; \tag{3.1}$$

its solution,

$$x_2 = c_1 \cos kx_1 + c_2 \sin kx_1\,, \qquad k = \sqrt{\frac{P}{A_{33}}}\,,$$

must satisfy the following boundary conditions:

(1) at $x_1 = 0\,,\quad x_2 = 0\,;$

(2) at $x_1 = l\,,\quad x_2 = 0\,.$

It is therefore concluded that $\sin kl = 0$, hence,

$$P_* = P_{\mathrm{cr}} = \frac{n^2\pi^2 A_{33}}{l^2}\,. \tag{3.2}$$

From this relation it follows that there exists an infinite set of critical forces corresponding to $n = 1, 2, 3, \ldots$. A particular form of the rod axis depends on n. Three mathematically admissible equilibrium shapes corresponding to $n = 1, 2, 3$ are shown in Fig. 3.1*b*.

The smallest critical force (for $n = 1$)

$$P_{\text{cr}} = P = \frac{\pi^2 A_{33}}{l^2} \tag{3.3}$$

is called the *Euler critical force*. Substituting (3.3) into (3.1), we arrive at

$$x_2 = c_1 \sin \frac{\pi x_1}{l}, \tag{3.4}$$

where c_1 is an arbitrary constant factor. The presence of such a factor is typical of solutions of linear differential equations governing deformation of a rod after loss of stability. In the linear approximation, the critical forces can be exactly determined while small deflections of the rod remain unknown. In many cases of practical interest, it is sufficient to determine the critical forces only.

Consider some examples (Figs. 3.2–3.5).

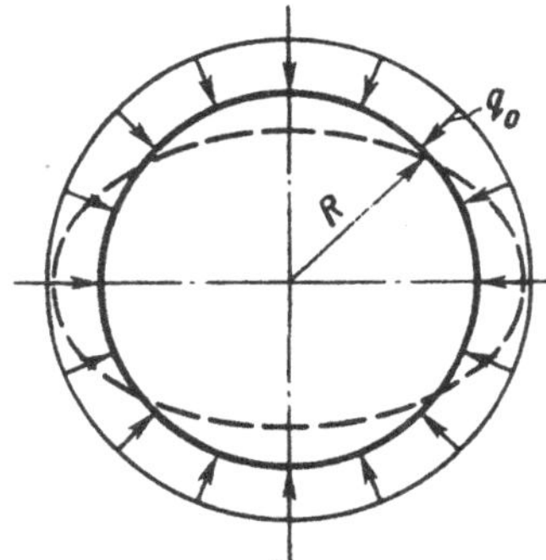

Fig. 3.2.

A ring subjected to a uniformly distributed radial load $\mathbf{q}_0$ is shown in Fig. 3.2. As soon as the critical value $|\mathbf{q}_0| = |\mathbf{q}_{\text{cr}}|$ is reached, a loss of stability of the ring happens. A new state of equilibrium is shown in Fig. 3.2 as a dashed line. Instability of the plane circular state may lead to a *spatial* equilibrium configuration (i.e. an out-of-plane deformation occurs).

Stability of preloaded rods is of great importance for applications. Consider a straight rod that had been loaded by a force $\mathbf{P}$ (dead or follower) and then was simply supported as shown in Fig. 3.3. In addition, a distributed load $\mathbf{q}$ (dead or follower) is applied to the rod. When designing such a structure, it is important to determine the critical value of $\mathbf{q}$. The dashed lines in Fig. 3.3 denote (qualitatively) possible equilibrium configurations of the rod after the loss of stability.

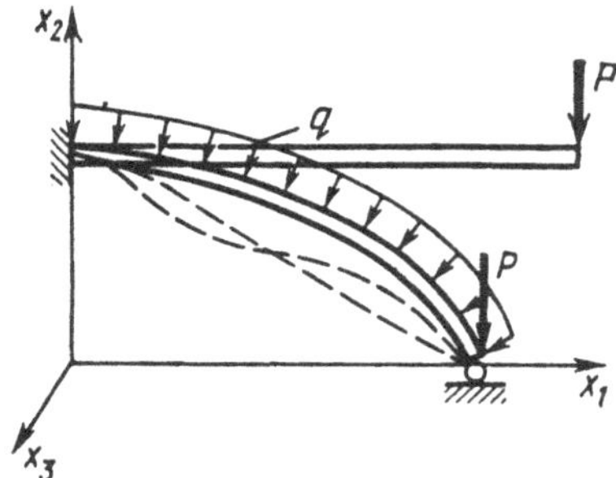

Fig. 3.3.

Figure 3.4 illustrates a plane spring of a clock balance wheel subjected to a bending moment **T**. The axis of the mainspring must remain plane. Hence, it requires that $|\mathbf{T}|$ does not exceed its critical value.

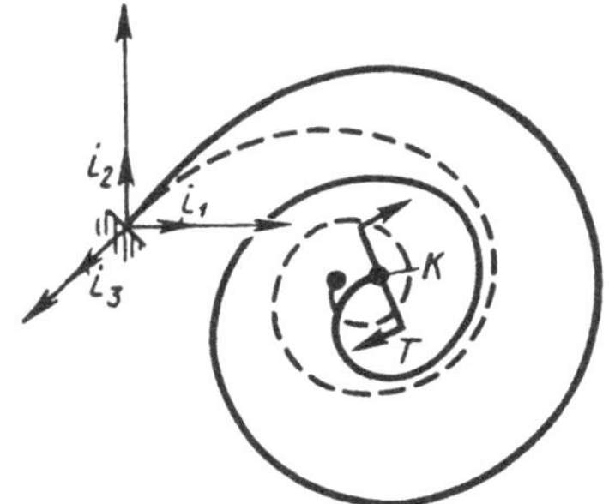

Fig. 3.4.

A cylindrical spring (e.g. an accelerometer spring) is shown in Fig. 3.5. The spring is subjected to a compressive force $\mathbf{P} = m\mathbf{a}$, where $\mathbf{a}$ is the acceleration of the object. Under a critical value of the acceleration, a loss of stability of the spring may result in an unwanted axial displacement of the mass point m. Consequently, the reading of the accelerometer may be inaccurate.

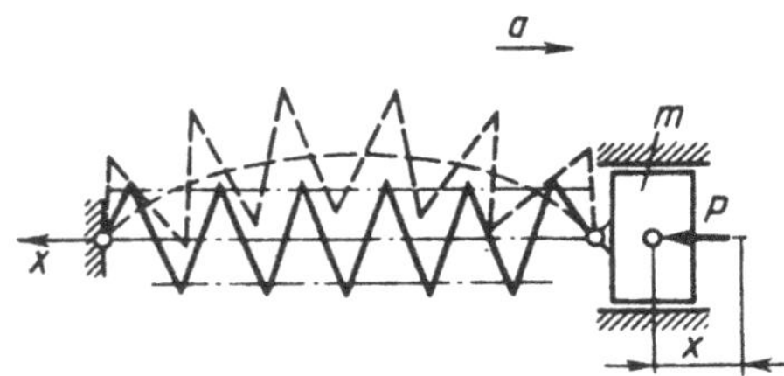

Fig. 3.5.

From here on, the configuration of a rod at the loss of stability will be called a *critical configuration.* This configuration is characterized by the *critical shape* and the *critical force.* The conservation of the original shape of an axial line until the loss of stability is characteristic of elementary problems like those concerned with the straight rod (Fig. 3.1) or the ring (Fig. 3.2). Unlike these problems, a loss of stability may occur with respect to a deformed configuration of a rod (Figs. 3.3 and 3.4).

3.2 Equilibrium Equations for a Rod After Loss of Stability

3.2.1 Vector Equilibrium Equations in the Attached Coordinate System

In this section, linear equilibrium equations for a rod after the loss of stability are derived. If the loss of stability *in the small* takes place, we can put $\mathbf{Q}_0 = \mathbf{Q}_* + \mathbf{Q}$, $\mathbf{M}_0 = \mathbf{M}_* + \mathbf{M}$, $\mathbf{q} = \mathbf{q}_* + \Delta\mathbf{q}$, $\boldsymbol{\mu} = \boldsymbol{\mu}_* + \Delta\boldsymbol{\mu}$, and $æ = æ_* + \Delta æ$, where the subscript '$*$' denotes the critical values; $\mathbf{Q}$, $\mathbf{M}$, $\Delta æ$, $\Delta\mathbf{q}$, and $\Delta\boldsymbol{\mu}$ are increments referred to the new state of equilibrium.

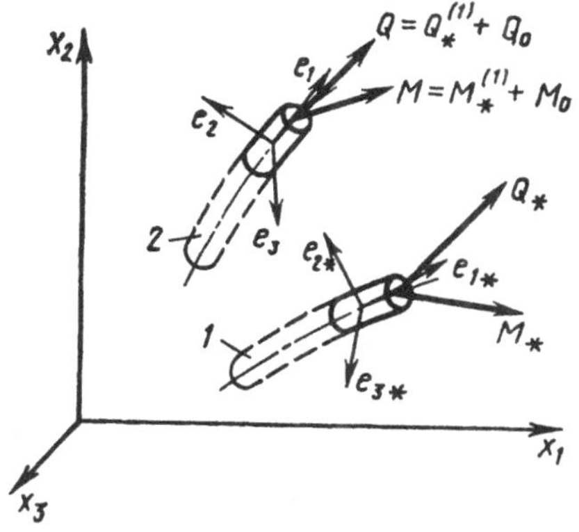

Fig. 3.6.

Since at the loss of stability in the small the new shape of equilibrium is assumed to be close to the critical one, the increments $\mathbf{Q}$, $\mathbf{M}$, $\Delta æ$, $\Delta\mathbf{q}$, $\Delta\mathbf{P}^{(i)}$, $\Delta\boldsymbol{\mu}$, and $\Delta\mathbf{T}^{(\nu)}$ can be regarded as small. Let us consider two configurations of a rod element (Fig. 3.6): critical state *1* and postcritical state *2*. In state *2*, i.e. in the basis $\{\mathbf{e}_i\}$, the asterisked vectors take the form

$$\mathbf{Q}_*^{(1)} = \sum_{j=1}^{3} Q_{*j}\mathbf{e}_j\,; \qquad \mathbf{M}_*^{(1)} = \sum_{j=1}^{3} M_{*j}\mathbf{e}_j\,,$$

where Q_{*j} and M_{*j} are projections in the basis $\{\mathbf{e}_{j*}\}$. Therefore, $\mathbf{Q}_*^{(1)} \neq \mathbf{Q}_*$ and $\mathbf{M}_*^{(1)} \neq \mathbf{M}_*$.

At the loss of stability, not only the vectors $\mathbf{Q}$ and $\mathbf{M}$ (these vectors characterize the stress state), but also the vectors that characterize the new shape of the rod (e.g. the vector $\boldsymbol{æ}$) are incremented. Therefore, to develop the equilibrium equations in terms of the increments, it is suitable to use vector equations referred to the attached basis $\{\mathbf{e}_i\}$. Let us retain in (1.57)–(1.61) terms linear in the increments. Then, omitting the equations corresponding to the critical equilibrium configuration, we obtain

$$\frac{\mathrm{d}\mathbf{Q}}{\mathrm{d}\eta} + \Delta\boldsymbol{æ} \times \mathbf{Q}_*^{(1)} + \boldsymbol{æ}_*^{(1)} \times \mathbf{Q} + \Delta\mathbf{P} = 0\,; \tag{3.5}$$

$$\frac{\mathrm{d}\mathbf{M}}{\mathrm{d}\eta} + \Delta\boldsymbol{æ} \times \mathbf{M}_*^{(1)} + \boldsymbol{æ}_*^{(1)} \times \mathbf{M} + \mathbf{e}_1 \times \mathbf{Q}_0 + \Delta\mathbf{T} = 0\,; \tag{3.6}$$

$$\mathbf{M} = \mathrm{A}\Delta\boldsymbol{æ}\,; \tag{3.7}$$

$$\frac{\mathrm{d}\boldsymbol{\vartheta}}{\mathrm{d}\eta} + \boldsymbol{æ}_*^{(1)} \times \boldsymbol{\vartheta} - \mathrm{A}^{-1}\mathbf{M} = 0\,; \tag{3.8}$$

$$\frac{\mathrm{d}\mathbf{u}}{\mathrm{d}\eta} + \boldsymbol{æ}_*^{(1)} \times \mathbf{u} + \mathrm{A}_1\boldsymbol{\vartheta} = 0\,. \tag{3.9}$$

The asterisked vectors can be determined from the following equations:

$$\frac{\mathrm{d}\mathbf{Q}_*}{\mathrm{d}\eta} + \boldsymbol{æ}_* \times \mathbf{Q}_* + \mathbf{P}_* = 0\,; \tag{3.10}$$

$$\frac{\mathrm{d}\mathbf{M}_*}{\mathrm{d}\eta} + \boldsymbol{æ}_* \times \mathbf{M}_* + \mathbf{e}_{1*} \times \mathbf{Q}_* + \mathbf{T}_* = 0\,; \tag{3.11}$$

$$\mathrm{L}_{1*}\frac{\mathrm{d}\boldsymbol{\vartheta}_*^{(1)}}{\mathrm{d}\eta} + \mathrm{L}_*\boldsymbol{æ}_0^{(1)} - \boldsymbol{æ}_* = 0\,; \tag{3.12}$$

$$\frac{\mathrm{d}\mathbf{u}_*}{\mathrm{d}\eta} + \boldsymbol{æ}_* \times \mathbf{u}_* + (l_{11*} - 1)\,\mathbf{e}_{1*} + l_{21*}\mathbf{e}_{2*} + l_{31*}\mathbf{e}_{3*} = 0\,; \tag{3.13}$$

$$\mathbf{M}_* = \mathrm{A}\,(\boldsymbol{æ}_* - \boldsymbol{æ}_0^{(1)})\,; \tag{3.14}$$

here $\mathbf{e}_{i*}$ are the base vectors of the basis $\{\mathbf{e}_{i*}\}$ corresponding to the critical configuration of the rod, L_{1*} and L_* are matrices whose elements are functions of the angles of rotation $\vartheta_{j*}^{(1)}$ of the basis $\{\mathbf{e}_{i*}\}$ with respect to the basis $\{\mathbf{e}_{i0}\}$.

Recall that the diagonal elements A_{ii} of the stiffness matrix A in nondimensional form are $A_{ii} = (A_{ii})_\mathrm{d}/(A_{33}(0))_\mathrm{d}$, where $(A_{ii})_\mathrm{d}$ are the dimensional stiffnesses and $(A_{33}(0))_\mathrm{d}$ is the dimensional bending stiffness $(A_{33})_\mathrm{d}$ at $\eta = 0$. In the case of rods of constant cross section, we have $A_{33} = 1$ and the nondimensional matrix A is as follows:

$$\mathrm{A} = \begin{bmatrix} A_{11} & 0 & 0 \\ 0 & A_{22} & 0 \\ 0 & 0 & 1 \end{bmatrix}.$$

It is important to note that the angles $\vartheta_{j*}^{(1)}$ and ϑ_j, being components of the vectors $\boldsymbol{\vartheta}_*^{(1)}$ and $\boldsymbol{\vartheta}_j$, do not coincide. At the loss of stability, the angles ϑ_j are not direct continuations of the angles $\vartheta_{j*}^{(1)}$.

Thus, the vectors $\mathbf{Q}_*$, $\mathbf{M}_*$, and $\boldsymbol{æ}_*$ in (3.5)–(3.9) can be found from the nonlinear equations (3.10)–(3.14). Application of numerical methods to (3.10)–(3.14) is discussed in Sect. 2.3.

The problem of loss of stability of a predeformed rod is comparatively complicated because when such a rod undergoes deformation, we do not know the critical configuration a priori. The best illustration to this fact is the spiral spring (Fig. 3.4): the critical shape of the spring shown as a dashed line differs sharply from its shape in the natural configuration. In general, the directions and the magnitudes of external forces change when a jump into another state of equilibrium takes place. Recall that $\Delta\mathbf{P}$ and $\Delta\mathbf{T}$ are functions of the increments of the external forces $\Delta\mathbf{P}^{(i)}$, $\Delta\mathbf{q}$, $\Delta\mathbf{T}^{(\nu)}$, and $\Delta\boldsymbol{\mu}$ (see (3.5) and (3.6)).

3.2.2 Increments of Forces and Moments

To solve (3.5)–(3.9) for critical loads, the explicit expressions for the force vectors $\Delta\mathbf{P}^{(i)}$ and $\Delta\mathbf{q}$ and the moment vectors $\Delta\mathbf{T}^{(\nu)}$ and $\Delta\boldsymbol{\mu}$) must be obtained. Let us consider different types of loads.

Follower Forces In this case, we have

$$\Delta\mathbf{P}^{(i)} = \Delta\mathbf{q} = \Delta\mathbf{T}^{(\nu)} = \Delta\boldsymbol{\mu} = 0\,,$$

hence, (3.5) and (3.6) are seriously simplified. Nevertheless, it is shown in the literature (Bolotin (1963) and Feodosiev (1996)) that the static loss of stability may not always take place. A loss of stability of an equilibrium configuration may occur when a system is set in motion with respect to this equilibrium state. In this case, as usual, the critical forces cannot be determined from the equilibrium equations. In problems like that, the equations of motion must be taken into consideration. (In more detail, this topic is covered in Chap. 3 of Svetlitsky (1987)).

Dead Forces In this case, we can write (see Appendix 1)

$$\Delta\mathbf{P}^{(i)} = \mathrm{A}_{P_*}^{(i)}\boldsymbol{\vartheta}\,(\eta_i)\,; \tag{3.15}$$

$$\Delta\mathbf{q} = \mathrm{A}_{q_*}\boldsymbol{\vartheta}\,(\eta)\,; \tag{3.16}$$

$$\Delta\mathbf{T}^{(\nu)} = \mathrm{A}_{T_*}^{(\nu)}\boldsymbol{\vartheta}\,(\eta_\nu)\,; \tag{3.17}$$

$$\Delta\boldsymbol{\mu} = \mathrm{A}_{\mu}\boldsymbol{\vartheta}\,(\eta) \tag{3.18}$$

Semi-Follower Forces Forces are called *semi-follower* if at a loss of stability they occupy an intermediate position between follower and dead forces. We have (for brevity, only the expression for $\Delta\mathbf{P}^{(i)}$ is written out)

$$\Delta\mathbf{P}^{(i)} = \mathrm{E}_P \mathrm{A}_{P_*}^{(i)} \boldsymbol{\vartheta}(\eta_i), \qquad \mathrm{E}_P = [(\alpha_{ii})_P], \qquad 0 \le (\alpha_{ii})_P \le 1, \tag{3.19}$$

where E_P is a diagonal matrix whose diagonal elements are equal to zero for follower forces and are equal to unity for dead forces.

Forces Which are Functions of Displacements, Their First Derivatives With Respect to η, and Angles of Rotation of Cross Sections This case is the most general one. The increments of forces and moments can be written in the form (1.99) as follows:

$$\Delta\mathbf{q} = \mathrm{C}_*^{(1)}\boldsymbol{\vartheta} + \mathrm{C}_*^{(2)}\mathbf{u} + \mathrm{C}_*^{(3)}\mathbf{u}'; \tag{3.20}$$

$$\Delta\mathbf{P}^{(i)} = \mathrm{B}_{i_*}^{(1)}\boldsymbol{\vartheta} + \mathrm{B}_{i_*}^{(2)}\mathbf{u} + \mathrm{B}_{i_*}^{(3)}\mathbf{u}'; \tag{3.21}$$

$$\Delta\boldsymbol{\mu} = \mathrm{C}_*^{(4)}\boldsymbol{\vartheta} + \mathrm{C}_*^{(5)}\mathbf{u} + \mathrm{C}_*^{(6)}\mathbf{u}'; \tag{3.22}$$

$$\Delta\mathbf{T}^{(\nu)} = \mathrm{B}_{\nu_*}^{(4)}\boldsymbol{\vartheta} + \mathrm{B}_{\nu_*}^{(5)}\mathbf{u} + \mathrm{B}_{\nu_*}^{(6)}\mathbf{u}'. \tag{3.23}$$

A more thorough derivation of relations for the increments of forces and moments is given in Sect. 3.4.

Examples of forces depending on the first derivative of the displacement vector $\mathbf{u}$ are presented in Chap. 6.

3.2.3 Equations in the Form Suitable for Integration

To present (3.5)–(3.9) in the form suitable for application of numerical methods, let us rewrite the vector products as follows:

$$\Delta\text{æ} \times \mathbf{Q}_*^{(1)} = \mathrm{A}_{Q_*}\Delta\text{æ}; \qquad \text{æ}_*^{(1)} \times \mathbf{Q} = \mathrm{A}_{\text{æ}_*}\mathbf{Q};$$

$$\Delta\text{æ} \times \mathbf{M}_*^{(1)} = \mathrm{A}_{M_*}\Delta\text{æ}; \qquad \text{æ}_* \times \mathbf{M} = \mathrm{A}_{\text{æ}_*}\mathbf{M};$$

$$\mathbf{e}_1 \times \mathbf{Q} = \mathrm{A}_1\mathbf{Q};$$

$$\text{æ}_*^{(1)} \times \mathbf{u} = \mathrm{A}_{\text{æ}_*}\mathbf{u}; \qquad \text{æ}_*^{(1)} \times \boldsymbol{\vartheta} = \mathrm{A}_{\text{æ}_*}\boldsymbol{\vartheta};$$

here

$$\mathrm{A}_{Q_*} = \begin{bmatrix} 0 & Q_{3*} & -Q_{2*} \\ -Q_{3*} & 0 & Q_{1*} \\ Q_{2*} & -Q_{1*} & 0 \end{bmatrix};$$

$$\mathrm{A}_{\text{æ}_*} = \begin{bmatrix} 0 & -\text{æ}_{3*} & \text{æ}_{2*} \\ \text{æ}_{3*} & 0 & -\text{æ}_{1*} \\ -\text{æ}_{2*} & \text{æ}_{1*} & 0 \end{bmatrix};$$

$$\mathrm{A}_{M_*} = \begin{bmatrix} 0 & M_{3*} & -M_{2*} \\ -M_{3*} & 0 & M_{1*} \\ M_{2*} & -M_{1*} & 0 \end{bmatrix}; \qquad \mathrm{A}_1 = \begin{bmatrix} 0 & 0 & 0 \\ 0 & 0 & -1 \\ 0 & 1 & 0 \end{bmatrix}.$$

On rearrangement, we get

$$\frac{d\mathbf{Q}}{d\eta} + A_{æ_*}\mathbf{Q} + A_{Q_*}A^{-1}\mathbf{M} + \Delta\mathbf{P} = 0\,; \tag{3.24}$$

$$\frac{d\mathbf{M}}{d\eta} + A_{æ_*}\mathbf{M} + A_{M_*}A^{-1}\mathbf{M} + A_1\mathbf{Q} + \Delta\mathbf{T} = 0\,; \tag{3.25}$$

$$\frac{d\boldsymbol{\vartheta}}{d\eta} + A_{æ_*}\boldsymbol{\vartheta} - A^{-1}\mathbf{M} = 0\,; \tag{3.26}$$

$$\frac{d\mathbf{u}}{d\eta} + A_{æ_*}\mathbf{u} + A_1\boldsymbol{\vartheta} = 0\,. \tag{3.27}$$

The system of equations (3.24)–(3.27) can be presented as a vector equation

$$\mathbf{Y}' + \mathbf{A}\mathbf{Y} + \mathbf{b}(\mathbf{u}, \mathbf{u}', \boldsymbol{\vartheta}) = 0\,, \tag{3.28}$$

where

$$\mathbf{Y} = \begin{bmatrix} \mathbf{Q} \\ \mathbf{M} \\ \boldsymbol{\vartheta} \\ \mathbf{u} \end{bmatrix}; \qquad \mathbf{b} = \begin{bmatrix} \Delta\mathbf{q} + \sum\limits_{i=1}^{n} \Delta\mathbf{P}^{(i)}\,\delta(\eta - \eta_i) \\ \Delta\boldsymbol{\mu} + \sum\limits_{\nu=1}^{\rho} \Delta\mathbf{T}^{(\nu)}\,\delta(\eta - \eta_\nu) \\ \mathbf{0} \\ \mathbf{0} \end{bmatrix};$$

$$A = \begin{bmatrix} A_{æ_*} & A_{Q_*}A^{-1} & 0 & 0 \\ A_1 & A_{M_*}A^{-1} + A_{æ_*} & 0 & 0 \\ 0 & -A^{-1} & A_{æ_*} & 0 \\ 0 & 0 & A_1 & A_{æ_*} \end{bmatrix}.$$

For arbitrary external loads, substitution of $\Delta\mathbf{P}^{(i)}$, $\Delta\mathbf{q}$, $\Delta\mathbf{T}^{(\nu)}$, and $\Delta\boldsymbol{\mu}$ into (3.28) transforms this inhomogeneous vector equation into a homogeneous equation. Methods for numerical determination of critical forces are discussed in Sect. 3.5.

In terms of projections on the attached coordinate axes, (3.24)–(3.27) are as follows:

$$\begin{aligned}
&\frac{dQ_1}{d\eta} + \frac{Q_{3*}}{A_{22}}M_2 - \frac{Q_{2*}}{A_{33}}M_3 - æ_{3*}Q_2 + æ_{2*}Q_3 + \Delta P_1 = 0\,;\\
&\frac{dQ_2}{d\eta} - \frac{Q_{3*}}{A_{11}}M_1 + \frac{Q_{1*}}{A_{33}}M_3 + æ_{3*}Q_1 - æ_{1*}Q_3 + \Delta P_3 = 0\,;\\
&\frac{dQ_3}{d\eta} + \frac{Q_{2*}}{A_{11}}M_1 - \frac{Q_{1*}}{A_{22}}M_2 - æ_{2*}Q_1 + æ_{1*}Q_2 + \Delta P_3 = 0\,;\\
&\frac{dM_1}{d\eta} + \frac{M_{3*}}{A_{22}}M_2 - \frac{M_{2*}}{A_{33}}M_3 - æ_{3*}M_2 + æ_{2*}M_3 + \Delta T_1 = 0\,;\\
&\frac{dM_2}{d\eta} - \frac{M_{3*}}{A_{11}}M_1 + \frac{M_{1*}}{A_{33}}M_3 + æ_{3*}M_1 - æ_{1*}M_3 - Q_3 + \Delta T_2 = 0\,;
\end{aligned} \tag{3.29}$$

$$\frac{\mathrm{d}M_3}{\mathrm{d}\eta}+\frac{M_{2*}}{A_{11}}M_1-\frac{M_{1*}}{A_{22}}M_2-æ_{2*}M_1+æ_{1*}M_2+Q_2+\Delta T_3=0\,; \tag{3.30}$$

$$\frac{\mathrm{d}\vartheta_1}{\mathrm{d}\eta}-\frac{M_1}{A_{11}}-æ_{3*}\vartheta_2+æ_{2*}\vartheta_3=0\,;$$
$$\frac{\mathrm{d}\vartheta_2}{\mathrm{d}\eta}-\frac{M_2}{A_{22}}+æ_{3*}\vartheta_1-æ_{1*}\vartheta_3=0\,;$$
$$\frac{\mathrm{d}\vartheta_3}{\mathrm{d}\eta}-\frac{M_3}{A_{33}}-æ_{2*}\vartheta_1+æ_{1*}\vartheta_2=0\,; \tag{3.31}$$

$$\frac{\mathrm{d}u_1}{\mathrm{d}\eta}-æ_{3*}u_2+æ_{2*}u_3=0\,;$$
$$\frac{\mathrm{d}u_2}{\mathrm{d}\eta}+æ_{3*}u_1-æ_{1*}u_3-\vartheta_3=0\,;$$
$$\frac{\mathrm{d}u_3}{\mathrm{d}\eta}-æ_{2*}u_1+æ_{1*}u_2+\vartheta_2=0\,. \tag{3.32}$$

(Recall that Q_j and M_j are small quantities.)

It should be emphasized that the system of equations (3.29)–(3.32) is of the most general type. It can be used to examine the static stability in the small of rods of constant ($A_{ii}=\text{const}$) or variable ($A_{ii}\neq\text{const}$) cross section for an arbitrary type of loading.

3.3 Plane Curvilinear Rods

3.3.1 Rods of Plane Axial Line Before Loss of Stability

If a rod axis during a loading remains a plane curve (e.g. a spiral spring), then we have $M_{1*}=M_{2*}=0$, $Q_{3*}=0$, and $æ_{1*}=æ_{2*}=0$; hence, the matrices A_{M_*}, A_{Q_*}, and $\mathrm{A}_{æ_*}$ in (3.24)–(3.27) are as follows:

$$\mathrm{A}_{Q_*}=\begin{bmatrix}0 & 0 & -Q_{2*}\\ 0 & 0 & Q_{1x}\\ Q_{2x} & -Q_{1*} & 0\end{bmatrix};\qquad \mathrm{A}_{M_*}=\begin{bmatrix}0 & M_{3*} & 0\\ -M_{3*} & 0 & 0\\ 0 & 0 & 0\end{bmatrix};$$

$$\mathrm{A}_{æ_*}=\begin{bmatrix}0 & -æ_{3*} & 0\\ æ_{3*} & 0 & 0\\ 0 & 0 & 0\end{bmatrix}.$$

If a plane configuration of the rod is no longer stable, the governing equations, a special case of (3.29)–(3.32), are as follows:

$$\frac{\mathrm{d}Q_1}{\mathrm{d}\eta}-\frac{Q_{2*}}{A_{33}}M_3-æ_{3*}Q_2+\Delta P_1=0\,;$$
$$\frac{\mathrm{d}Q_2}{\mathrm{d}\eta}+\frac{Q_{1*}}{A_{33}}M_3+æ_{3*}Q_1+\Delta P_2=0\,;$$

$$\frac{dQ_3}{d\eta} + \frac{Q_{2*}}{A_{11}} M_1 - \frac{Q_{1*}}{A_{22}} M_2 + \Delta P_3 = 0; \tag{3.33}$$

$$\frac{dM_1}{d\eta} - æ_{3*} M_2 + \frac{M_{3*}}{A_{22}} M_2 + \Delta T_1 = 0;$$

$$\frac{dM_2}{d\eta} + æ_{3*} M_1 - \frac{M_{3*}}{A_{11}} M_1 - Q_3 + \Delta T_2 = 0;$$

$$\frac{dM_3}{d\eta} + Q_2 + \Delta T_3 = 0; \tag{3.34}$$

$$\frac{d\vartheta_1}{d\eta} - \frac{M_1}{A_{11}} - æ_{3*}\vartheta_2 = 0;$$

$$\frac{d\vartheta_2}{d\eta} - \frac{M_2}{A_{22}} + æ_{3*}\vartheta_1 = 0;$$

$$\frac{d\vartheta_3}{d\eta} - \frac{M_3}{A_{33}} = 0; \tag{3.35}$$

$$\frac{du_1}{d\eta} - æ_{3*} u_2 = 0;$$

$$\frac{du_2}{d\eta} + æ_{3*} u_1 - \vartheta_3 = 0;$$

$$\frac{du_3}{d\eta} + \vartheta_2 = 0. \tag{3.36}$$

If the increments of external loads ΔP_j and ΔT_j depend on u_j and ϑ_j, $j = 1, 2, 3$, then the systems of equations (3.33)–(3.36) are coupled. Hence, when a loss of stability happens, the plane axial line becomes a spatial curve. If the load increments Δq_1, Δq_2, $\Delta P_1^{(i)}$, $\Delta P_2^{(i)}$, $\Delta\mu_3$, and $\Delta T_3^{(\nu)}$, which appear in the expressions for ΔP_j and ΔT_j, are functions only of u_1, u_2, and ϑ_3 while Δq_3, $\Delta P_3^{(i)}$, $\Delta\mu_1$, $\Delta\mu_2$, $\Delta T_1^{(\nu)}$, and $\Delta T_2^{(\nu)}$ are functions only of u_3, ϑ_1, and ϑ_2, the system of equations (3.33)–(3.36) may be decomposed into two independent systems, that is,

$$\frac{dQ_1}{d\eta} - \frac{Q_{2*}}{A_3} M_3 - æ_{3*} Q_2 + \Delta q_1 + \sum_{i=1}^{n} \Delta P_1^{(i)}\, \delta(\eta - \eta_i) = 0;$$

$$\frac{dQ_2}{d\eta} + \frac{Q_{1*}}{A_3} M_3 - æ_{3*} Q_1 + \Delta q_2 + \sum_{i=1}^{n} \Delta P_2^{(i)}\, \delta(\eta - \eta_i) = 0;$$

$$\frac{dM_3}{d\eta} + Q_2 + \Delta\mu_3 + \sum_{\nu=1}^{\rho} \Delta T_3^{\nu}\, \delta(\eta - \eta_\nu) = 0;$$

$$\frac{d\vartheta_3}{d\eta} - \frac{M_3}{A_3} = 0;$$

$$\frac{du_1}{d\eta} - æ_{3*} u_2 = 0;$$

$$\frac{du_2}{d\eta} + æ_{3*}u_1 - \vartheta_3 = 0\,; \tag{3.37}$$

$$\frac{dQ_3}{d\eta} + \frac{Q_{2*}}{A_{11}}M_1 - \frac{Q_{1*}}{A_{22}}M_2 + \Delta q_3 + \sum_{i=1}^{n}\Delta P_3^{(i)}\,\delta(\eta - \eta_i) = 0\,;$$

$$\frac{dM_1}{d\eta} - æ_{3*}M_2 + \frac{M_{3*}}{A_{22}}M_2 + \Delta\mu_1 + \sum_{\nu=1}^{\rho}\Delta T_1^{(\nu)}\,\delta(\eta - \eta_\nu) = 0\,;$$

$$\frac{dM_2}{d\eta} + æ_{3*}M_1 - \frac{M_{3*}}{A_{11}}M_1 - \Delta Q_3 + \Delta\mu_2 + \sum_{\nu=1}^{\rho}\Delta T_2^{(\nu)}\,\delta(\eta - \eta_\nu) = 0\,;$$

$$\frac{d\vartheta_1}{d\eta} - \frac{M_1}{A_1} - æ_{3*}\vartheta_2 = 0\,;$$

$$\frac{d\vartheta_2}{d\eta} - \frac{M_2}{A_2} + æ_{3*}\vartheta_1 = 0\,;$$

$$\frac{du_3}{d\eta} + \vartheta_2 = 0\,. \tag{3.38}$$

The critical load corresponding to the in-plane loss of stability can be found from (3.37); the critical load corresponding to the out-of-plane loss of stability can be found from (3.38).

Suppose that the magnitudes of loads applied to a rod vary linearly with a factor μ, i.e. the loads are μq_i, $\mu P_j^{(i)}$, etc., where q_i, $P_j^{(i)}$, etc. are known quantities. Thus, the loss of stability is governed only by μ. Denote the value of μ that corresponds to the in-plane loss of stability by μ_{*1} and the value of μ that corresponds to the out-of-plane loss of stability by μ_{*2}. The external loads do not result in a loss of stability provided the inequalities $\mu_{*1} > 1$ and $\mu_{*2} > 1$ hold.

If the smaller of the coefficients μ_{*1} and μ_{*2} is less than unity and the applied load does not exceed its critical value, then either the out-of-plane loss of stability ($\mu_{*2} < 1$) or the in-plane loss of stability ($\mu_{*1} < 1$) may occur.

For example, let us consider a critical moment $|\mathbf{T}_*|$ for a spiral spring (see Fig. 3.4). Suppose that one end of the spring, the cross section K, is free. For simplicity, we assume that $\mathbf{Q}_* = 0$.

It is essential to know the behavior of the applied moment $\mathbf{T}$ after the loss of stability. Suppose that the moment $\mathbf{T}$ is dead, i.e. the equality $\mathbf{T} = T_3\mathbf{i}_3$ holds; then, we get

$$\Delta\mathbf{T} = B_*^{(4)}\boldsymbol{\vartheta}\delta\,(\eta - 1) = -T_{3*}\vartheta_2(1)\,\mathbf{e}_1 + T_{3*}\vartheta_1(1)\,\mathbf{e}_2\,. \tag{3.39}$$

Using (3.29)–(3.32), we obtain the following system of equations for a spiral rod of variable cross section:

$$\frac{dQ_1}{d\eta} - æ_{3*}Q_2 = 0\,;$$

$$\frac{dQ_2}{d\eta} + æ_{3*}Q_1 = 0\,;$$
$$\frac{dQ_3}{d\eta} = 0\,; \tag{3.40}$$
$$\frac{dM_1}{d\eta} + \left(\frac{M_{3*}}{A_{22}} - æ_{3*}\right) M_2 - T_{3*}\vartheta_2\delta(\eta - 1) = 0\,;$$
$$\frac{dM_2}{d\eta} - \left(\frac{M_{3*}}{A_{11}} - æ_{3*}\right) M_1 - Q_3 + T_{3*}\vartheta_1\delta(\eta - 1) = 0\,;$$
$$\frac{dM_3}{d\eta} + Q_2 = 0\,; \tag{3.41}$$
$$\frac{d\vartheta_1}{d\eta} - \frac{M_1}{A_{11}} - æ_{3*}\vartheta_2 = 0\,;$$
$$\frac{d\vartheta_2}{d\eta} - \frac{M_2}{A_{22}} + æ_{3*}\vartheta_1 = 0\,;$$
$$\frac{d\vartheta_3}{d\eta} - \frac{M_3}{A_{33}} = 0\,; \tag{3.42}$$
$$\frac{du_1}{d\eta} - æ_{3*}u_2 = 0\,;$$
$$\frac{du_2}{d\eta} + æ_{3*}u_1 - \vartheta_3 = 0\,;$$
$$\frac{du_3}{d\eta} + \vartheta_2 = 0\,. \tag{3.43}$$

For rods of constant cross section, we have $A_{33} = 1$. Note that the critical values of the parameters M_{3*} and $æ_{3*}$ in (3.40)–(3.43) depend on the bending moment $\mathbf{T}_{3*}$. If we apply a follower moment, then $\Delta\mathbf{T} = 0$.

3.3.2 Stability of Plane Configuration of a Ring

Equations (3.33)–(3.36) are useful in the analysis of out-of-plane deformations of a rod. As an example, consider a ring subjected to a uniformly distributed load (see Fig. 3.2). We assume that the ring is of constant cross section, hence $A_{33} = 1$. Let the bending stiffnesses be different, $A_{22} \neq A_{33}$. We have $Q_{3*} = Q_{2*} = 0$, $M_{1*} = M_{2*} = M_{3*} = 0$, $æ_{3*} = 1/R^0$, and $æ_{1*} = æ_{2*} = 0$, where $R^0 = R/l = 1/(2\pi)$ is a nondimensional radius. Thus, (3.29)–(3.32) take the following form:

$$\frac{dQ_1}{d\eta} - \frac{1}{R^0} Q_2 = -\Delta q_1\,;$$
$$\frac{dQ_2}{d\eta} + \frac{1}{R^0} Q_1 + Q_{1*}M_3 = -\Delta q_2\,;$$
$$\frac{dQ_3}{d\eta} - \frac{Q_{1*}}{A_{22}} M_2 = -\Delta q_3\,; \tag{3.44}$$

$$\frac{dM_1}{d\eta} - \frac{1}{R^0} M_2 = 0 ;$$
$$\frac{dM_2}{d\eta} + \frac{1}{R^0} M_1 - Q_3 = 0 ;$$
$$\frac{dM_3}{d\eta} + Q_2 = 0 ; \qquad (3.45)$$
$$\frac{d\vartheta_1}{d\eta} - \frac{M_1}{A_{11}} - \frac{1}{R^0} \vartheta_2 = 0 ;$$
$$\frac{d\vartheta_2}{d\eta} - \frac{M_2}{A_{22}} + \frac{1}{R^0} \vartheta_1 = 0 ;$$
$$\frac{d\vartheta_3}{d\eta} - M_3 = 0 ; \qquad (3.46)$$
$$\frac{du_1}{d\eta} - \frac{1}{R^0} u_2 = 0 ;$$
$$\frac{du_2}{d\eta} + \frac{1}{R^0} u_1 - \vartheta_3 = 0 ;$$
$$\frac{du_3}{d\eta} + \vartheta_2 = 0 . \qquad (3.47)$$

The axial force Q_{1*} in (3.44) may be found from (3.10) with $æ_* = 1/R^0$ and $\mathbf{P}_*^{(i)} = 0$, i.e. $Q_{1*} = -q_{2*} R^0$.

Equations (3.44)–(3.47) enable us to determine the critical load $\mathbf{q}_*$ for any type of external loads. In the case of follower loads, we put $\Delta q_i = 0$ and the system of equations (3.44)–(3.47) yields two uncoupled systems

$$\frac{dQ_1}{d\eta} - \frac{1}{R^0} Q_2 = 0 ;$$
$$\frac{dQ_2}{d\eta} + \frac{1}{R^0} Q_1 + Q_{1*} M_3 = 0 ;$$
$$\frac{dM_3}{d\eta} + Q_2 = 0 ;$$
$$\frac{d\vartheta_3}{d\eta} - M_3 = 0 ;$$
$$\frac{du_1}{d\eta} - \frac{1}{R^0} u_2 = 0 ;$$
$$\frac{du_2}{d\eta} + \frac{1}{R^0} u_1 - \vartheta_3 = 0 \qquad (3.48)$$

and

$$\frac{dQ_3}{d\eta} - \frac{Q_{1*}}{A_{22}} M_2 = 0 ;$$
$$\frac{dM_1}{d\eta} - \frac{1}{R^0} M_2 = 0 ;$$

$$\frac{\mathrm{d}M_2}{\mathrm{d}\eta} + \frac{1}{R^0} M_1 - Q_3 = 0 ;$$
$$\frac{\mathrm{d}\vartheta_1}{\mathrm{d}\eta} - \frac{M_1}{A_{11}} - \frac{1}{R^0} \vartheta_2 = 0 ;$$
$$\frac{\mathrm{d}\vartheta_2}{\mathrm{d}\eta} - \frac{M_2}{A_{22}} + \frac{1}{R^0} \vartheta_1 = 0 ;$$
$$\frac{\mathrm{d}u_3}{\mathrm{d}\eta} + \vartheta_2 = 0 . \tag{3.49}$$

The critical value q_{2*}, at which the in-plane loss of stability occurs, can be evaluated from (3.48). The form of the ring after the loss of stability is shown in Fig. 3.2 as a dashed line. Equations (3.49) enable us to determine the critical value of the load at which the loss of stability may result in an out-of-plane deformation of the ring.

To determine the critical value q_{2*}, it is sufficient to use the first three equations of (3.48). Eliminating Q_{01} and M_{03} from these equations, we arrive at an equation in one unknown Q_2,

$$\frac{\mathrm{d}^2 Q_2}{\mathrm{d}\eta^2} + \left[\frac{1}{(R^0)^2} - Q_{1*} \right] Q_2 = 0 . \tag{3.50}$$

Its solution is of the form

$$Q_2 = c_1 \cos k\eta + c_2 \sin k\eta , \qquad k^2 = \frac{1}{(R^0)^2} - Q_{1*} R^0 .$$

This solution must be a periodic function in η, i.e. $Q_2(0) = Q_2(1)$ and $Q_2'(0) = Q_2'(1)$. These yield the following system of homogeneous linear equations:

$$(\cos k - 1)\, c_1 + \sin k c_2 = 0 ;$$
$$-k \sin k c_1 + (\cos k - 1)\, c_2 = 0 . \tag{3.51}$$

The determinant of (3.51) must equal zero, hence $\cos k - 1 = 0$ and $k = 2\pi n$. In view of $Q_{1*} = -q_{2*} R^0$, we get

$$\frac{1}{(R^0)^2} + q_{2*} R^0 = 4\pi^2 n^2 .$$

The nondimensional value of the critical load is

$$q_{2*} = \frac{4\pi^2 n^2 (R^0)^2 - 1}{(R^0)^3} . \tag{3.52}$$

The dimensional value is

$$(q_{2*})_{\mathrm{d}} = \frac{q_{2*} A_{33}}{l^3} = \frac{(n^2 - 1)\, A_{33}}{R^3} , \qquad R = R^0 l = \frac{l}{2\pi} . \tag{3.53}$$

Setting $n = 2$, we obtain the critical value at which the loss of stability may lead to an out-of-plane deformation

$$(q_{2*})_{\mathrm{d}} = \frac{3A_{33}}{R^3}\,. \tag{3.54}$$

Eliminating M_{01} and Q_{03} from the first three equations of (3.49), we obtain

$$\frac{\mathrm{d}^2 M_2}{\mathrm{d}\eta^2} + \left(\frac{1}{(R^0)^2} - \frac{Q_{1*}}{A_{22}}\right) M_2 = 0\,. \tag{3.55}$$

Equation (3.55) is the same as (3.50). Hence, we obtain the dimensional critical value

$$(q_{2*})_{\mathrm{d}} = \frac{3A_{22}}{R^3}\,. \tag{3.56}$$

Comparing the critical values q_{2*} given by (3.54) and (3.56), we get the following conclusions:

(1) If $A_{33} < A_{22}$, then the in-plane loss of stability occurs and the axial line of the ring remains a plane curve.

(2) If $A_{33} > A_{22}$, then the loss of stability results in an out-of-plane deformation of the ring. The projections of the displacement of the points of the axial line on the original plane are equal to zero (i.e. the projection of the deformed axial line on the original plane is a circle).

(3) If $A_{33} = A_{22}$, then the in-plane and out-of-plane loss of stability take place simultaneously: all the components of the displacement vector $\mathbf{u}$ are different from zero.

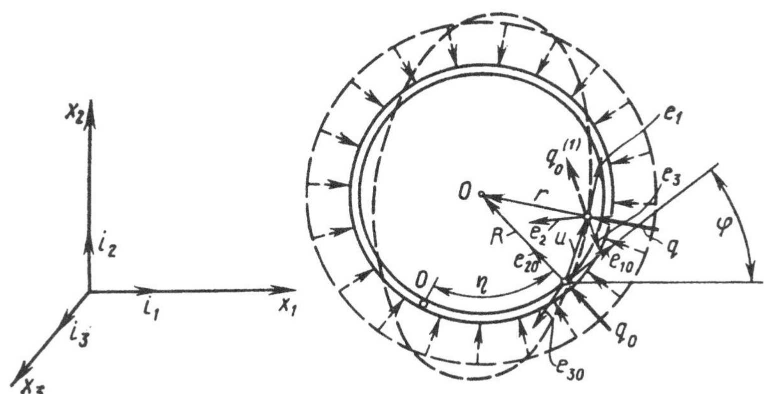

Fig. 3.7.

Let us assume that after the loss of stability the vector $\mathbf{q}$ remains parallel to its original direction (see Fig. 3.7). To evaluate the critical value $\mathbf{q}_*$, one must obtain the increment $\Delta\mathbf{q}$. If the direction and the magnitude of a vector $\mathbf{a}$ are constant, then the increments of its components are as follows (see (A.142)):

$$\Delta\mathbf{q} = q_{2*}\vartheta_3\mathbf{e}_1 - q_{2*}\vartheta_1\mathbf{e}_3 . \tag{3.57}$$

By use of the system of equations (3.44)–(3.47), which is similar to the system (3.48)–(3.49), we obtain two uncoupled systems of equations:

$$\begin{aligned}
&\frac{\mathrm{d}Q_1}{\mathrm{d}\eta} - \frac{1}{R^0}Q_2 + q_{2*}\vartheta_3 = 0 ;\\
&\frac{\mathrm{d}Q_2}{\mathrm{d}\eta} + \frac{1}{R^0}Q_1 + Q_{1*}M_3 = 0 ;\\
&\frac{\mathrm{d}M_3}{\mathrm{d}\eta} + Q_2 = 0 ;\\
&\frac{\mathrm{d}\vartheta_3}{\mathrm{d}\eta} - M_3 = 0 ;\\
&\frac{\mathrm{d}u_1}{\mathrm{d}\eta} - \frac{1}{R^0}u_2 = 0 ;\\
&\frac{\mathrm{d}u_2}{\mathrm{d}\eta} + \frac{1}{R^0}u_1 - \vartheta_3 = 0
\end{aligned} \tag{3.58}$$

and

$$\begin{aligned}
&\frac{\mathrm{d}Q_3}{\mathrm{d}\eta} - \frac{Q_{1*}}{A_{22}}M_2 - q_{2*}\vartheta_1 = 0 ;\\
&\frac{\mathrm{d}M_1}{\mathrm{d}\eta} - \frac{1}{R^0}M_2 = 0 ;\\
&\frac{\mathrm{d}M_2}{\mathrm{d}\eta} + \frac{1}{R^0}M_1 - Q_3 = 0 ;\\
&\frac{\mathrm{d}\vartheta_1}{\mathrm{d}\eta} - \frac{M_1}{A_{11}} - \frac{1}{R^0}\vartheta_2 = 0 ;\\
&\frac{\mathrm{d}\vartheta_2}{\mathrm{d}\eta} - \frac{M_2}{A_{22}} + \frac{1}{R^0}\vartheta_1 = 0 ;\\
&\frac{\mathrm{d}u_3}{\mathrm{d}\eta} + \vartheta_2 = 0 .
\end{aligned} \tag{3.59}$$

Eliminating Q_1, Q_2, and ϑ_3 from the first four equations of (3.58), we get an equation in one unknown M_3,

$$\frac{\mathrm{d}^4M_3}{\mathrm{d}\eta^4} + \left[\frac{1}{(R^0)^2} + q_{2*}R^0\right]\frac{\mathrm{d}^2M_3}{\mathrm{d}\eta^2} + \frac{q_{2*}}{R^0}M_3 = 0 . \tag{3.60}$$

Its solution must be a periodic function. Therefore, we put

$$M_3 = c \sin 2\pi n\eta\,. \tag{3.61}$$

Substituting (3.61) into (3.60), we arrive at

$$q_{2*} = \frac{n^2 A_{33}}{R^3}\,. \tag{3.62}$$

Thus, the value of the dimensional critical load is determined.

The variable n may equal either 1 or 2. If $n = 1$, then the ring moves as a rigid body (keeping its original form). If $n = 2$, then the loss of stability occurs and the ring takes the shape of an ellipse. The point O, however, does not change its original position.

Let us determine the critical value of the load at which an out-of-plane deformation of the ring may occur. We assume that after the loss of stability the vector $\mathbf{q}_{2*}$ remains parallel to its original direction. Eliminating Q_3 and M_1 from the first three equations of (3.59), we arrive at

$$\vartheta_1 = \frac{1}{q_{2*}}\left[M_2'' + \left(\frac{1}{(R^0)^2} - \frac{Q_{1*}}{A_{22}}\right)M_2\right]. \tag{3.63}$$

Eliminating M_1 and ϑ_2 from the second, the fourth, and the fifth equations of (3.59), we get

$$\vartheta_1'' - \left(\frac{1}{R^0 A_{11}} + \frac{1}{R^0 A_{22}}\right)M_2 + \frac{1}{(R^0)^2}\vartheta_1 = 0\,. \tag{3.64}$$

Substituting (3.63) into (3.64) and using the relation $Q_{1*} = -q_{2*}R^0$, we arrive at[1]

$$M_2^{\mathrm{IV}} + \left[\frac{2}{(R^0)^2} + \frac{q_{2*}R^0}{A_{22}}\right]M_2'' + \left[-q_{2*}\left(\frac{1}{R^0 A_{11}}\right) + \frac{1}{(R^0)^4}\right]M_2 = 0\,. \tag{3.65}$$

Let $M_2 = A\sin 2\pi n\eta$. Equation (3.65) can be solved for q_{2*} as follows:

$$q_{2*} = \frac{(n^2-1)^2 A_{22}}{\left(n^2 + \dfrac{A_{22}}{A_{11}}\right)(R^0)^3}\,. \tag{3.66}$$

The dimensional value of the critical load is

$$(q_{2*})_{\mathrm{d}} = \frac{q_{2*}(A_{33})_{\mathrm{d}}}{l^3} = \frac{(n^2-1)^2 (A_{22})_{\mathrm{d}}}{\left(n^2 + \dfrac{A_{22}}{A_{11}}\right)R^3}\,, \tag{3.67}$$

where $(A_{33})_{\mathrm{d}}$ and $(A_{22})_{\mathrm{d}}$ are dimensional bending stiffnesses and $R = R^0 l$ is a dimensional radius.

[1] The superscript 'IV' denotes the fourth derivative with respect to η.

The critical load (3.66) has a minimum at $n = 2$,

$$(q_{2*})_{\mathrm{d}} = \frac{9(A_{22})_{\mathrm{d}}}{\left(4 + \dfrac{A_{22}}{A_{11}}\right) R^3}. \tag{3.68}$$

The critical value q_{2*} can be obtained without reduction of the systems of equations (3.48), (3.49), (3.58), and (3.59) to one equation. Since all the variables, say, in (3.59), are periodic functions, we can write

$$\begin{aligned} Q_3 &= B_1 e^{i(2\pi n)\eta}; & M_1 &= B_2 e^{i(2\pi n)\eta}; & M_3 &= B_3 e^{i(2\pi n)\eta}; \\ \vartheta_1 &= B_4 e^{i(2\pi n)\eta}; & \vartheta_2 &= B_5 e^{i(2\pi n)\eta}; & u_3 &= B_6 e^{i(2\pi n)\eta}. \end{aligned} \tag{3.69}$$

Introduction of (3.69) into (3.59) yields a system of linear equations in the unknowns B_i. Matrix form of this system of equations is as follows:

$$\mathbf{HB} = 0;$$

here

$$\mathrm{H} = \begin{bmatrix} 2\pi ni & 0 & -\dfrac{Q_{1*}}{A_2} & -q_{2*} & 0 & 0 \\ 0 & 2\pi ni & -\dfrac{1}{R^0} & 0 & 0 & 0 \\ -1 & \dfrac{1}{R^0} & 2\pi ni & 0 & 0 & 0 \\ 0 & -\dfrac{1}{A_1} & 0 & 2\pi ni & -\dfrac{1}{R^0} & 0 \\ 0 & 0 & -\dfrac{1}{A_2} & \dfrac{1}{R^0} & 2\pi ni & 0 \\ 0 & 0 & 0 & 0 & 1 & 2\pi ni \end{bmatrix}; \quad \mathbf{B} = \begin{bmatrix} B_1 \\ B_2 \\ \vdots \\ B_6 \end{bmatrix}.$$

For this system to have nontrivial solutions, the equality $D = \det \mathrm{H} = 0$ must hold. Hence,

$$D = 2\pi ni \begin{vmatrix} 2\pi ni & -\dfrac{1}{R^0} & 0 & 0 \\ \dfrac{1}{R^0} & 2\pi ni & 0 & 0 \\ -\dfrac{1}{A_1} & 0 & 2\pi ni & -\dfrac{1}{R^0} \\ 0 & -\dfrac{1}{A_2} & \dfrac{1}{R^0} & 2\pi ni \end{vmatrix} - \begin{vmatrix} 0 & -\dfrac{Q_{1*}}{A_{22}} & -q_{2*} & 0 \\ 2\pi ni & -\dfrac{1}{R^0} & 0 & 0 \\ -\dfrac{1}{A_1} & 0 & 2\pi ni & -\dfrac{1}{R^0} \\ 0 & -\dfrac{1}{A_{22}} & \dfrac{1}{R^0} & 2\pi ni \end{vmatrix}$$

and, finally, we get

$$\begin{aligned} D = 2\pi ni \Bigg\{ & \left[(4\pi^2 n^2)^2 - \frac{2}{(R^0)^2}(4\pi^2 n^2) + \frac{1}{(R^0)^4} \right] \\ & - \frac{q_{2*}}{R^0} \left[\frac{4\pi^2 n^2 R^0}{A_{22}} + \frac{1}{A_{11}} \right] \Bigg\} = 0. \end{aligned} \tag{3.70}$$

The formula (3.66) can be rederived on the basis of (3.70).

3.3.3 Stability of a Plane Configuration of a Rod with Lateral Supports

Let us restrict our consideration to the case shown in Fig. 3.8. If the rod is bent in such a way that it lies in the plane x_1Ox_2, the elastic support plays no role. Due to the loss of stability, the rod undergoes an out-of-plane deformation, hence, a concentrated reaction force $\mathbf{R}$ applied to the cross section K occurs. Suppose that the rod is subjected to a follower bending moment $\mathbf{T} = T_3\mathbf{e}_3$. If displacements of the axial points are small, then the reaction $\mathbf{R}$ can be found in the following way. The reaction is a product of the stiffness of the elastic support c and the projection of the displacement of the point K on the direction of the vector $\mathbf{R}$ (this direction is assumed to be known). Thus, we have

$$|\mathbf{R}| = |\mathbf{e}_R \cdot \mathbf{u}_K|\, c\,, \tag{3.71}$$

where $\mathbf{e}_R$ is a unit vector such that $\mathbf{R} = |\mathbf{R}|\mathbf{e}_R$.

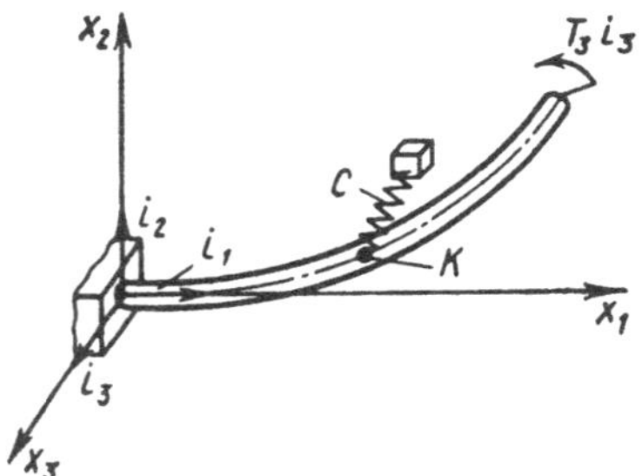

Fig. 3.8.

Suppose that the elastic support is a cylindrical spring whose axis is perpendicular to the plane x_1Ox_2. Introduction of $\mathbf{e}_R = -\mathbf{i}_3$ and $\mathbf{e}_3 = \mathbf{i}_3$ into (3.71) yields

$$|\mathbf{R}| = |-\mathbf{i}_3 \cdot \mathbf{u}_K|\, c = u_3 c\,,$$
$$\mathbf{R} = -u_3 c\mathbf{i}_3\,.$$

In the general case,

$$\mathbf{e}_R = \sum_{i=1}^{3} \beta_i \mathbf{i}_i\,,$$

where β_i are known quantities.

Since (3.5)–(3.9) are related to the attached system of coordinates, we must refer the vector $\mathbf{e}_R$ to this system. By use of (A.59), we obtain the expression for $\mathbf{e}_R$ in the basis $\{\mathbf{e}_i\}$,

$$\mathbf{e}_R^{(1)} = \mathrm{L}^{(1)}\mathbf{e}_R\,, \qquad \mathrm{L}^{(1)} = \mathrm{L}\mathrm{L}^0\,,$$

where $\mathrm{L}^{(1)}$ is the matrix of transformation from the basis $\{\mathbf{e}_j\}$ to the basis $\{\mathbf{e}_i\}$ and L^0 is the matrix of transformation from the basis $\{\mathbf{i}_j\}$ to the basis $\{\mathbf{e}_{j*}\}$. The elements of the matrix L^0 depend on the angles of rotation ϑ_{j*} corresponding to the instant of the loss of stability. In the attached coordinate system, the magnitude of the reaction force is

$$|\mathbf{R}| = c\,|\mathbf{e}_R^{(1)} \cdot \mathbf{u}_K|$$

or

$$|\mathbf{R}| = c\,|\mathrm{L}^{(1)}\mathbf{e}_R \cdot \mathbf{u}_K|\,, \qquad \mathbf{u}_K = \sum_{j=1}^{3} u_j \mathbf{e}_j\,.$$

In the basis $\{\mathbf{e}_i\}$, the vector $\mathbf{R}$ reads

$$\mathbf{R} = -c\,|\mathrm{L}^{(1)}\mathbf{e}_R \cdot \mathbf{u}_K|\,\mathrm{L}^{(1)}\mathbf{e}_R = -\sum_{j=1}^{3} R_j \mathbf{e}_j\,.$$

If we adopt the assumption that displacements of the axial points are small, then the components of the vector $\mathbf{R}$ are as follows:

$$R_i = -\sum_{j=1}^{3} \gamma_{ij} u_{Kj}(\eta_K)$$

or

$$R_i = -\left(\sum_{j=1}^{3} \gamma_{ij} u_j(\eta)\right) \delta\,(\eta - \eta_K)\,, \qquad i = 1\,,\,2\,,\,3\,.$$

For simplicity of algebraic manipulation and numerical analysis, it is convenient to present the vector $\mathbf{R}$ in the form

$$\mathbf{R} = -\mathrm{G}\,\mathbf{u}\,(\eta)\,\delta(\eta - \eta_K)\,. \tag{3.72}$$

The variables γ_{ij} are the elements of the square matrix G.

In the problem under consideration, we have $Q_{1*} = Q_{2*} = 0$ and $M_{3*} = T_{3*}$. Whatever the external load is, the postbuckling equilibrium equations are as follows:

$$\frac{\mathrm{d}Q_1}{\mathrm{d}\eta} - æ_{3*}Q_2 = R_1\,\delta\,(\eta - \eta_K)\,;$$
$$\frac{\mathrm{d}Q_2}{\mathrm{d}\eta} + æ_{3*}Q_1 = R_2\,\delta\,(\eta - \eta_K)\,;$$
$$\frac{\mathrm{d}Q_3}{\mathrm{d}\eta} = R_3\,\delta\,(\eta - \eta_K)\,. \tag{3.73}$$

Here we write out only the system of equations (3.40) for $\Delta D_i \neq 0$ because the systems of equations (3.41)–(3.43) remain the same.

Using the relation

$$æ_{3*} Q_i = æ_{3*} Q_i - æ_{30} Q_i + æ_{30} Q_i = \frac{M_{3*}}{A_{33}} Q_i + æ_{30} Q_i$$

and the equality $M_{3*} = T_*$, we get

$$\frac{dQ_1}{d\eta} - \frac{T_*}{A_{33}} Q_2 - æ_{30} Q_2 = \sum_{j=1}^{3} \gamma_{1j} u_j \, \delta(\eta - \eta_K);$$

$$\frac{dQ_2}{d\eta} + \frac{T_*}{A_{33}} Q_1 + æ_{30} Q_1 = \sum_{j=1}^{3} \gamma_{2j} u_j \, \delta(\eta - \eta_K);$$

$$\frac{dQ_3}{d\eta} = \sum_{j=1}^{3} \gamma_{3j} u_j \, \delta(\eta - \eta_K); \tag{3.74}$$

$$\frac{dM_1}{d\eta} - æ_{30} M_2 + \Delta T_1 \, \delta(\eta - 1) = 0;$$

$$\frac{dM_2}{d\eta} + æ_{30} M_1 - Q_3 + \Delta T_2 \, \delta(\eta - 1) = 0;$$

$$\frac{dM_3}{d\eta} + Q_2 + \Delta T_3 \, \delta(\eta - 1) = 0; \tag{3.75}$$

$$\frac{d\vartheta_1}{d\eta} - \frac{M_1}{A_1} - \frac{T_*}{A_{33}} \vartheta_2 - æ_{30} \vartheta_2 = 0;$$

$$\frac{d\vartheta_2}{d\eta} - \frac{M_2}{A_2} + \frac{T_*}{A_{33}} \vartheta_1 + æ_{30} \vartheta_1 = 0;$$

$$\frac{d\vartheta_3}{d\eta} - \frac{M_3}{A_{33}} = 0; \tag{3.76}$$

$$\frac{du_1}{d\eta} - \frac{T_*}{A_{33}} u_2 - æ_{30} u_2 = 0;$$

$$\frac{du_2}{d\eta} + \frac{T_*}{A_{33}} u_1 + æ_{30} u_1 - \vartheta_3 = 0;$$

$$\frac{du_3}{d\eta} + \vartheta_2 = 0. \tag{3.77}$$

It should be noted that (3.74) include terms dependent on the displacement of the point of application of the force **R**. Similar problems of statics of rods with elastic or rigid supports were discussed in Sect. 2.2. The prime difference between problems of statics of rods with intermediate supports and problems of static stability of rods with intermediate supports is that in the latter problems the critical values of the external forces are unknown. Numerical determination of the critical loads for rods with supports is presented in Sect. 3.5.

3.4 Increments of Loads at Loss of Stability

3.4.1 Forces Directed at a Fixed Point

In this subsection, we consider some examples of the determination of increments of concentrated forces directed at a fixed point O. Recall some examples of loss of stability of a ring discussed in the preceding sections. Before the loss of stability, the ring is subjected to a uniformly distributed normal load $\mathbf{q}_0 = q_2\mathbf{e}_{r0}$. After the loss of stability, this load either remains perpendicular to the axial line of the ring or remains parallel to its original direction. Now, we consider another situation assuming that the load $\mathbf{q}_0$ is pointed at the center of a ring (see Fig. 3.7). These three cases have one feature in common: the shape of the axial line remains unchanged before the loss of stability. For this reason, these cases are comparatively simple ones.

As it was mentioned above, a loss of stability may occur when a rod, being continuously deformed, takes a shape that differs considerably from its initial shape. This case was discussed in Sect. 1.2. Let us resume the discussion of the problem formulated above. In view of (1.48), we obtain

$$\begin{gathered}
\mathrm{A} = |\mathrm{L}^0\mathbf{R}| = R\,, \qquad \mathbf{R} = \mathbf{r}_0\,; \\
\mathrm{L}^0 = \begin{bmatrix} \cos\varphi & \sin\varphi & 0 \\ -\sin\varphi & \cos\varphi & 0 \\ 0 & 0 & 1 \end{bmatrix}; \quad \Delta\mathrm{L}_1 = \begin{bmatrix} 0 & \vartheta_3 & -\vartheta_2 \\ -\vartheta_3 & 0 & \vartheta_1 \\ \vartheta_2 & -\vartheta_1 & 0 \end{bmatrix}; \\
\mathrm{L}^0\mathbf{R} = \begin{bmatrix} 0 \\ R \\ 0 \end{bmatrix}; \qquad \mathbf{e}_{r0} = \begin{bmatrix} 0 \\ 1 \\ 0 \end{bmatrix}; \qquad \mathbf{u} = \begin{bmatrix} u_1 \\ u_2 \\ u_3 \end{bmatrix}.
\end{gathered} \tag{3.78}$$

On rearrangement, from (1.51) it follows that

$$\Delta\mathbf{q} = |\mathbf{q}_0|\left[\left(-\frac{u_1}{R} + \vartheta_3\right)\mathbf{e}_1 - \left(\frac{u_3}{R} + \vartheta_1\right)\mathbf{e}_3\right] = \Delta q_1\mathbf{e}_1 - \Delta q_3\mathbf{e}_3\,. \tag{3.79}$$

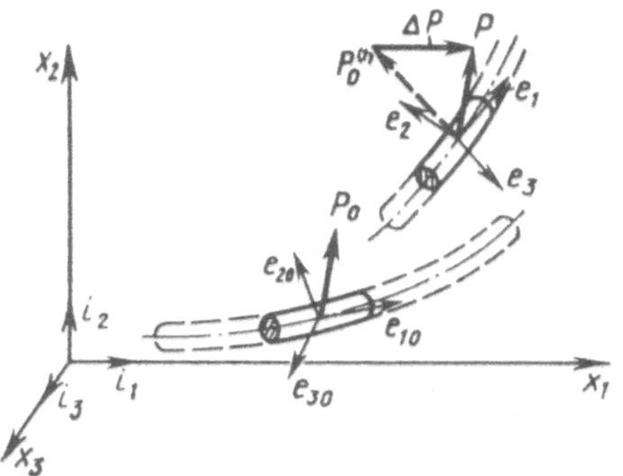

Fig. 3.9.

Assume now that the applied forces are dead (see Fig. 3.9). The components of the vector $\mathbf{P}_0$ depend on the angles of rotation of the attached basis, hence we have $\mathbf{P} = \mathrm{L}\mathbf{P}_0$. In the basis $\{\mathbf{e}_j\}$, we get

$$\Delta\mathbf{P} = \mathrm{L}\mathbf{P}_0 - \mathbf{P}_0 = (\mathrm{L} - \mathrm{E})\,\mathbf{P}_0\,. \tag{3.80}$$

If the attached basis rotates by small angles, then, in view of (1.46), we obtain

$$\begin{aligned}\Delta\mathbf{P} = \Delta\mathrm{L}_1\mathbf{P}_0 = [&(P_{02}\vartheta_3 - P_{03}\vartheta_2)\,\mathbf{e}_1 \\ &+ (-P_{01}\vartheta_3 + P_{03}\vartheta_1)\,\mathbf{e}_2 + (P_{01}\vartheta_2 - P_{02}\vartheta_1)\,\mathbf{e}_3]\,.\end{aligned} \tag{3.81}$$

For example, if the ring is subjected to a dead load $\mathbf{q}_0$, then we get

$$\Delta\mathbf{q} = q_2\vartheta_3\mathbf{e}_1 - q_2\vartheta_1\mathbf{e}_3\,; \tag{3.82}$$

this expression is identical to (3.57).

3.4.2 Forces Which Follow a Straight Line

Let us consider forces such that their vectors remain perpendicular to a fixed straight line during deformation. We assume that the forces are pointed at a certain straight line A-A (Fig. 3.10). Moreover, during deformation the forces lie in a plane that is perpendicular to this straight line. Examples of such forces are given in Figs. 3.11 and 3.12.

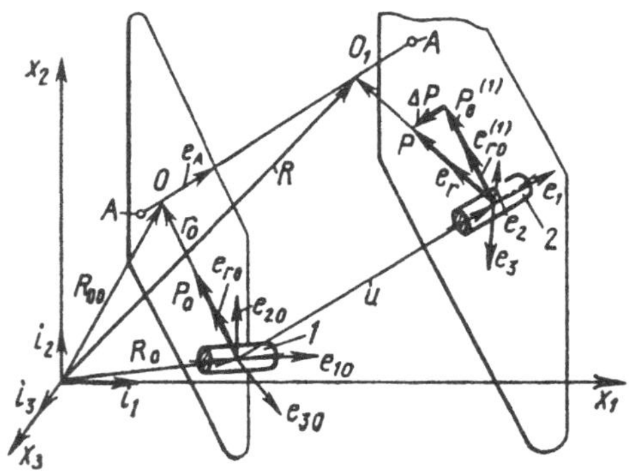

Fig. 3.10.

A rod that rotates about the axis x_2 is shown in Fig. 3.11.

When a loss of stability of the plane configuration of the rod happens, the distributed load $\mathbf{q}$ is perpendicular to the axis x_2.

A rod subjected to a magnetic field is shown in Fig. 3.12. Consider the case of small displacements of the axial points. After the loss of stability, the

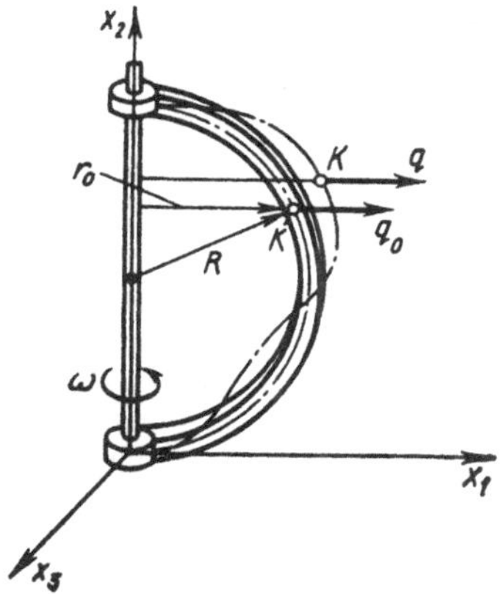

Fig. 3.11.

distributed magnetic force is nearly perpendicular to the line A-A. The distributed forces are directed oppositely with respect to the forces caused by rotation of the rod (see Fig. 3.11). Besides, the magnitudes of the forces change after the loss of stability because these magnitudes depend on the radius $\mathbf{r}$.

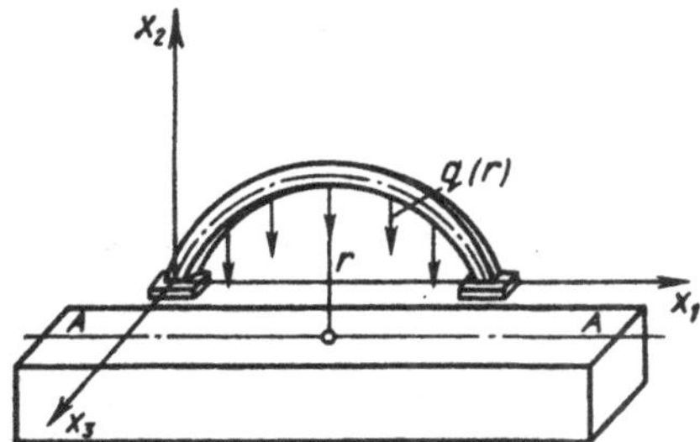

Fig. 3.12.

Let us obtain the increments of the components of a vector $\mathbf{P}_0$ that is pointed at the straight line A-A and lies in a plane perpendicular to A-A (see Fig. 3.10). The natural and a deformed configurations of an element of the rod are shown in Fig. 3.10 (configurations *1* and *2*, respectively). The force $\mathbf{P}_0$ causes the transition of the element from configuration *1* to configuration *2*. Hence, in the attached basis $\{\mathbf{e}_j\}$, the components of $\mathbf{P}_0$ are different in these configurations. The increment $\Delta\mathbf{P}$ satisfies the following relation:

$$\Delta\mathbf{P} = (\mathbf{P} - \mathbf{P}_0^{(1)}) = [|\mathbf{P}_0|(\mathbf{e}_r - \mathbf{e}_{r0}^{(1)})]; \qquad (3.83)$$

here

$$\mathbf{e}_r = \frac{\mathrm{LL}^0\mathbf{r}_0 + \mathbf{R}_{00_1}^{(1)} - \mathbf{u}}{|\mathrm{LL}^0\mathbf{r}_0 + \mathrm{LL}^0\mathbf{R}_{00_1} - \mathbf{u}|}; \qquad \mathbf{e}_{r0}^{(1)} = \frac{\mathrm{L}^0\mathbf{r}_0}{|\mathbf{r}_0|}.$$

An unknown vector $\mathbf{R}_{00_1}$ in the expression for the unit vector $\mathbf{e}_r$ depends on the vector $\mathbf{u}$. Since the vectors $\mathbf{r}_0$ and $\mathbf{r}$ are perpendicular to the line A-A, we see that the vector $\mathbf{R}_{00}$ can be written as (see Fig. 3.10)

$$\mathbf{R}_{00_1} = (\mathbf{e}_A \cdot \mathbf{u})\, \mathbf{e}_A$$

or, in the basis $\{\mathbf{e}_j\}$,

$$\mathbf{R}_{00_1}^{(1)} = (\mathrm{LL}^0\mathbf{e}_A \cdot \mathbf{u})\, \mathrm{LL}^0\mathbf{e}_A;$$

finally, we arrive at

$$\Delta\mathbf{P} = |\mathbf{P}_0| \left[\frac{\mathrm{LL}^0\mathbf{r}_0 + (\mathrm{LL}^0\mathbf{e}_A \cdot \mathbf{u})\, \mathrm{LL}^0\mathbf{e}_A - \mathbf{u}}{|\mathrm{LL}^0\mathbf{r}_0 + (\mathrm{LL}^0\mathbf{e}_A \cdot \mathbf{u})\, \mathrm{LL}^0\mathbf{e}_A - \mathbf{u}|} - \frac{\mathrm{L}^0\mathbf{r}_0}{|\mathbf{r}_0|} \right]. \tag{3.84}$$

The elements of the matrices L and L^0 in (3.84) depend on the coordinate η_K of the point of application of the force $\mathbf{P}_0$. Similarly, we can show that

$$\Delta\mathbf{q} = |\mathbf{q}_0| \left[\frac{\mathrm{LL}^0\mathbf{r}_0 + (\mathrm{LL}^0\mathbf{e}_A \cdot \mathbf{u})\, \mathrm{LL}^0\mathbf{e}_A - \mathbf{u}}{|\mathrm{LL}^0\mathbf{r}_0 + (\mathrm{LL}^0\mathbf{e}_A \cdot \mathbf{u})\, \mathrm{LL}^0\mathbf{e}_A - \mathbf{u}|} - \frac{\mathrm{L}^0\mathbf{r}_0}{|\mathbf{r}_0|} \right]; \tag{3.85}$$

$$\Delta\boldsymbol{\mu} = |\boldsymbol{\mu}_0| \left[\frac{\mathrm{LL}^0\mathbf{r}_0 + (\mathrm{LL}^0\mathbf{e}_A \cdot \mathbf{u})\, \mathrm{LL}^0\mathbf{e}_A - \mathbf{u}}{|\mathrm{LL}^0\mathbf{r}_0 + (\mathrm{LL}^0\mathbf{e}_A \cdot \mathbf{u})\, \mathrm{LL}^0\mathbf{e}_A - \mathbf{u}|} - \frac{\mathrm{L}^0\mathbf{r}_0}{|\mathbf{r}_0|} \right]; \tag{3.86}$$

$$\Delta\mathbf{T} = |\mathbf{T}_0| \left[\frac{\mathrm{LL}^0\mathbf{r}_0 + (\mathrm{LL}^0\mathbf{e}_A \cdot \mathbf{u})\, \mathrm{LL}^0\mathbf{e}_A - \mathbf{u}}{|\mathrm{LL}^0\mathbf{r}_0 + (\mathrm{LL}^0\mathbf{e}_A \cdot \mathbf{u})\, \mathrm{LL}^0\mathbf{e}_A - \mathbf{u}|} - \frac{\mathrm{L}^0\mathbf{r}_0}{|\mathbf{r}_0|} \right]. \tag{3.87}$$

3.4.3 Increments of Concentrated Forces Which Follow a Straight Line: Small Deflections of a Rod

The expression for $\Delta\mathbf{P}_0$ will be derived under the assumption that displacements of axial points and angles of rotation of the attached coordinate axes are small. As in the previous section, we assume that the force $\mathbf{P}_0$ is pointed at a certain straight line A-A and, during deformation, this force lies in a plane which is perpendicular to the straight line. Using (1.46) in (3.87) and retaining only first-order terms in $\Delta\mathrm{L}_1$ and $\mathbf{u}$, we obtain

$$\begin{aligned}\Delta\mathbf{P} = |\mathbf{P}_0| \Big[\Delta\mathrm{L}_1\mathbf{e}_{r_0} &+ \frac{1}{A}(\mathrm{L}^0\mathbf{e}_A \cdot \mathbf{u})\, \mathrm{L}^0\mathbf{e}_A - \frac{\mathbf{u}}{A} \\ &- \frac{1}{A}(\mathbf{e}_{r_0}\mathrm{L}^0\mathbf{e}_A)(\mathrm{L}^0\mathbf{e}_A \cdot \mathbf{u})\, \mathbf{e}_{r_0} + \frac{1}{A}(\mathbf{e}_{r_0}\mathbf{u})\, \mathbf{e}_{r_0} \Big].\end{aligned} \tag{3.88}$$

Consider an example shown in Fig. 3.13. A ring lies in the plane x_1Ox_2. A straight line A-A is perpendicular to this plane. Thus,

$$\mathbf{e}_A = \begin{bmatrix} 0 \\ 0 \\ 1 \end{bmatrix}; \qquad \mathbf{e}_{r_0} = \begin{bmatrix} 0 \\ 1 \\ 0 \end{bmatrix}.$$

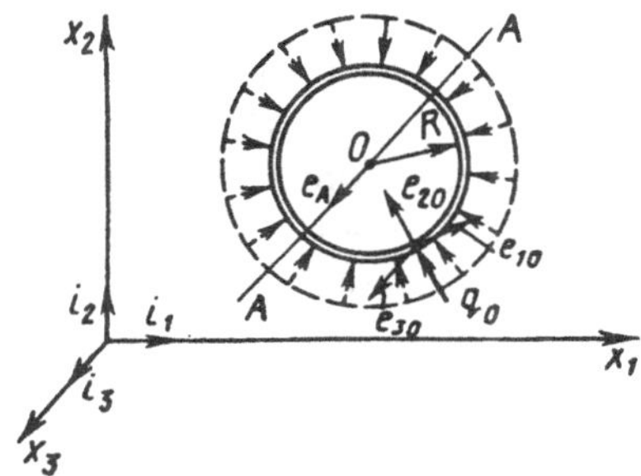

Fig. 3.13.

The matrices L^0 and ΔL_1 correspond to the matrices (3.78) from the previous example. It follows from (3.88) that

$$\Delta \mathbf{P} = |\mathbf{P}_0| \left[\left(\vartheta_3 - \frac{u_1}{R} \right) \mathbf{e}_1 - \vartheta_1 \mathbf{e}_3 \right] . \tag{3.89}$$

In the plane of the ring, the behavior of the force $\mathbf{q}_0$ is similar to the behavior of a force pointed at the center of the ring, therefore, the expression for Δq_1 is similar to the expression obtained in the previous section.

3.4.4 Increments of Concentrated Forces Directed at a Fixed Point: Large Deflections of a Rod

Suppose that a loss of stability of a rod occurs with respect to its deformed configuration. The deformed configuration is substantially different from the initial configuration. Restrict our consideration to the case when, during deformation, forces are of constant magnitude and directed at a fixed point O_1 (see Fig. 3.14). Since the absolute value of the force $\mathbf{P}_*^{(1)}$ remains the same after the loss of stability, we have $|\mathbf{P}_*^{(1)}| = |\mathbf{P}_*|$. Three positions of a rod element subjected to a concentrated force $\mathbf{P}_0$ are shown in Fig. 3.14. Our purpose is to determine $\Delta \mathbf{P}$. It follows from Fig. 3.14 that

$$\Delta \mathbf{P} = |\mathbf{P}_*| \left(\mathbf{e}_r - \mathbf{e}_{r_*}^{(1)} \right) . \tag{3.90}$$

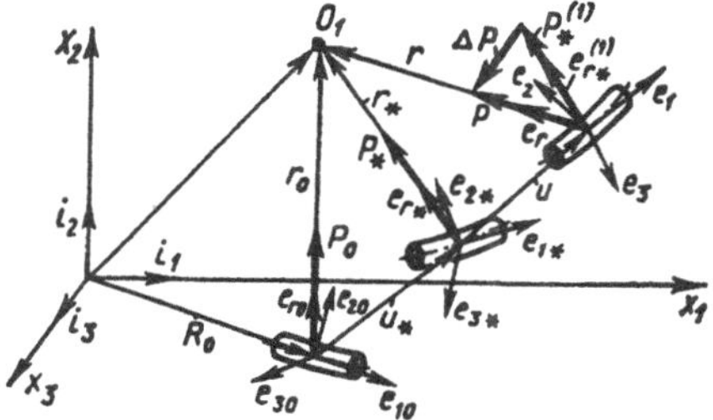

Fig. 3.14.

Equation (3.90) and the previously obtained relations for $\Delta\mathbf{P}$ are alike. In the previous examples, we did not take into consideration the deformations before the loss of stability. For large deformation of a rod before the loss of stability, the vector $\mathbf{e}_{r_*} = \mathbf{e}_{r_*}^{(1)}$ is to be found from the nonlinear equilibrium equations. After the loss of stability, the equilibrium of a rod is governed by (3.5)–(3.9), which are associated with the attached basis $\{\mathbf{e}_j\}$. Hence, it is necessary to obtain the components of the vector $\Delta\mathbf{P}$ in this basis. The unit vector $\mathbf{e}_{r_*}$ satisfies the relation

$$\mathbf{e}_{r_*} = \frac{\mathbf{r}_*}{|\mathbf{r}_*|}, \qquad \mathbf{r}_* = \mathrm{L}_0\mathrm{L}^0\mathbf{r}_0 - \mathbf{u}_*,$$

where L_0 is the matrix of transformation from the basis $\{\mathbf{e}_{j0}\}$ to the basis $\{\mathbf{e}_{j*}\}$ and L^0 is the matrix of transformation from the basis $\{\mathbf{i}_j\}$ to the basis $\{\mathbf{e}_{j0}\}$. The unit vector $\mathbf{e}_r$ satisfies the relation

$$\mathbf{e}_r = \frac{\mathbf{r}}{|\mathbf{r}|}, \qquad \mathbf{r} = \mathrm{L}\mathbf{r}_* - \mathbf{u},$$

where L is the matrix of transformation from the basis $\{\mathbf{e}_{j*}\}$ to the basis $\{\mathbf{e}_j\}$. As a result, we get

$$\Delta\mathbf{P} = |\mathbf{P}_*|\left[\frac{\mathrm{L}\,(\mathrm{L}_0\mathrm{L}^0\mathbf{r}_0 - \mathbf{u}_*) - \mathbf{u}}{|\mathrm{L}\,(\mathrm{L}_0\mathrm{L}^0\mathbf{r}_0 - \mathbf{u}_*) - \mathbf{u}|} - \frac{\mathrm{L}_0\mathrm{L}^0\mathbf{r}_0 - \mathbf{u}_*}{|\mathrm{L}_0\mathrm{L}^0\mathbf{r}_0 - \mathbf{u}_*|}\right]. \tag{3.91}$$

Let us consider now special cases of (3.91). Assume that after the loss of stability deflections with respect to the deformed state are small. Since the vector $\mathbf{u}_*$ is known, we put

$$\mathrm{L}_0\mathrm{L}^0\mathbf{r}_0 - \mathbf{u}_* = \mathbf{b}.$$

For small angles of rotation of the attached axes and small components of $\mathbf{u}$, we can rewrite the right-hand side of (3.91) as follows:

$$\Delta\mathbf{P} = |\mathbf{P}_*|\left[\frac{\mathbf{b} + \Delta\mathbf{a}}{|\mathbf{b} + \Delta\mathbf{a}|} - \frac{\mathbf{b}}{|\mathbf{a}|}\right], \qquad \Delta\mathbf{a} = \Delta\mathrm{L}_1\mathbf{b} - \mathbf{u}. \tag{3.92}$$

Assuming that the components of the vector $\Delta\mathbf{a}$ are small, we get

$$\frac{1}{|\mathbf{b} + \Delta\mathbf{a}|} = \frac{1}{A}\left[1 - \frac{1}{A^2}(\mathbf{b}\,\Delta\mathbf{a})\right], \qquad A = |\mathbf{b}|.$$

In view of (3.92), we obtain the sought increments

$$\Delta\mathbf{P} = |\mathbf{P}_*|\left[\Delta\mathrm{L}_1\mathbf{e}_b - \frac{\mathbf{u}}{A} - \mathbf{e}_b(\mathbf{e}_b\,\Delta\mathrm{L}_1\mathbf{e}_b) + \frac{1}{A}\,\mathbf{e}_b(\mathbf{e}_b\mathbf{u})\right], \qquad \mathbf{e}_b = \frac{\mathbf{b}}{|\mathbf{b}|}. \tag{3.93}$$

3.4.5 Increments of Concentrated Forces Directed at a Fixed Point: Small Deflections of a Rod

Let us assume that before the loss of stability the angles of rotation of the attached coordinate axes and displacements of axial points (i.e. the components of the vector $\mathbf{u}_*$) are small. The matrix L_0 is of the form $\mathrm{L}_0 = \mathrm{E}+\Delta\mathrm{L}_{10}$, where

$$\Delta\mathrm{L}_{10} = \begin{bmatrix} 0 & \vartheta_{30} & -\vartheta_{20} \\ -\vartheta_{30} & 0 & \vartheta_{10} \\ \vartheta_{20} & -\vartheta_{10} & 0 \end{bmatrix};$$

here ϑ_{i0} are small angles of rotation of the attached axes before the loss of stability. By use of (3.91), we obtain

$$\Delta\mathbf{P} = |\mathbf{P}_*| \left\{ \frac{(\mathrm{E}+\Delta\mathrm{L}_1)\,[\,(\mathrm{E}+\Delta\mathrm{L}_{10})\,\mathrm{L}^0\mathbf{r}_0 - \mathbf{u}_* - \mathbf{u}\,]}{|(\mathrm{E}+\Delta\mathrm{L}_1)\,[\,(\mathrm{E}+\Delta\mathrm{L}_{10})\,\mathrm{L}^0\mathbf{r}_0 - \mathbf{u}_* - \mathbf{u}\,]|} - \frac{(\mathrm{E}+\Delta\mathrm{L}_{10})\,\mathrm{L}^0\mathbf{r}_0 - \mathbf{u}_*}{|(\mathrm{E}+\Delta\mathrm{L}_{10})\,\mathrm{L}^0\mathbf{r}_0 - \mathbf{u}_*|} \right\};$$

on rearrangement, we get

$$\Delta\mathbf{P} = |\mathbf{P}_*| \left[\Delta\mathrm{L}_1\mathbf{e}_{r_0} - \frac{\mathbf{u}}{A} + \frac{1}{A}\,(\mathbf{e}_{r_0}\cdot\mathbf{u})\,\mathbf{e}_{r_0} \right]. \tag{3.94}$$

The variables $\Delta\mathrm{L}_{10}$ and $\mathbf{u}_*$ do not enter into the right-hand side of (3.94). Hence, in the case under consideration, the components of the critical force depend neither on small angles of rotation of the attached basis nor on small displacements of the axial points. Equation (3.94) for $\Delta\mathbf{P}_*$ coincides with (1.49) obtained for the case when the deformation of a rod before the loss of stability is not taken into consideration.

3.5 Computer-Oriented Methods

3.5.1 Natural and Critical Configurations Coincide

In this case, the critical load can be found numerically because the behavior of a rod is governed by linear equations.

The critical load can be found by sequential solution of the two linear equations:

$$\mathbf{Y}_*' + \mathrm{A}_*\mathbf{Y}_* = \mathbf{b}_*; \tag{3.95}$$

$$\mathbf{Y}' + \mathrm{A}\,(\mathbf{Y}_*)\,\mathbf{Y} + \mathbf{b}\,(\mathbf{Y}_*,\,\mathbf{Y}) = 0. \tag{3.96}$$

Equation (3.95) can be obtained by direct specialization of (3.10)–(3.14). Equation (3.96) is the same as (3.28). The vector $\mathbf{b}$ depends on the increments of external loads, which, in the general case, are functions of the vectors $\boldsymbol{\vartheta}$, $\mathbf{u}$, and $\mathbf{u}'$ (see (3.20)–(3.23)). Using (3.27), we can eliminate the vector $\mathbf{u}'$ from (3.20)–(3.23). Therefore, the increments of external loads depend on the vectors $\boldsymbol{\vartheta}$ and $\mathbf{u}$.

Eliminating the vector $\mathbf{u}'$ from (3.20)–(3.22), we arrive at

$$\Delta\mathbf{q} = \mathrm{C}_1\boldsymbol{\vartheta} + \mathrm{C}_2\mathbf{u}\,; \qquad \Delta\boldsymbol{\mu} = \mathrm{C}_3\boldsymbol{\vartheta} + \mathrm{C}_4\mathbf{u}\,,$$

where

$$\mathrm{C}_1 = \mathrm{C}_*^{(1)} - \mathrm{C}_*^{(3)}\mathrm{A}_1\,; \qquad \mathrm{C}_2 = \mathrm{C}_*^{(2)} - \mathrm{C}_*^{(3)}\mathrm{A}_{æ*}\,;$$
$$\mathrm{C}_3 = \mathrm{C}_*^{(4)} - \mathrm{C}_*^{(6)}\mathrm{A}_1\,; \qquad \mathrm{C}_5 = \mathrm{C}_*^{(5)} - \mathrm{C}_*^{(6)}\mathrm{A}_{æ*}\,.$$

The vector $\mathbf{b}$ in (3.96) can be written as

$$\mathbf{b} = \begin{bmatrix} 0 & 0 & \mathrm{C}_1 & \mathrm{C}_2 \\ 0 & 0 & \mathrm{C}_3 & \mathrm{C}_4 \\ 0 & 0 & 0 & 0 \\ 0 & 0 & 0 & 0 \end{bmatrix} \cdot \begin{bmatrix} \mathbf{Q}_0 \\ \mathbf{M}_0 \\ \boldsymbol{\vartheta} \\ \mathbf{u} \end{bmatrix} = \mathrm{C}\mathbf{Y}\,.$$

Thus, the following homogeneous equation is obtained:

$$\mathbf{Y}' + \mathrm{B}\mathbf{Y} = 0\,, \qquad \mathrm{B} = \mathrm{A} + \mathrm{C}\,. \tag{3.97}$$

The vector $\mathbf{Y}_*$ is a characteristic of stress-strain state of the rod in its critical configuration after the loss of stability. This vector can be found from (3.95). Substituting the vector $\mathbf{Y}_*$ into (3.96) and replacing $\mathbf{b}$ by its explicit expression (as a function of increments of external loads), we see that (3.96) becomes a homogeneous equation. The value of the load at which the solution of (3.96) satisfies the boundary conditions is the sought critical value.

Numerical analysis of linear equilibrium equations is discussed in Sect. 2.3. There we put the solution to be of the form

$$\mathbf{Y} = \mathrm{K}(\eta)\mathbf{C} + \mathbf{Y}_1\,,$$

where $\mathrm{K}(\eta)$ is the fundamental matrix to be determined.

In Sect. 2.1, we discussed methods for improvement of the accuracy of numerical determination of the matrix $\mathrm{K}(\eta)$. These methods can also be applied to (3.97).

To determine the critical loads from (3.96), one must evaluate the vector $\mathbf{Y}_*$. To obtain the required accuracy, we can use the procedure of correction of the matrix $\mathrm{K}(\eta)$ presented in Sect. 2.1. This procedure is time-consuming and we see little reason to follow it unless the matrix $\mathrm{K}(\eta)$ need also be obtained.

Consider the method of iterations. In view of (3.95), we get

$$\mathbf{Y}_* = -\int \mathrm{A}_*\mathbf{Y}_*\,\mathrm{d}h + \int \mathbf{b}_*\,\mathrm{d}h + \mathbf{Y}_{*0}\,. \tag{3.98}$$

If an approximate value of the vector $\mathbf{Y}_*$ is known, say, it can be $\mathbf{Y}_*^{(0)}$, then, in view of (3.98), we obtain

$$\mathbf{Y}_*^{(1)} = -\int \mathrm{A}_* \mathbf{Y}_*^{(0)}\, \mathrm{d}h + \int \mathbf{b}_*\, \mathrm{d}h + \mathbf{Y}_{*0}^{(1)}\,. \tag{3.99}$$

Some components of the vector $\mathbf{Y}_{*0}$ are equal to zero (this fact follows from the boundary conditions at $\eta = 0$). The other components are to be found from the boundary conditions at $\eta = 1$. The mth approximate value is

$$\mathbf{Y}_*^{(m)} = -\int \mathrm{A}_* \mathbf{Y}^{(m-1)}\, \mathrm{d}h^{(1)} + \int \mathbf{b}_*\, \mathrm{d}h^{(1)} + \mathbf{Y}_{*0}^{(m)}\,. \tag{3.100}$$

If the required accuracy is achieved at the mth step, that is,

$$\max_{\eta} \left| \frac{Y_{*j}^{(m)} - Y_{*j}^{(m-1)}}{Y_{*j}^{(m)}} \right| 100\% \leq \Delta\,,$$

then the vector $\mathbf{Y}_*^{(m)}$ can be taken as the sought solution of (3.95).

This brings up the question: What is the way for selecting the zeroth approximation $\mathbf{Y}_*^{(0)}$? For example, we can take the vector obtained by the method of initial parameters as the zeroth approximation, that is,

$$\mathbf{Y}_*^{(0)} = \mathrm{K}\,(\eta)\, \mathbf{C} + \mathbf{Y}_{10}\,. \tag{3.101}$$

The vector $\mathbf{Y}_*^{(0)}$ satisfies the boundary conditions. In this case, we have to determine the matrix $\mathrm{K}(\eta)$ just once. To clarify our line of reasoning, let us consider an example. Suppose that the applied loads vary linearly with a coefficient β. For each $\beta^{(i)}$, the vector $\mathbf{Y}_*$ can be found. Having determined $\mathbf{Y}_*$, we rewrite (3.96) as a homogeneous equation

$$\mathbf{Y}' + \mathbf{B}\mathbf{Y} = 0\,, \tag{3.102}$$

therefore,

$$\mathbf{Y} = \mathrm{K}\,(\eta\,, \beta^{(j)})\, \mathbf{C}\,, \qquad \mathrm{K}\,(0\,, \beta^{(j)}) = \mathrm{E}\,. \tag{3.103}$$

The matrix $\mathrm{K}(\eta)$ can be found more accurately in the following way:

$$\mathrm{K}^{(m)}(\eta\,, \beta^{(j)}) = -\int \mathrm{B}\mathrm{K}^{(m-1)}(h^{(1)}\,, \beta^{(j)})\, \mathrm{d}h^{(1)} + \mathrm{E}\,.$$

Once the matrix $\mathrm{K}(\eta\,, \beta^{(i)})$ of the required accuracy is obtained, then, from the boundary conditions at $\eta = 1$, we get the system of six homogeneous equations:

$$\sum_{j=1}^{6} k_{ij}(1\,, \beta^{(j)})\, c_j = 0\,, \qquad i = 1\,, 2\,, \ldots\,, 6\,. \tag{3.104}$$

The other six components of the vector $\mathbf{C}$ must equal zero to satisfy the boundary conditions at $\eta = 0$. Nontrivial solutions of (3.104) do exist provided

$$\det(k_{ij}) = 0\,. \tag{3.105}$$

The values $\beta^{(i)}$ found from (3.105) are the critical values of the factor β_*. The corresponding loads $\mathbf{q}_* = \beta_* \mathbf{q}_0$, $\boldsymbol{\mu}_* = \beta_* \boldsymbol{\mu}_0$, $\mathbf{P}_*^{(i)} = \beta_* \mathbf{P}_0^{(i)}$, and $\mathbf{T}_*^{(\nu)} = \beta_* \mathbf{T}_0^{(\nu)}$ are the critical loads. Under the action of the loads $\mathbf{q}_0$, $\boldsymbol{\mu}_0$, $\mathbf{P}_0^{(i)}$, and $\mathbf{T}_0^{(\nu)}$, the loss of stability will not occur provided the inequality $\beta_* > 1$ holds.

Let us consider in more detail the derivation of the formula (3.105) for the case of a cantilever beam subjected to a concentrated force $\mathbf{P}$ and a concentrated moment $\mathbf{T}$ applied at the end $\eta = 1$. The vector $\mathbf{Y}$ must satisfy the following boundary conditions:

$$\begin{aligned} &(1) \quad \text{at} \quad \eta = 0\,, \quad \boldsymbol{\vartheta} = \mathbf{u} = 0\,; \\ &(2) \quad \text{at} \quad \eta = 1\,, \quad \mathbf{Q}_0 = \Delta\mathbf{P} \quad \text{and} \quad \mathbf{M}_0 = \Delta\mathbf{T}\,; \end{aligned}$$

if $\Delta\mathbf{P}$ and $\Delta\mathbf{T}$ are functions only of $\mathbf{u}(1)$ and $\boldsymbol{\vartheta}(1)$, then we have

$$\mathbf{Q}_0(1) = \mathrm{B}_*^{(2)}\mathbf{u}(1) + \mathrm{B}_*^{(1)}\boldsymbol{\vartheta}(1)\,; \qquad \mathbf{M}_0(1) = \mathrm{B}_*^{(4)}\boldsymbol{\vartheta}(1) + \mathrm{B}_*^{(5)}\mathbf{u}(1)\,.$$

Considering the homogeneous boundary conditions at $\eta = 0$, we obtain $c_7 = 0$, $c_8 = 0$, ..., $c_{12} = 0$. The other six constants can be determined from the boundary conditions at $\eta = 1$. Let us consider the corresponding equations in detail. Introduction of the obtained values of c_7, ..., c_{12} into (3.103) yields

$$\mathbf{Y} = \begin{bmatrix} \mathbf{Q}_0 \\ \mathbf{M}_0 \\ \boldsymbol{\vartheta} \\ \mathbf{u} \end{bmatrix} = \begin{bmatrix} \mathrm{K}^{(11)} & \mathrm{K}^{(12)} & \mathrm{K}^{(13)} & \mathrm{K}^{(14)} \\ \mathrm{K}^{(21)} & \mathrm{K}^{(22)} & \mathrm{K}^{(23)} & \mathrm{K}^{(24)} \\ \mathrm{K}^{(31)} & \mathrm{K}^{(32)} & \mathrm{K}^{(33)} & \mathrm{K}^{(34)} \\ \mathrm{K}^{(41)} & \mathrm{K}^{(42)} & \mathrm{K}^{(43)} & \mathrm{K}^{(44)} \end{bmatrix} \cdot \begin{bmatrix} \mathbf{C}^{(1)} \\ \mathbf{C}^{(2)} \\ \mathbf{0} \\ \mathbf{0} \end{bmatrix}, \tag{3.106}$$

where $\mathrm{K}^{(ij)}$ is a 3×3 block matrix and

$$\mathbf{C}^{(1)} = \begin{bmatrix} c_1 \\ c_2 \\ c_3 \end{bmatrix}; \qquad \mathbf{C}^{(2)} = \begin{bmatrix} c_4 \\ c_5 \\ c_6 \end{bmatrix}.$$

Using (3.106), we get

$$\begin{aligned} \mathbf{Q}_0 &= \mathrm{K}^{(11)}\mathbf{C}^{(1)} + \mathrm{K}^{(12)}\mathbf{C}^{(2)}\,; \\ \mathbf{M}_0 &= \mathrm{K}^{(21)}\mathbf{C}^{(1)} + \mathrm{K}^{(22)}\mathbf{C}^{(2)}\,; \\ \boldsymbol{\vartheta} &= \mathrm{K}^{(31)}\mathbf{C}^{(1)} + \mathrm{K}^{(32)}\mathbf{C}^{(2)}\,; \\ \mathbf{u} &= \mathrm{K}^{(41)}\mathbf{C}^{(1)} + \mathrm{K}^{(42)}\mathbf{C}^{(2)}\,. \end{aligned} \tag{3.107}$$

From (3.107) it follows that the critical loads can be found provided the first six columns of the fundamental matrix $\mathrm{K}(\eta)$ are known. We should successively solve (3.102) for the following initial vectors:

$$\mathbf{Y}^{(1)} = \begin{bmatrix} 1 \\ 0 \\ 0 \\ 0 \\ 0 \\ 0 \\ 0 \\ \vdots \\ 0 \end{bmatrix}; \qquad \mathbf{Y}^{(2)} = \begin{bmatrix} 0 \\ 1 \\ 0 \\ 0 \\ 0 \\ 0 \\ 0 \\ \vdots \\ 0 \end{bmatrix}; \ldots; \qquad \mathbf{Y}^{(6)} = \begin{bmatrix} 0 \\ 0 \\ 0 \\ 0 \\ 0 \\ 1 \\ 0 \\ \vdots \\ 0 \end{bmatrix}.$$

Considering the boundary conditions at $\eta = 1$, we obtain two vector equations

$$\begin{aligned}
&(\mathrm{K}^{(11)} - \mathrm{B}_*^{(2)}\mathrm{K}^{(41)} - \mathrm{B}_*^{(1)}\mathrm{K}^{(31)})\,\mathbf{C}^{(1)} \\
&\qquad + (\mathrm{K}^{(12)} - \mathrm{B}_*^{(1)}\mathrm{K}^{(32)} - \mathrm{B}_*^{(2)}\mathrm{K}^{(42)})\,\mathbf{C}^{(2)} = 0\,; \\
&(\mathrm{K}^{(21)} - \mathrm{B}_*^{(4)}\mathrm{K}^{(31)} - \mathrm{B}_*^{(5)}\mathrm{K}^{(41)})\,\mathbf{C}^{(1)} \\
&\qquad + (\mathrm{K}^{(22)} - \mathrm{B}_*^{(4)}\mathrm{K}^{(32)} - \mathrm{B}_*^{(5)}\mathrm{K}^{(42)})\,\mathbf{C}^{(2)} = 0
\end{aligned}$$

or

$$\begin{bmatrix} \mathrm{D}_{11} & \mathrm{D}_{12} \\ \mathrm{D}_{21} & \mathrm{D}_{22} \end{bmatrix} \cdot \begin{bmatrix} \mathbf{C}^{(1)} \\ \mathbf{C}^{(2)} \end{bmatrix} = 0\,,$$

where

$$\begin{aligned}
\mathrm{D}_{11} &= \mathrm{K}^{(11)} - \mathrm{B}_*^{(1)}\mathrm{K}^{(31)} - \mathrm{B}_*^{(2)}\mathrm{K}^{(41)}\,; \\
\mathrm{D}_{12} &= \mathrm{K}^{(12)} - \mathrm{B}_*^{(1)}\mathrm{K}^{(32)} - \mathrm{B}_*^{(2)}\mathrm{K}^{(42)}\,; \\
\mathrm{D}_{21} &= \mathrm{K}^{(21)} - \mathrm{B}_*^{(4)}\mathrm{K}^{(31)} - \mathrm{B}_*^{(5)}\mathrm{K}^{(41)}\,; \\
\mathrm{D}_{22} &= \mathrm{K}^{(22)} - \mathrm{B}_*^{(4)}\mathrm{K}^{(32)} - \mathrm{B}_*^{(5)}\mathrm{K}^{(42)}\,.
\end{aligned}$$

The critical values of β_* (and, thus, the critical loads) can be found from the relation

$$D = \det \begin{bmatrix} \mathrm{D}_{11} & \mathrm{D}_{12} \\ \mathrm{D}_{21} & \mathrm{D}_{22} \end{bmatrix} = 0\,.$$

3.5.2 Natural and Critical Configurations Differ

The key feature of loss of stability of curvilinear rods with respect to a deformed configuration is that the critical strain-stress state is not known a priori. In particular, the shape of the axial line corresponding to the critical state may differ greatly from the natural configuration of the axial line. For example, when we determine the critical load for a rod that is straight in its natural configuration, we assume that the rod remains straight in the critical state. We also used this assumption in the determination of the critical loads for a ring, i.e. we assumed that the critical configuration coincide with the natural configuration. Under this assumption, all the geometrical characteristics of the axial line, $æ_{j0}(\eta)$, and the location of the principal axes, $\vartheta_{j0}(\eta)$, are known. The internal loads $Q_{j0}(\eta)$ and $M_{j0}(\eta)$ can be found from the equilibrium equations of the zeroth approximation.

When the loss of stability takes place with respect to a deformed configuration, the critical equilibrium configuration must first be determined (see (3.10)–(3.14)). The parameters $\boldsymbol{æ}_*$, $\mathbf{Q}_*$, and $\mathbf{M}_*$, which characterize this configuration, will enter into the linear equilibrium equations after the loss of stability (see (3.24)–(3.27) or (3.28)). Since the critical configuration is unknown a priori, we are to examine the stability of the equilibrium states that occur while the rod is subjected to a continuously increasing loading (step-by-step loading). In Chap. 2, we dealt with nonlinear equilibrium equations while the loads applied to a rod were known. Hence, the corresponding vector characteristics of the strain-stress state of the rod can be obtained numerically (say, using the method of step-by-step loading).

In the analysis of stability of a rod, loads are to be found either from the nonlinear equilibrium equations (3.10)–(3.14) or from the linear equations (3.24)–(3.27) under the homogeneous boundary conditions. Numerical analysis of (3.10)–(3.14) for each step of loading is given in Sect. 2.3. Types of loading are as follows: (1) a load grows continuously; (2) a step-by-step loading takes place. Of course, more complicated types of loading exist, e.g. a rod is subjected to a set of distributed or concentrated forces and moments. During a loading, it is possible to select one particular load such that continuous increasing of this load will take the rod to a critical state. All the above is essential for numerical analysis when the eigenvalues (the critical loads) for the boundary-value problem are to be found.

Let us consider the determination of the critical loads assuming that they vary linearly with a coefficient β. If the required accuracy of the solution of (3.10)–(3.14) is achieved at the mth step, then we can find the fundamental matrix $\mathrm{K}(\eta)$ for (3.100) and check the condition (3.103). Then, we determine the strain-stress state of the rod at the mth step from (3.96) and check the validity of the equality (3.105). If this equality holds, then the corresponding value $\beta^{(m)} = \beta_0 m = \beta_*$ is the critical one and, thus, the critical load is found. If the determinant in (3.104) does not equal zero, then we have to proceed with the $(m+1)$th step. If at the $(m+1)$th and mth steps the determinant in (3.104) has opposite signs, then the sought critical value lies between $\beta^{(m)}$ and $\beta^{(m-1)}$.

The critical value β_* can also be found by the method of successive approximations.

3.5.3 Concentrated Loads Applied to Arbitrary Cross Sections: Determination of Critical Loads

Consider a rod subjected to a distributed load. Let a concentrated force **P** and a concentrated moment **T** be applied to the cross sections η_P and η_T, respectively. Eliminating $\mathbf{u}'$ from (3.21) and (3.23), we obtain

$$\Delta\mathbf{P} = \mathrm{B}_1\boldsymbol{\vartheta} + \mathrm{B}_2\mathbf{u}; \qquad \Delta\mathbf{T} = \mathrm{B}_3\boldsymbol{\vartheta} + \mathrm{B}_4\mathbf{u},$$

where

$$\mathrm{B}_1 = \mathrm{B}_*^{(1)} - \mathrm{B}_*^{(3)}\mathrm{A}_1; \qquad \mathrm{B}_2 = \mathrm{B}_*^{(2)} - \mathrm{B}_*^{(3)}\mathrm{A}_{æ_*};$$
$$\mathrm{B}_3 = \mathrm{B}_*^{(4)} - \mathrm{B}_*^{(6)}\mathrm{A}_1; \qquad \mathrm{B}_4 = \mathrm{B}_*^{(5)} - \mathrm{B}_*^{(6)}\mathrm{A}_{æ_*}.$$

The vector **b** satisfies the relation

$$\mathbf{b} = \mathbf{C}\mathbf{Y} + \mathrm{C}_P\mathbf{Y}\,\delta(\eta - \eta_P) + \mathrm{C}_T\mathbf{Y}\,\delta(\eta - \eta_T),$$

where

$$\mathbf{C}_P = \begin{bmatrix} 0 & 0 & \mathrm{B}_1 & \mathrm{B}_2 \\ 0 & 0 & 0 & 0 \\ 0 & 0 & 0 & 0 \\ 0 & 0 & 0 & 0 \end{bmatrix}; \qquad \mathbf{C}_T = \begin{bmatrix} 0 & 0 & 0 & 0 \\ 0 & 0 & \mathrm{B}_3 & \mathrm{B}_4 \\ 0 & 0 & 0 & 0 \\ 0 & 0 & 0 & 0 \end{bmatrix}.$$

In view of the presence of the concentrated load, (3.96) takes the form

$$\mathbf{Y}' + \mathbf{B}\mathbf{Y} = -\mathrm{C}_P\mathbf{Y}\,\delta(\eta - \eta_P) - \mathrm{C}_T\mathbf{Y}\,\delta(\eta - \eta_T). \tag{3.108}$$

Its solution is

$$\mathbf{Y} = \mathrm{K}(\eta)\,\mathbf{C} + \mathrm{F}_1\mathbf{Y}(\eta_P)\,H(\eta - \eta_P) + \mathrm{F}_2\mathbf{Y}(\eta_T)\,H(\eta - \eta_T), \tag{3.109}$$

where

$$\mathrm{F}_1 = \mathrm{K}(\eta)\,\mathrm{K}^{-1}(\eta_P)\,\mathrm{C}_P; \qquad \mathrm{F}_2 = \mathrm{K}(\eta)\,\mathrm{K}^{-1}(\eta_T)\,\mathrm{C}_T.$$

It is more convenient to present the term $\mathrm{K}(\eta)\mathbf{C}$ as follows:

$$\mathrm{K}(\eta)\mathbf{C} = \begin{bmatrix} \mathrm{K}^{(11)} & \dots & \mathrm{K}^{(14)} \\ \vdots & \ddots & \vdots \\ \mathrm{K}^{(41)} & \dots & \mathrm{K}^{(44)} \end{bmatrix} \cdot \begin{bmatrix} \mathbf{C}^{(1)} \\ \vdots \\ \mathbf{C}^{(4)} \end{bmatrix};$$

here K^{ij} are 3×3 matrices.

By use of block-matrix notation, the matrices F_1 and F_2 can be presented in similar form. The solution (3.109) must satisfy the boundary conditions at $\eta = 0$ and $\eta = 1$ and the following additional conditions:

$$\begin{aligned}
&(1) \quad \text{at} \quad \eta = \eta_P\,, \quad \boldsymbol{\vartheta} = \boldsymbol{\vartheta}\,(\eta_P)\,; \quad \mathbf{u} = \mathbf{u}\,(\eta_P)\,;\\
&(2) \quad \text{at} \quad \eta = \eta_T\,, \quad \boldsymbol{\vartheta} = \boldsymbol{\vartheta}\,(\eta_T)\,; \quad \mathbf{u} = \mathbf{u}\,(\eta_T)\,.
\end{aligned} \tag{3.110}$$

As an illustration, consider a cantilever beam subjected to the following boundary conditions:

$$\begin{aligned}
&(1) \quad \text{at} \quad \eta = 0\,, \quad \boldsymbol{\vartheta} = \mathbf{u} = 0\,;\\
&(2) \quad \text{at} \quad \eta = 1\,, \quad \mathbf{Q}_0 = \mathbf{M}_0 = 0\,.
\end{aligned}$$

From the boundary conditions at $\eta = 0$, we obtain $\mathbf{C}^{(3)} = \mathbf{C}^{(4)} = 0$. In view of the boundary conditions at $\eta = 1$, we have

$$\begin{aligned}
&\mathrm{K}^{(11)}(1)\,\mathbf{C}^{(1)} + \mathrm{K}^{(12)}(1)\,\mathbf{C}^{(2)} + \mathrm{F}_1^{(13)}(1)\,\boldsymbol{\vartheta}\,(\eta_P)\\
&\qquad + \mathrm{F}_1^{(14)}(1)\,\mathbf{u}\,(\eta_P) + \mathrm{F}_2^{(13)}(1)\,\boldsymbol{\vartheta}\,(\eta_T) + \mathrm{F}_2^{(14)}(1)\,\mathbf{u}\,(\eta_T) = 0\,;\\
&\mathrm{K}^{(21)}(1)\,\mathbf{C}^{(1)} + \mathrm{K}^{(22)}(1)\,\mathbf{C}^{(2)} + \mathrm{F}_1^{(23)}(1)\,\boldsymbol{\vartheta}\,(\eta_P)\\
&\qquad + \mathrm{F}_1^{(24)}(1)\,\mathbf{u}\,(\eta_P) + \mathrm{F}_2^{(23)}(1)\,\boldsymbol{\vartheta}\,(\eta_T) + \mathrm{F}_2^{(24)}(1)\,\mathbf{u}\,(\eta_T) = 0\,.
\end{aligned} \tag{3.111}$$

Besides (3.111), the relations (3.110) must be satisfied (for definiteness, assume that $\eta_P < \eta_T$):

at $\eta = \eta_P$,

$$\begin{aligned}
&\boldsymbol{\vartheta}\,(\eta_P) = \mathrm{K}^{(31)}(\eta_P)\,\mathbf{C}^{(1)} + \mathrm{K}^{(32)}(\eta_P)\,\mathbf{C}^{(2)}\,;\\
&\mathbf{u}\,(\eta_P) = \mathrm{K}^{(41)}(\eta_P)\,\mathbf{C}^{(1)} + \mathrm{K}^{(42)}(\eta_P)\,\mathbf{C}^{(2)}\,;
\end{aligned} \tag{3.112}$$

at $\eta = \eta_T$,

$$\begin{aligned}
&\boldsymbol{\vartheta}\,(\eta_T) = \mathrm{K}^{(31)}(\eta_T)\,\mathbf{C}^{(1)}\\
&\qquad + \mathrm{K}^{(32)}(\eta_T)\,\mathbf{C}^{(2)} + \mathrm{F}_1^{(33)}(\eta_T)\,\boldsymbol{\vartheta}\,(\eta_P) + \mathrm{F}_1^{(34)}(\eta_T)\,\mathbf{u}\,(\eta_P)\,;\\
&\mathbf{u}\,(\eta_T) = \mathrm{K}^{(41)}(\eta_T)\,\mathbf{C}^{(1)}\\
&\qquad + \mathrm{K}^{(42)}(\eta_T)\,\mathbf{C}^{(2)} + \mathrm{F}_1^{(43)}(\eta_T)\,\boldsymbol{\vartheta}\,(\eta_P) + \mathrm{F}_1^{(44)}(\eta_T)\,\mathbf{u}\,(\eta_P)\,.
\end{aligned} \tag{3.113}$$

Equations (3.111)–(3.113) is a homogeneous-equation system in the six unknown vectors: $\mathbf{C}^{(1)}$, $\mathbf{C}^{(2)}$, $\boldsymbol{\vartheta}(\eta_P)$, $\mathbf{u}(\eta_P)$, $\boldsymbol{\vartheta}(\eta_T)$, and $\mathbf{u}(\eta_T)$. This system can be rewritten as a vector equation

$$\mathrm{F}\,\mathbf{s} = 0\,, \tag{3.114}$$

where

$$\mathbf{s} = \mathbf{s}\,(\,\mathbf{C}^{(1)}\,,\,\mathbf{C}^{(2)}\,,\,\boldsymbol{\vartheta}\,(\eta_P)\,,\,\mathbf{u}\,(\eta_P)\,,\,\boldsymbol{\vartheta}\,(\eta_T)\,,\,\mathbf{u}\,(\eta_T)\,)\,.$$

The values of the variables q_{*j}, μ_{*j}, P_{*j}, and T_{*j} at which the determinant of the matrix F is equal to zero are the sought critical loads.

If the load varies linearly, the critical value of the coefficient of proportionality and the corresponding critical loads can be found from the relation $\det[\mathrm{F}] = 0$.

Numerical determination of critical loads is also discussed in Chapters 4, 5, and 6.

3.6 Problems

3.1. An elastic spiral is mounted on an object moving with an acceleration **a** (Fig. 3.15). The vector **a** is parallel to the axis x_1. Derive equations from which the critical loads (the critical value of $|\mathbf{a}|$) can be found.

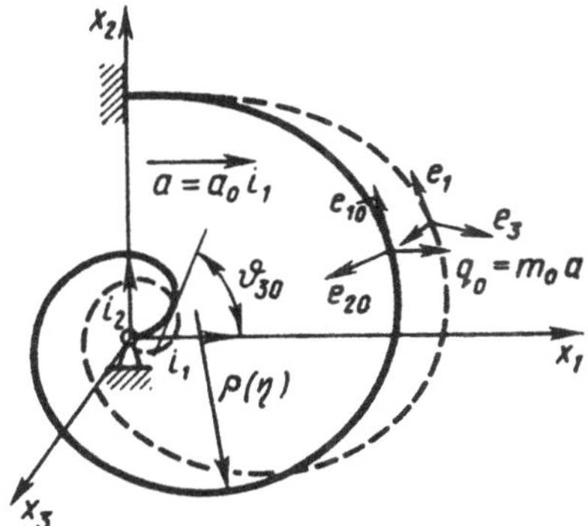

Fig. 3.15.

3.2. A spiral rod rotates with an angular velocity ω (Fig. 3.16). Derive linear equilibrium equations for the following cases: (1) at the loss of stability, the shape of the rod is close to its natural shape; (2) these shapes differ greatly.

3.3. A straight rod of constant cross section (Fig. 3.17*a*) is deformed into a ring (Fig. 3.17*b*). The ring is subjected to a follower distributed load *q*. The bending stiffnesses of the ring are different ($A_{22} \neq A_{33}$). The nondimensional stiffness A_{33} equals unity. Does the critical value of $|\mathbf{q}|$ depend on the initial stress state of the ring? This stress state is characterized by the bending moment M_{30}.

3.4. Determine the critical load **q** (see Fig. 3.17*b*) assuming that after the loss of stability the vector **q** remains parallel to its original direction. Take into consideration the initial stress-strain state. Examine both in-plane and of out-of-plane loss of stability.

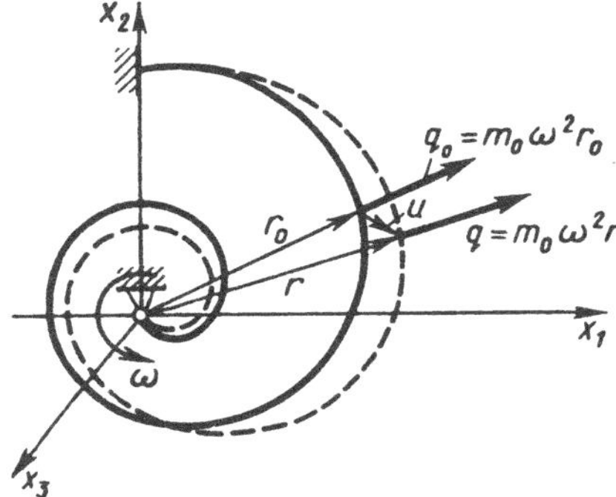

Fig. 3.16.

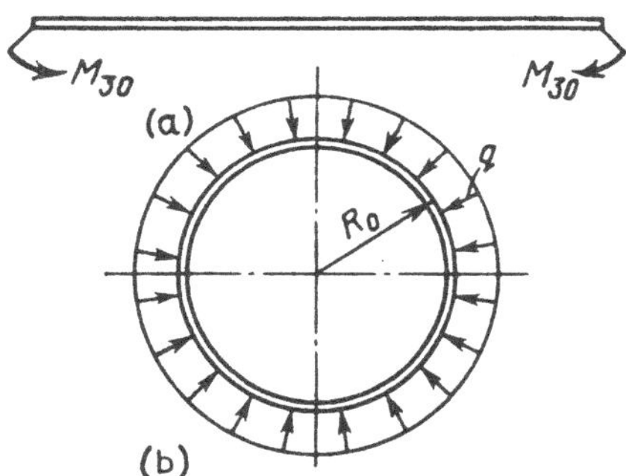

Fig. 3.17.

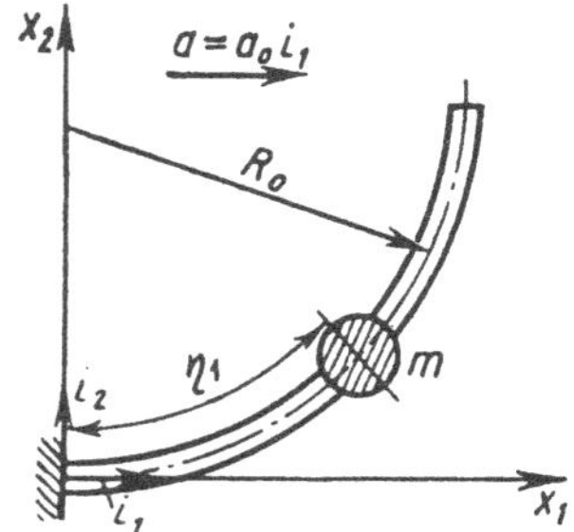

Fig. 3.18.

3.5. A circular rod is mounted on an object moving with an acceleration **a** (Fig. 3.18). The vector **a** is parallel to the axis x_1. A mass point m is rigidly connected to the rod. Derive the equilibrium equations for the rod after the loss of stability. Assume that the critical shape of the rod coincides with its natural shape, i.e. $æ_{3*} = 1/R^0$, where $R^0 = l/R_0$.

4. Straight Rods

This chapter is devoted to applied problems of statics of straight rods. A great quantity of structural elements of mechanisms and devices could be treated as straight rods. Straight rods find wide application in mechanical engineering. Some illustrative examples are given in Introduction.

Most design problems of mechanics of rods cannot be solved exactly (e.g. problems of statics and dynamics of rods of varying cross section, nonlinear problems, etc.). These problems are usually solved by means of numerical or approximate methods. Moreover, usually the explicit representation of the exact solution (if we managed to develop it) is too complicated and useless for mathematical and physical interpretation. In general, the solution must be simplified through the use of numerical methods. The most-used methods are methods based on variational principles of mechanics.

Exact and approximate methods for analysis of equilibrium equations are discussed in this chapter. The approximate methods are based on the theory of distributions; these methods allow us to find solutions for problems of statics of straight rods subjected to concentrated forces and interacting with additional lateral supports.

4.1 Rods of Straight Natural Configuration

4.1.1 Traditional Routines of Derivation of Equilibrium Equations

Suppose that a rod axis in the natural configuration is a straight line; suppose also that the axial line remains a plane curve during deformation. Let the loads applied to the rod be dead. Under these conditions, the equilibrium equations for the rod can be easily derived within the framework of university courses of the strength of materials and structural mechanics. If a rod (see Fig. 0.21) in its natural configuration is twisted and subjected to loads of complicated nature (e.g. forces and moments which follow the normal to the axial line or are permanently directed at a fixed point or depend on the displacement of the axial points), then the equilibrium equations could hardly be derived by use of the standard methods. It is more convenient to obtain these equations by direct specialization of the general equilibrium equations (1.31)–(1.35) (or (1.57)–(1.61)) for the case of rods of straight initial configuration.

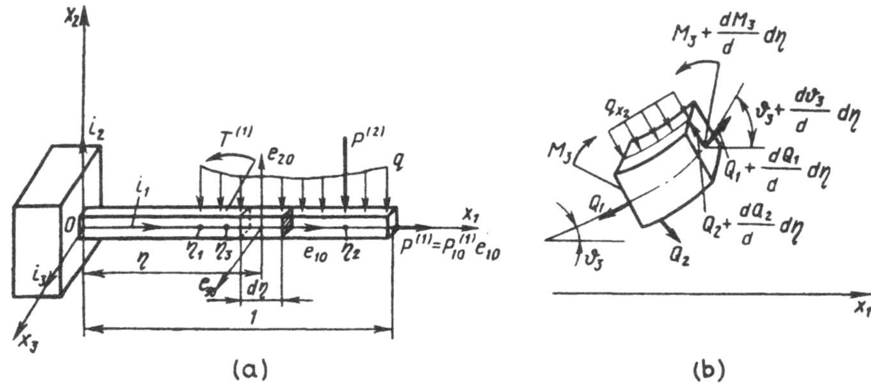

Fig. 4.1.

Figure 4.1 *a* illustrates a rod of rectangular cross section subjected to a distributed load $\mathbf{q} = -q_2\mathbf{e}_{20}$ applied to a part of the rod $\eta_1 \leq \eta \leq 1$. In addition, concentrated forces $\mathbf{P}^{(1)} = P_{10}^{(1)}\mathbf{e}_{10}$ and $\mathbf{P}^{(2)} = -P_2^{(2)}\mathbf{e}_{20}$ and a concentrated moment $\mathbf{T}^{(1)} = T_3\mathbf{e}_{30}$ are applied to the rod. The base vectors $\mathbf{e}_{20}$ and $\mathbf{e}_{30}$ of the attached coordinate system are directed along the principal axes of the cross section. In the natural configuration, we have $\mathbf{e}_{20}||\mathbf{i}_2$ and $\mathbf{e}_{30}||\mathbf{i}_3$. Hence, during deformation, the rod axis lies in the plane x_1Ox_2.

Suppose that displacements of the axial points are small. Let us consider an element of the rod and the forces and the moments acting on this element (see Fig. 4.1 *b*). The element is assumed to be in equilibrium. Since in the Cartesian basis the components of the applied loads remain unchanged, it is more convenient to represent the equilibrium equations in this basis. Besides, we assume that the plane cross sections of the rod orthogonal to its axial line remain plane and orthogonal to the axial line after deformation. Hence, shear strains may be neglected.

From summation of forces in the x_1 and x_2 directions, we obtain

$$\left(Q_1 + \frac{\mathrm{d}Q_1}{\mathrm{d}\eta}\,\mathrm{d}\eta\right)\cos(\vartheta_3 + \mathrm{d}\vartheta_3) - \left(Q_2 + \frac{\mathrm{d}Q_2}{\mathrm{d}\eta}\,\mathrm{d}\eta\right)\sin(\vartheta_3 + \mathrm{d}\vartheta_3)$$
$$- Q_1\cos\vartheta_3 + Q_2\sin\vartheta_3 + P_{x_1}^{(1)}\,\delta(\eta - 1)\,\mathrm{d}\eta = 0\,; \tag{4.1}$$
$$\left(Q_1 + \frac{\mathrm{d}Q_1}{\mathrm{d}\eta}\,\mathrm{d}\eta\right)\sin(\vartheta_3 + \mathrm{d}\vartheta_3) + \left(Q_2 + \frac{\mathrm{d}Q_2}{\mathrm{d}\eta}\,\mathrm{d}\eta\right)\cos(\vartheta_3 + \mathrm{d}\vartheta_3)$$
$$- Q_1\sin\vartheta_3 - Q_2\cos\vartheta_3 - q_{x_2}\,\mathrm{d}\eta - P_{x_2}^{(2)}\,\delta(\eta - \eta_2)\,\mathrm{d}\eta = 0\,. \tag{4.2}$$

The concentrated forces enter into (4.1) and (4.2) though these forces are not applied to the element under consideration. A concentrated force $\mathbf{q}_P$ can

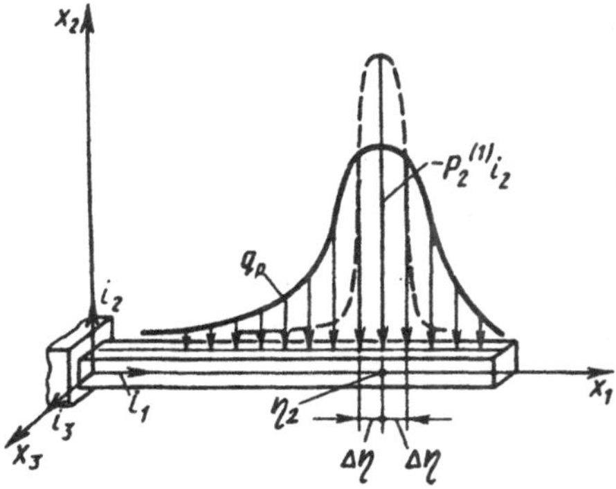

Fig. 4.2.

be regarded as a distributed force with a narrow zone of application (equal to $2\,\Delta\eta$) as illustrated in Fig. 4.2. Such a representation of concentrated forces is discussed in Appendix 4 in greater detail. For small ϑ_3, from (4.1)–(4.2) it follows that

$$\frac{\mathrm{d}Q_1}{\mathrm{d}\eta} - \frac{\mathrm{d}}{\mathrm{d}\eta}(Q_2\vartheta_3) + P_{x_1}^{(1)}\,\delta(\eta-1) = 0\,; \tag{4.3}$$

$$\frac{\mathrm{d}Q_2}{\mathrm{d}\eta} + \frac{\mathrm{d}}{\mathrm{d}\eta}(Q_1\vartheta_3) - q_{x_2} - P_{x_2}^{(2)}\,\delta(\eta-\eta_2) = 0\,, \tag{4.4}$$

where

$$\vartheta_3 = \frac{\mathrm{d}u_2}{\mathrm{d}\eta}\,.$$

These equations are nonlinear because they contain products of the sought functions $Q_1\vartheta_3$ and $Q_2\vartheta_3$. Recall that (4.3) and (4.4) are the equilibrium equations in the Cartesian basis while Q_1 and Q_2 are the components of the vector $\mathbf{Q}$ in the attached basis. To obtain linear equations, we are to adopt additional assumptions.

From (4.3) it follows that

$$Q_1 = Q_2\vartheta_3 - P_{x_1}^{(1)}H(\eta-1) + Q_{10}\,. \tag{4.5}$$

If the forces $\mathbf{P}^{(2)}$ and $\mathbf{q}$ result in $\mathbf{Q}_2$ such that $\mathbf{P}^{(1)}$ and $\mathbf{Q}_2$ are of the same order of magnitude, then the term $Q_2\vartheta_3$ can be neglected because of the smallness of ϑ_3; as a result, we get

$$Q_1 \approx -P_{x_1}^{(1)}H(\eta-1) + Q_{10}\,. \tag{4.6}$$

Since the axial force Q_1 equals zero to the right of the cross section of application of the force $P_{x_1}^{(1)}$, from (4.6) we obtain $Q_{10} = P_{x_1}^{(1)}$, i.e. $Q_1 = Q_{10} = \text{const}$ (provided the axial force $q_{x_1} = 0$).

Knowing Q_1, we can represent (4.4) as a linear equation

$$\frac{\mathrm{d}Q_2}{\mathrm{d}\eta} + P_{x_1}^{(1)} \frac{\mathrm{d}\vartheta_3}{\mathrm{d}\eta} - q_{x_2} - P_{x_2}^{(2)} \delta(\eta - \eta_2) = 0 . \tag{4.7}$$

The moments must satisfy the relation

$$M_3 + \frac{\mathrm{d}M_3}{\mathrm{d}\eta}\mathrm{d}\eta - M_3 + \left(Q_2 + \frac{\mathrm{d}Q_2}{\mathrm{d}\eta}\mathrm{d}\eta\right)\mathrm{d}\eta + T_3^{(1)}(\eta - \eta_3)\,\mathrm{d}\eta = 0 .$$

It gives

$$\frac{\mathrm{d}M_3}{\mathrm{d}\eta} + Q_2 + T_3^{(1)} \delta(\eta - \eta_3) = 0 . \tag{4.8}$$

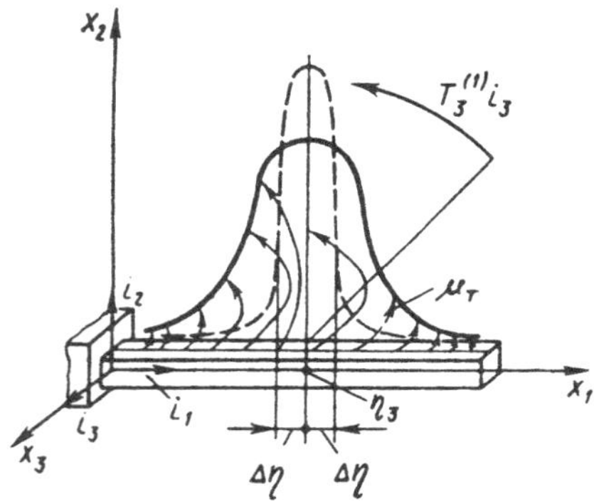

Fig. 4.3.

Equation (4.8) includes the concentrated moment applied at the cross section η_3. Arguing as in the case of the force $P_{x_2}^{(2)}$, we can treat the moment $T_3^{(1)}$ as the limit (at the cross section η_3) of a series of distributed moments μ_3 as indicated in Fig. 4.3.

Differentiating (4.8) and eliminating the term $\mathrm{d}Q_2/\mathrm{d}\eta$ by use of (4.4), we get

$$\frac{\mathrm{d}^2 M_3}{\mathrm{d}\eta^2} - \frac{\mathrm{d}}{\mathrm{d}\eta}\left(Q_1 \frac{\mathrm{d}u_2}{\mathrm{d}\eta}\right) + q_{x_2} + P_{x_2}^{(2)} \delta(\eta - \eta_2) + T_3^{(1)} \delta(\eta - \eta_3) = 0 . \tag{4.9}$$

Recall that

$$M_3 = A_{33}\,\Delta æ_3 ; \qquad \Delta æ_3 = \frac{\mathrm{d}\vartheta_3}{\mathrm{d}\eta} = \frac{\mathrm{d}^2 u_2}{\mathrm{d}\eta^2} .$$

Considering small displacements of the axial points, we obtain the following equilibrium equation:

$$\frac{\mathrm{d}^2}{\mathrm{d}\eta^2}\left(A_{33} \frac{\mathrm{d}^2 u_2}{\mathrm{d}\eta^2}\right) - P_{x_1}^{(1)} \frac{\mathrm{d}^2 u_2}{\mathrm{d}\eta} + q_{x_2} + P_{x_2}^{(2)} \delta(\eta - \eta_2) + T_3^{(1)} \delta(\eta - \eta_3) = 0 . \tag{4.10}$$

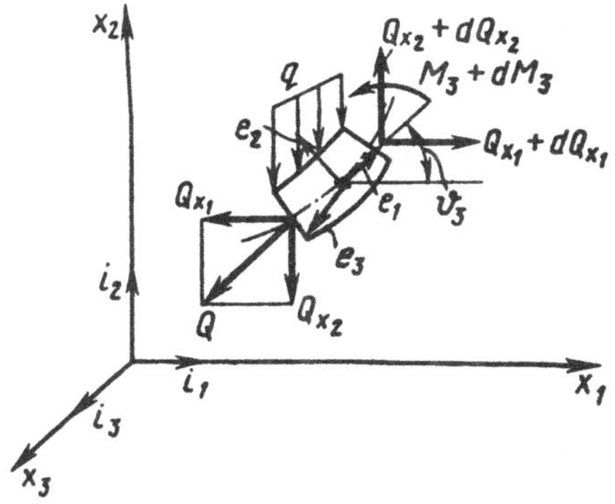

Fig. 4.4.

Another form of the equilibrium equations can be obtained if we refer the vector $\mathbf{Q}$ to the Cartesian basis, i.e. $\mathbf{Q} = Q_{x_1}\mathbf{i}_1 + Q_{x_2}\mathbf{i}_2$ (see Fig. 4.4). In this case, we have two linear equations

$$b_{x_1} = \frac{dQ_{x_1}}{d\eta} + P_{x_1}^{(1)}\,\delta(\eta-1) = 0\,, \qquad Q_{x_1} = Q_1\cos\vartheta_3 - Q_2\sin\vartheta_3\,; \tag{4.11}$$

$$b_{x_2} = \frac{dQ_{x_2}}{d\eta} - q_{x_2} - P_{x_2}^{(2)}\,\delta(\eta-\eta_3) = 0\,, \qquad Q_{x_2} = Q_1\sin\vartheta_3 + Q_2\cos\vartheta_3\,. \tag{4.12}$$

From (4.8) a nonlinear equation follows:

$$\frac{dM_3}{d\eta} + Q_{x_2} - Q_{x_1}\vartheta_3 + T_3^{(1)}\,\delta(\eta-\eta_3) = 0\,, \qquad Q_2 = Q_{x_2} - Q_{x_1}\vartheta_3\,. \tag{4.13}$$

Equation (4.11) is uncoupled from the others. This equation can be solved for Q_{x_1}. As a result, we get

$$Q_{x_1} = -P_{x_1}^{(1)}H(\eta-1) + Q_{x_1 0}\,, \qquad Q_{x_1 0} = P_{x_1}^{(1)}$$

or

$$Q_{x_1} = P_{x_1}^{(1)}\,.$$

We see that (4.13) becomes a linear equation. Further, eliminating Q_{x_2}, M_3, and ϑ_3 from (4.13), we obtain an equation, which is the same as (4.10). Solving for u_2 from (4.10) and determining its first, second, and third derivatives, we find ϑ_3, M_3, Q_1, and Q_2.

In Chap. 1, the following equilibrium equation in the form (1.31) was obtained:

$$\frac{d\mathbf{Q}}{d\eta} + \mathbf{q} + \sum_{i=1}^{n}\mathbf{P}^{(i)}\,\delta(\eta-\eta_i) = 0\,. \tag{4.14}$$

The equilibrium equations (4.11) and (4.12) for a rod element (see Fig. 4.4) are almost the same as (4.14) but written in terms of projections in the Cartesian frame.

There exists another form of representation of the equilibrium equations. In the attached basis, the basis $(\mathbf{e}_1, \mathbf{e}_2)$ in Fig. 4.4, the equilibrium equation reads

$$\frac{\mathrm{d}\mathbf{Q}}{\mathrm{d}\eta} + \text{æ} \times \mathbf{Q} + \mathbf{q} + \sum_{i=1}^{2} \mathbf{P}^{(i)}\, \delta(\eta - \eta_i) = 0\,. \tag{4.15}$$

In the basis $\{\mathbf{e}_i\}$, the force vectors are

$$\begin{aligned} \mathbf{q} &= -q_{x_2} \sin\vartheta_3 \mathbf{e}_1 - q_{x_2} \cos\vartheta_3 \mathbf{e}_2\,; \\ \mathbf{P}^{(2)} &= -P^{(2)}_{x_2} \sin\vartheta_3 \mathbf{e}_1 - P^{(2)}_{x_2} \cos\vartheta_3 \mathbf{e}_2\,; \\ \mathbf{P}^{(1)} &= P^{(1)}_{x_1} \cos\vartheta_3 \mathbf{e}_1 - P^{(1)}_{x_1} \sin\vartheta_3 \mathbf{e}_2\,. \end{aligned} \tag{4.16}$$

For small values of the angle ϑ_3, we have

$$\begin{aligned} \mathbf{q} &= -q_{x_2}\vartheta_3 \mathbf{e}_1 - q_{x_2} \mathbf{e}_2\,; \\ \mathbf{P}^{(2)} &= -P^{(2)}_{x_2}\vartheta_3 \mathbf{e}_1 - P^{(2)}_{x_2} \mathbf{e}_2\,; \\ \mathbf{P}^{(1)} &= P^{(1)}_{x_1} \mathbf{e}_1 - P^{(1)}_{x_1}\vartheta_3 \mathbf{e}_2\,. \end{aligned} \tag{4.17}$$

Since

$$\text{æ} = \frac{\mathrm{d}\vartheta_3}{\mathrm{d}\eta}\, \mathbf{i}_3\,,$$

we obtain two equations in the attached basis,

$$b_1 = \frac{\mathrm{d}Q_1}{\mathrm{d}\eta} - Q_2 \frac{\mathrm{d}\vartheta_3}{\mathrm{d}\eta} - q_{x_2}\vartheta_3 + P^{(1)}_{x_1}\, \delta(\eta - 1) - P^{(2)}_{x_2}\vartheta_3\, \delta(\eta - \eta_2) = 0\,; \tag{4.18}$$

$$b_2 = \frac{\mathrm{d}Q_2}{\mathrm{d}\eta} + Q_1 \frac{\mathrm{d}\vartheta_2}{\mathrm{d}\eta} - q_{x_2} - P^{(2)}_{x_2}\, \delta(\eta - \eta_2) - P^{(1)}_{x_1}\vartheta_3\, \delta(\eta - 1) = 0\,. \tag{4.19}$$

If the external forces are follower, i.e. $\mathbf{q} = -q_2\mathbf{e}_2$, $\mathbf{P}^{(1)} = P_1^{(1)}\mathbf{e}_1$, and $\mathbf{P}^{(2)} = -P_2^{(2)}\mathbf{e}_2$, then the equilibrium equations (4.18) and (4.19) are as follows:

$$\begin{aligned} &\frac{\mathrm{d}Q_1}{\mathrm{d}\eta} - Q_2 \frac{\mathrm{d}\vartheta_3}{\mathrm{d}\eta} + P_1^{(1)}\, \delta(\eta - \eta_1) = 0\,; \\ &\frac{\mathrm{d}Q_2}{\mathrm{d}\eta} + Q_1 \frac{\mathrm{d}\vartheta_3}{\mathrm{d}\eta} - q_2 - P_2^{(2)}\, \delta(\eta - \eta_2) = 0\,. \end{aligned}$$

To derive (4.18) and (4.19) on the basis of (4.11) and (4.12), one should use the formulas of transformation of coordinates. In vector notation, (4.11) and (4.12) are as follows:

$$\mathbf{b} = b_{x_1}\mathbf{i}_1 + b_{x_2}\mathbf{i}_2 = 0 .$$

For small values of the angle ϑ_3, we have

$$\mathbf{i}_1 = 1 \cdot \mathbf{e}_1 - \vartheta_3\mathbf{e}_2 ; \qquad \mathbf{i}_2 = \vartheta_3\mathbf{e}_1 + 1 \cdot \mathbf{e}_2 ;$$

on rearrangement, in view of (4.11) and (4.12), we get

$$\mathbf{b} = b_1\mathbf{e}_1 + b_2\mathbf{e}_2 = 0$$

or

$$b_1 = b_{x_1} + b_{x_2}\vartheta_3 = 0 ; \qquad b_2 = b_{x_2} + b_{x_1}\vartheta_3 = 0 .$$

For numerical analysis, it is more convenient to represent the equilibrium equations in the form of a system of first-order differential equations,

$$\frac{\mathrm{d}Q_2}{\mathrm{d}\eta} + P_{x_1}^{(1)}\frac{M_3}{A_{33}} - q_{x_2} - P_{x_2}^{(2)}\,\delta\left(\eta - \eta_2\right) = 0 ; \tag{4.20}$$

$$\frac{\mathrm{d}M_3}{\mathrm{d}\eta} + Q_2 + T_3^{(1)}\,\delta\left(\eta - \eta_3\right) = 0 ; \tag{4.21}$$

$$\frac{\mathrm{d}\vartheta_3}{\mathrm{d}\eta} - \frac{M_3}{A_{33}} = 0 ; \tag{4.22}$$

$$\frac{\mathrm{d}u_2}{\mathrm{d}\eta} - \vartheta_3 = 0 . \tag{4.23}$$

From here on, we will use either the fourth-order equilibrium equation (4.10) or the system of first-order equations (4.20)–(4.23) depending on the features of a particular problem.

In some problems of statics of rods, a rod axis may become a spatial curve, e.g. if a rod of twisted natural configuration (see Fig. 0.20) is subjected to a load which is perpendicular to the rod axis. The analysis of such problems may be performed by use of equilibrium equations of more complicated nature.

4.1.2 General Equilibrium Equations in the Case of Straight Rods

As mentioned above, it is often more convenient to derive equilibrium equations on the basis of the general nonlinear equilibrium equations.

These general equilibrium equations referred to various coordinate systems are presented in Chap. 1. These equations allow us to account for the peculiarities of the structure of a rod and the behavior of external loads.

If the principal axes of a cross section of a straight rod are directed along the Cartesian axes, we have

$$æ_0^{(1)} = 0 \qquad (i.e.\ æ_{10} = æ_{20} = æ_{30} = 0) ; \qquad \mathrm{L}^{(0)} = \mathrm{E} .$$

If a straight rod in its natural configuration is twisted (e.g. a drilling bit), then we have

$$
æ_{10} = \frac{\mathrm{d}\vartheta_{10}}{\mathrm{d}\eta}\,; \qquad æ_{20} = æ_{30} = 0\,;
$$

$$
\mathrm{L}^{(0)} = \begin{bmatrix} 1 & 0 & 0 \\ 0 & \cos\vartheta_{10} & \sin\vartheta_{10} \\ 0 & -\sin\vartheta_{10} & \cos\vartheta_{10} \end{bmatrix},
$$

where $\vartheta_{10}(\eta)$ is a known function. The case $\mathrm{d}\vartheta_{10}/\mathrm{d}\eta \neq 0$ is considered in Sect. 4.3.

Nonlinear Equilibrium Equations for a Rod whose Axial Line in a Deformed Configuration is a Spatial Curve In the case of large deflections of the rod axis, the equilibrium equations can be obtained on the basis of (1.57)–(1.61).

In the attached axes, we have

$$
\frac{\mathrm{d}\mathbf{Q}}{\mathrm{d}\eta} + \boldsymbol{æ} \times \mathbf{Q} + \mathbf{P} = 0\,; \tag{4.24}
$$

$$
\frac{\mathrm{d}\mathbf{M}}{\mathrm{d}\eta} + \boldsymbol{æ} \times \mathbf{M} + \mathbf{e}_1 \times \mathbf{Q} + \mathbf{T} = 0\,; \tag{4.25}
$$

$$
\mathbf{M} = A\boldsymbol{æ}\,, \qquad \boldsymbol{æ} = \left(æ_1 + \frac{\mathrm{d}\vartheta_{10}}{\mathrm{d}\eta}\right)\mathbf{e}_1 + æ_2\mathbf{e}_2 + æ_3\mathbf{e}_3\,; \tag{4.26}
$$

$$
\mathrm{L}_1\,\frac{\mathrm{d}\boldsymbol{\vartheta}}{\mathrm{d}\eta} - A^{-1}\mathbf{M} = 0\,; \tag{4.27}
$$

$$
\frac{\mathrm{d}\mathbf{u}}{\mathrm{d}\eta} + \boldsymbol{æ} \times \mathbf{u} + (l_{11} - 1)\,\mathbf{e}_1 + l_{21}\mathbf{e}_2 + l_{31}\mathbf{e}_3 = 0\,. \tag{4.28}
$$

This system could be also obtained from the general equilibrium equations (1.57)–(1.61) with $\boldsymbol{æ}_0^{(1)} = 0$.

For naturally twisted rods, we get

$$
M_1 = A_{11}\left(æ_1 + \frac{\mathrm{d}\vartheta_{10}}{\mathrm{d}\eta}\right); \qquad M_2 = A_{22}æ_2\,; \qquad M_3 = A_{33}æ_3\,, \tag{4.29}
$$

where the angle $\vartheta_{10}(\eta)$ describes the location of the principal axes with respect to the attached coordinate axes.

In the attached basis, the following equations, a special case of (1.64)–(1.68), can be obtained:

$$
\frac{\mathrm{d}Q_1}{\mathrm{d}\eta} + Q_3æ_2 - Q_2æ_3 + P_1 = 0\,;
$$

$$
\frac{\mathrm{d}Q_2}{\mathrm{d}\eta} + Q_1æ_3 - Q_3æ_1 + P_2 = 0\,;
$$

$$
\frac{\mathrm{d}Q_3}{\mathrm{d}\eta} + Q_2æ_1 - Q_1æ_2 + P_3 = 0\,; \tag{4.30}
$$

$$\frac{\mathrm{d}M_1}{\mathrm{d}\eta} + M_3 æ_2 - M_2 æ_3 + T_1 = 0\,;$$
$$\frac{\mathrm{d}M_2}{\mathrm{d}\eta} + M_1 æ_3 - M_3 æ_1 - Q_3 + T_2 = 0\,;$$
$$\frac{\mathrm{d}M_3}{\mathrm{d}\eta} + M_2 æ_1 - M_1 æ_2 + Q_2 + T_3 = 0\,; \tag{4.31}$$
$$M_i = A_{ii} æ_i\,, \qquad i = 1\,,2\,,3\,; \tag{4.32}$$
$$æ_1 = \frac{\mathrm{d}\vartheta_1}{\mathrm{d}\eta}\cos\vartheta_2\cos\vartheta_3 - \frac{\mathrm{d}\vartheta_3}{\mathrm{d}\eta}\sin\vartheta_2\,;$$
$$æ_2 = \frac{\mathrm{d}\vartheta_2}{\mathrm{d}\eta} - \frac{\mathrm{d}\vartheta_1}{\mathrm{d}\eta}\sin\vartheta_3\,;$$
$$æ_1 = \frac{\mathrm{d}\vartheta_3}{\mathrm{d}\eta}\cos\vartheta_2 + \frac{\mathrm{d}\vartheta_1}{\mathrm{d}\eta}\sin\vartheta_2\cos\vartheta_3\,; \tag{4.33}$$
$$\frac{\mathrm{d}u_1}{\mathrm{d}\eta} + u_3 æ_2 - u_2 æ_3 + l_{11} - 1 = 0\,;$$
$$\frac{\mathrm{d}u_2}{\mathrm{d}\eta} + u_1 æ_3 - u_3 æ_1 + l_{21} = 0\,;$$
$$\frac{\mathrm{d}u_3}{\mathrm{d}\eta} + u_2 æ_1 - u_1 æ_2 + l_{31} = 0\,, \tag{4.34}$$

where

$$P_j = q_j + \sum_{i=1}^{n} P_j^{(i)}\,\delta(\eta - \eta_i)\,;$$
$$T_j = \mu_j + \sum_{\nu=1}^{\rho} T_j^{(\nu)}\,\delta(\eta - \eta_\nu)\,, \qquad j = 1\,,2\,,3\,.$$

In the Cartesian frame, the sought equations, a special case of (1.84)–(1.88), are as follows:

$$\frac{\mathrm{d}\mathbf{Q}_x}{\mathrm{d}\eta} + \mathbf{P}_x = 0\,; \tag{4.35}$$
$$\frac{\mathrm{d}\mathbf{M}_x}{\mathrm{d}\eta} + (\mathrm{L}^{\mathrm{T}}\mathbf{i}_1) \times \mathbf{Q}_x + \mathbf{T}_x = 0\,; \tag{4.36}$$
$$\mathbf{M}_x = \mathrm{L}^{\mathrm{T}}\mathrm{A}æ\,; \tag{4.37}$$
$$\frac{\mathrm{d}\boldsymbol{\vartheta}}{\mathrm{d}\eta} - \mathrm{L}_1^{-1}\mathrm{A}^{-1}\mathrm{L}\mathbf{M}_x = 0\,; \tag{4.38}$$
$$\frac{\mathrm{d}\mathbf{u}_x}{\mathrm{d}\eta} - (\mathrm{L}^{\mathrm{T}} - \mathrm{E})\,\mathbf{i}_1 = 0\,; \tag{4.39}$$

here vectors denoted by the subscript 'x' are referred to the Cartesian basis, i.e. $\mathbf{a}_x = a_{x_1}\mathbf{i}_1 + a_{x_2}\mathbf{i}_2 + a_{x_3}\mathbf{i}_3$.

Nonlinear Equilibrium Equations for a Rod whose Axial Line in a Deformed Configuration is a Plane Curve In the attached basis, these equations are a special case of (4.30)–(4.34):

$$\frac{dQ_1}{d\eta} - Q_2 æ_3 + P_1 = 0\,;$$

$$\frac{dQ_2}{d\eta} + Q_1 æ_3 + P_2 = 0\,; \tag{4.40}$$

$$\frac{dM_3}{d\eta} + Q_2 + T_3 = 0\,; \tag{4.41}$$

$$M_3 = A_{33} æ_3\,; \tag{4.42}$$

$$\frac{d\vartheta_3}{d\eta} - \frac{M_3}{A_{33}} = 0\,; \tag{4.43}$$

$$\frac{du_1}{d\eta} - u_2 æ_3 + l_{11} - 1 = 0\,, \qquad l_{11} = \cos\vartheta_3\,;$$

$$\frac{du_2}{d\eta} + u_1 æ_3 + l_{21} = 0\,, \qquad l_{21} = -\sin\vartheta_3\,. \tag{4.44}$$

In the Cartesian basis, we have

$$\frac{dQ_{x_1}}{d\eta} + P_{x_1} = 0\,;$$

$$\frac{dQ_{x2}}{d\eta} + P_{x_2} = 0\,; \tag{4.45}$$

$$\frac{dM_{x3}}{d\eta} + \cos\vartheta_3 Q_{x2} - \sin\vartheta_3 Q_{x_1} + T_{x_3} = 0\,; \tag{4.46}$$

$$\frac{d\vartheta_3}{d\eta} - \frac{M_{x3}}{A_{33}} = 0\,; \tag{4.47}$$

$$\frac{du_{x_1}}{d\eta} - \cos\vartheta_3 + 1 = 0\,;$$

$$\frac{du_{x2}}{d\eta} - \sin\vartheta_3 = 0\,. \tag{4.48}$$

For a rod of constant cross section, we must put $A_{33} = 1$.

4.2 Equilibrium Equations for Small Displacements and Angles of Rotation

4.2.1 Vector Equations

If the displacements u_i and the angles of rotation ϑ_i are small, then the curvature increments $\Delta æ_j$ can also be considered small. Since for straight rods $æ_0 = 0$, we have $æ = \Delta æ$. In view of (4.24)–(4.28), we arrive at

$$\frac{d\mathbf{Q}}{d\eta} + \Delta æ \times \mathbf{Q} + \mathbf{P}_0 + \Delta\mathbf{P} = 0\,; \tag{4.49}$$

$$\frac{d\mathbf{M}}{d\eta} + \Delta æ \times \mathbf{M} + \mathbf{e}_1 \times \mathbf{Q} + \mathbf{T}_0 + \Delta\mathbf{T} = 0\,; \tag{4.50}$$

$$\mathbf{M} = \mathrm{A}\,\Delta æ\,; \tag{4.51}$$

$$\frac{d\boldsymbol{\vartheta}}{d\eta} - \mathrm{A}^{-1}\mathbf{M} = 0\,; \tag{4.52}$$

$$\frac{d\mathbf{u}}{d\eta} + \mathrm{A}_1\boldsymbol{\vartheta} = 0\,, \tag{4.53}$$

where

$$\Delta\mathbf{P} = \Delta\mathbf{q} + \sum_{i=1}^{n} \Delta\mathbf{P}^{(i)}\,\delta(\eta - \eta_i)\,;$$

$$\Delta\mathbf{T} = \Delta\boldsymbol{\mu} + \sum_{\nu=1}^{\rho} \Delta\mathbf{T}^{(\nu)}\,\delta(\eta - \eta_\nu)\,. \tag{4.54}$$

In general, the increments $\Delta\mathbf{q}$, $\Delta\mathbf{P}^{(i)}$, $\Delta\boldsymbol{\mu}$, and $\Delta\mathbf{T}^{(\nu)}$ are functions of $\mathbf{u}$, $\mathbf{u}'$, and $\boldsymbol{\vartheta}$. Eliminating $\mathbf{u}'$ from (3.20)–(3.23), we get

$$\Delta\mathbf{q} = \mathbf{C}_1\boldsymbol{\vartheta} + \mathbf{C}_2\mathbf{u}\,; \qquad \Delta\mathbf{P}^{(i)} = \mathbf{B}_1^{(1)}\boldsymbol{\vartheta} + \mathbf{B}_2^{(i)}\mathbf{u}\,;$$

$$\Delta\boldsymbol{\mu} = \mathbf{C}_3\boldsymbol{\vartheta} + \mathbf{C}_4\mathbf{u}\,; \qquad \Delta\mathbf{T}^{(\nu)} = \mathbf{B}_3^{(\nu)}\boldsymbol{\vartheta} + \mathbf{B}_4^{(\nu)}\mathbf{u}\,. \tag{4.55}$$

Substituting (4.55) into (4.49) and (4.50), we obtain the system of the five vector equations (4.49)–(4.53) in the five unknown vectors $\mathbf{Q}$, $\mathbf{M}$, $\Delta æ$, $\boldsymbol{\vartheta}$, and $\mathbf{u}$.

In the general case, the components of the vectors $\mathbf{Q}$ and $\mathbf{M}$ (unlike the vectors $\mathbf{u}$, $\Delta æ$, and $\boldsymbol{\vartheta}$) cannot be considered small. Therefore, even for small $\mathbf{u}$, $\boldsymbol{\vartheta}$, and $\Delta æ$, (4.49)–(4.53) are nonlinear differential equations. The equations of the zeroth approximation were obtained in Sect. 1.4 under the assumption $æ \approx æ_0^{(1)}$ or $\Delta æ = 0$. Therefore, all the terms containing $\Delta æ$ are assumed to be equal to zero: $\Delta æ \times \mathbf{Q} = \Delta æ \times \mathbf{M} \approx 0$. The latter assumption is reasonable for curvilinear rods because $æ_{j0} \gg \Delta æ_j$. For straight rods, we have $æ_{j0} = 0$, hence we need additional argumentation to omit the terms $\Delta æ \times \mathbf{Q}$ and $\Delta æ \times \mathbf{M}$ in the vector equations without argumentation.

4.2.2 Equilibrium Equations in the Attached Coordinate System

First, let us estimate the nonlinear terms in the vector equations (4.49) and (4.50). To do this, consider (4.49)–(4.53) in terms of projections on the attached axes. We have

$$\frac{dQ_1}{d\eta} + \Delta æ_2 Q_3 - \Delta æ_3 Q_2 + P_{10} + \Delta P_1 = 0\,;$$
$$\frac{dQ_2}{d\eta} + \Delta æ_3 Q_1 - \Delta æ_1 Q_3 + P_{20} + \Delta P_2 = 0\,;$$
$$\frac{dQ_3}{d\eta} + \Delta æ_1 Q_2 - \Delta æ_2 Q_1 + P_{30} + \Delta P_3 = 0\,; \tag{4.56}$$
$$\frac{dM_1}{d\eta} + \Delta æ_2 M_3 - \Delta æ_3 M_2 + T_{10} + \Delta T_1 = 0\,;$$
$$\frac{dM_2}{d\eta} + \Delta æ_3 M_1 - \Delta æ_1 M_3 - Q_3 + T_{20} + \Delta T_2 = 0\,;$$
$$\frac{dM_3}{d\eta} + \Delta æ_1 M_2 - \Delta æ_2 M_1 + Q_2 + T_{30} + \Delta T_3 = 0\,; \tag{4.57}$$
$$\frac{d\vartheta_1}{d\eta} - \frac{M_1}{A_{11}} = 0\,;$$
$$\frac{d\vartheta_2}{d\eta} - \frac{M_2}{A_{22}} = 0\,;$$
$$\frac{d\vartheta_3}{d\eta} - \frac{M_3}{A_{33}} = 0\,; \tag{4.58}$$
$$\frac{du_1}{d\eta} = 0\,;$$
$$\frac{du_2}{d\eta} - \vartheta_3 = 0\,;$$
$$\frac{du_3}{d\eta} + \vartheta_2 = 0\,; \tag{4.59}$$
$$M_i = A_{ii}\,\Delta æ_i\,, \tag{4.60}$$

where

$$P_{j0} = q_{j0} + \sum_{i=1}^{n} P_{j0}^{(i)}\,\delta(\eta - \eta_i)\,;$$
$$\Delta P_j = \Delta q_j + \sum_{i=1}^{n} \Delta P_j^{(i)}\,\delta(\eta - \eta_i)\,; \tag{4.61}$$
$$T_{j0} = \mu_{j0} + \sum_{\nu=1}^{\rho} T_{j0}^{(\nu)}\,\delta(\eta - \eta_\nu)\,;$$
$$\Delta T_j = \Delta \mu_j + \sum_{\nu=1}^{\rho} \Delta T_j^{(\nu)}\,\delta(\eta - \eta_\nu)\,. \tag{4.62}$$

Equations (4.56)–(4.60) need be treated together with (1.67) and (1.68); we must put $æ_{j0} = 0$ in (1.67) and replace $\boldsymbol{æ}_j$ by $\Delta\boldsymbol{æ}_j$ in (1.68). From (4.56) it follows that

$$Q_1 = -\int_0^\eta (\Delta æ_3 Q_2 - \Delta æ_2 Q_3 + \Delta P)\, \mathrm{d}\zeta - \int_0^\eta P_{10}\, \mathrm{d}\zeta + Q_{100}$$

or

$$Q_1 = Q_{10} + \Delta Q_1 ,$$

where ΔQ_1 is a small quantity. If the force projections on the tangent to the rod axis equal zero ($P_{10} = 0$), then the axial force Q_1 can be considered small and the terms $\Delta æ_3 Q_1$ and $\Delta æ_2 Q_1$ in (4.56) may be neglected. If $P_{10} \neq 0$, then the nonlinear terms can be rearranged as follows:

$$\Delta æ_i (Q_{10} + \Delta Q_1) = \Delta æ_i Q_{10} , \qquad i = 2, 3 .$$

Since the quantities Q_2 and Q_3 cannot be considered small, the terms $\Delta æ_2 Q_3$ and $\Delta æ_3 Q_2$ in (4.56) are of the first order of smallness. Therefore, these terms cannot be neglected unless they are small as compared to the other terms in (4.56) (e.g. P_{10}). The term ΔP_1 depends linearly on u_j and ϑ_j, which are small quantities. Hence, if $P_{10} = 0$, then all the terms in (4.56) are of the same order of magnitude. Thus, the terms $\Delta æ_2 Q_3$ and $\Delta æ_3 Q_2$ cannot be neglected.

Suppose that the applied forces are follower ($\Delta P_1 = 0$) and, moreover, $P_{10} = 0$; then the first equation of the system (4.56) is uncoupled from the other equations. Solving for $\Delta æ_2$, $\Delta æ_3$, Q_2, and Q_3, we arrive at

$$Q_1 = \Delta Q_1 = \Delta Q_{100} + \int_0^\eta (\Delta æ_3 Q_2 - \Delta æ_2 Q_3)\, \mathrm{d}\zeta . \tag{4.63}$$

From this it follows that if $P_{10} = \Delta P_1 = 0$, then Q_1 is of the first order of smallness. Therefore, the products in (4.56) containing ΔQ_1 can be neglected.

Suppose that the applied forces are not follower ($\Delta P_1 \neq 0$). Assume that ΔP_1 depends linearly on u_j and ϑ_j. In this case, the first equation of the system (4.56) is coupled with the other equations. In this case, the first equation of (4.56) is nonlinear because the nonlinear terms $\Delta æ_2 Q_3$ and $\Delta æ_3 Q_2$ are of the same order of magnitude as the term ΔP_1 and, therefore, they cannot be neglected. We could drop out the terms $\Delta æ_2 Q_3$ and $\Delta æ_3 Q_2$ provided the following inequality holds:

$$\Delta æ_2 Q_3 \approx \Delta æ_3 Q_2 \ll \Delta P_1 (u_j , \vartheta_j) . \tag{4.64}$$

To check the validity of (4.64), one should solve the system (4.56)–(4.60) in which only linear terms are retained. If the inequality (4.64) is satisfied, then the nonlinear terms in (4.56) may be neglected; otherwise, the system of equations (4.56)–(4.60) is nonlinear even for small deflections.

Consider those nonlinear terms in the second and third equations of the system (4.56) which contain $\Delta æ_1$. From (4.57), in view of (4.62), we get

$$M_1 = A_{11}\,\Delta æ_1 = -\int_0^\eta [(A_{33}-A_{22})\,\Delta æ_2\,\Delta æ_3 - \Delta T_1]\,\mathrm{d}\zeta - \int_0^\eta T_{10}\,\mathrm{d}\zeta + M_{100}$$

or

$$M_1 = A_{11}\,\Delta æ_1 = M_{10} + \Delta M_1^{(1)} + \Delta M_1^{(2)}\,, \tag{4.65}$$

where

$$M_{10} = -\int_0^\eta T_{10}\,\mathrm{d}\zeta\,; \qquad M_1^{(1)} = -\int_0^\eta \Delta T_1\,\mathrm{d}\zeta\,;$$
$$\Delta M_1^{(2)} = -\int_0^\eta (A_{33}-A_{22})\,\Delta æ_2\,\Delta æ_3\,\mathrm{d}\zeta\,.$$

The quantities $\Delta M_1^{(1)}$ and $\Delta M_1^{(2)}$ are of the first and the second order of smallness, respectively. Therefore, we can neglect the term $\Delta M_1^{(2)}$ when deriving the linear equations (4.57). For straight rods we can put

$$M_1 = A_{11}\,\Delta æ_1 = M_{10} + \Delta M_1^{(1)}\,.$$

Consequently, the terms containing $\Delta æ_1$ in the second and third equations of the system (4.56 can be rewritten as follows:

$$\Delta æ_1 Q_j = \left(\frac{M_{10}}{A_{11}} + \frac{\Delta M_1}{A_{11}}\right) Q_j \approx \frac{M_{10}}{A_{11}}\,Q_j\,.$$

Suppose that the projection T_{10} of the moment $\mathbf{T}_0$ on the tangent to the rod axis equals zero and the moment $\mathbf{T}_0$ is a follower one ($M_{10} = 0$ and $\Delta T_1 = 0$). Using (4.65), we get

$$\Delta æ_1 = \frac{1}{A_{11}}\,\Delta M_1 = -\frac{1}{A_{11}}\int_0^\eta (A_{33}-A_{22})\,\Delta æ_3\,\Delta æ_3\,\mathrm{d}\zeta\,. \tag{4.66}$$

From this relation it is seen that $\Delta æ_1$ is of the second order of smallness. Therefore, the terms $\Delta æ_1 Q_2$ and $\Delta æ_1 Q_3$ in (4.56) can be neglected. Particularly, for a rod of circular cross section ($A_{22} = A_{33}$), we have $\Delta æ_1 \equiv 0$.

If the applied moment is not a follower one ($\Delta T_1 \neq 0$), then, even under the condition $T_{10} = 0$, the nonlinear terms $\Delta æ_1 Q_j$ cannot be neglected. Therefore, the second and third equations of the system (4.56) remain nonlinear.

Using (4.60), we can rewrite the first two equations of the system (4.57) as follows:

$$\frac{\mathrm{d}M_2}{\mathrm{d}\eta} + (A_{11}-A_{33})\,\Delta æ_1\,\Delta æ_3 - Q_3 + T_{20} + \Delta T_2 = 0\,;$$
$$\frac{\mathrm{d}M_3}{\mathrm{d}\eta} + (A_{22}-A_{11})\,\Delta æ_1\,\Delta æ_3 + Q_2 + T_{30} + \Delta T_3 = 0\,. \tag{4.67}$$

If $T_{10} = \Delta T_1 = 0$, then the term $\Delta æ_1$ defined by (4.66) is of the second order of smallness. Therefore, the terms containing $\Delta æ_1\, \Delta æ_2$ and $\Delta æ_1\, \Delta æ_3$ in (4.67) are of the third order of smallness and thus can be neglected.

In the case of a follower moment with a nonzero projection on the tangent to the rod axis ($T_{10} \neq 0$), we have $\Delta æ_1 = M_{10}/A_{11}$; consequently, the terms containing the products $\Delta æ_1\, \Delta æ_2$ and $\Delta æ_1\, \Delta æ_3$ need be retained.

If the applied moments are not follower but still $T_{10} = 0$, then we have $\Delta T_1 \neq 0$ and $M_1 \neq 0$. Nevertheless, the moment M_1 can be considered small, i.e. $M_1 = \Delta M_1 = A_{11}\, \Delta æ_1$. Thus, we get (see (4.57))

$$\Delta æ_3 M_1 - \Delta æ_1 M_3 = \frac{A_{11} - A_{33}}{A_{11}}\, \Delta æ_3\, \Delta M_1 \approx 0\,;$$

$$\Delta æ_3 M_1 - \Delta æ_1 M_3 = \frac{A_{11} - A_{33}}{A_{11}}\, \Delta æ_3\, \Delta M_1 \approx 0\,.$$

From these relations we obtain

$$\frac{\mathrm{d}\Delta M_1}{\mathrm{d}\eta} + \frac{A_{33} - A_{12}}{A_{22}A_{33}}\, M_2 M_3 + \Delta T_1 = 0\,;$$

$$\frac{\mathrm{d}M_2}{\mathrm{d}\eta} - Q_3 + T_{20} + \Delta T_2 = 0\,;$$

$$\frac{\mathrm{d}M_3}{\mathrm{d}\eta} + Q_2 + T_{30} + \Delta T_3 = 0\,. \tag{4.68}$$

Finally, let $T_{10} \neq 0$ and $\Delta T_1 \neq 0$. Then we have

$$\Delta æ_1 = \frac{1}{A_{11}}\,(M_{10} + \Delta M_1)\,,$$

and

$$\frac{1}{A_{11}}\,(A_{11} - A_{33})\,\Delta æ_3 (M_{10} + \Delta M_1) \approx \frac{1}{A_{11}}\,(A_{11} - A_{33})\, M_{10}\, \Delta æ_3\,; \tag{4.69}$$

$$\frac{1}{A_{11}}\,(A_{22} - A_{11})\,\Delta æ_2 (M_{10} + \Delta M_1) \approx \frac{1}{A_{11}}\,(A_{22} - A_{11})\, M_{10}\, \Delta æ_2\,. \tag{4.70}$$

From the particular cases of expressions for $\Delta æ_1$ considered above, it is seen that the terms containing the products $\Delta æ_1 M_j$ in (4.57) can be neglected; the only exception is the case $T_{10} \neq 0$. Therefore, if $T_{10} = 0$, then (4.57) are linear equations.

For $T_{10} \neq 0$ ($M_{10} = T_{10}$), we have $\Delta æ_1 = M_{10}/A_{11}$. Hence, $\Delta æ_1$ is known and the terms $\Delta æ_1 M_j$ are linear in the unknown projections M_1 and M_2.

Let us consider two cases:

(1) $Q_{10} = M_{10} = 0$;

(2) $Q_{10} \neq 0$ and $M_{10} \neq 0$.

In the first case, the equilibrium equations (4.56)–(4.60) are as follows:

$$\frac{dQ_1}{d\eta} + \Delta æ_2 Q_3 - \Delta æ_3 Q_2 + \Delta P_1 = 0\,;$$
$$\frac{dQ_2}{d\eta} + \Delta æ_3 Q_1 - \Delta æ_1 Q_3 + P_{20} + \Delta P_2 = 0\,;$$
$$\frac{dQ_3}{d\eta} + \Delta æ_1 Q_2 - \Delta æ_2 Q_1 + P_{30} + \Delta P_3 = 0\,; \tag{4.71}$$
$$\frac{dM_1}{d\eta} + \Delta æ_2 M_3 - \Delta æ_3 M_2 + \Delta T_1 = 0\,;$$
$$\frac{dM_2}{d\eta} + \Delta æ_3 M_1 - \Delta æ_1 M_3 - Q_3 + T_{20} + \Delta T_0 = 0\,;$$
$$\frac{dM_3}{d\eta} + \Delta æ_1 M_2 - \Delta æ_2 M_1 + Q_2 + T_{30} + \Delta T_3 = 0\,. \tag{4.72}$$

The presence of nonlinear terms in these equations substantially complicates the analysis. If $T_{10} = M_{10} = 0$, the terms $\Delta æ_j M_i$ can be neglected. (We will omit these terms in the derivation of the equations of the zeroth approximation.) To derive the equations of the first and higher-order approximation, we should retain the nonlinear terms in (4.72).

Equations (4.71) and (4.72) with (4.51)–(4.53) can be solved by use of approximate methods; to do this, the equations of the zeroth and the first approximation for curvilinear rods should be used (see Sect. 1.4).

Recall that the equilibrium equations of the zeroth approximation do not contain increments of the vector æ, therefore, we can put $\Delta æ_i = 0$ in (4.71) and (4.72). As a result, we obtain the equations of the zeroth approximation

$$\frac{dQ_1^{(0)}}{d\eta} + \Delta P_1^{(0)} = 0\,;$$
$$\frac{dQ_2^{(0)}}{d\eta} + P_{20} + \Delta P_2^{(0)} = 0\,;$$
$$\frac{dQ_3^{(0)}}{d\eta} + P_{30} + \Delta P_3^{(0)} = 0\,; \tag{4.73}$$
$$\frac{dM_1^{(0)}}{d\eta} + \Delta T_1^{(0)} = 0\,;$$
$$\frac{dM_2^{(0)}}{d\eta} - Q_3^{(0)} + T_{20} + \Delta T_2^{(0)} = 0\,;$$
$$\frac{dM_3^{(0)}}{d\eta} + Q_2^{(0)} + T_{30} + \Delta T_3^{(0)} = 0\,, \tag{4.74}$$

where

$$\Delta P_j^{(0)} = \Delta q_j^{(0)} + \sum_{i=1}^{n} \Delta P_j^{(i)(0)}\, \delta(\eta - \eta_i)\,;$$

$$\Delta T_j^{(0)} = \Delta\mu_j^{(0)} + \sum_{\nu=1}^{\rho} \Delta T_j^{(\nu)(0)}\,\delta(\eta - \eta_\nu)\,.$$

For straight rods, the increments $\Delta q_j^{(0)}$, $\Delta\mu_j^{(0)}$, $\Delta P_j^{(i)(0)}$, and $\Delta T_j^{(\nu)(0)}$ depend, in general, on u_2, u_3, ϑ_2, and ϑ_3 and do not depend on ϑ_1 and u_1 ($u_1 = 0$). Consequently, these increments can be written as follows:

$$\begin{aligned}
\Delta q_j^{(0)} &= c_{j2}^{(2)} u_2^{(0)} + c_{j3}^{(1)} u_3^{(0)} + c_{j2}^{(2)} \vartheta_2^{(0)} + c_{j3}^{(1)} \vartheta_3^{(0)}\,;\\
\Delta P_j^{(i)(0)} &= (b_{j2}^{(2)})_i\, u_2^{(0)} + (b_{j3}^{(2)})_i\, u_3^{(0)} + (b_{j2}^{(1)})_i\, \vartheta_2^{(0)} + (b_{j3}^{(1)})_i\, \vartheta_3^{(0)}\,;\\
\Delta \mu_j^{(0)} &= c_{j2}^{(4)} u_2^{(0)} + c_{j3}^{(4)} u_3^{(0)} + c_{j2}^{(3)} \vartheta_2^{(0)} + c_{j3}^{(3)} \vartheta_3^{(0)}\,;\\
\Delta T_j^{(i)(0)} &= (b_{j2}^{(4)})_\nu\, u_2^{(0)} + (b_{j3}^{(4)})_\nu\, u_3^{(0)} + (b_{j2}^{(3)})_\nu\, \vartheta_2^{(0)} + (b_{j3}^{(3)})_\nu\, \vartheta_3^{(0)}\,.
\end{aligned} \tag{4.75}$$

Equations (4.73) and (4.74) are to be solved together with the equations of the zeroth approximation (4.51)–(4.53). The latter equations are of the form

$$\mathbf{M}^{(0)} = \mathrm{A}\,\Delta\text{æ}^{(0)}\,; \tag{4.76}$$

$$\frac{\mathrm{d}\boldsymbol{\vartheta}^{(0)}}{\mathrm{d}\eta} - \mathrm{A}^{-1}\mathbf{M}^{(0)} = 0\,; \tag{4.77}$$

$$\frac{\mathrm{d}\mathbf{u}^{(0)}}{\mathrm{d}\eta} + \mathrm{A}_1\boldsymbol{\vartheta}^{(0)} = 0\,. \tag{4.78}$$

The system of equations (4.73), (4.74) and (4.76)–(4.78) can be represented as two uncoupled systems,

$$\begin{aligned}
&\frac{\mathrm{d}Q_1^{(0)}}{\mathrm{d}\eta} + \Delta P_1^{(0)} = 0\,;\\
&\frac{\mathrm{d}M_1^{(0)}}{\mathrm{d}\eta} + \Delta T_1^{(0)} = 0\,;\\
&M_1^{(0)} = A_{11}\,\Delta\text{æ}_1^{(0)}\,;\\
&\frac{\mathrm{d}\vartheta_1^{(0)}}{\mathrm{d}\eta} - \frac{M_1^{(0)}}{A_{11}} = 0\,;\\
&\frac{\mathrm{d}u_1^{(0)}}{\mathrm{d}\eta} = 0
\end{aligned} \tag{4.79}$$

and

$$\begin{aligned}
&\frac{\mathrm{d}Q_2^{(0)}}{\mathrm{d}\eta} + P_{20} + \Delta P_2^{(0)} = 0\,;\\
&\frac{\mathrm{d}Q_3^{(0)}}{\mathrm{d}\eta} + P_{30} + \Delta P_3^{(0)} = 0\,;
\end{aligned}$$

$$\begin{aligned}
&\frac{\mathrm{d}M_2^{(0)}}{\mathrm{d}\eta} - Q_3^{(0)} + T_{20} + \Delta T_2^{(0)} = 0\,;\\
&\frac{\mathrm{d}M_3^{(0)}}{\mathrm{d}\eta} + Q_2^{(0)} + T_{30} + \Delta T_3^{(0)} = 0\,,\\
&\frac{\mathrm{d}\vartheta_2^{(0)}}{\mathrm{d}\eta} - \frac{M_2^{(0)}}{A_{22}} = 0\,;\\
&\frac{\mathrm{d}\vartheta_3^{(0)}}{\mathrm{d}\eta} - \frac{M_3^{(0)}}{A_{33}} = 0\,;\\
&\frac{\mathrm{d}u_2^{(0)}}{\mathrm{d}\eta} - \vartheta_3^{(0)} = 0\,;\\
&\frac{\mathrm{d}u_3^{(0)}}{\mathrm{d}\eta} + \vartheta_2^{(0)} = 0\,;\\
&M_2^{(0)} = A_{22}\,\Delta æ_2^{(0)}\,;\\
&M_3^{(0)} = A_{33}\,\Delta æ_3^{(0)}\,.
\end{aligned} \tag{4.80}$$

Taking into account (4.75), we can rewrite (4.80) in vector form

$$\begin{aligned}
&\frac{\mathrm{d}\mathbf{Q}^{(0)}}{\mathrm{d}\eta} + \mathbf{P}_0 + \Delta\mathbf{P}^{(0)} = 0\,;\\
&\frac{\mathrm{d}\mathbf{M}^{(0)}}{\mathrm{d}\eta} + \mathrm{A}_1\mathbf{Q}^{(0)} + \mathbf{T}_0 + \Delta\mathbf{T}^{(0)} = 0\,;\\
&\frac{\mathrm{d}\boldsymbol{\vartheta}^{(0)}}{\mathrm{d}\eta} - \mathrm{A}^{-1}\mathbf{M}^{(0)} = 0\,;\\
&\frac{\mathrm{d}\mathbf{u}^{(0)}}{\mathrm{d}\eta} + \mathrm{A}_1\,\boldsymbol{\vartheta}^{(0)} = 0\,,
\end{aligned} \tag{4.81}$$

where

$$\mathrm{A}_1 = \begin{bmatrix} 0 & -1 \\ 1 & 0 \end{bmatrix}; \qquad \mathrm{A} = \begin{bmatrix} A_{22} & 0 \\ 0 & A_{33} \end{bmatrix}; \qquad \mathbf{Q}^{(0)} = \begin{bmatrix} Q_2^{(0)} \\ Q_3^{(0)} \end{bmatrix};$$

$$\mathbf{M}^{(0)} = \begin{bmatrix} M_2^{(0)} \\ M_3^{(0)} \end{bmatrix}; \qquad \boldsymbol{\vartheta}^{(0)} = \begin{bmatrix} \vartheta_2^{(0)} \\ \vartheta_3^{(0)} \end{bmatrix}; \qquad \mathbf{u}^{(0)} = \begin{bmatrix} u_2^{(0)} \\ u_3^{(0)} \end{bmatrix};$$

$$\mathbf{P}_0 = \begin{bmatrix} P_{20} \\ P_{30} \end{bmatrix}; \qquad \mathbf{T}_0 = \begin{bmatrix} T_{20} \\ T_{30} \end{bmatrix}.$$

The variables $u_2^{(0)}$, $u_3^{(0)}$, $\vartheta_2^{(0)}$, and $\vartheta_3^{(0)}$ in the formulas for $\Delta P_1^{(0)}$ and $\Delta T_1^{(0)}$ (see (4.75)) can be found from (4.81). Then, in view of (4.79), we can determine $Q_1^{(0)}$, $M_1^{(0)}$, $\Delta æ_1$, and $\vartheta_1^{(0)}$ ($u_1^{(0)} \equiv 0$).

In view of (4.75), (4.81) can be written as follows:

$$\mathbf{Y}^{(0)\prime} + \mathrm{A}^{(0)}\mathbf{Y}^{(0)} = \mathbf{f}^{(0)}\,; \tag{4.82}$$

here

$$\mathbf{Y}^{(0)} = \begin{bmatrix} \mathbf{Q}^{(0)} \\ \mathbf{M}^{(0)} \\ \boldsymbol{\vartheta}^{(0)} \\ \mathbf{u}^{(0)} \end{bmatrix}; \qquad \mathrm{A}^{(0)} = \begin{bmatrix} 0 & 0 & C_1 & C_2 \\ \mathrm{A}_1 & 0 & C_3 & C_4 \\ 0 & -\mathrm{A}^{-1} & 0 & 0 \\ 0 & 0 & \mathrm{A}_1 & 0 \end{bmatrix};$$

$$\mathbf{f}^{(0)} = \begin{bmatrix} -\mathbf{P}^{(0)} - \mathrm{B}_1\boldsymbol{\vartheta}^{(0)} - \mathrm{B}_2\mathbf{u}^{(0)} \\ -\mathbf{T}^{(0)} - \mathrm{B}_3\boldsymbol{\vartheta}^{(0)} - \mathrm{B}_4\mathbf{u}^{(0)} \\ \mathbf{0} \\ \mathbf{0} \end{bmatrix}.$$

The elements of the matrices B_1, B_2, B_3, and B_4 contain the factors $\delta(\eta-\eta_i)$ and $\delta(\eta-\eta_\nu)$. For this reason, these terms are rearranged to the right-hand side of (4.82).

If the zeroth-approximation solution is not satisfactory or the increments $\Delta\mathbf{P}^{(0)}$ and $\Delta\mathbf{T}^{(0)}$ depend on the angle ϑ_1, then (4.73), (4.74) and (4.76)–(4.78) cannot be decomposed into two systems of equations. In this general case, by use of vector notation, we obtain

$$\begin{aligned}
&\frac{\mathrm{d}\mathbf{Q}^{(0)}}{\mathrm{d}\eta} + \mathrm{A}_{10}\mathbf{P}_0 + \Delta\mathbf{P}^{(0)} = 0\,;\\
&\frac{\mathrm{d}\mathbf{M}^{(0)}}{\mathrm{d}\eta} + \mathrm{A}_1\mathbf{Q}^{(0)} + \mathrm{A}_{10}\mathbf{T}^{(0)} + \Delta\mathbf{T}^{(0)} = 0\,;\\
&\mathbf{M}^{(0)} = \mathrm{A}\,\Delta\boldsymbol{æ}^{(0)}\,;\\
&\frac{\mathrm{d}\boldsymbol{\vartheta}^{(0)}}{\mathrm{d}\eta} - \mathrm{A}^{-1}\mathbf{M}^{(0)} = 0\,;\\
&\frac{\mathrm{d}\mathbf{u}^{(0)}}{\mathrm{d}\eta} + \mathrm{A}_1\,\Delta\boldsymbol{\vartheta}^{(0)} = 0
\end{aligned} \tag{4.83}$$

or, in view of the formulas for $\Delta\mathbf{P}^{(0)}$ and $\Delta\mathbf{T}^{(0)}$, this system can be written as

$$\mathbf{Y}^{(0)\prime} + \mathrm{A}^{(0)}\mathbf{Y}^{(0)} = \mathbf{f}^{(0)}\,,$$

where

$$\mathbf{Y}^{(0)} = \begin{bmatrix} \mathbf{Q}^{(0)} \\ \mathbf{M}^{(0)} \\ \boldsymbol{\vartheta}^{(0)} \\ \mathbf{u}^{(0)} \end{bmatrix}; \qquad \mathbf{f}^{(0)} = \begin{bmatrix} -\mathrm{A}_{10}\mathbf{P}_0 - \mathrm{B}_1\boldsymbol{\vartheta}^{(0)} - \mathrm{B}_2\mathbf{u}^{(0)} \\ -\mathrm{A}_{10}\mathbf{T}_0 - \mathrm{B}_3\boldsymbol{\vartheta}^{(0)} - \mathrm{B}_4\mathbf{u}^{(0)} \\ \mathbf{0} \\ \mathbf{0} \end{bmatrix};$$

$$\mathrm{A} = \begin{bmatrix} 0 & 0 & C_1 & C_2 \\ \mathrm{A}_1 & 0 & C_3 & C_4 \\ 0 & -\mathrm{A}^{-1} & 0 & 0 \\ 0 & 0 & \mathrm{A}_1 & 0 \end{bmatrix}; \qquad \mathrm{A}_{10} = \begin{bmatrix} 0 & 0 & 0 \\ 0 & 1 & 0 \\ 0 & 0 & 1 \end{bmatrix}. \tag{4.84}$$

Unlike (4.82), the vectors $\mathbf{Q}^{(0)}$, $\mathbf{M}^{(0)}$, $\boldsymbol{\vartheta}^{(0)}$, and $\mathbf{u}^{(0)}$ in (4.84) have three components.

Let us derive the equations of the first approximation on the basis of (4.71) and (4.72). For this purpose, we put $Q_i = Q_i^{(0)} + Q_i^{(1)}$, $\Delta æ_i = \Delta æ_i^{(0)} + \Delta æ_i^{(1)}$, $\Delta P_i = \Delta P_i^{(0)} + \Delta P_i^{(1)}$, $M_i = M_i^{(0)} + M_i^{(1)}$, and $\Delta T_i = \Delta T_i^{(0)} + \Delta T_i^{(1)}$. Omitting terms of the second order of smallness, we arrive at

$$\begin{aligned}
&\frac{\mathrm{d}Q_1^{(1)}}{\mathrm{d}\eta} + \Delta æ_2^{(0)} Q_3^{(1)} - \Delta æ_3^{(0)} Q_2^{(1)} + \Delta æ_2^{(1)} Q_3^{(0)} \\
&\qquad - \Delta æ_3^{(1)} Q_2^{(0)} + \Delta P_1^{(1)} = \Delta æ_3^{(0)} Q_2^{(0)} - \Delta æ_2^{(0)} Q_3^{(0)}\,; \\
&\frac{\mathrm{d}Q_2^{(1)}}{\mathrm{d}\eta} + \Delta æ_3^{(0)} Q_1^{(1)} - \Delta æ_1^{(0)} Q_3^{(1)} + \Delta æ_3^{(1)} Q_1^{(0)} \\
&\qquad - \Delta æ_1^{(1)} Q_3^{(0)} + \Delta P_2^{(1)} = -\Delta æ_3^{(0)} Q_1^{(0)} + \Delta æ_1^{(0)} Q_3^{(0)}\,; \\
&\frac{\mathrm{d}Q_3^{(1)}}{\mathrm{d}\eta} + \Delta æ_1^{(0)} Q_2^{(1)} - \Delta æ_2^{(0)} Q_1^{(1)} + \Delta æ_1^{(1)} Q_2^{(0)} \\
&\qquad - \Delta æ_2^{(1)} Q_1^{(0)} + \Delta P_3^{(1)} = -\Delta æ_2^{(0)} Q_1^{(0)} - \Delta æ_1^{(0)} Q_2^{(0)}\,;
\end{aligned} \tag{4.85}$$

$$\begin{aligned}
&\frac{\mathrm{d}M_1^{(1)}}{\mathrm{d}\eta} + \Delta æ_2^{(0)} M_3^{(1)} - \Delta æ_3^{(0)} M_2^{(1)} + \Delta æ_2^{(1)} M_3^{(0)} \\
&\qquad - \Delta æ_3^{(1)} M_2^{(0)} + \Delta T_1 = \Delta æ_3^{(0)} M_2^{(0)} - \Delta æ_2^{(0)} M_3^{(0)}\,; \\
&\frac{\mathrm{d}M_2^{(1)}}{\mathrm{d}\eta} + \Delta æ_3^{(0)} M_1^{(1)} - \Delta æ_1^{(0)} M_3^{(1)} + \Delta æ_3^{(1)} M_1^{(0)} \\
&\qquad - \Delta æ_1^{(1)} M_3^{(0)} - Q_3^{(1)} + \Delta T_2^{(1)} = -\Delta æ_3^{(0)} M_1^{(0)} + \Delta æ_1^{(0)} M_3^{(0)}\,; \\
&\frac{\mathrm{d}M_3^{(1)}}{\mathrm{d}\eta} + \Delta æ_1^{(0)} M_2^{(1)} - \Delta æ_2^{(0)} M_1^{(1)} + \Delta æ_1^{(1)} M_2^{(0)} \\
&\qquad - \Delta æ_2^{(1)} M_1^{(1)} + Q_2^{(2)} + \Delta T_3 = -\Delta æ_2^{(0)} M_1^{(0)} - \Delta æ_1^{(0)} M_2^{(0)}\,.
\end{aligned} \tag{4.86}$$

The vector equations (4.85) and (4.86) together with (4.51)–(4.53) represent the complete system of equilibrium equations of the first approximation

$$\begin{aligned}
&\frac{\mathrm{d}\mathbf{Q}^{(1)}}{\mathrm{d}\eta} + \mathrm{A}_æ^{(0)} \mathbf{Q}^{(1)} + \mathrm{A}_Q^{(0)} \mathrm{A}^{-1} \mathbf{M}^{(1)}\, \Delta \mathbf{P}^{(1)} \\
&= -\mathrm{A}^{-1} \mathbf{M}^{(0)} \times \mathbf{Q}^{(0)}\,; \\
&\frac{\mathrm{d}\mathbf{M}^{(1)}}{\mathrm{d}\eta} + \mathrm{A}_æ^{(0)} \mathbf{M}^{(1)} + \mathrm{A}_M^{(0)} \mathrm{A}^{-1} \mathbf{M}^{(1)} + \mathrm{A}_1 \mathbf{Q}^{(1)} + \Delta \mathbf{T}^{(1)} \\
&= \mathrm{A}^{-1} \mathbf{M}^{(0)} \times \mathbf{M}^{(0)}\,; \\
&\mathbf{M}^{(1)} = \mathrm{A}\, \Delta æ^{(1)}\,; \\
&\frac{\mathrm{d}\boldsymbol{\vartheta}^{(1)}}{\mathrm{d}\eta} - \mathrm{A}^{-1} \mathbf{M}^{(1)} = 0\,;
\end{aligned}$$

$$\frac{\mathrm{d}\mathbf{u}^{(1)}}{\mathrm{d}\eta} + \mathrm{A}_1\,\Delta\boldsymbol{\vartheta}^{(1)} = 0\,, \tag{4.87}$$

where

$$\mathrm{A}_{æ}^{(0)} = \begin{bmatrix} 0 & -\Delta æ_3^{(0)} & \Delta æ_2^{(0)} \\ \Delta æ_3^{(0)} & 0 & -\Delta æ_1^{(0)} \\ -\Delta æ_2^{(0)} & \Delta æ_1^{(0)} & 0 \end{bmatrix};$$

$$\mathrm{A}_{Q}^{(0)} = \begin{bmatrix} 0 & Q_3^{(0)} & -Q_2^{(0)} \\ -Q_3^{(0)} & 0 & Q_1^{(0)} \\ Q_2^{(0)} & -Q_1^{(0)} & 0 \end{bmatrix};$$

$$\mathrm{A}_{M}^{(0)} = \begin{bmatrix} 0 & M_3^{(0)} & -M_2^{(0)} \\ -M_3^{(0)} & 0 & M_1^{(0)} \\ M_2^{(0)} & -M_1^{(0)} & 0 \end{bmatrix}; \qquad \mathrm{A}_1 = \begin{bmatrix} 0 & 0 & 0 \\ 0 & 0 & -1 \\ 0 & 1 & 0 \end{bmatrix}.$$

Consider now the equilibrium equations (4.56)-(4.61) with $Q_{10} \neq 0$ and $M_{10} \neq 0$. From (4.56) and (4.59) it follows that

$$Q_1 = Q_{10} + \Delta Q_1\,; \qquad M_1 = M_{10} + \Delta M_1\,,$$

where

$$\begin{aligned} Q_{10} &= -\int_0^{\eta} P_{10}\,\mathrm{d}\zeta + Q_{100}\,; \\ \Delta Q_1 &= -\int_0^{\eta} (\Delta æ_2 Q_3 - \Delta æ_3 Q_2)\,\mathrm{d}\zeta + \Delta Q_{10}\,; \\ M_{10} &= -\int_0^{\eta} T_{10}\,\mathrm{d}\zeta + M_{100}\,; \\ \Delta M_1 &= -\int_0^{\eta} (\Delta æ_2 M_3 - \Delta æ_3 M_2)\,\mathrm{d}\zeta + \Delta M_{10}\,. \end{aligned}$$

Since ΔQ_1 and ΔM_1 could be considered small, we replace Q_1 and M_1 in (4.57), (4.58) and (4.60), (4.61) by Q_{10} and M_{10} defined by the relations

$$\begin{aligned} &\frac{\mathrm{d}Q_{10}}{\mathrm{d}\eta} + P_{10} = 0\,; \\ &\frac{\mathrm{d}M_{10}}{\mathrm{d}\eta} + T_{10} = 0\,, \end{aligned} \tag{4.88}$$

where

$$\begin{aligned} P_{10} &= q_{10} + \sum_{i=1}^{n} P_0^{(i)}\,\delta(\eta - \eta_i)\,; \\ T_{10} &= \mu_{10} + \sum_{\nu=1}^{\rho} T_{10}^{(\nu)}\,\delta(\eta - \eta_\nu)\,. \end{aligned} \tag{4.89}$$

Integrating (4.88), we get

$$Q_{10} = -\int_0^{\eta} q_{10}\,\mathrm{d}\zeta - \sum_{i=1}^{n} P_{10}^{(i)} H(\eta - \eta_i) + Q_{100}\,; \tag{4.90}$$

$$M_{10} = -\int_0^{\eta} \mu_{10}\,\mathrm{d}\zeta - \sum_{\nu=1}^{\rho} T_{10}^{(\nu)} H(\eta - \eta_\nu) + M_{100}\,. \tag{4.91}$$

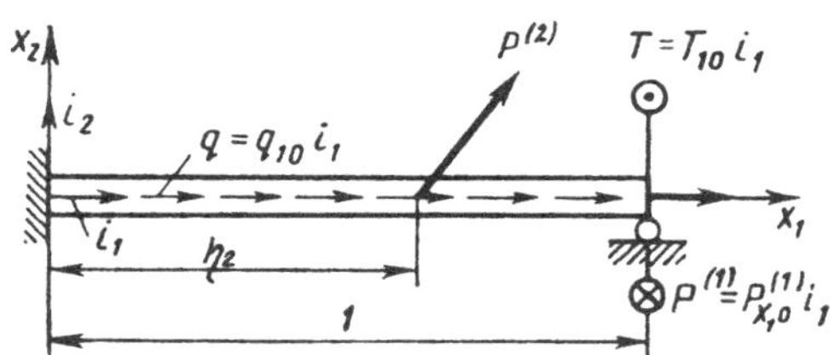

Fig. 4.5.

The arbitrary constants Q_{100} and M_{100} are defined by the end fixity conditions. For example, in the case shown in Fig. 4.5, we have

$$Q_{10}(0) = Q_{100} = \int_0^1 q_{10}\,\mathrm{d}\zeta + P_{x_2 0}^{(2)} + P_{x_1 0}^{(1)}\,;$$
$$M_{10}(0) = M_{100} = T_{x_1 0}\,.$$

For $M_{10} \neq 0$, we have $æ_{10} = M_{10}/A_{11}$. By use of (4.57), (4.58), (4.60), and (4.61), we obtain the equations of the zeroth approximation

$$\frac{\mathrm{d}Q_1^{(0)}}{\mathrm{d}\eta} + P_{10} + \Delta P_1^{(0)} = 0\,;$$
$$\frac{\mathrm{d}Q_2^{(0)}}{\mathrm{d}\eta} + \Delta æ_3^{(0)} Q_{10} - \frac{M_{10}}{A_{11}} Q_3^{(0)} + P_{20} + \Delta P_2^{(0)} = 0\,;$$
$$\frac{\mathrm{d}Q_3^{(0)}}{\mathrm{d}\eta} + \frac{M_{10}}{A_{11}} Q_2^{(0)} - \Delta æ_2^{(0)} Q_{10} + P_{30} + \Delta P_3^{(0)} = 0\,; \tag{4.92}$$
$$\frac{\mathrm{d}M_1^{(0)}}{\mathrm{d}\eta} + T_{10} + \Delta T_1^{(0)} = 0\,;$$
$$\frac{\mathrm{d}M_2^{(0)}}{\mathrm{d}\eta} + \Delta æ_3^{(0)} M_{10} - \frac{M_{10}}{A_{11}} M_3^{(0)} - Q_3^{(0)} + T_{20} + \Delta T_2^{(0)} = 0\,;$$
$$\frac{\mathrm{d}M_3^{(0)}}{\mathrm{d}\eta} + \frac{M_{10}}{A_{11}} M_2^{(0)} - \Delta æ_2^{(0)} M_{10} + Q_2^{(0)} + T_{30} + \Delta T_3^{(0)} = 0\,. \tag{4.93}$$

If $\Delta P_i^{(0)}$ and $\Delta T_j^{(0)}$ do not depend on $\vartheta_1^{(0)}$ ($u_1^{(0)} \equiv 0$), then, using (4.92), (4.93) and (4.76)–(4.78), we obtain two systems of equations similar to (4.79) and (4.80)

$$\frac{dQ_1^{(0)}}{d\eta} + P_{10} + \Delta P_1^{(0)} = 0\,;$$
$$\frac{dM_1^{(0)}}{d\eta} + T_{10} + \Delta T_1^{(0)} = 0\,;$$
$$M_1^{(0)} = A_{11}\,\Delta æ_1^{(0)}\,, \qquad \Delta æ_1^{(0)} = æ_{10}\,;$$
$$\frac{d\vartheta_1^{(0)}}{d\eta} - \frac{M_1^{(0)}}{A_{11}} = 0\,;$$
$$\frac{du_1^{(0)}}{d\eta} = 0\,; \tag{4.94}$$
$$\frac{dQ_2^{(0)}}{d\eta} + \Delta æ_3^{(0)} Q_{10} - \frac{M_{10}}{A_{11}}\, Q_3^{(0)} + P_{20} + \Delta P_2^{(0)} = 0\,;$$
$$\frac{dQ_3^{(0)}}{d\eta} - \Delta æ_2^{(0)} Q_{10} + \frac{M_{10}}{A_{11}}\, Q_2^{(0)} + P_{30} + \Delta P_3^{(0)} = 0\,;$$
$$\frac{dM_2^{(0)}}{d\eta} + \Delta æ_3^{(0)} M_{10} - \frac{M_{10}}{A_{11}}\, M_3^{(0)} - Q_3^{(0)} + T_{20} + \Delta T_2^{(0)} = 0\,;$$
$$\frac{dM_3^{(0)}}{d\eta} - \Delta æ_2^{(0)} M_{10} + \frac{M_{10}}{A_{11}}\, M_2^{(0)} + Q_2^{(0)} + T_{30} + \Delta T_3^{(0)} = 0\,;$$
$$\frac{d\vartheta_2^{(0)}}{d\eta} - æ_{10}\vartheta_3^{(0)} - \frac{M_2^{(0)}}{A_{22}} = 0\,;$$
$$\frac{d\vartheta_3^{(0)}}{d\eta} + æ_{10}\vartheta_2^{(0)} - \frac{M_3^{(0)}}{A_{33}} = 0\,;$$
$$\frac{du_2^{(0)}}{d\eta} - æ_{10}u_3^{(0)} - \vartheta_3^{(0)} = 0\,;$$
$$\frac{du_3^{(0)}}{d\eta} + æ_{10}u_2^{(0)} + \vartheta_2^{(0)} = 0\,, \tag{4.95}$$

where

$$M_2^{(0)} = A_{22}\,\Delta æ_2^{(0)}\,; \qquad \Delta æ_2^{(0)} = \frac{d\vartheta_2^{(0)}}{d\eta}\,;$$
$$M_3^{(0)} = A_{33}\,\Delta æ_3^{(0)}\,; \qquad \Delta æ_3^{(0)} = \frac{d\vartheta_3^{(0)}}{d\eta}\,.$$

The expressions for $\Delta P_j^{(0)}$ and $\Delta T_j^{(0)}$ are identical to those in (4.73) and (4.74). By use of (4.75), these expressions can be rewritten as follows (for brevity, the relations only for one concentrated force and one concentrated moment are written out):

$$\Delta \mathbf{P}^{(0)} = \mathrm{C}_1\boldsymbol{\vartheta}^{(0)} + \mathrm{C}_2\mathbf{u}^{(0)} + \mathrm{B}_1\boldsymbol{\vartheta}^{(0)} + \mathrm{B}_2\mathbf{u}^{(0)}\,;$$
$$\Delta \mathbf{T}^{(0)} = \mathrm{C}_3\boldsymbol{\vartheta}^{(0)} + \mathrm{C}_4\mathbf{u}^{(0)} + \mathrm{B}_3\boldsymbol{\vartheta}^{(0)} + \mathrm{B}_4\mathbf{u}^{(0)}\,; \tag{4.96}$$

here the vector $\boldsymbol{\vartheta}^{(0)}$ has two components $\vartheta_2^{(0)}$ and $\vartheta_3^{(0)}$, the vector $\mathbf{u}^{(0)}$ has two components $u_2^{(0)}$ and $u_2^{(0)}$; C_i and B_i are 2×2 matrices.

Taking into account (4.96), we can represent (4.95) in vector form (the same vector notation was used in (4.81))

$$\begin{aligned}
&\frac{\mathrm{d}\mathbf{Q}^{(0)}}{\mathrm{d}\eta} - \mathrm{A}_1 Q_{10}\, \Delta æ^{(0)} + \mathrm{A}_1 \frac{M_{10}}{A_{11}} \mathbf{Q}^{(0)} \\
&\qquad + \mathbf{P}^{(0)} + \mathrm{C}_1 \boldsymbol{\vartheta}^{(0)} + \mathrm{C}_2 \mathbf{u}^{(0)} = -\mathrm{B}_1 \boldsymbol{\vartheta}^{(0)} - \mathrm{B}_2 \mathbf{u}^{(0)}\,; \\
&\frac{\mathrm{d}\mathbf{M}^{(0)}}{\mathrm{d}\eta} + \mathrm{A}_1 \mathbf{Q}^{(0)} - \mathrm{A}_1 M_{10}\, \Delta æ^{(0)} + \mathrm{A}_1 \frac{M_{10}}{A_{11}} \mathbf{M}^{(0)} \\
&\qquad + \mathbf{T}^{(0)} + \mathrm{C}_3 \boldsymbol{\vartheta}^{(0)} + \mathrm{C}_4 \mathbf{u}^{(0)} = -\mathrm{B}_3 \boldsymbol{\vartheta}^{(0)} - \mathrm{B}_4 \mathbf{u}^{(0)}\,; \\
&\frac{\mathrm{d}\boldsymbol{\vartheta}^{(0)}}{\mathrm{d}\eta} - \mathrm{A}^{-1} \mathbf{M}^{(0)} = 0\,; \\
&\frac{\mathrm{d}\mathbf{u}^{(0)}}{\mathrm{d}\eta} + \mathrm{A}_1 \boldsymbol{\vartheta}^{(0)} = 0\,; \\
&\mathbf{M}^{(0)} = \mathrm{A}\, \Delta æ^{(0)}\,.
\end{aligned} \tag{4.97}$$

The elements of the matrices B_1, B_2, B_3, and B_4 contain δ-function; the terms containing these matrices are rearranged to the right-hand side of the equations. Equations (4.97) are a system of inhomogeneous linear differential equations. Methods for analysis of equations containing δ-function are presented in Chap. 2.

4.2.3 Equilibrium Equations in the Cartesian Coordinate System

Using the expressions for $\Delta \mathbf{P}_x$ and $\Delta \mathbf{T}_x$, we can rewrite (1.84) and (1.85) as follows:

$$\frac{\mathrm{d}\mathbf{Q}_x}{\mathrm{d}\eta} + \mathbf{P}_x + \Delta \mathbf{P}_x = 0\,; \tag{4.98}$$

$$\begin{aligned}
&\frac{\mathrm{d}\mathbf{M}_x}{\mathrm{d}\eta} - (\vartheta_3 Q_{x_3} + \vartheta_2 Q_{x_2})\, \mathbf{i}_1 + (\vartheta_2 Q_{x_1} - Q_{x_3})\, \mathbf{i}_2 \\
&\qquad + (Q_{x_2} - \vartheta_3 Q_{x_1})\, \mathbf{i}_3 + \mathbf{T}_{x_0} + \Delta \mathbf{T}_x = 0\,;
\end{aligned} \tag{4.99}$$

$$\Delta æ = \mathrm{A}^{-1} \mathrm{L} \mathbf{M}_x\,; \tag{4.100}$$

$$\frac{\mathrm{d}\boldsymbol{\vartheta}}{\mathrm{d}\eta} - \mathrm{A}^{-1} \mathrm{L} \mathbf{M}_x = 0\,; \tag{4.101}$$

$$\frac{\mathrm{d}\mathbf{u}_x}{\mathrm{d}\eta} + \boldsymbol{\vartheta}_3 \mathbf{i}_2 - \boldsymbol{\vartheta}_2 \mathbf{i}_3 = 0\,; \tag{4.102}$$

here

$$\mathrm{L}=\begin{bmatrix} 1 & -\vartheta_3 & \vartheta_2 \\ \vartheta_3 & 1 & -\vartheta_1 \\ -\vartheta_2 & \vartheta_1 & 1 \end{bmatrix}.$$

Although (4.100) and (4.101) are referred to the attached basis, they contain the components of the vector **M** in the Cartesian basis. As noted in Sect. 1.3, the use of the two bases $\{\mathbf{i}\}_j$ and $\{\mathbf{e}_j\}$ may be very advantageous for numerical analysis and algebraic manipulation. For dead forces, we have

$$\Delta\mathbf{q}_x=\Delta\boldsymbol{\mu}_x=\Delta\mathbf{P}_x^{(i)}=\Delta\mathbf{T}_x^{(\nu)}=0\,.$$

By use of (4.98)–(4.102), we obtain the following equations in terms of projections on the Cartesian axes:

$$\begin{aligned}
&\frac{\mathrm{d}Q_{x1}}{\mathrm{d}\eta}+P_{x_10}+\Delta P_{x_1}=0\,;\\
&\frac{\mathrm{d}Q_{x2}}{\mathrm{d}\eta}+P_{x_20}+\Delta P_{x_2}=0\,;\\
&\frac{\mathrm{d}Q_{x3}}{\mathrm{d}\eta}+P_{x_30}+\Delta P_{x_3}=0\,;
\end{aligned}\tag{4.103}$$

$$\begin{aligned}
&\frac{\mathrm{d}M_{x_1}}{\mathrm{d}\eta}-\vartheta_3 Q_{x_3}-\vartheta_2 Q_{x_2}+T_{x_10}+\Delta T_{x_1}=0\,;\\
&\frac{\mathrm{d}M_{x_2}}{\mathrm{d}\eta}+\vartheta_2 Q_{x_1}-Q_{x_3}+\mu_{x20}+\Delta\mu_{x_2}+\sum_{\nu=1}^{\rho}(T_{x20}^{(\nu)}+\Delta T_{x_2}^{(\nu)})\,\delta(\eta-\eta_\nu)=0\,;\\
&\frac{\mathrm{d}M_{x3}}{\mathrm{d}\eta}+Q_{x_2}-\vartheta_3 Q_{x_1}+\mu_{x_30}+\Delta\mu_{x_3}+\sum_{\nu=1}^{\rho}(T_{x_30}^{(\nu)}+\Delta T_{x_3}^{(\nu)})\,\delta(\eta-\eta_\nu)=0\,;
\end{aligned}\tag{4.104}$$

$$\begin{aligned}
&\Delta æ_1=\frac{1}{A_{11}}(M_{x_1}-\vartheta_3 M_{x_2}+\vartheta_2 M_{x_3})\,;\\
&\Delta æ_2=\frac{1}{A_{22}}(M_{x_2}+\vartheta_3 M_{x_1}-\vartheta_1 M_{x_3})\,;\\
&\Delta æ_3=\frac{1}{A_{33}}(M_{x_3}-\vartheta_2 M_{x_1}+\vartheta_1 M_{x_2})\,;
\end{aligned}\tag{4.105}$$

$$\begin{aligned}
&\frac{\mathrm{d}\vartheta_1}{\mathrm{d}\eta}-\frac{1}{A_{11}}(M_{x_1}-\vartheta_3 M_{x_2}+\vartheta_2 M_{x_3})=0\,;\\
&\frac{\mathrm{d}\vartheta_2}{\mathrm{d}\eta}-\frac{1}{A_{22}}(M_{x_2}+\vartheta_3 M_{x_1}-\vartheta_1 M_{x_3})=0\,;\\
&\frac{\mathrm{d}\vartheta_3}{\mathrm{d}\eta}-\frac{1}{A_{33}}(M_{x_3}-\vartheta_2 M_{x_1}+\vartheta_1 M_{x_2})=0\,;
\end{aligned}\tag{4.106}$$

$$\frac{\mathrm{d}u_{x1}}{\mathrm{d}\eta}=0\,;$$

$$\frac{\mathrm{d}u_{x2}}{\mathrm{d}\eta}-\vartheta_3=0\,;$$

$$\frac{du_{x3}}{d\eta} + \vartheta_2 = 0 . \tag{4.107}$$

If the increments Δq_{x_j} and $\Delta P_{x_j}^{(i)}$ are linear in u_{x_k} and ϑ_k, then (4.103) are linear. Equations (4.104)–(4.107) are nonlinear because they contain the products $\vartheta_j Q_{x_i}$ and $\vartheta_j M_{x_i}$, where Q_{x_j} and M_{x_j} cannot be considered small. Therefore, we need to estimate the nonlinear terms. From (4.103) it follows that

$$Q_{x_j} = Q_{x_j 0} + \Delta Q_{x_j} ,$$

where ΔQ_{x_j} is small, hence, the relations

$$\begin{aligned} \vartheta_3 Q_{x_3} &= \vartheta_3 Q_{x_3 0} ; \qquad & \vartheta_2 Q_{x_2} &= \vartheta_2 Q_{x_2 0} ; \\ \vartheta_1 Q_{x_1} &= \vartheta_1 Q_{x_1 0} ; \qquad & \vartheta_3 Q_{x_1} &= \vartheta_3 Q_{x_1 0} \end{aligned}$$

are to be substituted into (4.104).

Suppose that the products $\vartheta_i M_{x_j}$ are small as compared to M_{x_k}; then we can represent (4.105) and (4.106) as linear equations. For $\mu_{x_1 0} \neq 0$ and $T_{x_1 0}^{(\nu)} \neq 0$, the terms $\vartheta_2 Q_{x_2}$, $\vartheta_2 Q_{x_3}$, $\Delta\mu_{x_1}$, and $\Delta T_{x_1}^{(\nu)}$ in the first equation of (4.104) are small as compared to $\mu_{x_1 0}$ and $T_{x_1 0}$. Thus these increments can be omitted and the equation becomes uncoupled from the other equations of the system (4.103)–(4.107). If the increments Δq_{x_1} and $\Delta P_{x_1}^{(i)}$ are dropped out, the first equation of the system (4.103) is uncoupled from the others. Hence, for any set of boundary conditions, we get $Q_{x_1} = Q_{x_1 0}$. Finally, using (4.103)–(4.107), we obtain two uncoupled systems of equations:

$$\begin{aligned}
&\frac{dQ_{x_1 0}}{d\eta} + P_{x_1 0} = 0 ; \\
&\frac{dM_{x_1 0}}{d\eta} + T_{x_1 0} = 0 , \qquad M_{x_1 0} = \Delta_{11}\, \Delta æ_1 ; \\
&\frac{d\vartheta_1}{d\eta} - \frac{M_{x_1 0}}{A_{11}} = 0 ; \\
&\frac{du_1}{d\eta} = 0 ; \qquad\qquad (4.108) \\
&\frac{dQ_{x_2}}{d\eta} + q_{x_2 0} + \Delta q_{x_2} + \sum_{i=1}^{n} (P_{x_2 0}^{(i)} + \Delta P_{x_2}^{(i)})\, \delta(\eta - \eta_i) = 0 ; \\
&\frac{dQ_{x_3}}{d\eta} + q_{x_3 0} + \Delta q_{x_3} + \sum_{i=1}^{n} (P_{x_3 0}^{(i)} + \Delta P_{x_3}^{(i)})\, \delta(\eta - \eta_i) = 0 ; \\
&\frac{dM_{x_2}}{d\eta} + \vartheta_2 Q_{x_1 0} - Q_{x_3} + \mu_{x_2 0} + \Delta\mu_{x_2} + \sum_{\nu=1}^{\rho} (T_{x_2 0}^{(\nu)} + \Delta T_{x_2}^{(\nu)})\, \delta(\eta - \eta_\nu) = 0 ; \\
&\frac{dM_{x_3}}{d\eta} + Q_{x_2} - \vartheta_3 Q_{x_1 0} + \mu_{x_3 0} + \Delta\mu_{x_3} + \sum_{\nu=1}^{\rho} (T_{x_3 0}^{(\nu)} + \Delta T_{x_3}^{(\nu)})\, \delta(\eta - \eta_\nu) = 0 ;
\end{aligned}$$

$$
\begin{aligned}
&\frac{\mathrm{d}\vartheta_2}{\mathrm{d}\eta} - \frac{M_{x2}}{A_{22}} = 0\,; \\
&\frac{\mathrm{d}\vartheta_3}{\mathrm{d}\eta} - \frac{M_{x3}}{A_{33}} = 0\,; \\
&\frac{\mathrm{d}u_{x2}}{\mathrm{d}\eta} + \vartheta_3 = 0\,; \\
&\frac{\mathrm{d}u_{x3}}{\mathrm{d}\eta} - \vartheta_2 = 0\,.
\end{aligned} \tag{4.109}
$$

Equations (4.109) enable us to determine the components of the unknown vectors $\mathbf{Q}$, $\mathbf{M}$, $\Delta\text{æ}$, $\boldsymbol{\vartheta}$, and $\mathbf{u}$ in the Cartesian coordinate system.

4.3 Naturally Twisted Straight Rods

4.3.1 Nonlinear Vector Equations of Equilibrium

For naturally twisted rods, we have

$$
\begin{aligned}
&\text{æ}_0^{(1)} = \text{æ}_{10}\mathbf{e}_1 = \frac{\mathrm{d}\vartheta_{10}}{\mathrm{d}\eta}\,\mathbf{e}_1\,; \\
&\text{æ}_0^{(1)} \times \mathbf{Q} = \text{æ}_{10}(\mathbf{e}_1 \times \mathbf{Q})\,; \\
&\text{æ}_0^{(1)} \times \mathbf{M} = \text{æ}_{10}(\mathbf{e}_1 \times \mathbf{M})\,.
\end{aligned}
$$

In view of the general equilibrium equations (1.57)–(1.61), we get

$$
\frac{\mathrm{d}\mathbf{Q}}{\mathrm{d}\eta} + \text{æ} \times \mathbf{Q} + \mathbf{P} = 0\,; \tag{4.110}
$$

$$
\frac{\mathrm{d}\mathbf{M}}{\mathrm{d}\eta} + \text{æ} \times \mathbf{M} + \mathbf{e}_1 \times \mathbf{Q} + \mathbf{T} = 0\,; \tag{4.111}
$$

$$
\mathbf{M} = \mathrm{A}\,(\text{æ} - \text{æ}_{10}\mathbf{e}_1)\,; \tag{4.112}
$$

$$
\mathrm{L}_1\frac{\mathrm{d}\boldsymbol{\vartheta}}{\mathrm{d}\eta} + \mathrm{L}_2\text{æ}_{10}\mathbf{e}_1 - \mathrm{A}^{-1}\mathbf{M} = 0\,; \tag{4.113}
$$

$$
\frac{\mathrm{d}\mathbf{u}}{\mathrm{d}\eta} + \text{æ} \times \mathbf{u} + (l_{11} - 1)\,\mathbf{e}_1 + l_{21}\mathbf{e}_2 + l_{31}\mathbf{e}_3 = 0\,. \tag{4.114}
$$

4.3.2 Linear Vector Equations of Equilibrium

When a rod slightly deflects with respect to its natural straight configuration, we have

$$
\text{æ} = \text{æ}_0^{(1)} + \Delta\text{æ}\,; \qquad \mathbf{P} = \mathbf{P}_0 + \Delta\mathbf{P}\,; \qquad \mathbf{T} = \mathbf{T}_0 + \Delta\mathbf{T}\,.
$$

Therefore, using (4.110)–(4.114), we arrive at

$$\frac{d\mathbf{Q}}{d\eta} + æ_{10}(\mathbf{e}_1 \times \mathbf{Q}) + \Delta æ \times \mathbf{Q} + \mathbf{P}_0 + \Delta \mathbf{P} = 0\,; \tag{4.115}$$

$$\frac{d\mathbf{M}}{d\eta} + æ_{10}(\mathbf{e}_1 \times \mathbf{M}) + \Delta æ \times \mathbf{M} + \mathbf{e}_1 \times \mathbf{Q} + \mathbf{T}_0 + \Delta \mathbf{T} = 0\,; \tag{4.116}$$

$$\mathbf{M} = \mathrm{A}\,\Delta æ\,; \tag{4.117}$$

$$\frac{d\boldsymbol{\vartheta}}{d\eta} + æ_{10}\mathrm{A}_1\boldsymbol{\vartheta} - \mathrm{A}^{-1}\mathbf{M} = 0\,; \tag{4.118}$$

$$\frac{d\mathbf{u}}{d\eta} + æ_{10}\mathrm{A}_1\mathbf{u} + \mathrm{A}_1\boldsymbol{\vartheta} = 0\,, \qquad æ_{10} = \frac{d\vartheta_{10}}{d\eta}\,. \tag{4.119}$$

The increments $\Delta\mathbf{P}$ and $\Delta\mathbf{T}$ in (4.115) and (4.116) defined by (4.54) depend on $\mathbf{u}$ and $\boldsymbol{\vartheta}$ (see (4.55)). Consider the nonlinear terms $\Delta æ \times \mathbf{Q}$ and $\Delta æ \times \mathbf{M}$ in (4.115) and (4.116). If the projections of the vectors $\mathbf{P}_0$ and $\mathbf{T}_0$ on the tangent to the rod axis do not equal zero, we get

$$\Delta æ \times \mathbf{Q} = Q_{10}(\Delta æ \times \mathbf{e}_1) + \Delta æ \times (Q_2\mathbf{e}_2 + Q_3\mathbf{e}_3)\,; \tag{4.120}$$

$$\Delta æ \times \mathbf{M} = M_{10}(\Delta æ \times \mathbf{e}_1) + \Delta æ \times (M_2\mathbf{e}_2 + M_3\mathbf{e}_3)\,, \tag{4.121}$$

where $Q_{10} = P_{10}$ and $M_{10} = T_{10}$. To obtain linear equilibrium equations, we set $\Delta æ \times (Q_2\mathbf{e}_2 + Q_3\mathbf{e}_3) = 0$ and $\Delta æ \times (M_2\mathbf{e}_2 + M_3\mathbf{e}_3) = 0$. Hence, the following approximate relations are valid:

$$\Delta æ \times \mathbf{Q} \approx Q_{10}(\Delta æ \times \mathbf{e}_1)\,; \qquad \Delta æ \times \mathbf{M} \approx M_{10}(\Delta æ \times \mathbf{e}_1)\,.$$

A more detailed discussion of the role of the nonlinear terms in the equilibrium equations for small deflections is given in Sect. 4.2.

Finally, in view of (4.115) and (4.116), we obtain the following linear equations:

$$\frac{d\mathbf{Q}}{d\eta} + æ_{10}(\mathbf{e}_1 \times \mathbf{Q}) + Q_{10}(\Delta æ \times \mathbf{e}_1) + \mathbf{P}_0 + \Delta \mathbf{P} = 0\,; \tag{4.122}$$

$$\frac{d\mathbf{M}}{d\eta} + æ_{10}(\mathbf{e}_1 \times \mathbf{M}) + M_{10}(\Delta æ \times \mathbf{e}_1) + \mathbf{e}_1 \times \mathbf{Q} + \mathbf{T}_0 + \Delta \mathbf{T} = 0\,. \tag{4.123}$$

4.3.3 Equilibrium Equations in the Attached Coordinate System

In terms of projections on the attached axes, the system of equations (4.122), (4.123), (4.118), and (4.119) for naturally twisted rods can be rewritten as follows:

$$\frac{dQ_1}{d\eta} + P_{10} + \Delta P_1 = 0\,;$$

$$\frac{dQ_2}{d\eta} - æ_{10}Q_3 + \Delta æ_3 Q_{10} + P_{20} + \Delta P_2 = 0\,;$$

$$\frac{dQ_3}{d\eta} + æ_{10}Q_2 - \Delta æ_2 Q_{10} + P_{30} + \Delta P_3 = 0\,; \tag{4.124}$$

$$\frac{dM_1}{d\eta} + T_{10} + \Delta T_1 = 0\,;$$
$$\frac{dM_2}{d\eta} - æ_{10} M_3 + \Delta æ_3 M_{10} - Q_3 + T_{20} + \Delta T_2 = 0\,;$$
$$\frac{dM_3}{d\eta} + æ_{10} M_2 - \Delta æ_2 M_{10} + Q_2 + T_{30} + \Delta T_3 = 0\,; \tag{4.125}$$
$$\frac{d\vartheta_1}{d\eta} - \frac{M_1}{A_{11}} = 0\,;$$
$$\frac{d\vartheta_2}{d\eta} - æ_{10}\vartheta_3 - \frac{M_2}{A_{22}} = 0\,;$$
$$\frac{d\vartheta_3}{d\eta} + æ_{20}\vartheta_2 - \frac{M_3}{A_{33}} = 0\,; \tag{4.126}$$
$$\frac{du_1}{d\eta} = 0\,;$$
$$\frac{du_2}{d\eta} - æ_{10} u_3 - \vartheta_3 = 0\,;$$
$$\frac{du_3}{d\eta} + æ_{10} u_2 + \vartheta_2 = 0\,. \tag{4.127}$$

Being of the most general form, the system of equations (4.124)–(4.127) is useful in many applications.

Let us derive the equilibrium equations for a twisted drilling bit (see Fig. 0.21) after a loss of stability. Assume that the force $\mathbf{P}_{10}$ and the moment $\mathbf{T}_{10}$ are follower and that the static mode of loss of stability may take place. In view of (4.124)–(4.127), we get

$$\frac{dQ_1}{d\eta} - P_{10}\,\delta\,(\eta - 1) = 0\,; \tag{4.128}$$
$$\frac{dQ_2}{d\eta} - æ_{10} Q_3 - \Delta æ_3 P_{10} = 0\,; \tag{4.129}$$
$$\frac{dQ_3}{d\eta} + æ_{10} Q_2 + \Delta æ_2 P_{10} = 0\,; \tag{4.130}$$
$$\frac{dM_1}{d\eta} + T_{10}\,\delta\,(\eta - 1) = 0\,; \tag{4.131}$$
$$\frac{dM_2}{d\eta} - æ_{10}^{(1)} M_3 - Q_3 = 0\,; \tag{4.132}$$
$$\frac{dM_3}{d\eta} + æ_{10}^{(1)} M_2 + Q_2 = 0\,; \tag{4.133}$$
$$æ_{10}^{(1)} = \frac{d\vartheta_{10}}{d\eta} - \frac{T_{10}}{A_{11}}\,. \tag{4.134}$$

(Equations (4.126) and (4.127) remain unchanged.)

From (4.128) and (4.131) it follows that $Q_1 = -P_{10}$ and $M_1 = M_{10}$. These equalities are used in (4.129), (4.130), and (4.132)–(4.134). From the first equations of the systems (4.126) and (4.127), we get $\vartheta_1' = M_{10}/A_{11} = \text{const}$ and $u_1 = 0$. Let us eliminate $\Delta æ_2$ and $\Delta æ_3$ from the other eight equations and represent these system of equations in vector form as follows:

$$\frac{d\mathbf{Y}}{d\eta} + \mathbf{A}\mathbf{Y} = 0\,; \tag{4.135}$$

here

$$\mathbf{Y} = (Q_2\,,\, Q_3\,,\, M_2\,,\, M_3\,,\, \vartheta_2\,,\, \vartheta_3\,,\, u_2\,,\, u_3)^{\mathrm{T}}\,;$$

$$A = \begin{bmatrix} 0 & -æ_{10} & 0 & -\dfrac{P_{10}}{A_{33}} & 0 & 0 & 0 & 0 \\ æ_{10} & 0 & \dfrac{P_{10}}{A_{22}} & 0 & 0 & 0 & 0 & 0 \\ 0 & -1 & 0 & \dfrac{T_{10}}{A_{33}} - æ_{10} & 0 & 0 & 0 & 0 \\ 1 & 0 & æ_{10} - \dfrac{T_{10}}{A_{22}} & 0 & 0 & 0 & 0 & 0 \\ 0 & 0 & -\dfrac{1}{A_{22}} & 0 & 0 & -æ_{10} & 0 & 0 \\ 0 & 0 & 0 & -\dfrac{1}{A_{33}} & æ_{10} & 0 & 0 & 0 \\ 0 & 0 & 0 & 0 & 0 & -1 & 0 & -æ_{10} \\ 0 & 0 & 0 & 0 & 0 & 0 & æ_{10} & 0 \end{bmatrix}.$$

4.4 Straight Rods on Elastic Foundation

4.4.1 Forces Acting on a Rod

The problems of rods contacting with elastic foundations, elastic layers, etc. (see Fig. 0.19) are of great practical importance. Interaction between a rod and a medium and between a rod and a force field is illustrated in Figs. 4.6 and 4.7, respectively.

A transmission pipeline is shown in Fig. 4.6. The pipe rests on elastic soil. As the pipe bends, a force of interaction between the pipe and the soil appears. This force depends on the displacement of the pipe in the vertical direction u_{x_2}.

The key feature of problems of interaction of rods with an elastic medium is as follows: when a rod deflects from its initial configuration (no matter straight or curvilinear), distributed forces $\mathbf{q}$ appear. These forces are functions of the displacement vector $\mathbf{u}$, i.e. $\mathbf{q} = \mathbf{q}(\mathbf{u})$. In the case of a linearly deformed elastic medium, we have

$$\mathbf{q} = -k u_{x_2} \mathbf{i}_2\,.$$

For a nonlinearly deformed medium, we get

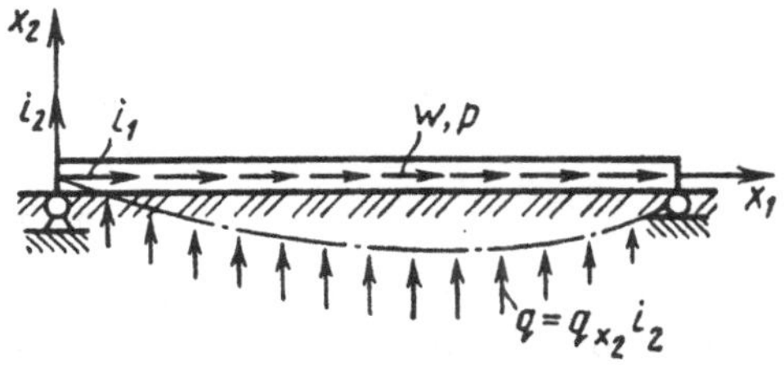

Fig. 4.6.

$$\mathbf{q} = -f(u_{x_2}) \operatorname{sign} u_{x_2} \mathbf{i}_2 .$$

As shown in Chapters 1–3, in most applied problems we deal with forces and moments which depend on displacements of the axial points and angles of rotation of the attached coordinate axes, i.e. $\mathbf{q} = \mathbf{q}(\mathbf{u}, \boldsymbol{\vartheta})$ and $\boldsymbol{\mu}=\boldsymbol{\mu}(\mathbf{u}, \boldsymbol{\vartheta})$. If displacements and angles of rotation are small, then the expressions for $\mathbf{q}$ and $\boldsymbol{\mu}$ take the following form (see (4.55)):

$$\mathbf{q} = \mathrm{C}_1\boldsymbol{\vartheta} + \mathrm{C}_2\mathbf{u}; \qquad \boldsymbol{\mu} = \mathrm{C}_3\boldsymbol{\vartheta} + \mathrm{C}_4\mathbf{u}. \tag{4.136}$$

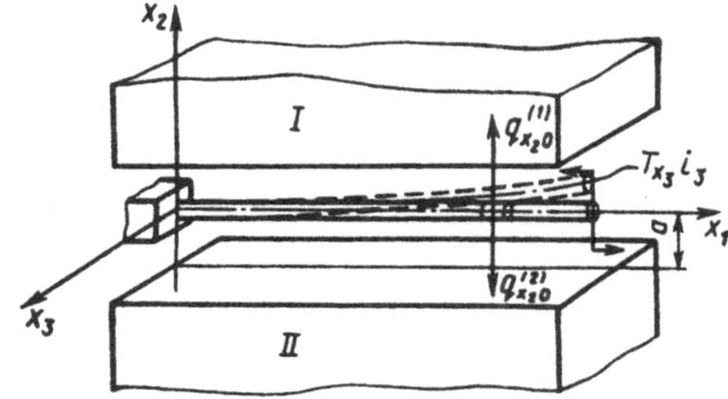

Fig. 4.7.

Hence, problems of statics of rods on an elastic foundation are special cases of the problems of statics considered above with $\mathrm{C}_2 = \mathrm{C}_3 = \mathrm{C}_4 = 0$ and

$$\mathrm{C}_1 = \begin{bmatrix} 0 & 0 & 0 \\ 0 & -k & 0 \\ 0 & 0 & 0 \end{bmatrix}.$$

A rod attracted by two magnets, *I* and *II*, is shown in Fig. 4.7. The magnitudes of the distributed attracting forces are denoted by $q^{(1)}_{x_2}$ and $q^{(2)}_{x_2}$, respectively.

Suppose that the rod in its natural configuration lies right in the middle between the magnets, hence, the attracting forces are equal $q^{(1)}_{x_2 0} = q^{(2)}_{x_2 0}$. If we apply an additional load to the rod, say, a concentrated moment T_{x_3} as indicated in Fig. 4.7, then the rod deflects and the attracting forces are no longer equal ($q_{x_2}^{(1)} > q_{x_2}^{(2)}$). Moreover, if $u_{x_2} \ll a$, the total attracting force acting on the rod is linear in u_{x_2}, that is,

$$\mathbf{q}_{x_2} = \mathbf{q}_{x_2}^{(1)} - \mathbf{q}_{x_2}^{(2)} = k_1 u_{x_2} \mathbf{i}_2 \,. \tag{4.137}$$

From this example we see that when the rod deflects from the initial equilibrium configuration, the additional distributed force is directed away from this initial equilibrium configuration. In the preceding example (Fig. 4.6), on the contrary, the distributed force caused by the deformation of the pipe is directed towards the initial equilibrium configuration of the pipe.

4.4.2 Equilibrium Equations

Consider a rod of varying cross section lying on an elastic foundation. Suppose that displacements of the axial points are small. Setting $T_{x_1 0} = 0$, $\mu_{x_1 0} = 0$, $q_{x_1 0} = 0$, $P_{x_1 0} \neq 0$, and $\Delta q_{x_2} = -k u_{x_2}$ in (4.98)–(4.102), we obtain the following equilibrium equations:

$$\frac{\mathrm{d}Q_{x_2}}{\mathrm{d}\eta} + q_{x_2 0} H(\eta - \eta_q) - k u_{x_2} + P_{x_2 0}\,\delta(\eta - \eta_P) = 0\,; \tag{4.138}$$

$$\frac{\mathrm{d}M_{x_3}}{\mathrm{d}\eta} + Q_{x_2} - \vartheta_3 P_{x_1 0} + \mu_{x_3 0} H(\eta - \eta_\mu) + T_{x_3 0}^{(1)}\,\delta(\eta - \eta_T) = 0\,; \tag{4.139}$$

$$\frac{\mathrm{d}\vartheta_3}{\mathrm{d}\eta} - \frac{M_{x_3 0}}{A_{33}} = 0\,; \tag{4.140}$$

$$\frac{\mathrm{d}u_{x_2}}{\mathrm{d}\eta} - \vartheta_3 = 0\,; \tag{4.141}$$

here $k = k_0 l^4 / A_{33}^{(0)}$ is a nondimensional coefficient (k_0 is the dimensional elastic stiffness of the foundation).

Equations (4.138)–(4.141) can be rewritten as one equation as follows:

$$\begin{aligned}(A_{33} u_{x_2}{''})'' - P_{x_1 0} u_{x_2}'' + k u_{x_2} &= q_{x_2 0} H(\eta - \eta_q) + P_{x_2 0}\,\delta(\eta - \eta_P) \\ &\quad - T_{x_3 0}^{(1)}\,\delta'(\eta - \eta_T) - \mu_{x_3 0}\,\delta(\eta - \eta_\mu)\,.\end{aligned} \tag{4.142}$$

For rods of constant cross section, the nondimensional bending stiffness $A_{33} = 1$. Hence, (4.142) becomes a linear differential equation with constant coefficients

$$u_{x_2}^{\mathrm{IV}} - Q_{10} u_{x_2}'' + k u_{x_2} = b\,, \tag{4.143}$$

where

$$b = -q_{x_2 0} H - P_{x_2 0}\,\delta_P - T_{x_3 0}\,\delta_T' - \mu_{x_3 0}\,\delta_\mu\,.$$

Equations (4.142) and (4.143) are useful in the derivation of approximate solutions. Approximating methods for analysis of equilibrium equations of straight rods are discussed in Sect. 4.5.

4.4.3 Krylov's Functions

Let us consider a simple problem of statics of rods. A straight rod of constant cross section rests on an elastic foundation. Suppose that the axial components of external forces are equal to zero; then, setting $P_{x_1 0} = 0$ in (4.143), we obtain

$$u_{x_2}^{\mathrm{IV}} + 4\alpha_1^4 u_{x_2} = b\,. \tag{4.144}$$

We seek the solution of the form $u_{x_2} = u_0 \exp(\lambda\eta)$. The corresponding characteristic equation $\lambda^4 + 4\alpha_1^4 = 0$ has the roots $\lambda_{1,2} = \pm(i-1)\,\alpha_1$ and $\lambda_{3,4} = \pm(i+1)\,\alpha_1$. The solution of the homogeneous equation (4.144) ($b = 0$) takes the form

$$\begin{aligned} u_{x_2} &= c_1\, e^{-\alpha_1\eta}\cos\alpha_1\eta + c_2\, e^{\alpha_1\eta}\sin\alpha_1\eta \\ &\quad + c_3\, e^{\alpha_1\eta}\cos\alpha_1\eta + c_4\, e^{\alpha_1\eta}\sin\alpha_1\eta \\ &= c_1 k_{11}^{(0)}(\eta) + c_2 k_{12}^{(0)}(\eta) + c_3 k_{13}^{(0)}(\eta) + c_4 k_{14}^{(0)}(\eta)\,. \end{aligned}$$

For straight rods of constant cross section ($Q_{10} = 0$), we have

$$Q_{x_2} = -u_{x_2}''' \,; \qquad M_{x_3} = u_{x_2}'' \,; \qquad \vartheta_3 = u_{x_2}' \,.$$

Hence, (4.144) is equivalent to a system of four ordinary differential equations

$$\begin{aligned} &Q_{x_2}' + 4\alpha_1^4 u_{x_2} = b_1\,; \\ &M_{x_3}' + Q_{x_2} = b_2\,; \\ &\vartheta_3' - M_{x_3} = 0\,; \\ &u_{x_2}' - \vartheta_3 = 0\,. \end{aligned} \tag{4.145}$$

The fundamental matrix for the homogeneous system of equations (4.145) ($b_1 = b_2 = 0$) is as follows:

$$\mathrm{K}^{(0)}(\eta) = \begin{bmatrix} k_{11}^{(0)'''} & k_{12}^{(0)'''} & k_{13}^{(0)'''} & k_{14}^{(0)'''} \\ k_{11}^{(0)''} & k_{12}^{(0)''} & k_{13}^{(0)''} & k_{14}^{(0)''} \\ k_{11}^{(0)'} & k_{12}^{(0)'} & k_{13}^{(0)'} & k_{14}^{(0)'} \\ k_{11}^{(0)} & k_{12}^{(0)} & k_{13}^{(0)} & k_{14}^{(0)} \end{bmatrix}.$$

The matrix $\mathrm{K}^{(0)}(\eta)$ for $\eta = 0$ is not a unit matrix. To avoid further difficulties, we should select the partial solutions such that the equality $\mathrm{K}^{(0)}(0) = \mathrm{E}$ holds; to do this, let us compose a linear combination of the partial solutions

$$k_{ij}(\eta) = \sum_{\nu=1}^{n} k_{ij}^{(0)}(\eta)\, b_{\nu i} \tag{4.146}$$

and choose coefficients $b_{\nu i}$ in such a way that the matrix K(0) with the components $k_{ij}(0)$ becomes the unit matrix.

The coefficients $b_{\nu i}$ can be found in the following way. For any matrix B with constant elements, the matrix $\mathrm{K} = \mathrm{K}^{(0)}(\eta)\mathrm{B}$ satisfies the homogeneous system of equations corresponding to (4.145). Evidently, the matrix B can be so chosen as to satisfy the equality $\mathrm{K}(0) = \mathrm{E}$. If the matrix B is equal to the inverse of the matrix $\mathrm{K}^{(0)}(0)$, we get

$$\mathrm{K}\,(\eta) = \mathrm{K}^0(\eta)\,[\,\mathrm{K}^0(0)\,]^{-1}\,. \tag{4.147}$$

On rearrangement, we have

$$\mathrm{K}\,(\eta) = [\,k_{ij}\,] = \begin{bmatrix} K_1 & -\alpha_1 K_4 & -2\alpha_1^2 K_3 & -2\alpha_1^3 K_2 \\ \dfrac{K_2}{2\alpha_1} & K_1 & -\alpha_1 K_4 & -2\alpha_1^2 K_3 \\ \dfrac{K_3}{2\alpha_1^2} & \dfrac{K_2}{2\alpha_1} & K_1 & -\alpha_1 K_4 \\ \dfrac{K_4}{4\alpha_1^3} & \dfrac{K_3}{2\alpha_1^2} & \dfrac{K_2}{2\alpha_1} & K_1 \end{bmatrix}, \tag{4.148}$$

where K_i are the following Krylov's functions:

$$\begin{aligned} K_1 &= \cosh\alpha_1\eta\cos\alpha_1\eta\,; \\ K_2 &= \cosh\alpha_1\eta\sin\alpha_1\eta + \sinh\alpha_1\eta\cos\alpha_1\eta\,; \\ K_3 &= \sinh\alpha_1\eta\sin\alpha_1\eta\,; \\ K_4 &= \cosh\alpha_1\eta\sin\alpha_1\eta\,. \end{aligned} \tag{4.149}$$

Krylov's functions are tabulated.

The derivatives of the functions K_i are

$$\begin{aligned} &K_1{}' = -\alpha_1 K_4\,; && K_1{}'' = -2\alpha_1^2 K_3\,; && K_1{}''' = -2\alpha_1^3 K_2\,; \\ &K_2{}' = 2\alpha_1 K_1\,; && K_2{}'' = -2\alpha_1^2 K_4\,; && K_2{}''' = -4\alpha_1^3 K_3\,; \\ &K_3{}' = \alpha_1 K_2\,; && K_3{}'' = 2\alpha_1^2 K_1\,; && K_3{}''' = -2\alpha_1^3 K_4\,; \\ &K_4{}' = 2\alpha_1 K_3\,; && K_4{}'' = 2\alpha_1^2 K_2\,; && K_4{}''' = 4\alpha_1^3 K_1\,. \end{aligned} \tag{4.150}$$

4.4.4 Equilibrium Equations for Rods of Constant Cross Section

To use computer-oriented methods, it is convenient to represent (4.115) in vector form

$$\mathbf{Y}' + \mathrm{A}\mathbf{Y} = \mathbf{b}\,, \tag{4.151}$$

where

$$\mathbf{Y} = \begin{bmatrix} Q_{x2} \\ M_{x3} \\ \vartheta_3 \\ u_{x2} \end{bmatrix} = \begin{bmatrix} -u'''_{x2} \\ u''_{x2} \\ u'_{x2} \\ u_{x2} \end{bmatrix}; \qquad \mathrm{A} = \begin{bmatrix} 0 & 0 & 0 & -4\alpha_1^4 \\ 1 & 0 & 0 & 0 \\ 0 & -1 & 0 & 0 \\ 0 & 0 & -1 & 0 \end{bmatrix};$$

$$\mathbf{b} = \begin{bmatrix} b_1 \\ b_2 \\ 0 \\ 0 \end{bmatrix} = \begin{bmatrix} -q_{x20} H(\eta - \eta_q) - P_{x20}\,\delta(\eta - \eta_P) \\ -T_{x30}\,\delta(\eta - \eta_T) - \mu_{x30} H(\eta - \eta_\mu) \\ 0 \\ 0 \end{bmatrix}. \tag{4.152}$$

Let us consider the general approach to the analysis of the inhomogeneous equation (4.151). The corresponding homogeneous equation has a solution

$$\mathbf{Y}^{(0)} = \mathbf{Y}_0\, e^{\lambda\eta}. \tag{4.153}$$

The characteristic equation $\det[\mathrm{E}\lambda + \mathrm{A}] = 0$ (or $\lambda^4 + 4\alpha_1^4 = 0$) has the following roots:

$$\lambda_{1,2} = \pm(i-1)\,\alpha_1; \qquad \lambda_{3,4} = \pm(i+1)\,\alpha_1. \tag{4.154}$$

The four independent vectors are thus obtained

$$\begin{aligned} &\mathbf{Y}^{(1)} = \mathbf{Y}_0^{(1)}\, e^{\lambda_1\eta}; && \mathbf{Y}^{(2)} = \mathbf{Y}_0^{(2)}\, e^{\lambda_2\eta}; \\ &\mathbf{Y}^{(3)} = \mathbf{Y}_0^{(3)}\, e^{\lambda_3\eta}; && \mathbf{Y}^{(4)} = \mathbf{Y}_0^{(4)}\, e^{\lambda_4\eta}. \end{aligned} \tag{4.155}$$

The components of the vectors $\mathbf{Y}_0^{(i)}$ can be found from the system of algebraic equations $[\mathrm{E}\lambda_i + \mathrm{A}]\,\mathbf{Y}_0^{(i)} = 0$. The expanded form of these equations is as follows:

$$\begin{aligned} &\lambda_i Y_{01}^{(i)} - Y_{02}^{(i)} = 0; && \lambda_i Y_{02}^{(i)} - Y_{03}^{(i)} = 0; \\ &\lambda_i Y_{03}^{(i)} + Y_{04}^{(i)} = 0; && -4\alpha_1^4 Y_{01}^{(i)} + \lambda_i Y_{04}^{(i)} = 0. \end{aligned} \tag{4.156}$$

According to (4.156), the vectors $Y_{02}^{(i)}$, $Y_{03}^{(i)}$, and $Y_{04}^{(i)}$ are related to $Y_{01}^{(i)}$ as follows:

$$Y_{02}^{(i)} = \lambda_i Y_{01}^{(i)}; \qquad Y_{03}^{(i)} = (\lambda_i)^2 Y_{01}^{(i)}; \qquad Y_{04}^{(i)} = -(\lambda_i)^3 Y_{01}^{(i)}. \tag{4.157}$$

The factor $Y_{01}^{(i)}$ may be prescribed arbitrarily. (This factor occurs in each component of the vectors $\mathbf{Y}_0^{(i)}$.) For definiteness, we put $Y_{01}^{(i)} = 1 + i$. Taking into account the third equation of (4.157), we see that the fourth equation of (4.156) becomes an identity. Since the roots λ_i are complex, we see that the solutions of (4.156) are also complex, that is, $Y_{0j}^{(i)} = (Y_{0j}^{(i)})_1 + i\,(Y_{0j}^{(i)})_2$.

From (4.157) it follows that

$$\mathbf{Y}^{(1)} = \begin{bmatrix} 1 \\ -2\alpha_1 \\ 2\alpha_1^2 \\ 0 \end{bmatrix} + i \begin{bmatrix} 1 \\ 0 \\ -2\alpha_1^2 \\ -4\alpha_1^3 \end{bmatrix}; \qquad \mathbf{Y}^{(2)} = \begin{bmatrix} 1 \\ 2\alpha_1 \\ 2\alpha_1^2 \\ 0 \end{bmatrix} + i \begin{bmatrix} 1 \\ 0 \\ -2\alpha_1^2 \\ -4\alpha_1^3 \end{bmatrix};$$

$$\mathbf{Y}^{(3)} = \begin{bmatrix} 1 \\ 0 \\ -2\alpha_1^2 \\ 4\alpha_1^3 \end{bmatrix} + i \begin{bmatrix} 1 \\ 2\alpha_1 \\ 2\alpha_1^2 \\ 0 \end{bmatrix}; \qquad \mathbf{Y}^{(4)} = \begin{bmatrix} 1 \\ 0 \\ -2\alpha_1^2 \\ 4\alpha_1^3 \end{bmatrix} + i \begin{bmatrix} 1 \\ -2\alpha_1 \\ 2\alpha_1^2 \\ 0 \end{bmatrix}. \tag{4.158}$$

Substituting (4.157) into (4.155), we obtain the complex eigenvectors

$$\mathbf{Y}^{(i)} = (\mathbf{Y}^{(i)})_1 + i\,(\mathbf{Y}^{(i)})_2\,, \tag{4.159}$$

whose real and imaginary parts satisfy the homogeneous equation (4.151).

If the eigenvectors $\mathbf{Y}^{(i)}$ are known, we can get two fundamental matrices whose columns are the real parts and the imaginary parts of the vectors $\mathbf{Y}^{(j)}$, respectively. These matrices differ only in the arrangement of their columns. Finally, for the homogeneous equations (4.151) ($\mathbf{b} = 0$), the fundamental matrix $\mathrm{K}\,(\eta)$ takes the form

$$\mathrm{K}(\eta) = [K_{ij}(\eta)]\,, \tag{4.160}$$

where

$$\begin{aligned}
K_{11} &= e^{-\alpha_1\eta}(\cos\alpha_1\eta - \sin\alpha_1\eta)\,; \qquad K_{12} = e^{\alpha_1\eta}(\cos\alpha_1\eta + \sin\alpha_1\eta)\\
K_{13} &= e^{\alpha_1\eta}(\cos\alpha_1\eta - \sin\alpha_1\eta)\,; \qquad K_{14} = e^{-\alpha_1\eta}(\cos\alpha_1\eta + \sin\alpha_1\eta)\,;\\
K_{21} &= -2\alpha_1\, e^{-\alpha_1\eta}\cos\alpha_1\eta\,; \qquad K_{22} = 2\alpha_1\, e^{\alpha_1\eta}\cos\alpha_1\eta\,;\\
K_{23} &= 2\alpha_1\, e^{\alpha_1\eta}\sin\alpha_1\eta\,; \qquad K_{24} = -2\alpha_1\, e^{-\alpha_1\eta}\sin\alpha_1\eta\,;\\
K_{31} &= 2\alpha_1^2\, e^{-\alpha_1\eta}\,(\cos\alpha_1\eta + \sin\alpha_1\eta)\,;\\
K_{32} &= 2\alpha_1^2\, e^{\alpha_1\eta}\,(\cos\alpha_1\eta - \sin\alpha_1\eta)\,;\\
K_{33} &= -2\alpha_1^2\, e^{\alpha_1\eta}\,(\cos\alpha_1\eta + \sin\alpha_1\eta)\,;\\
K_{34} &= 2\alpha_1^2\, e^{-\alpha_1\eta}\,(\sin\alpha_1\eta - \cos\alpha_1\eta)\,;\\
K_{41} &= 4\alpha_1^3\, e^{-\alpha_1\eta}\sin\alpha_1\eta\,; \qquad K_{42} = -4\alpha_1^3\, e^{\alpha_1\eta}\sin\alpha_1\eta\,;\\
K_{43} &= 4\alpha_1^3\, e^{\alpha_1\eta}\cos\alpha_1\eta\,; \qquad K_{44} = -4\alpha_1^3\, e^{-\alpha_1\eta}\cos\alpha_1\eta\,.
\end{aligned}$$

The elements of the matrix $\mathrm{K}(\eta)$ are linear combinations of the partial solutions $k_{1j}^{(0)}(\eta)$ obtained above. If the matrix $\mathrm{K}(\eta)$ (4.160) is not the unit matrix at $\eta = 0$, then, using the transformation (4.147), we can define a fundamental matrix $\mathrm{K}_1(\eta)$ such that the equality $\mathrm{K}_1(0) = E$ is satisfied. Hence, in what follows, we assume that the fundamental matrix K is the unit matrix at $\eta = 0$. For example, the matrix defined by (4.148) is such a matrix.

The solution of the inhomogeneous equation (4.151) can be written as

$$\mathbf{Y}^{(0)} = \mathbf{Y}_0 + \mathbf{Y}_H = \mathrm{K}(\eta)\,\mathbf{C} + \int_0^{\eta} \mathrm{K}(\eta - \zeta)\,\mathbf{b}\,\mathrm{d}\zeta\,. \tag{4.161}$$

The components of the vector $\mathbf{Y}_H$ are as follows:

$$\begin{aligned} Y_{Hj} &= -\int_0^{\eta} k_{j1} H(\zeta - \eta_q)\, q_{x_2 0}\,\mathrm{d}\zeta - \int_0^{\eta} k_{j1}\,\delta(\zeta - \eta_P)\, P_{x_2 0}\,\mathrm{d}\zeta \\ &\quad - \int_0^{\eta} k_{j2} H(\zeta - \eta_\mu)\,\mu_{x_3 0}\,\mathrm{d}\zeta - \int_0^{\eta} k_{j2}\,\delta(\zeta - \eta_T)\, T_{x_3 0}\,\mathrm{d}\zeta \end{aligned} \tag{4.162}$$

or

$$Y_{Hj} = (J_{j1})_q + (J_{j1})_P + (J_{j2})_\mu + (J_{j2})_T\,, \qquad j = 1\,,2\,,3\,,\ldots\,, \tag{4.163}$$

where

$$\begin{aligned} (J_{j1})_q &= -\int_0^{\eta} k_{j1} H(\zeta - \eta_q)\, q_{x_2 0}\,\mathrm{d}\zeta\,; \\ (J_{j1})_P &= -\int_0^{\eta} k_{j1}\,\delta(\zeta - \eta_P)\, P_{x_2 0}\,\mathrm{d}\zeta\,; \\ (J_{j2})_\mu &= -\int_0^{\eta} k_{j2} H(\zeta - \eta_\mu)\,\mu_{x_3 0}\,\mathrm{d}\zeta\,; \\ (J_{j2})_T &= -\int_0^{\eta} k_{j2}\,\delta(\zeta - \eta_T)\, T_{x_3 0}\,\mathrm{d}\zeta\,. \end{aligned} \tag{4.164}$$

Formulas (4.164) containing concentrated forces and moments can be rewritten as follows:

$$\begin{aligned} (J_{j1})_P &= -k_{j1}(\alpha_1 \eta_P)\, P_{x_2 0} H(\eta - \eta_P)\,; \\ (J_{j2})_T &= -k_{j2}(\alpha_1 \eta)\, T_{x_3 0} H(\eta - \eta_T)\,; \end{aligned} \tag{4.165}$$

introducing Krylov's functions, we arrive at

$$\begin{aligned} (J_{11})_P &= -P_{x_2 0} K_1(\alpha_1 \eta_P)\, H(\eta - \eta_P)\,; \\ (J_{21})_P &= -\frac{P_{x_2 0}}{2\alpha_1}\, K_2(\alpha_1 \eta_P)\, H(\eta - \eta_P)\,; \\ (J_{31})_P &= -\frac{P_{x_2 0}}{2\alpha_1^2}\, K_3(\alpha_1 \eta_P)\, H(\eta - \eta_P)\,; \\ (J_{41})_P &= -\frac{P_{x_2 0}}{4\alpha_1^3}\, K_4(\alpha_1 \eta_P)\, H(\eta - \eta_P)\,; \end{aligned} \tag{4.166}$$

$$\begin{aligned} (J_{12})_T &= \alpha T_{x_3 0} K_4(\alpha_1 \eta_T)\, H(\eta - \eta_T)\,; \\ (J_{22})_T &= -T_{x_3 0} K_1(\alpha_1 \eta_T)\, H(\eta - \eta_T)\,; \\ (J_{32})_T &= -\frac{T_{x_3 0}}{2\alpha_1}\, K_2(\alpha_1 \eta_T)\, H(\eta - \eta_T)\,; \\ (J_{42})_T &= -\frac{T_{x_3 0}}{2\alpha_1}\, K_3(\alpha_1 \eta_T)\, H(\eta - \eta_T)\,. \end{aligned} \tag{4.167}$$

Recall that (4.150) relate Krylov's functions to their derivatives. For $q_{x_2 0} = \text{const}$ and $\mu_{x_3 0} = \text{const}$, (4.162) take the form

$$
\begin{aligned}
(J_{11})_q &= -\frac{q_{x_2 0}}{2\alpha_1^2}\,[\,K_2(\alpha_1\eta) - K_2(\alpha_1\eta_q)\,]\,H\,(\eta-\eta_q)\,;\\
(J_{21})_q &= -\frac{q_{x_2 0}}{2\alpha_1^3}\,[\,K_3(\alpha_1\eta) - K_3(\alpha_1\eta_q)\,]\,H\,(\eta-\eta_q)\,;\\
(J_{31})_q &= -\frac{q_{x_2 0}}{4\alpha_1^4}\,[\,K_4(\alpha_1\eta) - K_4(\alpha_1\eta_q)\,]\,H\,(\eta-\eta_q)\,;\\
(J_{41})_q &= -\frac{q_{x_2 0}}{4\alpha_1^5}\,[\,K_1(\alpha_1\eta) - K_1(\alpha_1\eta_q)\,]\,H\,(\eta-\eta_q)\,;\\
(J_{12})_\mu &= -\frac{\mu_{x_3 0}}{\alpha_1}\,[\,K_1(\alpha_1\eta) - K_1(\alpha_1\eta_\mu)\,]\,H\,(\eta-\eta_\mu)\,;\\
(J_{22})_\mu &= -\frac{\mu_{x_3 0}}{2\alpha_1^2}\,[\,K_2(\alpha_1\eta) - K_2(\alpha_1\eta_\mu)\,]\,H\,(\eta-\eta_\mu)\,;\\
(J_{32})_\mu &= -\frac{\mu_{x_3 0}}{2\alpha_1^3}\,[\,K_3(\alpha_1\eta) - K_3(\alpha_1\eta_\mu)\,]\,H\,(\eta-\eta_\mu)\,;\\
(J_{42})_\mu &= -\frac{\mu_{x_3 0}}{4\alpha_1^4}\,[\,K_4(\alpha_1\eta) - K_4(\alpha_1\eta_\mu)\,]\,H\,(\eta-\eta_\mu)\,.
\end{aligned}
\tag{4.168}
$$

Finally, assuming that there are several concentrated forces $\mathbf{P}^{(i)}$ and moments $\mathbf{T}^{(\nu)}$ applied to the rod, we can write the general solution of (4.151) in expanded form as follows:

$$
\begin{aligned}
Y_1 = Q_{x_2} &= \sum_{j=1}^{4} k_{ij}c_j + (J_{11})_q + \sum_{i=1}^{n}(J_{11})_P^{(i)} + (J_{12})_\mu + \sum_{\nu=1}^{\rho}(J_{12})_T^{(\nu)}\,;\\
Y_2 = M_{x_3} &= \sum_{j=1}^{4} k_{2j}c_j + (J_{21})_q + \sum_{i=1}^{n}(J_{21})_P^{(i)} + (J_{22})_\mu + \sum_{\nu=1}^{\rho}(J_{22})_T^{(\nu)}\,;\\
Y_3 = \nu_3 &= \sum_{j=1}^{4} k_{3j}c_j + (J_{31})_q + \sum_{i=1}^{n}(J_{31})_P^{(i)} + (J_{32})_\mu + \sum_{\nu=1}^{\rho}(J_{32})_T^{(\nu)}\,;\\
Y_4 = u_{x_2} &= \sum_{j=1}^{4} k_{4j}c_j + (J_{41})_q + \sum_{i=1}^{n}(J_{41})_P^{(i)} + (J_{42})_\mu + \sum_{\nu=1}^{\rho}(J_{42})_T^{(\nu)}\,;
\end{aligned}
\tag{4.169}
$$

here k_{ij} are the elements of the fundamental matrix (4.148) of the vector equation (4.151); these elements are expressed in terms of Krylov's functions.

Arbitrary constants c_j in (4.169) are to be determined from the boundary conditions at $\eta = 0$ and $\eta = 1$. As an example, consider a case shown in Fig. 0.19. Let the terms in (4.169) depending on external loads be denoted by Y_{Hj}, $j = 1, 2, 3, 4$. From (4.163) we can see that $Y_{Hj} = 0$ at $\eta = 0$, hence, $Y_2(0) = 0$, $Y_4(0) = 0$, and $c_2 = c_4 = 0$. Since $Y_2(1) = 0$ and $Y_4(1) = 0$, we see that the constants c_1 and c_3 satisfy the relations

$$k_{21}(\alpha_1)\,c_1 + k_{23}(\alpha_1)\,c_3 = -Y_{H2}(1)\,;$$
$$k_{41}(\alpha_1)\,c_1 + k_{43}(\alpha_1)\,c_3 = -Y_{H4}(1)\,. \tag{4.170}$$

4.4.5 Equilibrium Equations for Rods with Lateral Supports

A reaction force R_1 applied to a support (see Fig. 4.8) can be treated as an additional external force applied to the rod. The rod is fixed as shown in Fig. 4.8. Consider the boundary conditions. At $\eta = 0$, we have $c_3 = c_4 = 0$; at $\eta = 1$, (4.169) yield the following relations:

$$k_{21}(\alpha_1)\,c_1 + k_{22}(\alpha_1)\,c_2 = -\frac{q_{x_20}}{2\alpha_1^3}\,[\,K_3(\alpha_1) - K_3(0.5\alpha_1)\,] - \frac{R_1}{2\alpha_1}\,K_2(0.5\alpha_1)\,;$$
$$k_{41}(\alpha_1)\,c_1 + k_{42}(\alpha_1)\,c_2 = \frac{q_{x_20}}{4\alpha_1^5}\,[\,K_1(\alpha_1) - K_1(0.5\alpha_1)\,] - \frac{R_1}{4\alpha_1^3}\,K_4(0.5\alpha_1)\,; \tag{4.171}$$

here

$$k_{21} = \frac{K_2}{2\alpha_1}\,; \qquad k_{22} = K_1\,; \qquad k_{41} = \frac{K_4}{4\alpha_1^3}\,; \qquad k_{42} = \frac{K_3}{2\alpha_1^3}\,.$$

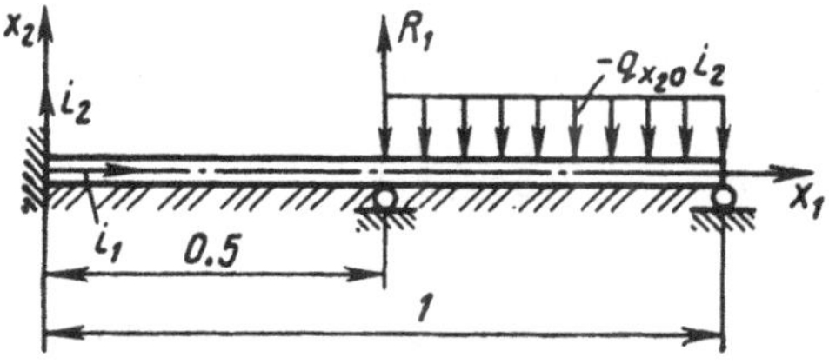

Fig. 4.8.

The two equations (4.171) contain three unknowns c_1, c_2, and R_1. The third equation can be obtained through the condition $u_{x_2} = 0$ at $\eta = 0.5$:

$$Y_4(0.5) = k_{41}(0.5\alpha_1)\,c_1 + k_{42}(0.5\alpha_1)\,c_2 = 0\,.$$

This approach to the solution of the equilibrium equations for a rod on an elastic foundation with a lateral hinge support is efficient and can be applied in the case of several supports of this kind. Moreover, this approach can be generalized to the case of elastic supports. In the latter case, the reaction at the support is

$$R^{(i)} = -c u_{x_2}\,\delta\,(\eta - \eta_i)\,.$$

Then, additional terms of the form

$$(J_{j1})_R^{(i)} = \int_0^\eta k_{j1} c u_{x_2}\, \delta(\zeta - \eta_i)\, d\zeta$$

appear in the general solution (4.169). These terms are similar to those in the second relation (4.164). On rearrangement, we get

$$\begin{aligned}
(J_{11})_R^{(i)} &= c u_{x_2}(\eta_i)\, K_1(\alpha_1 \eta_i)\, H(\eta - \eta_i);\\
(J_{21})_R^{(i)} &= c u_{x_2}(\eta_i)\, K_2(\alpha_1 \eta_i)\, H(\eta - \eta_i);\\
(J_{31})_R^{(i)} &= c u_{x_2}(\eta_i)\, K_3(\alpha_1 \eta_i)\, H(\eta - \eta_i);\\
(J_{41})_R^{(i)} &= c u_{x_2}(\eta_i)\, K_4(\alpha_1 \eta_i)\, H(\eta - \eta_i),
\end{aligned}$$

where

$$u_{x_2}(\eta_i) = \sum_{j=1}^{4} k_{4j}(\eta_i)\, c_j .$$

If we replace the hinge support (see Fig. 4.8) by an elastic support, then $R_1 = -c u_{x_2}\, \delta(\eta - 0.5)$. The constants c_1 and c_2 can be determined from the relations

$$\begin{aligned}
k_{21}(\alpha_1)\, c_1 + k_{22}(\alpha_1)\, c_2 &= -\frac{q_{x_2 0}}{2\alpha_1^3}\left[K_3(\alpha_1) - K_3(0.5\alpha_1)\right]\\
&\quad - \frac{c}{2\alpha_1} K_2(0.5\alpha_1)\left[k_{41}(0.5\alpha_1)\, c_1 + k_{42}(0.5\alpha_1)\, c_2\right];\\
k_{41}(\alpha_1)\, c_1 + k_{42}(\alpha_1)\, c_2 &= \frac{q_{x_2 0}}{4\alpha_1^5}\left[K_1(\alpha_1) - K_1(0.5\alpha_1)\right]\\
&\quad - \frac{c}{4\alpha_1^3} K_4(0.5\alpha_1)\left[k_{41}(0.5\alpha_1)\, c_1 + k_{42}(0.5\alpha_1)\, c_2\right].
\end{aligned}$$

4.4.6 Equilibrium Equations for Rods of Varying Cross Section

Let us consider the equilibrium equations (4.138)–(4.141) for a rod of varying cross section. These equations can be solved only by use of numerical methods. As noted, it is convenient to represent the equations in vector form

$$\mathbf{Y}' + \mathrm{A}(\eta)\,\mathbf{Y} = \mathbf{b}, \tag{4.172}$$

where

$$\mathbf{Y} = \begin{bmatrix} Q_{x_2} \\ M_{x_3} \\ \vartheta_3 \\ u_{x_2} \end{bmatrix}; \qquad \mathrm{A} = \begin{bmatrix} 0 & 0 & 0 & -k \\ 1 & 0 & -P_{x_1 0} & 0 \\ 0 & \dfrac{1}{A_{33}(\eta)} & 0 & 0 \\ 0 & 0 & 1 & 0 \end{bmatrix};$$

$$\mathbf{b} = \begin{bmatrix} b_1 \\ b_2 \\ 0 \\ 0 \end{bmatrix} = \begin{bmatrix} -q_{x_2 0} H(\eta - \eta_q) - P_{x_2 0}\, \delta(\eta - \eta_P) \\ -T_{x_3 0}\, \delta(\eta - \eta_T) - \mu_{x_3 0} H(\eta - \eta_\mu) \\ 0 \\ 0 \end{bmatrix}.$$

Recall the basic results obtained in Sect. 1.4. The general solution of (4.172) is

$$\mathbf{Y} = \mathrm{K}(\eta)(\mathbf{C}) + \int_0^{\eta} \mathrm{K}(\eta)\,\mathrm{K}^{-1}(\zeta)\,\mathbf{b}(\zeta)\,\mathrm{d}\zeta = \mathbf{Y}^{(0)} + \mathbf{Y}_H\,.$$

This formula is valid for any vector $\mathbf{b}$. The vector $\mathbf{b}$ can be written as follows:

$$\mathbf{b} = \mathbf{b}_q H(h - \eta_q) + \mathbf{b}_\mu H(\zeta - \eta_\mu) + \mathbf{b}_P\,\delta(\zeta - \eta_P) + \mathbf{b}_T\,\delta(\zeta - \eta_T)\,,$$

Hence,

$$\mathbf{Y}_H = \int_{\eta_q}^{\eta} \mathrm{G}(\eta, \zeta)\,\mathbf{b}_q\,\mathrm{d}\zeta + \int_{\eta_\mu}^{\eta} \mathrm{G}(\eta, \zeta)\,\mathbf{b}_\mu\,\mathrm{d}\zeta + \mathrm{G}(\eta, \eta_P)\,\mathbf{b}_P H(\eta - \eta_P)$$
$$+ \mathrm{G}(\eta, \eta_T)\,\mathbf{b}_T H(\eta - \eta_T)\,,$$

where

$$\mathrm{G}(\eta, \zeta) = \mathrm{K}(\eta)\,\mathrm{K}^{-1}(\zeta)\,;$$
$$\mathrm{G}(\eta, \eta_P) = \mathrm{K}(\eta)\,\mathrm{K}^{-1}(\eta_P)\,;$$
$$\mathrm{G}(\eta, \eta_T) = \mathrm{K}(\eta)\,\mathrm{K}^{-1}(\eta_T)\,.$$

Computer-oriented methods for determination of the matrix $\mathrm{K}(\eta)$ are discussed in Chap. 2.

4.5 Application of Approximate Methods

4.5.1 Principle of Virtual Displacements

Methods based on the principle of virtual displacements are useful in analysis of linear and, what is more important, of nonlinear problems. In theoretical mechanics, this principle is formulated as follows:

A mechanical system with ideal constraints is in equilibrium if and only if the work done by external forces through any virtual displacement of the system is equal to zero.

Constraints are said to be *ideal* if the work done by the reaction forces through any virtual displacement equals zero.

The principle of virtual displacements can be written as follows:

$$\delta A = \sum_{i=1}^{n} (\mathbf{F}_i \cdot \delta \mathbf{r}_i) = 0\,; \tag{4.173}$$

here δA is the work done by the forces $\mathbf{F}_i$ through the displacement $\delta \mathbf{r}_i$. Let the forces $\mathbf{F}_i$ be conservative, hence, they can be represented in the following form:

$$F_{x_j}^{(i)} = -\frac{\partial U}{\partial x_j^{(i)}}, \qquad j = 1, 2, 3, \quad i = 1, 2, \ldots, n;$$

the function $U(x_j^{(i)})$ is called the *potential energy of the system.*

From (4.173) it follows that

$$\delta A = -\sum_{i=1}^{n} \left(\frac{\partial U}{\partial x_1^{(i)}} \, \delta x_1^{(i)} + \frac{\partial U}{\partial x_2^{(i)}} \, \delta x_2^{(i)} + \frac{\partial U}{\partial x_3^{(i)}} \, \delta x_3^{(i)} \right) = -\delta U = 0$$

or

$$\delta U = 0, \tag{4.174}$$

where δU is the variation of the potential energy. The relation (4.174) is the necessary condition for the potential energy to have an extremum in an equilibrium configuration. Consequently, we see that the necessary and sufficient condition for a conservative system with ideal constraints to be in equilibrium are the same as the necessary condition (not a sufficient one) for the potential energy to attain an extremum.

Along with traditional methods based on analysis of differential equilibrium equations, the principle of virtual displacements is probably the most powerful tool of structural analysis, since it allows us to obtain approximate solutions. When dealing with elastic elements (rods, plates, and shells), we should take into consideration not only the work done by the external forces but also the work done by the internal forces as the structure deflects with respect to its original location. Now, let us discuss the concept of *virtual displacement* within the context of the theory of rods. Any small displacement of rod points away from the initial configuration that can be performed without breakdown of the constraints is termed *virtual* or *allowable.* As an example, consider a rod illustrated in Fig. 4.9. The shape of the axial line is described by a function $y(\eta)$. Suppose that a function $\delta\, y(\eta)$ is continuous in η and $\delta\, y(0) = \delta\, y(1) = 0$; then, this function defines a virtual-displacement field of the axial points.

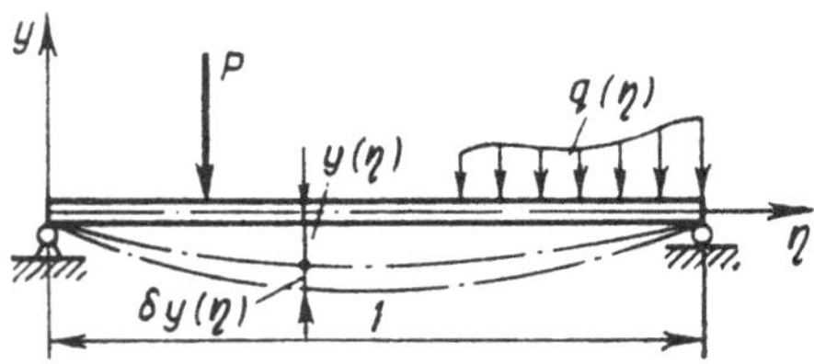

Fig. 4.9.

Consider the formulation of the principle of virtual displacements for a straight rod subjected to an arbitrary set of loads (see Fig. 4.9). Under the action of the loads **P**, **T**, and **q**, the rod deforms, therefore the work done by the loads is converted into the strain energy. Neglecting the energy lost in internal friction, we get $U = A$, where U is the strain energy and A is the work done by the external loads. Thus, for deformable systems, the principle of virtual displacements can be formulated as follows:

If a deformable system subjected to external forces is in equilibrium, then the work done by the external forces through any virtual displacement is equal to the work done by the internal forces through the same displacement, that is,

$$\delta \mathbf{A} = \delta U\,. \tag{4.175}$$

The work δA is the work done by the generalized external forces through virtual displacements of the points of application of these forces.

Upon a virtual displacement of the system, the external forces remain unchanged. Hence, the work done by an external force is equal to the magnitude of the force times the virtual displacement, i.e.,

$$\delta A_Q = \sum_{k=1}^{n} (\mathbf{Q}_k\, \delta \mathbf{y}_k)\,,$$

where $\mathbf{Q}_k$ is a generalized force and $\delta \mathbf{y}_k$ is a generalized virtual displacement. Here the concentrated forces and moments $\mathbf{P}^{(i)}$ and $\mathbf{T}^{(\nu)}$ are termed *generalized forces*, the linear displacements $\delta \mathbf{u}^{(i)}$ and angles of rotation $\delta \vartheta^{(\nu)}$ of the attached axes are called *generalized displacements* $\delta \mathbf{y}_k$. For distributed forces and moments, the work done through virtual displacements is as follows:

$$\delta A_q = \int_{\eta_{q1}}^{\eta_{q2}} (\mathbf{q}\, \delta \mathbf{u})\, \mathrm{d}\zeta\,; \qquad \delta A_\mu = \int_{\eta_{\mu 1}}^{\eta_{\mu 2}} (\boldsymbol{\mu}\, \delta \boldsymbol{\vartheta})\, \mathrm{d}\zeta\,; \tag{4.176}$$

here η_{q_j} and η_{μ_j} are the axial coordinates of the endpoints of the interval of application of the external loads.

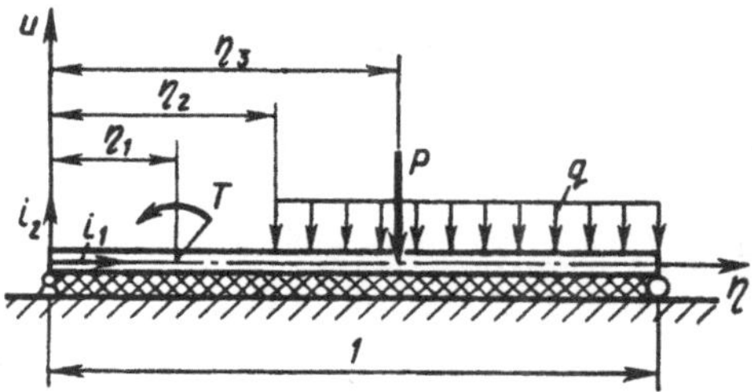

Fig. 4.10.

As an example, consider a rod on an elastic foundation (see Fig. 4.10). Let us determine the work done by the external forces. As in the previous chapter, we denote nondimensional deflections by u_j. The work done by the external forces through a virtual displacement can be written as follows:

$$\delta A = -P\,\delta u\,(\eta_3) + T\,\delta u'\,(\eta_1) - \int_{\eta_2}^{1} q\,\delta u\,\mathrm{d}\zeta - \int_0^1 q_\mathrm{f}\,\delta u\,\mathrm{d}\zeta\,, \qquad q_\mathrm{f} = k^4 u\,.$$

The equilibrium equations can be obtained from (4.175). Let us apply a tensile axial force Q_{10}. The potential energy of the rod and the variation of the energy are

$$U = \frac{1}{2}\int_0^1 A_{33}u''^2\,\mathrm{d}\zeta\,; \tag{4.177}$$

$$\delta U = \int_0^1 A_{33}u''\,\delta u''\,\mathrm{d}\zeta\,. \tag{4.178}$$

The work done by the external forces through virtual displacements is as follows:

$$\delta A = -P\,\delta u\,(\eta_3) + T\,\delta u'\,(\eta_1) - \int_0^1 q\,\delta u H\,(\zeta - \eta_2)\,\mathrm{d}\zeta$$
$$- \int_0^1 k^4 u\,\delta u\,\mathrm{d}\zeta - \delta\left(Q_{10}\cdot\frac{1}{2}\int_0^1 u'^2\,\mathrm{d}\zeta\right)\,.$$

Introducing the δ-function, we have

$$\delta A = -P\int_0^1 \delta u\,\delta\,(\zeta - \eta_3)\,\mathrm{d}\zeta - T\int_0^1 \delta u'\,\delta\,(\zeta - \eta_1)\,\mathrm{d}\zeta$$
$$- \int_0^1 qH\,(\zeta - \eta_2)\,\delta u\,\mathrm{d}\zeta - \int_0^1 k^4 u\,\delta u\,\mathrm{d}\zeta - Q_{10}\int_0^1 u'\,\delta u'\,\mathrm{d}\zeta\,. \tag{4.179}$$

Integrating (4.178) by parts, we get

$$\int_0^1 A_{33}u''\,\delta u''\,\mathrm{d}h = A_{33}u''\,\delta u'|_0^1 - (A_{33}u'')'\,\delta u\Big|_0^1 + \int_0^1 (A_{33}u'')''\,\delta u\,\mathrm{d}\zeta\,.$$

In view of the boundary conditions, the first and the second terms in the right-hand side of this relation vanish.

Integration of the last term in the right-hand side of (4.179) yields

$$Q_{10}\int_0^1 u'\,\delta u'\,\mathrm{d}\zeta = Q_{10}(u'\,\delta u)|_0^1 - \int_0^1 u''\,\delta u\,\mathrm{d}\zeta\,.$$

Equating δU and δA, we get

$$\int_0^1 \Big[(A_{33}u'')'' + k^4 u + P\,\delta\,(\zeta - \eta_3) + T\,\delta'\,(\zeta - \eta_1) + qH\,(\zeta - \eta_2) - Q_{10}u'' \Big]\,\delta u\,\mathrm{d}\zeta = 0\,. \tag{4.180}$$

Taking into account the fact that δu is an arbitrary nonzero function, from (4.180), we get

$$(A_{33}u'')'' - Q_{10}u'' + k^4 u + P\,\delta\,(\eta - \eta_3) + T\,\delta'\,(\eta - \eta_1) + qH\,(\eta - \eta_2) = 0\,. \tag{4.181}$$

Equation (4.181) is the equilibrium equation for the case illustrated in Fig. 4.10.

In more general and compact notation, the relation (4.180) reads

$$\int_0^1 L\,(u)\,\delta u\,\mathrm{d}\zeta = 0\,, \tag{4.182}$$

where $L(u)$ is the left-hand side of the equilibrium equation (e.g. (4.181)). If u is an exact solution of the equilibrium equation, then $L(u) = 0$. If u is an approximate solution, then it does not satisfy the equilibrium equation; as a result, we can regard (4.182) as an additional condition (besides the boundary conditions) that u must satisfy. Suppose that u_{a} is an approximate solution. Substituting u_{a} into the equilibrium equation $\mathrm{L}(u) = 0$, we have $L(u_{\mathrm{a}}) = q_{\mathrm{a}} \neq 0$, where q_{a} has the dimensions of distributed load. From (4.182) it follows that

$$\int_0^1 q_{\mathrm{a}}\,\delta u\,\mathrm{d}\zeta = 0\,,$$

i.e. the work done by the unbalanced distributed load q_{a} through virtual displacements must equal zero.

The relation (4.182) is the basic formula for determination of approximate solutions. Let us represent the displacement u as a series

$$u_{\mathrm{a}} = \sum_{i=1}^{n} a_i v_i(\eta)\,, \tag{4.183}$$

where a_i are constants and $v_i(\eta)$ are known functions, which satisfy the geometrical and the physical boundary conditions. We will seek the virtual displacements of the axial points in the form

$$\delta u = \sum_{i=1}^{n} \delta b_i v_i(\eta)\,, \tag{4.184}$$

where δb_i are independent arbitrary quantities.

Substitution of (4.183) and (4.184) into (4.182) yields

$$\sum_{i=1}^{n} \delta b_i \int_0^1 L\,(a_i v_i)\, v_j \,\mathrm{d}\zeta = 0\,. \tag{4.185}$$

Recall that δb_i are independent functions; in view of (4.185), we have

$$\int_0^1 L\,(a_i v_i)\, v_j \,\mathrm{d}\zeta = 0\,. \tag{4.186}$$

Integration of (4.186) yields the following linear system of algebraic equations in the unknowns a_i:

$$\sum_{i=1}^{n} \alpha_{ij} a_i = c_j\,, \qquad j = 1\,,\, 2\,,\, \ldots\,,\, n\,.$$

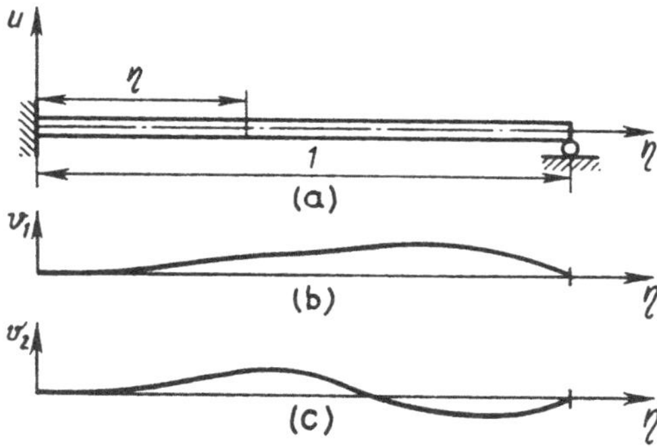

Fig. 4.11.

During the determination of an approximate solution, we used the relation (4.183), which contains the functions $v_i(\eta)$. As these functions we can take a set of mutually orthogonal polynomials which satisfy the boundary conditions is very effective in determination of approximate solutions. Let us obtain such polynomials for a rod shown in Fig. 4.11*a*. It is reasonable to represent the nondimensional displacement v in the form of a finite power series

$$v_1(\eta) = a_0 + a_1\eta + a_2\eta^2 + a_3\eta^3 + a_4\eta^4\,,$$

where the number n of the coefficients a_i exceeds by one the number of the boundary conditions. In view of the boundary conditions, setting $a_4 = 1$, we get

$$v_1 = \eta^4 - \frac{5}{2}\eta^3 + \frac{3}{2}\eta^2\,. \tag{4.187}$$

This function is positive for $\eta \in (0\,,\,1)$ (see Fig. 4.11*b*). The degree of the polynomial for $v_2(\eta)$ exceeds by one the degree of the polynomial for $v_1(\eta)$; thus,

$$v_2 = a_0 + a_1\eta + a_2\eta^2 + a_3\eta^3 + a_4\eta^4 + a_5\eta^5 .$$

Let $a_5 = 1$, then

$$v_2 = \eta^5 + a_4\eta^4 - \left(\frac{9}{2} + \frac{5}{2}a_4\right)\eta^3 + \left(\frac{7}{2} + \frac{3}{2}a_4\right)\eta^2 . \tag{4.188}$$

From the orthogonality condition for v_1 and v_2 on the interval $(0, 1)$ we have

$$\int_0^1 v_1 v_2 \,\mathrm{d}\zeta = c_1 a_4 + b_1 = 0 ;$$

as a result, we get $a_4 = -2.2535$.

The further calculations yield

$$v_2 = \eta^5 - 2.2535\eta^4 + 1.1337\eta^3 + 0.1198\eta^2 . \tag{4.189}$$

As η runs from 0 to 1, the function $v_2(\eta)$ equals zero just once as illustrated in Fig. 4.11*b*. The expression for $v_3(\eta)$ consists of seven summands and contains two free parameters a_4 and a_5 (if we put $a_6 = 1$). These parameters are to be found from the following orthogonality conditions:

$$\int_0^1 v_1 v_3 \,\mathrm{d}\zeta = c_{11} a_4 + c_{12} a_5 + b_1 = 0 ;$$

$$\int_0^1 v_2 v_3 \,\mathrm{d}\zeta = c_{21} a_4 + c_{22} a_5 + b_2 = 0 .$$

For a particular problem, the procedure of evaluation of the functions v_i can be easily realized on computer. Let us approximate the nondimensional displacement by a finite sum,

$$u_{\mathrm{a}}(\eta) = \sum_{i=1}^{n} a_i v_i(\eta) , \tag{4.190}$$

where a_i are arbitrary constants. Now, we can estimate the accuracy of the solution and the minimum number of terms in (4.190) to be taken to satisfy the required accuracy. The absolute error cannot be estimated because the exact solution is unknown. Hence, the only thing we can do is to compare the values of the left-hand side of (4.190) for different n.

Let $u_{\mathrm{a}1} = a_1 v_1$ and $u_{\mathrm{a}2} = a_1 v_1 + a_2 v_2$ be approximate solutions. If for any η such that $\eta \in [0, 1]$ we have $\max|\Delta u_{\mathrm{a}1}| \le \Delta u_{\mathrm{al}}$, where Δu_{al} is the allowable error, then v_2 can be regarded as the sought solution. If $\max|\Delta u_{\mathrm{a}1}| > \Delta u_{\mathrm{al}}$, then we should seek the solution in the form $u_{\mathrm{a}3} = a_1 v_1 + a_2 v_2 + a_3 v_3$ and check the validity of the inequality

$$\max|\Delta u_{\mathrm{a}2}| = \max|u_{\mathrm{a}3} - u_{\mathrm{a}2}| \le \Delta u_{\mathrm{al}} .$$

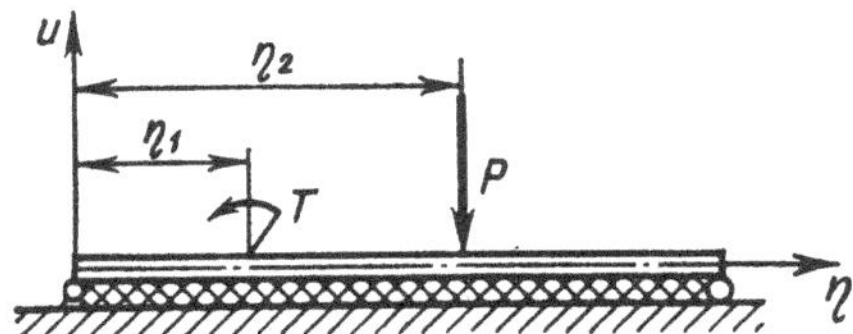

Fig. 4.12.

Finally, a function u_{ak} such that

$$\max |\Delta u_{a(k-1)}| = \max |u_{ak} - u_{a(k-1)}| \leq \Delta u_{al} \tag{4.191}$$

can be taken as the sought deflection of the rod.

As an illustration, consider a simply supported rod on a linearly elastic foundation as shown in Fig. 4.12. Let us take the functions $v_i(\eta)$ in the form

$$v_i(\eta) = \sin i\pi\eta .$$

If $A_{33} = 1$, the equilibrium equation is as follows:

$$u^{\mathrm{IV}} + k^4 u + T\,\delta'(\eta - \eta_1) + P\delta\,(\eta - \eta_2) = 0 . \tag{4.192}$$

We seek an approximate solution as a finite sum

$$u_a = \sum_{i=1}^{n} a_i \sin i\pi\eta .$$

Let virtual displacements of the points of the rod axis be of the form

$$\delta u = \sum_{i=1}^{n} \delta b_i \sin i\pi\eta .$$

According to the principle of virtual displacements, we get

$$\int_0^1 L\,(u_a) \sin j\pi\eta \,\mathrm{d}\eta = 0 , \qquad j = 1 , 2 , \ldots , n ,$$

or

$$\frac{a_j}{2}\,[\,(j\pi)^4 + k^4\,] - Tj\pi \cos j\pi\eta_1 - P \sin j\pi\eta_2 = 0 .$$

Consequently,

$$a_j = 2\,\frac{P \sin j\pi\eta_2 + Tj\pi \cos j\pi\eta_1}{(j\pi)^4 + k^4} .$$

The two-term approximate solution of (4.192) is as follows:

$$u_a = \frac{2\left(P \sin \pi\eta_2 + T \cos \pi\eta_1\right)}{\pi^4 + k^4} \sin \pi\eta + \frac{2\left(P \sin 2\pi\eta_2 + T \cos 2\pi\eta_1\right)}{16\pi^4 + k^4} \sin 2\pi\eta .$$

The method presented above implies reduction of a system of the first-order differential equilibrium equations to one fourth-order equation. This reduction requires a great deal of calculation. It is especially so for rods of varying cross section or when force increments Δq, $\Delta \mu$, ΔP_i, and ΔT are not linear in the displacement u and the angle ϑ_3. For example, a rod of varying cross section is illustrated in Fig. 4.10. The rod is subjected to an axial force $\mathbf{P}_1 = P_{10}\mathbf{i}_1$ ($Q_{10} = P_{10} \neq 0$). In the case of follower forces, the equilibrium equations of the zeroth approximation are

$$\begin{aligned}
&\frac{\mathrm{d}Q_2^{(0)}}{\mathrm{d}\eta} + \frac{Q_{10}}{A_{33}} M_3^{(0)} + k^4 u_2^{(0)} - qH\left(\eta - \eta_2\right) - P\,\delta\left(\eta - \eta_3\right) = 0\,;\\
&\frac{\mathrm{d}M_3^{(0)}}{\mathrm{d}\eta} + Q_2^{(0)} + T\,\delta\left(\eta - \eta_1\right) = 0\,;\\
&\frac{\mathrm{d}\vartheta_3^{(0)}}{\mathrm{d}\eta} - \frac{M_3^{(0)}}{A_{33}} = 0\,;\\
&\frac{\mathrm{d}u_2^{(0)}}{\mathrm{d}\eta} - \vartheta_3^{(0)} = 0\,.
\end{aligned} \tag{4.193}$$

Let $M_{10} = 0$. For follower forces, the first two equations of the system (4.193) are special cases of (4.92) and (4.93), respectively. Eliminating successively $\vartheta_3^{(0)}$, $M_3^{(0)}$, and $Q_3^{(0)}$, we obtain (see (4.181))

$$\left(A_{33} u_2^{(0)\prime}\right)'' - Q_{10} u_2^{(0)\prime\prime} + k^4 u_2^{(0)} + P\,\delta\left(\eta - \eta_3\right) + T\,\delta'\left(\eta - \eta_1\right) + qH\left(\eta - \eta_2\right) = 0\,.$$

Recall that (4.181) was used as an illustrative example during the derivation of (4.182).

Now let us discuss the derivation of approximate solutions of the equilibrium equations (4.193) without reduction to one differential equation. In vector notation, we have

$$\mathbf{L}\left(\mathbf{Y}\right) = \mathbf{Y}' + \mathbf{A}\mathbf{Y} + \mathbf{b} = 0\,, \tag{4.194}$$

where

$$\mathbf{Y} = \begin{bmatrix} Q_2^{(0)} \\ M_3^{(0)} \\ \vartheta_3^{(0)} \\ u_2^{(0)} \end{bmatrix}; \qquad \mathbf{b} = \begin{bmatrix} -qH\left(\eta - \eta_2\right) - P\,\delta\left(\eta - \eta_3\right) \\ T\,\delta\left(\eta - \eta_1\right) \\ 0 \\ 0 \end{bmatrix};$$

$$\mathrm{A} = \begin{bmatrix} 0 & \dfrac{Q_{10}}{A_{33}} & 0 & 0 \\ 1 & 0 & 0 & 0 \\ 0 & -\dfrac{1}{A_{33}} & 0 & 0 \\ 0 & 0 & -1 & 0 \end{bmatrix}.$$

Consider a one-term approximate solution

$$\mathbf{Y}_{\mathrm{a}}(\eta) = a_1 \mathbf{v}^{(1)}(\eta), \tag{4.195}$$

where a_1 is an arbitrary constant and $\mathbf{v}^{(1)}(\eta)$ is a vector whose components satisfy the boundary conditions. For example, the function $v_1(\eta)$ defined by (4.187) and its derivatives satisfy the boundary conditions (see Fig. 4.11). Consequently, under the same boundary conditions, an approximate solution of (4.193) may be taken in the form

$$\mathbf{v}^{(1)} = \begin{bmatrix} v_1^{(1)} \\ v_2^{(1)} \\ v_3^{(1)} \\ v_4^{(1)} \end{bmatrix} = \begin{bmatrix} -v_1''' \\ v_1'' \\ v_1' \\ v_1 \end{bmatrix}.$$

To improve the accuracy of the solution, we can take the vector $\mathbf{Y}$ as a sum

$$\mathbf{Y}_{\mathrm{a}} = \sum_{i=1}^{m} a_i \mathbf{v}^{(i)}, \tag{4.196}$$

where

$$\mathbf{v}^{(i)} = \begin{bmatrix} -v_i''' \\ v_i'' \\ v_i' \\ v_i \end{bmatrix}.$$

Let virtual displacements be of the form

$$\delta \mathbf{v} = \delta b_1 \mathbf{v}^{(1)},$$

where δb_1 is an arbitrary factor. Consider the scalar product

$$J = (\mathbf{L} \cdot \mathrm{E}_0\, \delta \mathbf{v}^{(1)}) = \delta b_1\, (\mathbf{L} \cdot \mathrm{E}_0 \mathbf{v}^{(1)}), \tag{4.197}$$

where

$$\mathrm{E}_0 = \begin{bmatrix} 0 & 0 & 0 & 1 \\ 0 & 0 & 1 & 0 \\ 0 & 1 & 0 & 0 \\ 1 & 0 & 0 & 0 \end{bmatrix}.$$

All the summands in the scalar product (4.197) have the dimension of work.

Let the work be equal to zero, that is,

$$\int_0^1 J\,\mathrm{d}\zeta = \delta b_1 \int_0^1 (\mathbf{L}\cdot \mathrm{E}_0\mathbf{v}^{(1)})\,\mathrm{d}\zeta = 0$$

or

$$\int_0^1 [\,(\mathbf{v}^{(1)\prime}\mathrm{E}_0\mathbf{v}^{(1)} + \mathrm{A}\mathbf{v}^{(1)}\mathrm{E}_0\mathbf{v}^{(1)})\,a_1 - \mathbf{b}\mathrm{E}_0\mathbf{v}^{(1)}\,]\,\mathrm{d}\zeta\,. \tag{4.198}$$

From (4.198) we get

$$a_1 = \frac{b_1}{b_2}\,,$$

where

$$b_1 = \int_0^1 \mathbf{b}\mathrm{E}_0\mathbf{v}^{(1)}\,\mathrm{d}\zeta\,; \qquad b_2 = \int_0^1 (\mathbf{v}^{(1)\prime}\mathrm{E}_0\mathbf{v}^{(1)} + \mathrm{A}\mathbf{v}^{(1)}\mathrm{E}_0\mathbf{v}^{(1)})\,\mathrm{d}\zeta\,.$$

Using a two-term approximation

$$\mathbf{Y}_{\mathrm{a}} = a_1\mathbf{v}^{(1)} + a_2\mathbf{v}^{(2)}\,,$$

we can obtain a system of two linear inhomogeneous algebraic equations in the unknowns a_1 and a_2.

Let us consider a more complicated problem assuming that a rod axis is not, in general, a plane curve. This is usually so when a torsional moment M_{10} is applied to a rod. The corresponding equilibrium equations were obtained in Sect. 4.2 (see (4.95)). If the applied forces are follower, then these equations can be rewritten as two vector equations

$$\mathbf{L}_1 = \mathbf{Y}_1' + \mathrm{A}_1\mathbf{Y}_1 + \mathrm{B}_1\mathbf{Y}_2 + \mathbf{b}_1 = 0\,; \tag{4.199}$$

$$\mathbf{L}_2 = \mathbf{Y}_2' + \mathrm{A}_2\mathbf{Y}_2 + \mathrm{B}_2\mathbf{Y}_1 + \mathbf{b}_2 = 0\,, \tag{4.200}$$

where

$$\mathbf{Y}_1 = \begin{bmatrix} Q_2^{(0)} \\ M_3^{(0)} \\ \vartheta_3^{(0)} \\ u_2^{(0)} \end{bmatrix}; \qquad \mathrm{A}_1 = \begin{bmatrix} 0 & \dfrac{Q_{10}}{A_{33}} & 0 & 0 \\ 1 & 0 & 0 & 0 \\ 0 & -\dfrac{1}{A_{33}} & 0 & 0 \\ 0 & 0 & -1 & 0 \end{bmatrix};$$

$$\mathbf{b}_1 = \begin{bmatrix} P_{20} \\ T_{30} \\ 0 \\ 0 \end{bmatrix}; \qquad \mathrm{B}_1 = \begin{bmatrix} -\dfrac{M_{10}}{A_{11}} & 0 & 0 & 0 \\ 0 & \dfrac{M_{10}(A_{22}-A_{11})}{A_{11}A_{22}} & 0 & 0 \\ 0 & 0 & 0 & 0 \\ 0 & 0 & 0 & 0 \end{bmatrix};$$

$$\mathbf{Y}_2 = \begin{bmatrix} Q_3^{(0)} \\ M_2^{(0)} \\ \vartheta_2^{(0)} \\ u_3^{(0)} \end{bmatrix}; \qquad \mathrm{A}_2 = \begin{bmatrix} 0 & -\dfrac{Q_{10}}{A_{22}} & 0 & 0 \\ -1 & 0 & 0 & 0 \\ 0 & -\dfrac{1}{A_{22}} & 0 & 0 \\ 0 & 0 & 1 & 0 \end{bmatrix};$$

$$\mathbf{b}_2 = \begin{bmatrix} P_{30} \\ T_{20} \\ 0 \\ 0 \end{bmatrix}; \qquad \mathrm{B}_2 = \begin{bmatrix} \dfrac{M_{10}}{A_{11}} & 0 & 0 & 0 \\ 0 & \dfrac{M_{10}(A_{33}-A_{11})}{A_{33}A_{11}} & 0 & 0 \\ 0 & 0 & 0 & 0 \\ 0 & 0 & 0 & 0 \end{bmatrix}.$$

We restrict our consideration to a one-term approximation. Let

$$\mathbf{Y}_1 = a_1 \mathbf{v}^{(1)}; \qquad \mathbf{Y}_2 = b_1 \mathbf{u}^{(1)}, \tag{4.201}$$

where

$$\mathbf{v}^{(1)} = \begin{bmatrix} -v_1''' \\ v_1'' \\ v_1' \\ v_1 \end{bmatrix}; \qquad \mathbf{u}^{(1)} = \begin{bmatrix} u_1''' \\ u_1'' \\ u_1' \\ u_1 \end{bmatrix}.$$

The functions v_1 and u_1 as well as their derivatives satisfy the end fixity conditions in the vertical plane x_1Ox_2 and in the horizontal plane x_1Ox_3, respectively. Let generalized virtual displacements be of the form

$$\delta \mathbf{v} = \delta c_1 \mathbf{v}^{(1)}; \qquad \delta \mathbf{u} = \delta q_1 \mathbf{u}^{(1)}.$$

Then, we substitute the vectors (4.201) into (4.199) and multiply both sides of (4.199) by the vector $\mathrm{E}^{(1)}\,\delta c_1\,\mathbf{v}^{(1)}$. The following equality must hold:

$$\delta c_1 \int_0^1 (\mathbf{L}_1 \cdot \mathrm{E}_0 \mathbf{v}^{(1)})\, \mathrm{d}\zeta = 0$$

or

$$a_1 \int_0^1 (\mathbf{v}^{(1)\prime} \cdot \mathrm{E}_0 \mathbf{v}^{(1)} + \mathrm{A}_1 \mathbf{v}^{(1)} \cdot \mathrm{E}_0 \mathbf{v}^{(1)})\, \mathrm{d}\zeta$$
$$+ b_1 \int_0^1 (\mathrm{B}_1 \mathbf{u}^{(1)} \cdot \mathrm{E}_0 \mathbf{v}^{(1)})\, \mathrm{d}\zeta + \int_0^1 (\mathbf{b}_1 \cdot \mathrm{E}_0 \mathbf{v}^{(1)})\, \mathrm{d}\zeta = 0. \tag{4.202}$$

Multiplying both sides of (4.200) by $E^{(1)}\,\delta q_1\,\mathbf{u}^{(1)}$, we get

$$\delta q_1 \int_0^1 (\mathbf{L}_2 \cdot \mathrm{E}_0 \mathbf{u}^{(1)})\, \mathrm{d}\zeta = 0$$

or

$$a_1 \int_0^1 (\mathbf{B}_2 \mathbf{v}^{(1)} \cdot \mathbf{E}_0 \mathbf{u}^{(1)})\, \mathrm{d}\zeta + b_1 \int_0^1 (\mathbf{u}^{(1)\prime} \cdot \mathbf{E}_0 \mathbf{u}^{(1)}$$

$$+ \mathbf{A}_2 \mathbf{u}^{(1)} \cdot \mathbf{E}_0 \mathbf{u}^{(1)})\, \mathrm{d}\zeta + \int_0^1 (\mathbf{b}_2 \cdot \mathbf{E}_0 \mathbf{u}^{(1)})\, \mathrm{d}\zeta = 0\,. \tag{4.203}$$

As a result, two equations (4.202) and (4.203) in the two unknowns a_1 and b_1 are obtained.

4.5.2 Principle of Minimum of Potential Energy

One of the widely used approximating methods for statics of elastic structures is based on the following principle:

An elastic structure subjected to static external forces takes an equilibrium configuration (among the variety of admissible ones) such that the potential energy attains a local minimum:

$$J = U = \min\,. \tag{4.204}$$

As the structure deforms, the external forces do perform the work. This work is converted into the strain energy of the structure; as a result, we have an additional condition

$$U = A\,, \tag{4.205}$$

where A is the work done by the external forces. The relation (4.205) is valid only for elastic systems because in this case the strain energy is not dissipated by irreversible processes. For a rod lying on an elastic foundation (see Fig. 4.12), the potential energy of the system and the work of the external forces may be written as follows:

$$U = \frac{1}{2} \int_0^1 (A_{33} u'^2 + k^4 u^2)\, \mathrm{d}\zeta\,; \tag{4.206}$$

$$A = \frac{1}{2} P u\,(\eta_2) + \frac{1}{2} T u'\,(\eta_1)\,. \tag{4.207}$$

Thus we arrive at a conditional-extremum problem of calculus of variations. Instead of (4.204), let us introduce the functional

$$J_1 = U + \lambda\,(U - A) = \mathrm{extr}\,, \tag{4.208}$$

where λ is a Lagrangian multiplier. If an elastic system obeys Hooke's law, the potential energy can be represented as a quadratic form in displacements u_i of the points of application of the external forces, that is,

$$U = \sum_{i=1}^{n} \sum_{j=1}^{n} \frac{1}{2} c_{ij} u_i u_j\,, \tag{4.209}$$

where u_i are generalized displacements (linear and angular) of the cross sections of application of the external forces and moments, c_{ij} are stiffness parameters, and n is the total number of external forces and moments.

In the case of static loading, the work done by external forces is

$$A = \frac{1}{2} \sum_{i=1}^{n} R_i u_i \,, \tag{4.210}$$

where the components of the forces and the moments are denoted by R_i.

The relation (4.210) is valid provided the forces R_i and the displacements have the same directions. In the general case, the work done by external forces can be written as follows:

$$A = \frac{1}{2} \sum_{i=1}^{n} (\mathbf{R}_i \cdot \mathbf{u}_i) \,.$$

It is important to note that the potential energy of a system is a homogeneous function of degree two while the work done by external forces is a homogeneous function of degree one.

Recall some basic properties of homogeneous functions. Here we restrict our consideration to polynomials. A function is said to be *homogeneous* if it can be represented as a sum of terms of the same degree. For example, the function

$$f\,(x\,,\,y) = x^2 + 2xy - 3y^2$$

is a homogeneous polynomial of degree two. The substitution of tx and ty for x and y, respectively, is equivalent to the multiplication of the polynomial by t^2. Similarly, for homogeneous functions of degree m, the substitution of tx for x and ty for y is equivalent to the multiplication by t^m, that is,

$$f\,(tx\,,\,ty) = t^m f(x\,,\,y) \,.$$

Suppose that a homogeneous function $f\,(x\,,\,y\,,\,z)$ of degree m has continuous partial derivatives with respect to x, y, and z. Then, for fixed values of the arguments, we have

$$f\,(tx_0\,,\,ty_0\,,\,tz_0) = t^m f(x_0\,,\,y_0\,,\,z_0) \,.$$

Differentiation with respect to t yields

$$\frac{\partial f}{\partial x}\,x_0 + \frac{\partial f}{\partial y}\,y_0 + \frac{\partial f}{\partial z}\,z_0 = t^{m-1} m f\,(x_0\,,\,y_0\,,\,z_0) \,. \tag{4.211}$$

Setting $t = 1$ in (4.211) and using the Euler theorem, we arrive at

$$\frac{\partial f}{\partial x}\,x_0 + \frac{\partial f}{\partial y}\,y_0 + \frac{\partial f}{\partial z}\,z_0 = m f\,(x_0\,,\,y_0\,,\,z_0) \,. \tag{4.212}$$

The necessary conditions for the functional J_1 (see (4.208)) to attain a local extremum are as follows:

$$\frac{\partial J_1}{\partial u_i} = (1+\lambda)\,\frac{\partial U}{\partial u_i} - \lambda\,\frac{\partial A}{\partial u_i} = 0\,, \qquad i = 1\,,2\,,\ldots\,,n\,. \tag{4.213}$$

Multiplying (4.213) by u_i and summing the modified equations, we get

$$(1+\lambda)\left(\sum_{i=1}^{n}\frac{\partial U}{\partial u_i}\,u_i\right) - \lambda\left(\sum_{i=1}^{n}\frac{\partial A}{\partial u_i}\,u_i\right) = 0\,. \tag{4.214}$$

According to the Euler theorem,

$$\sum_{i=1}^{n}\frac{\partial U}{\partial u_i}\,u_i = 2U\,; \qquad \sum_{i=1}^{n}\frac{\partial A}{\partial u_i}\,u_i = A\,,$$

therefore, from (4.214) we get

$$(1+\lambda)\,2U - \lambda A = 0\,. \tag{4.215}$$

By use of the equality $A = U$, we obtain the Lagrangian multiplier $\lambda = -2$. For elastic systems governed by Hooke's law and subjected to external forces satisfying (4.209) (e.g. conservative forces), we have $\lambda = -2$. Consequently, we have an extremum problem (not a conditional-extremum problem) for the functional,

$$J_1 = 2A - U = \text{extr}\,. \tag{4.216}$$

We can prove that the extremum of the functional (4.216) is a local maximum, hence the functional

$$J_1 = U - 2A \tag{4.217}$$

attain a minimum. The equilibrium equations can be obtained on the basis of the principle of minimum of the potential energy (see functional (4.217)) as well as on the basis of the principle of virtual displacements.

The equality $J_1 = U - 2A = \min$ is efficient for derivation of approximate solutions of problems of statics. Differential equations that appear in the problems of calculus of variations (e.g. equilibrium equations) can hardly be integrated. Hence, it is important to develop methods that do not reduce the analysis of functionals to the analysis of systems of differential equations. Such methods for variational problems are termed *direct methods*.

4.5.3 Ritz Method

For straight rods, the functional J_1 depends on a rod deflection u and its derivatives u' and u'':

$$J_1 = J_1(u\,,\, u'\,,\, u'') = \min\,,$$

Let us represent u as a series

$$u = \sum a_j v_j(\eta)\,,$$

where a_j are constants to be determined; it is sufficient that the known functions $v_j(\eta)$ satisfy only the geometrical boundary conditions. (Recall that when we used the methods based on the principle of virtual displacements, it was needed that the approximating functions $v_j(\eta)$ satisfy both geometrical and physical boundary conditions.) As such functions, we can take orthogonal polynomials similar to (4.187) and (4.188). Substituting u into the functional J_1 (4.217), we get

$$J_1 = J_1(a_j) = \min\,;$$

as a result, we arrive at the following equations in the unknown coefficients a_j:

$$\frac{\partial J_1}{\partial a_j} = 0\,, \qquad j = 1\,,\, 2\,,\, \ldots\,,\, n\,. \tag{4.218}$$

Solving these equations for a_j, we readily get an approximate expression for u.

As an example, let us find deflections of the rod shown in Fig. 4.12. Using (4.206) and (4.207), we get

$$J_1 = \frac{1}{2}\int_0^1 (A_{33}u^{2''} + k^4u^2)\,\mathrm{d}\zeta - Pu\,(\eta_2) - Tu'\,(\eta_1)\,. \tag{4.219}$$

Restrict our consideration to two-term approximation

$$u = a_1 \sin\pi\eta + a_2 \sin 2\pi\eta\,. \tag{4.220}$$

Substituting (4.220) into (4.219) and using (4.218), we get two independent equations in the unknowns a_1 and a_2:

$$\frac{1}{2}\,(\pi^4 + k^4)\,a_1 = P\sin\pi\eta_2 + T\pi\cos\pi\eta_1\,;$$
$$\frac{1}{2}\,(16\pi^4 + k^4)\,a_2 = P\sin 2\pi\eta_2 + T2\pi\cos 2\pi\eta_1$$

or

$$a_1 = \frac{2\,(P\sin\pi\eta_2 + T\pi\cos\pi\eta_1)}{\pi^4 + k^4}\,;$$
$$a_2 = \frac{2\,(P\sin 2\pi\eta_2 + T2\pi\cos 2\pi\eta_1)}{16\pi^4 + k^4}\,.$$

4.5.4 Approximating Methods Based on Lagrangian Multipliers

Let us consider the method based on the principle of minimum of the potential energy when approximating functions satisfy just a part of geometrical boundary conditions. For a rod shown in Fig. 4.13, we seek an approximate solution as a finite sum

$$u = \sum_{i=1}^{k} a_i \sin i\pi\eta \,. \tag{4.221}$$

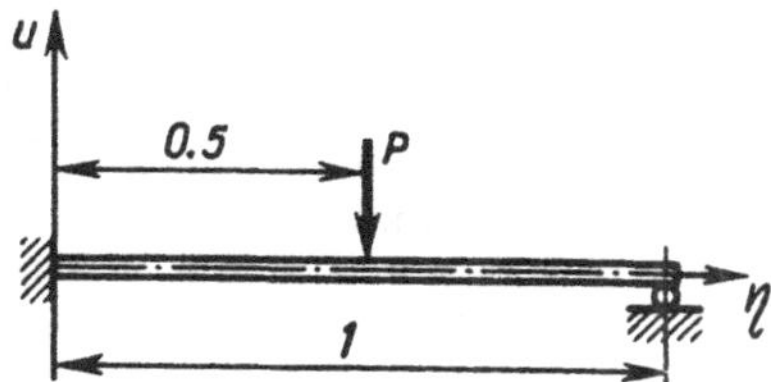

Fig. 4.13.

These functions satisfy only two boundary conditions, $u(0) = 0$ and $u(1) = 0$, but they do not satisfy the third condition, $u'(0) = 0$. Of course, we could choose the approximating functions such that all the three boundary conditions are satisfied (e.g. the functions defined by (4.187) and (4.189)). Anyway, we use the trigonometric functions (4.221) because they are convenient for algebraic manipulation. To account for the third boundary condition, we write

$$u'(0) = \sum_{i=1}^{k} a_i \pi \cos \pi i \eta|_{\eta=0} = 0 \,, \qquad i = 1\,, 2\,, \dots\,, k\,. \tag{4.222}$$

To take advantage of the method of Lagrangian multipliers, let us consider a functional

$$J_2 = J_1 + \lambda u'(0)\,,$$

where λ is a Lagrangian multiplier. The unknown coefficients a_i can be found from the equations

$$\frac{\partial J_2}{\partial a_i} = \frac{\partial J_1}{\partial a_i} + \lambda \frac{\partial u'(0)}{\partial a_i} = 0\,.$$

The coefficients a_i are linear in the multiplier λ, that is, $a_i = a_{i0}\lambda + a_{i1}$. Substituting these expressions into (4.222), we determine λ and a_i.

Suppose that a rod is not simply supported but clamped at the end $\eta = 1$ (Fig. 4.13). In this case, the approximating functions satisfy only one of the three boundary conditions. The function (4.221) must satisfy two additional conditions: $u'(0) = 0$ and $u(1) = 0$. The functional J_2 takes the form

$$J_2 = J_1 + \lambda_1 u'(0) + \lambda_2 u'(1)\,.$$

Thus, the use of Lagrangian multipliers substantially extends the class of approximating functions that can be used in the methods based on the principle of minimum of potential energy.

4.6 Stability of Compressed-Twisted Rods

In the preceding sections of this chapter when considering a straight rod in compression-tension, we assumed that the rod remained straight, i.e. an axial force of any direction does not change the straight-line shape of the rod. It is a well-known fact that this assumption is not always valid.

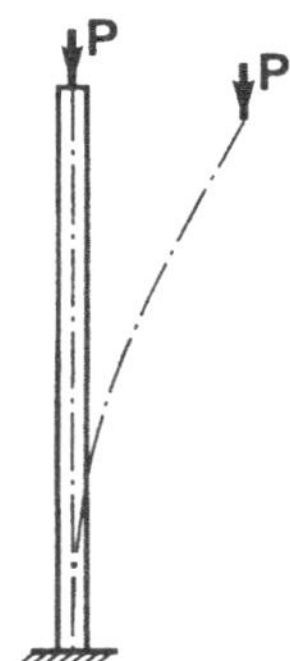

Fig. 4.14.

When a compressive force reaches a certain value, the straight-line configuration of a rod becomes unstable and the rod jumps into another equilibrium configuration (see Fig. 4.14). In applications, most structural rod-like elements are loaded by forces of more complicated nature than those in the Euler problem.

A rod immersed into a steady flow of liquid or gas (e.g. a blade, an aerofoil or a wing) is subjected to distributed forces $\mathbf{q}$ and a torsional moment μ_1 as shown in Fig. 4.15. These loads depend on the angle of rotation of the cross section ϑ_1. Therefore, for a critical value of the flow velocity, a loss of bending-twisting stability of the rod (a divergence) may occur.

A rod subjected to a magnetic field is shown in Fig. 4.16. When the rod is in the straight-line configuration, the distributed attractive forces of the magnets are of the same magnitude and perpendicular to the rod axis. If the rod deflects from the straight-line configuration, then the attractive forces become different and the resultant force acting on the rod is directed away

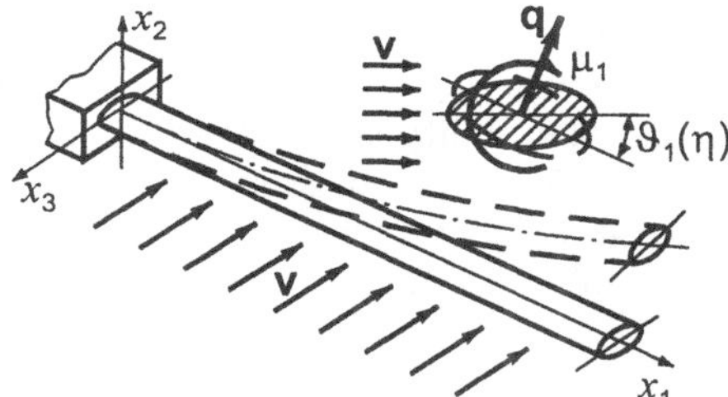

Fig. 4.15.

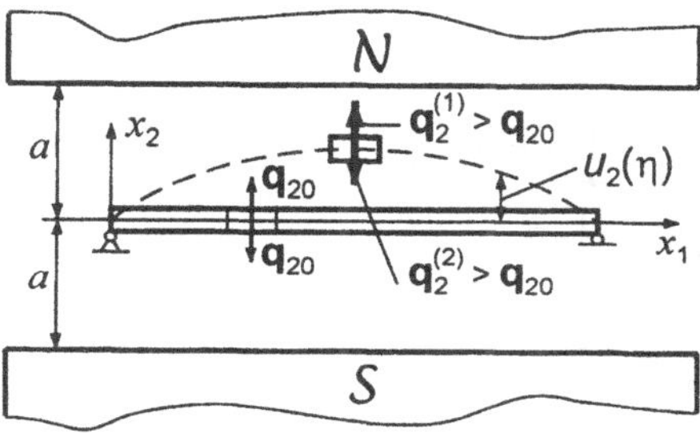

Fig. 4.16.

from the initial location of the rod. The magnitude of the resultant force depends on the component u_2 of the displacement vector **u**. Under a critical value of the strength of the magnetic field, the rod loses its stability. These examples illustrate feedback distributed forces and moments; $q_2^{(j)}$ depends only on the component u_2 of the displacement vector while μ_1 depends on the angle of rotation ϑ_1 of the cross sections.

The two examples considered above, Figs. 4.15 and 4.16, clearly show us that the loss of stability of a straight rod may occur not necessarily under the action of compressive forces.

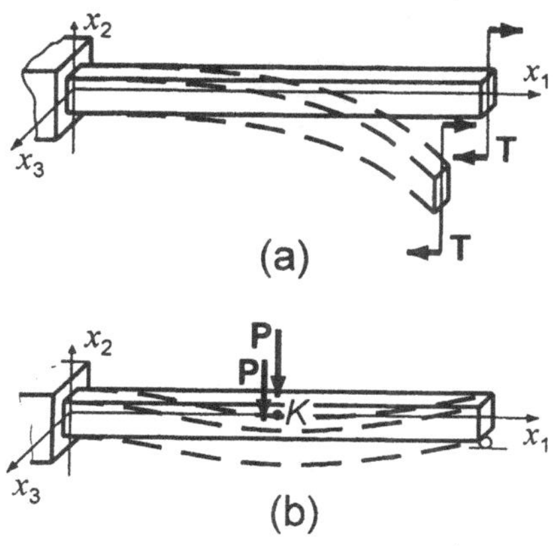

Fig. 4.17.

Straight rods subjected to a dead bending moment **T** and a dead force **P** are shown in Fig. 4.17*a* and Fig. 4.17*b*, respectively. Although no internal axial forces arise in the straight rods, a loss of stability is still possible. With an increase of the absolute values $|\mathbf{P}|$ and $|\mathbf{T}|$, the rod axis bends into a curve lying in the plane x_1Ox_2. Under the critical values P_* and T_*, the plane configuration of the rod is no longer stable and, hence, the out-of-plane loss of stability of the rod takes place.

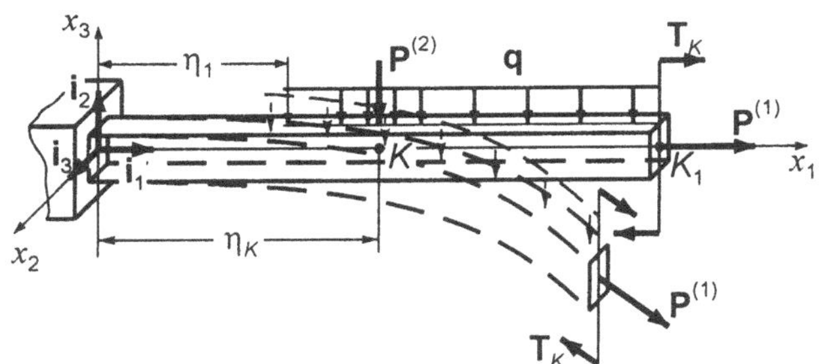

Fig. 4.18.

In structural design, more complicated cases of loading exist, e.g. a rod subjected to a lateral forces **q** and $\mathbf{P}^{(2)}$ and an axial force $\mathbf{P}^{(1)}$ is illustrated in Fig. 4.18. The force $\mathbf{P}^{(1)}$ may be either compressive or tensile. It should be noted that a nonplane mode of loss of stability of a straight rod may be caused by both compressive and tensile forces.

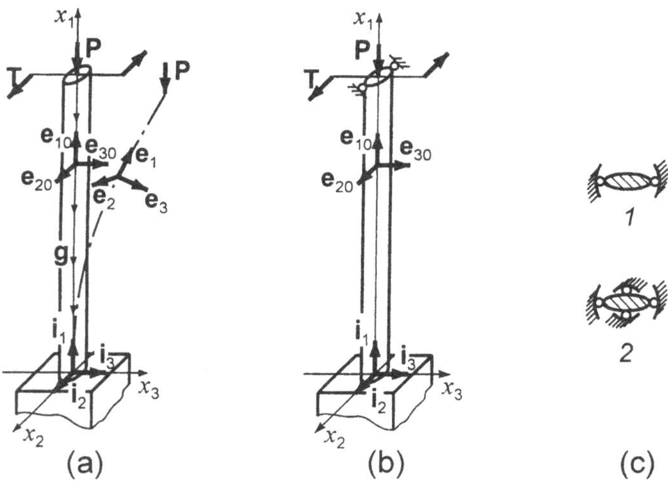

Fig. 4.19.

In the examples considered above, the equilibrium equations are referred to the critical state of the rods, thus the equilibrium equations governing the behavior of the rods after the loss of stability can be obtained by direct specialization of the general equilibrium equations (3.5)–(3.9) and (3.10)–(3.14) (see Chap. 3).

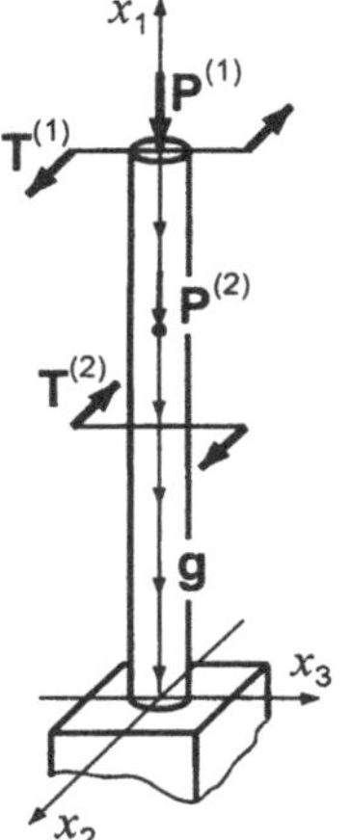

Fig. 4.20.

To illustrate numerical determination of critical loads, let us consider a straight compressed-twisted rod whose bending stiffnesses are different ($A_{22} \neq A_{33}$). Some other examples of numerical determination of critical loads are presented in Sect. 5.4. Straight rods subjected to a compressive force **P**, a distributed axial force **g**, and a concentrated moment **T** are shown in Figs. 4.19*a*-*c* and 4.20. In the general case, each rod has different bending stiffnesses and arbitrary end fixity conditions (Fig. 4.19*c*). Bits for deep-hole drilling (Fig. 4.21) and pipe strings used in boring (Fig. 4.22) can be mathematically treated as rods subjected to compressive forces and torsional moments.

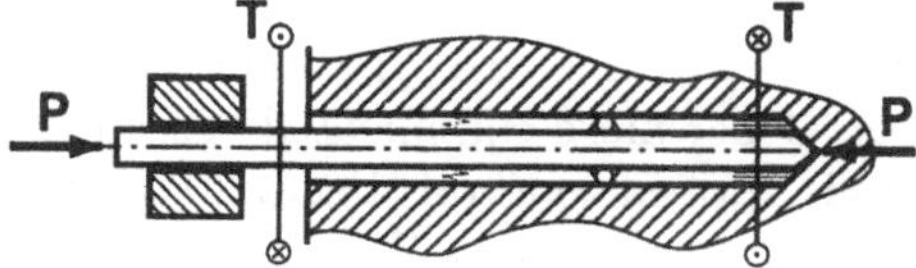

Fig. 4.21.

In spite of a great amount of publications devoted to problems of stability of straight rods, many problems in the field of static stability of rods have not been yet discussed, these being, problems of stability of compressed-twisted rods of different bending stiffnesses. In most papers on problems of stability of straight compressed-twisted rods, the authors assume that the natural (before the loss of stability) twist of a rod can be neglected. This assumption implies that the torsional stiffness of the rod tends to infinity, while in practice the torsional and bending stiffnesses of a rod are of the same order of magnitude. Hence, torsional stiffness must be taken into consideration.

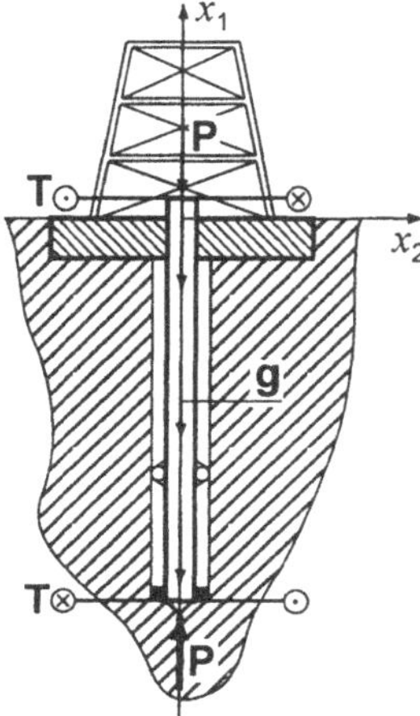

Fig. 4.22.

The loss of static stability is one of the most undesirable phenomena in applications. Consider a rod shown in Fig. 4.19*b*. Before the loss of stability, the rod remains straight. Hence, the state vector $\mathbf{Z}_*$, which characterizes the critical configuration of the rod, has the following components:

$$\begin{aligned}
&\mathbf{Q}_* = -Q_{1*}\mathbf{e}_1\,, &&\text{i.e.}\quad Q_{1*} = P_*\,; &&Q_{2*} = Q_{3*} = 0\,;\\
&\mathbf{M}_* = M_{1*}\mathbf{e}_1\,, &&\text{i.e.}\quad M_{1*} = T_*\,; &&M_{2*} = M_{3*} = 0\,;\\
&\boldsymbol{\ae}_* = \ae_{1*}\mathbf{e}_1\,; &&\text{i.e.}\quad \ae_{2*} = \ae_{3*} = 0\,;
\end{aligned}$$

here P_* and T_* are the critical values of compressive force and torsional moment, respectively.

Suppose that the rod is also subjected to a distributed axial force, say, the gravity force $\mathbf{g}$ as indicated in Fig. 4.19*a*. In this case, the internal axial force depends on the axial coordinate η. Other more complicated cases may appear, e.g. a set of forces and moments is applied to a rod (Fig. 4.20). Some structures may include rigid and elastic local constraints (Fig. 4.21). These constraints may seriously affect the load critical values.

Let us analyze the stability of a compressed-twisted rod subjected to a concentrated force $\mathbf{P}$ and an end torsional moment $\mathbf{T}$. Several types of end fixity conditions will be considered (see Fig. 4.19*a*-*c*). In addition, the rod may be loaded by the distributed axial force $\mathbf{g}$.

The behavior of the rod after the loss of stability can be studied on the basis of (3.29)–(3.32). On rearrangement, we arrive at the following equilibrium equations referred to the attached coordinate system:

$$\begin{aligned}
&\frac{dQ_2}{d\eta} - \frac{Q_{1*}}{A_{33}} M_3 - æ_{1*} Q_3 + \Delta g_2 = 0 ; \\
&\frac{dQ_3}{d\eta} + \frac{Q_{1*}}{A_{22}} M_2 + æ_{1*} Q_2 + \Delta g_3 = 0 ; \\
&\frac{dM_2}{d\eta} + \left(\frac{M_{1*}}{A_{33}} - æ_{1*} \right) M_3 - Q_3 = 0 ; \\
&\frac{dM_3}{d\eta} + \left(æ_{1*} - \frac{M_{1*}}{A_{22}} \right) M_2 + Q_2 = 0 ; \\
&\frac{d\vartheta_2}{d\eta} - \frac{M_2}{A_{22}} - æ_{1*} \vartheta_3 = 0 ; \\
&\frac{d\vartheta_3}{d\eta} - \frac{M_3}{A_{33}} + æ_{1*} \vartheta_2 = 0 ; \\
&\frac{du_2}{d\eta} - \vartheta_3 - æ_{1*} u_3 = 0 ; \\
&\frac{du_3}{d\eta} + \vartheta_2 + æ_{1*} u_2 = 0 .
\end{aligned} \tag{4.223}$$

For a straight rod (Fig. 4.19a), the critical values of the parameters Q_{1*}, M_{1*}, and $æ_{1*}$ are

$$Q_{1*} = P_* ; \qquad M_{1*} = T_* ; \qquad æ_{1*} = \frac{T_*}{A_{11}} .$$

Hence, Q_{1*} and M_{1*} in (4.223) can be replaced by P_* and T_*.

When, besides the forces mentioned above, the rod is subjected to a distributed axial load q_{x_1} and a distributed torsional moment μ_1, then, in general, Q_{1*} and M_{1*} are not equal to P_* and T_*, respectively. That is why the coefficients in (4.223) are represented in terms of Q_{1*} and M_{1*}. This representation of the equations is useful for analysis of more complicated problems.

The terms Δg_2 and Δg_3 in (4.223) are functions of the distributed axial load $\mathbf{g} = -g_{x_1} \mathbf{i}_1$. These terms allows us to account for the rod weight in the stability analysis.

Recall that the equilibrium equations (4.223) were obtained under assumption that the strain-stress state of the rod before and after a loss of stability obeys the generalized Hooke's law. For $M_{1*} = 0$, the system of equations (4.223) decomposes into two uncoupled systems. Each system consists of four equations and governs rod equilibrium in the planes x_1Ox_2 and x_1Ox_3, respectively.

Unlike the commonly used equilibrium equations, the initial twisting is taken into consideration in (4.223).

As noted above, the equality $æ_{1*} = 0$ is used in most publications. Since

$$æ_{1*} = \frac{M_{1*}}{A_{11}}, \tag{4.224}$$

we can see that if $M_{1*} \neq 0$, then the condition $æ_{1*} = 0$ is equivalent to the assumption $A_{11} = \infty$. The latter equality is in contradiction with common sense because all the three stiffnesses of a real rod (the two bending stiffnesses A_{22} and A_{33} and the torsional stiffness A_{11}) are of the same order of magnitude.

It should be emphasized that the system of equations (4.223) is valid for rods of varying and constant cross section. For rods of varying cross section, the bending stiffnesses A_{22} and A_{33} and the torsional stiffness A_{11} are functions of the axial coordinate η. For rods of constant cross section, the nondimensional stiffness A_{33} is equal to unity (see (1.29)).

To account for the force of gravity $\mathbf{g}$, we include the terms Δg_2 and Δg_3 into (4.223). Let us obtain expressions for these terms. Since the gravity force is a dead force (see Fig. 4.19*a*), we get

$$\mathbf{g} = -g_{x_1}\mathbf{i}_1 \,. \tag{4.225}$$

Assume that the angles of rotation are small. In the attached basis, we have

$$\mathbf{i}_1 = \mathbf{e}_1 - \vartheta_3\mathbf{e}_2 + \vartheta_2\mathbf{e}_3 \,,$$

hence,

$$\mathbf{g} = -g_{x_1}\mathbf{e}_1 + g_{x_1}\vartheta_3\mathbf{e}_2 - g_{x_1}\vartheta_2\mathbf{e}_3 \,. \tag{4.226}$$

In the case of critical load, we arrive at

$$\mathbf{g} = \mathbf{g}_* + \Delta\mathbf{g} \,,$$

where

$$\mathbf{g}_* = -g_{x_{1*}}\mathbf{e}_1 \,; \qquad \Delta\mathbf{g} = g_{x_{1*}}\vartheta_3\mathbf{e}_2 - g_{x_{1*}}\vartheta_2\mathbf{e}_3 \,.$$

Finally, we get

$$\Delta\mathbf{g}_2 = g_{x_1}\vartheta_3 \,; \qquad \Delta\mathbf{g}_3 = -g_{x_1}\vartheta_2 \,. \tag{4.227}$$

Because of the gravity force, the internal compressive force Q_{1*} varies along the rod as follows:

$$Q_{1*}(\eta) = Q_{1*}^{(0)} + (1-\eta)\, g_{x_{1*}} \,, \qquad Q_{1*}^{(0)} = P_* \,. \tag{4.228}$$

The first two equations of the system (4.223) take the form

$$\frac{dQ_2}{d\eta} - \frac{Q_{1*}}{A_{33}} M_3 - æ_{1*} Q_3 + g_{x_{1*}} \vartheta_3 = 0\,;$$
$$\frac{dQ_3}{d\eta} + \frac{Q_{1*}}{A_{22}} M_2 + æ_{1*} Q_2 - g_{x_{1*}} \vartheta_2 = 0\,. \tag{4.229}$$

The other equations of the system (4.223) remain unchanged.

Later, when solving (4.223) numerically, we put

$$g_{x_{1*}} = \gamma Q_{1*}^{(0)}\,, \qquad \gamma \geq 0\,,$$

hence,

$$Q_{1*} = Q_{1*}^{(0)} \left[1 + \gamma (1 - \eta) \right] \tag{4.230}$$

Using the formulas (4.227) for Δg_2 and Δg_3, we can represent (4.223) in vector form

$$\frac{d\mathbf{Y}}{d\eta} + \mathrm{A}\mathbf{Y} = 0\,, \tag{4.231}$$

where

$$\mathbf{Y} = (Y_1\,, Y_2\,, \ldots\,, Y_8)^{\mathrm{T}} = (Q_2\,, Q_3\,, M_2\,, M_3\,, \vartheta_2\,, \vartheta_3\,, u_2\,, u_3)^{\mathrm{T}}\,;$$

$$\mathrm{A} = \begin{bmatrix} 0 & -æ_{1*} & 0 & -\dfrac{Q_{1*}}{A_{33}} & 0 & g_{x_{1*}} & 0 & 0 \\ æ_{1*} & 0 & \dfrac{Q_{1*}}{A_{22}} & 0 & -g_{x_{1*}} & 0 & 0 & 0 \\ 0 & -1 & 0 & \dfrac{M_{1*}}{A_{33}} - æ_{1*} & 0 & 0 & 0 & 0 \\ 1 & 0 & æ_{1*} - \dfrac{M_{1*}}{A_{22}} & 0 & 0 & 0 & 0 & 0 \\ 0 & 0 & -\dfrac{1}{A_{22}} & 0 & 0 & -æ_{1*} & 0 & 0 \\ 0 & 0 & 0 & -\dfrac{1}{A_{33}} & æ_{1*} & 0 & 0 & 0 \\ 0 & 0 & 0 & 0 & 0 & -1 & 0 & -æ_{1*} \\ 0 & 0 & 0 & 0 & 1 & 0 & æ_{1*} & 0 \end{bmatrix}.$$

The solution of (4.231) is of the form

$$\mathbf{Y} = \mathrm{K}(\eta)\,\mathbf{C}\,, \qquad \mathrm{K}(0) = \mathrm{E}\,, \tag{4.232}$$

where $\mathrm{K}(\eta)$ is the fundamental matrix of the homogeneous system of equations (4.231) and $\mathbf{C} = (C_1\,, C_2\,, \ldots\,, C_8)^{\mathrm{T}}$ is a vector with arbitrary constant components.

Let us consider a cantilever beam shown in Fig. 4.19*a*. Suppose that the beam is clamped at $\eta = 0$, i.e. $Y_5(0) = Y_6(0) = Y_7(0) = Y_8(0) = 0$. Hence, $C_5 = C_6 = C_7 = C_8 = 0$. The boundary conditions at $\eta = 1$ depend on the type of loading. For a dead force $\mathbf{P}_* = -P_*\mathbf{i}_1$ and a dead torsional moment $\mathbf{T}_* = T_*\mathbf{i}_1$, we have

$$
\begin{aligned}
&Q_2(1) = (\mathbf{P}_* \cdot \mathbf{e}_2(1)) = -P_*(\mathbf{i}_1 \cdot \mathbf{e}_2(1))\,;\\
&Q_3(1) = (\mathbf{P}_* \cdot \mathbf{e}_3(1)) = -P_*(\mathbf{i}_1 \cdot \mathbf{e}_3(1))\,;\\
&M_2(1) = (\mathbf{T}_* \cdot \mathbf{e}_2(1)) = T_*(\mathbf{i}_1 \cdot \mathbf{e}_2(1))\,;\\
&M_3(1) = (\mathbf{T}_* \cdot \mathbf{e}_3(1)) = T_*(\mathbf{i}_1 \cdot \mathbf{e}_3(1))\,.
\end{aligned}
\tag{4.233}
$$

If the angles of rotation are small, the matrix of transformation from the basis $\{\mathbf{i}_j\}$ to the basis $\{\mathbf{e}_j\}$ is as follows (see Appendix I, (A.46)):

$$
\mathrm{L} = \begin{bmatrix} 1 & \vartheta_3 & -\vartheta_2 \\ -\vartheta_3 & 1 & \vartheta_1 \\ \vartheta_2 & -\vartheta_1 & 1 \end{bmatrix}.
$$

Consequently, the boundary conditions (4.223) take the form

$$
\begin{aligned}
&Q_2(1) - P_*\vartheta_3(1) = 0\,; \qquad Q_3(1) - P_*\vartheta_2(1) = 0\,;\\
&M_2(1) + T_*\vartheta_3(1) = 0\,; \qquad M_3(1) - T_*\vartheta_2(1) = 0
\end{aligned}
\tag{4.234}
$$

or, in terms of the components of the state vector $\mathbf{Y}$,

$$
\begin{aligned}
&Y_1(1) - P_*Y_6(1) = 0\,; \qquad Y_2(1) + T_*Y_6(1) = 0\,;\\
&Y_3(1) + P_*Y_5(1) = 0\,; \qquad Y_4(1) - T_*Y_5(1) = 0\,.
\end{aligned}
\tag{4.235}
$$

For a follower load, the boundary conditions at $\eta = 1$ are

$$
Y_1(1) = Y_2(1) = Y_3(1) = Y_4(1) = 0\,.
$$

In the case of a dead force $\mathbf{P}$ and a follower moment $\mathbf{T}_* = T_*\mathbf{e}_1(1)$, the boundary conditions at $\eta = 1$ are as follows:

$$
\begin{aligned}
&Y_1(1) - P_*Y_6(1) = 0\,;\\
&Y_2(1) + P_*Y_5(1) = 0\,;\\
&Y_3(1) = Y_4(1) = 0\,.
\end{aligned}
\tag{4.236}
$$

In view of the boundary conditions at $\eta = 0$, the solution of (4.232) in coordinate form is

$$
Y_j = \sum_{i=1}^{4} k_{ji}C_i\,, \qquad j = 1\,, 2\,, \ldots\,, 8\,.
\tag{4.237}
$$

Therefore, we can rewrite the boundary conditions (4.232) for dead loads as follows:

$$\sum_{j=1}^{4}(k_{1j} - P_* k_{6j})\, C_j = 0\,; \qquad \sum_{j=1}^{4}(k_{2j} + P_* k_{5j})\, C_j = 0\,;$$
$$\sum_{j=1}^{4}(k_{3j} + T_* k_{6j})\, C_j = 0\,; \qquad \sum_{j=1}^{4}(k_{4j} - T_* k_{5j})\, C_j = 0\,. \tag{4.238}$$

For the linear algebraic system (4.238) to have nontrivial solutions, the determinant of the coefficients C_i must vanish, that is,

$$D\,(P_*\,,\, T_*) = 0\,. \tag{4.239}$$

This relation allows us to determine P_{*j} as functions of T_*. The roots of (4.239) are $P_{*j}(T_*)$. These roots are termed *eigenvalues* of the boundary-value problem. The least root is called the *critical force*. Equation (4.239) has an infinite set of roots; in general, these roots can be determined only by use of numerical methods.

Further, we will give examples of determination of critical loads for various types of external loads. Emphasize is on the analysis of the influence of the initial twist $æ_{1*} \neq 0$. Equations (4.223) are written in terms of nondimensional quantities. Hence, P_{*J} and T_* are also nondimensional. To evaluate the corresponding dimensional force and moment, we should multiply P_{*j} and T_* by $A_{33}(0)/l^2$ and $A_{33}(0)/l$, respectively (here, $A_{33}(0)$ is the dimensional bending stiffness and l is the rod length).

As a numerical example, consider a steel cantilever beam of constant cross section the following parameters: $l = 0.7\,\mathrm{m}$, $h = 0.7 \cdot 10^{-2}\,\mathrm{m}$, $b = 0.5 \cdot 10^{-2}\,\mathrm{m}$, $E = 2 \cdot 10^{11}\,\mathrm{N} \cdot \mathrm{m}^{-2}$, and $G = 8 \cdot 10^{10}\,\mathrm{N} \cdot \mathrm{m}^{-2}$. The values of nondimensional stiffnesses are $A_{11} = 0.577$, $A_{22} = 0.5$, and $A_{33} = 1$.

The plots of the first two eigenvalues P_{*1} versus the magnitude of the torsional moment $T = |\mathbf{T}|$ are shown in Fig. 4.23*a*. The force $\mathbf{P}$ and the moment $\mathbf{T}$ are dead. The plots of P_{*3} and P_{*4} versus T are illustrated in Fig. 4.23*b*.

Similar plots can be obtained for other pairs of eigenvalues. The common property of the plots is that as the magnitude of $\mathbf{T}$ is increased, the curves get closer and finally merge at $T = T_{nj}$. In general, for different pairs, the corresponding values T_{nj} are also different. It should be noted that any two successive eigenvalues $P_{*(2j-1)}$ and P_{*2j}, $j = 1\,,\,2\,,\ldots$, are related in such a manner.

The plots for P_{*3} and P_{*4} are of theoretical rather than of practical interest. Nevertheless, the eigenvalues P_{*j}, $j = 2\,,\,3\,,\ldots$, are useful because they allow us to determine the eigenvectors $\mathbf{z}^{(j)}(\eta)$. These vector-functions satisfy the boundary conditions and can be used in the derivation of approximate solutions of linear and nonlinear problems of statics and dynamics.

The least eigenvalue $P_{*1}^{(0)}$ corresponds to the loss of stability in the plane x_1Ox_3. This plane in its turn corresponds to the least bending stiffness of the rod (see Fig. 4.19*a*). If the rod is subjected to a compressive force P of

P
$P_{*2}^{(0)}$
2.4
$P_{*2}(T_*)$
$æ_{1*} \neq 0$
$æ_{1*} = 0$
2.0
II
1
2
1.6
$P_{*2}(T_*)$
III
$P_{*1}^{(0)}$
1.2
I
$\frac{A_{22}}{A_{33}} = 0.5$
0.8
0
0.2
0.4
T_{n2}
T_{n1}
T
P
T
(a)

P
$P_{*2}^{(0)}$
22
20
18
16
14
12
10
$æ_{1*} = 0$
$æ_{1*} \neq 0$
$P_{*2}^{(0)}$
T_{n1}
T_{n2}
0
0.5
1.0
1.5
2.0
T
(b)

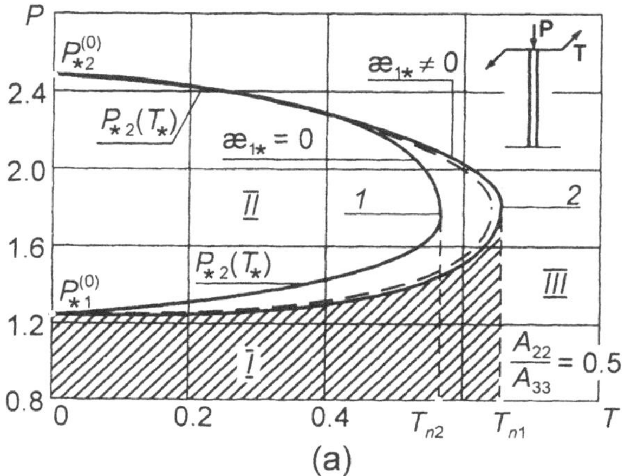

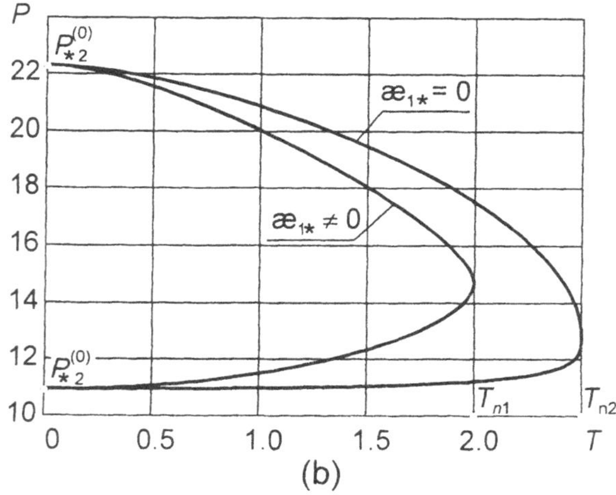

Fig. 4.23.

slow growth, then, as soon as P reaches the value $P_{*1}^{(0)}$, the loss of stability occurs. If after the loss of stability the force P continues its growth, the rod will bend in the plane x_1Ox_3. Therefore, the loss of stability of a cantilever beam in the plane x_1Ox_2, which is the plane of the largest bending stiffness, can hardly take place in practice.

Let us consider in detail some numerical results (see Fig. 4.23*a*). If $æ_{1*} \neq 0$, then a region of values of P and T at which straight line configuration of the rod is stable is shown by hatching (region *I*). The boundary points of the region correspond to the critical state of the rod when straight-line configuration is no longer stable. For $P > P_{*1}$ and some fixed value of T, the rod jumps (bifurcates) into another equilibrium configuration (region *II*).

Linear equilibrium equations allow us to determine the eigenvalues P_{*j} of a boundary-value problem; however, the parameters of a deformed configuration of the rod after the loss of stability (the displacements u_2 and u_3 and the angles of rotation ϑ_2 and ϑ_3 of the attached axes) cannot be obtained from linear equations. That is why, to investigate the strain-stress state of a rod after the loss of stability, one should use nonlinear equilibrium equations, say, (4.40)–(4.44).

As noted, for each pair of eigenvalues $(P_{*(2j-1)}, P_{*2j})$ there exists a certain limit moment T_{nj}, say, T_{n1} in Fig. 4.23*a*. If $T < T_{n1}$, (4.239) has a nonempty set of roots, i.e. static loss of stability is possible. If $T > T_{n1}$ (region *III*), there is no eigenvalues such that $P_{*1} \leq P \leq P_{*2}$. Hence, for T and P from region *III*, the static loss of stability can never appear, i.e. the rod has no equilibrium configurations close to the straight-line configuration. Therefore, for P and T belonging to region *III*, the behavior of the rod can be described by equations of small oscillations; at that, a dynamic loss of stability is possible. In greater detail, the outlined problems are discussed in Svetlitsky (1987) devoted to the dynamics of rods.

Plot *1* in Fig. 4.23*a* corresponds to the case of an infinitely large torsional stiffness ($æ_{1*} = 0$). Plot *2* corresponds to the case of a finite value of torsional stiffness ($æ_{1*} \neq 0$). The dashed curve corresponds to the case of a dead force **P** and a follower moment **T**. For cantilever beams, the values of critical load are insensitive to the behavior of follower moments.

The fact that the real properties of a rod ($æ_{1*} \neq 0$) are taken into consideration enables us to determine the critical loads with a higher accuracy. For example, for $T = 0.56$, the critical value P_{*1} determined for the case $æ_{1*} \neq 0$ is 10% less than the critical value obtained under the condition $æ_{1*} = 0$. Of special note is the fact that the value of the critical force for $æ_{1*} = 0$ exceeds the value of the critical force for a real rod ($æ_{1*} \neq 0$). Such errors are impermissible in design of optimal structures, in which operating characteristics of the elements are selected to be very close to their critical values.

The plots of P_{*1} and P_{*2} versus $T = |\mathbf{T}|$ for $A_{22}/A_{33} = 0.125$ are shown in Fig. 4.24. Recall that $P = |\mathbf{P}|$. For P and T from the hatched region, the static loss of stability of a rod do not occur. Taking into consideration

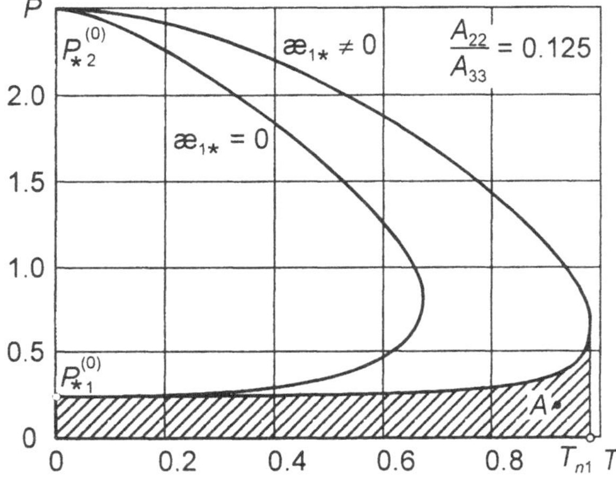

Fig. 4.24.

the initial twist, we can essentially enlarge the region of static stability, i.e. a rod remains stable under torsional moments which considerably exceed the previously obtained critical value. Indeed, the critical moment T_{n1} for $æ_{1*} \neq 0$ differs by 30% from the moment T_{n2} for $æ_{1*} = 0$. For instance, from Fig. 4.24 it follows that for the values of T and P corresponding to the point A the rods keeps its straight configuration while computations not accounting for the initial twist show us that for these values of T and P the loss of stability has already happened.

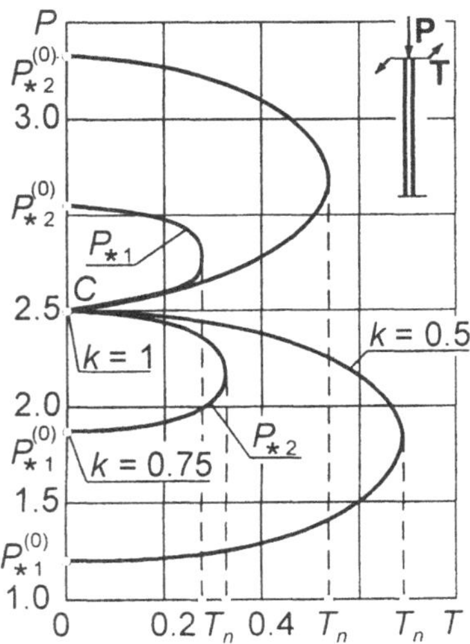

Fig. 4.25.

Now let us consider the relationship between the stability of compressed-twisted rods and the ratio of bending stiffnesses $A_{22}/A_{33} = k$. For dead loads $\mathbf{P}$ and $\mathbf{T}$, the plots of eigenvalues of $P = |\mathbf{P}|$ versus $T = |\mathbf{T}|$ for some values of k are illustrated in Fig. 4.25. An interesting feature of the plots is that as k approaches unity, the eigenvalues $P_{*1}^{(0)}$ and $P_{*2}^{(0)}$ get closer and merge at $k = 1$ while the hatched regions in Fig. 4.24 collapse into a point C. The critical value of the moment decreases continually as k tends to unity and vanishes for $k = 1$. Let us examine the situation for $k = 1$ in greater detail. In this case, the stiffnesses in the planes x_1Ox_2 and x_1Ox_2 are equal, hence, the eigenvalues are also equal. Suppose that the rod is subjected to a torsional moment T such that the magnitude of the moment is as small as is wished; then, for $k = 1$, the rod has no static equilibrium configurations. Indeed, any infinitesimally small torsional moment does exceed the limit value because it equals zero (see Fig. 4.25). Consequently, (4.239) has no roots and thus there is no adjacent equilibrium configurations. In this case, the loss of dynamic stability of the rod occurs.

The special case $k = 1$, when a rod is subjected to a compressive force and a torsional moment, is known as the *Nikolai paradox* (Bolotin (1963)). Actually, the *paradox* can be explained if we assume that the concept ***stability*** implies not only the transition from an unstable equilibrium state to a stable one but also the transition from an unstable equilibrium state to a motion with respect to this equilibrium state.

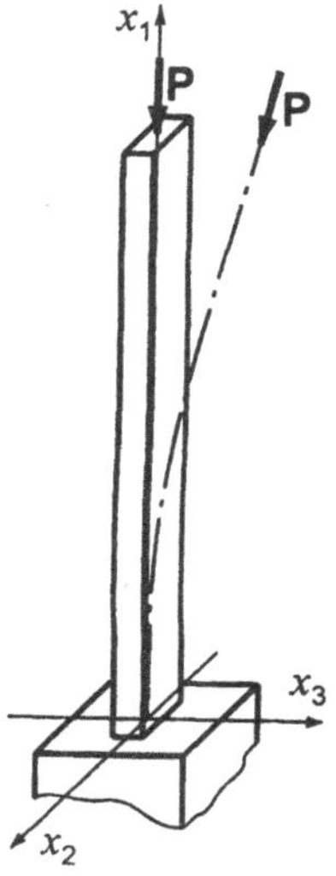

Fig. 4.26.

A similar paradox appears in the stability analysis of a straight cantilever beam of an arbitrary ratio k subjected to a compressive follower force $\mathbf{P}$ (see Fig. 4.26). The equation, which is a counterpart of (4.239), has no roots.

As mentioned above, the roots (eigenvalues) are characteristics of possible equilibrium configurations. Thus, we have proved that the straight-line equilibrium configuration is stable irrespective of the follower force **P**. Considering small oscillations of a rod subjected to a follower compressive force, we can show that that at a certain value of the force the straight-line configuration becomes unstable and the rod breaks into oscillation, i.e. the loss of dynamic stability occurs. An overview of nonstandard problems of elastic stability of rods is presented in Bolotin (1963) devoted to the theory of dynamic stability of rods.

For some nondimensional values of the dead distributed load g_{x_1} (see Fig. 4.19*a*), the plots of the first two eigenvalues P_{*1} and P_{*2} are illustrated in Fig. 4.27. The curves obtained with regard for the initial twist are shown as solid lines and those obtained with no regard for the initial twist are shown as dashed lines. The loads **P** and **T** are dead.

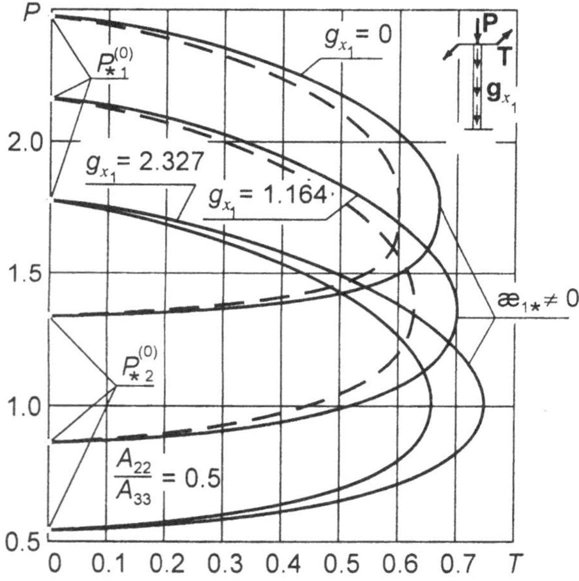

Fig. 4.27.

Solving for the eigenvalues P_{*j}, we obtain the corresponding eigenvectors $\mathbf{Y}_0^{(j)}$. For each eigenvalue P_{*j}, we get the following equation:

$$\frac{\mathrm{d}\mathbf{Y}_0^{(j)}}{\mathrm{d}\eta} + A\left[\eta\,,\ P_{*j}(T_*)\right]\mathbf{Y}_0^{(j)} = 0\,; \tag{4.240}$$

here $\mathbf{Y}_0^{(j)}$ is the corresponding eigenvector.

The eigenvalues P_{*j} are determined from (4.239), which is the characteristic equation for (4.238). Substituting the roots $P_{*j}(T_*)$ into (4.238), we get

$$\sum_{\nu=1}^{4}(k_{1\nu}^{(j)} - P_{*j}k_{6\nu}^{(j)})\,C_\nu^{(j)} = 0\,; \qquad \sum_{\nu=1}^{4}(k_{2\nu}^{(j)} + P_{*j}k_{5\nu}^{(j)})\,C_\nu^{(j)} = 0\,;$$
$$\sum_{\nu=1}^{4}(k_{3\nu}^{(j)} + T_{*j}k_{6\nu}^{(j)})\,C_\nu^{(j)} = 0\,; \qquad \sum_{\nu=1}^{4}(k_{4\nu}^{(j)} - T_{*j}k_{5\nu}^{(j)})\,C_\nu^{(j)} = 0\,, \qquad (4.241)$$

where $k_{\rho\nu}^{(1)}$ are the elements of the fundamental matrix of the vector equation (4.240). In view of the first three equations of the system (4.241), we can write

$$C_1^{(j)} = \alpha_1 C_4^{(j)}\,; \qquad C_2^{(j)} = \alpha_2 C_4^{(j)}\,;$$
$$C_3^{(j)} = \alpha_3 C_4^{(j)}\,; \qquad C_4^{(j)} = 1\,. \qquad (4.242)$$

Setting $C_4^{(j)} = 1$ and using (4.237), we have

$$Y_{0i}^{(j)}(\eta) = \sum_{\nu=1}^{4} k_{i\nu}^{(j)}(\eta)\,\alpha_\nu\,, \qquad i = 1\,, 2\,, \ldots\,, 8\,. \qquad (4.243)$$

This procedure of determination of eigenvectors can be applied to any boundary conditions and loads under which the static loss of stability is possible.

Rods Simply Supported at $\eta = 1$ A rod can be simply supported at $\eta = 1$ in two ways as indicated in Fig. 4.19*c*. Restrict our consideration to case *1*, i.e. let us study an in-plane loss of stability of a rod which is simply supported in the plane x_1Ox_2; let the rod be subjected to a dead load. We have, then, the following boundary-condition equations at $\eta = 1$:

$$Q_3(1) + P_*\vartheta_2(1) = 0\,; \qquad M_2(1) + T_*\vartheta_3(1) = 0\,;$$
$$M_3(1) - T_*\vartheta_2(1) = 0\,; \qquad u_2(1) = 0$$

or

$$Y_2(1) + P_*Y_5(1) = 0\,; \qquad Y_3(1) + T_*Y_6(1) = 0\,;$$
$$Y_4(1) - T_*Y_5(1) = 0\,; \qquad Y_7(1) = 0\,. \qquad (4.244)$$

Plots of the first two eigenvalues P_{*j} versus $T = |\mathbf{T}|$ for $A_{22}/A_{33} = 0.5$ are shown in Fig. 4.28 for the cases $æ_{1*} = 0$ and $æ_{1*} \neq 0$. We clearly see that these plots $P_{*j} = P_{*j}(T)$ differ qualitatively from those for a cantilever beam (see Fig. 4.23*a*). If a continuously growing torsional moment $\mathbf{T}$ is applied to the rod, then the first eigenvalue P_{*1} also increases at the beginning but then starts decreasing and, finally, at a certain $T_{*1}^{(0)}$, vanishes. The value of $T_{*1}^{(0)}$ that corresponds to the case $æ_{1*} = 0$ is different from that in the case $æ_{1*} \neq 0$. Thus, the loss of stability may be caused by the torsional moment $T_{*1}^{(0)}$ alone (for $P = 0$). The region of values of P and T which do not cause the loss of stability is hatched (see Fig. 4.28). If the initial twist is not taken

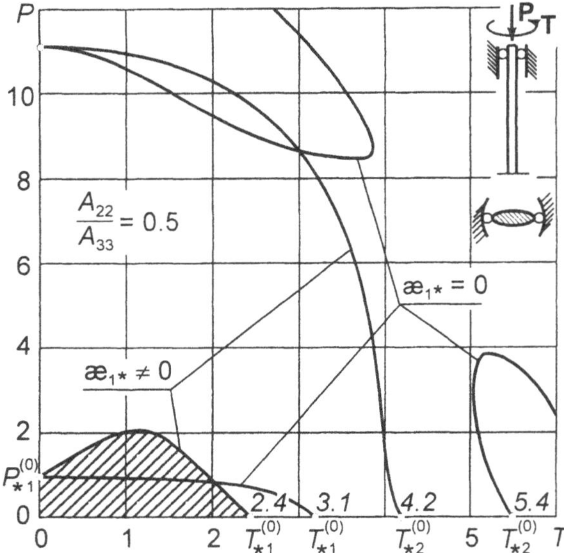

Fig. 4.28.

into consideration, then the corresponding critical value of the moment is 25% above the real critical value (for $P = 0$). Another qualitative difference between the plots in Figs. 4.28 and 4.23*a* is that in Fig. 4.23*a* the eigenvalues get closer as T increases and merge at $T = T_n$. The fact is that a torsional moment alone cannot cause a loss of stability of a cantilever beam.

4.7 Stability of Straight Rods with Local Constraints

A straight rod with elastic and hinge local constraints is shown in Fig. 4.29. Some basic types of local constraints are illustrated in Fig. 4.29*b*. For simplicity, we will consider a rod with only two hinged cross sections η_1 and η_2 (variant *1*). The loss of stability results in reaction forces $\mathbf{R}^{(1)}$ and $\mathbf{R}^{(2)}$. The directions of the reaction forces are unknown a priori. Nevertheless, for small angles of rotation of the attached coordinate axes, the reaction forces are assumed to lie in planes orthogonal to the rod axis after the loss of stability. To account for the reaction forces $\mathbf{R}^{(1)}$ and $\mathbf{R}^{(2)}$ in the equilibrium equations (4.231) after the loss of stability, we will use the δ-functions $\delta_j = \delta(\eta - \eta_j)$. Then, we can write

$$\frac{\mathrm{d}\mathbf{Y}}{\mathrm{d}\eta} + A\mathbf{Y} = \mathbf{b}\,, \tag{4.245}$$

where

$$\mathbf{b} = (-\mathbf{R}^{(1)}\delta_1 - \mathbf{R}^{(2)}\delta_2\,, 0\,, 0\,, 0\,, 0\,, 0\,, 0)^{\mathrm{T}}\,;$$

$$\mathbf{R}^{(1)} = R_2^{(1)}\mathbf{e}_2 + R_3^{(1)}\mathbf{e}_3\,; \qquad \mathbf{R}^{(2)} = R_2^{(2)}\mathbf{e}_2 + R_3^{(2)}\mathbf{e}_3\,.$$

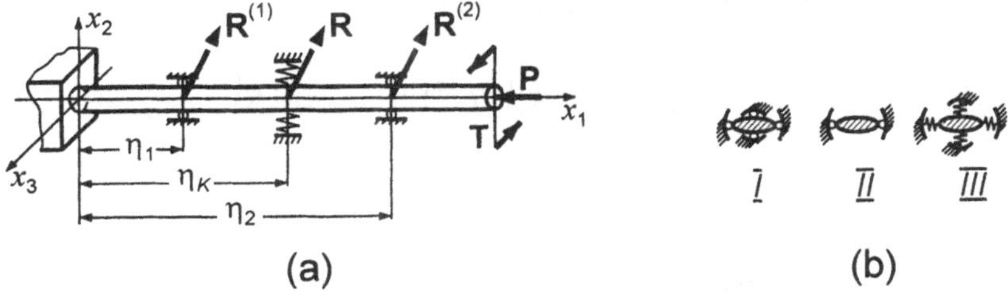

Fig. 4.29.

Equation (4.245) has a solution

$$\mathbf{Y} = \mathrm{K}(\eta)\,\mathbf{C} + \int_0^{\eta} \mathrm{G}(\eta, \zeta)\,\mathbf{b}(\zeta)\,\mathrm{d}\zeta, \tag{4.246}$$

where $\mathrm{G}(\eta, \zeta)$ is a Green matrix

$$\mathrm{G}(\eta, \zeta) = \mathrm{K}(\eta)\,\mathrm{K}^{-1}(\zeta), \qquad \mathrm{K}(0) = \mathrm{E}.$$

For rods of constant cross section, the Green matrix takes the form

$$\mathrm{G}(\eta, \zeta) = \mathrm{K}(\eta - \zeta). \tag{4.247}$$

Let us consider in detail the term containing the vector $\mathbf{b}$ in the right-hand side of (4.246),

$$\mathbf{Y}_H = \int_0^{\eta} \mathrm{K}(\eta - \zeta)\,\mathbf{b}(\zeta)\,\mathrm{d}\zeta. \tag{4.248}$$

The right-hand side of (4.245) can be represented as a sum

$$\mathbf{b} = \mathbf{b}_1\,\delta_1 + \mathbf{b}_2\,\delta_2, \tag{4.249}$$

where

$$\mathbf{b}_1 = (-R_2^{(1)}, -R_3^{(1)}, 0, 0, 0, 0, 0, 0)^{\mathrm{T}};$$
$$\mathbf{b}_2 = (-R_2^{(2)}, -R_3^{(2)}, 0, 0, 0, 0, 0, 0)^{\mathrm{T}}.$$

Substituting the vector $\mathbf{b}$ into (4.248) and integrating, we have

$$\mathbf{Y}_H = \mathrm{K}(\eta - \eta_1)\,\mathbf{b}_1 H_1 + \mathrm{K}(\eta - \eta_2)\,\mathbf{b}_2 H_2, \tag{4.250}$$

where $H_j = H(\eta - \eta_j)$ is a Heaviside function.

As a result, the vector $\mathbf{Y}$ takes the form

$$\mathbf{Y} = \mathrm{K}(\eta)\,\mathbf{C} + \mathrm{K}(\eta - \eta_1)\,\mathbf{b}_1 H_1 + \mathrm{K}(\eta - \eta_2)\,\mathbf{b}_2 H_2. \tag{4.251}$$

For brevity, let us denote the elements of the matrices $\mathrm{K}(\eta)$, $\mathrm{K}(\eta - \eta_1)$, and $\mathrm{K}(\eta - \eta_2)$ by k_{ij}, $k_{ij}^{(1)}$, and $k_{ij}^{(2)}$, respectively.

The components of the vector $\mathbf{Y}$ are as follows:

$$Y_j(\eta) = \sum_{\nu=1}^{8} k_{j\nu} C_\nu - (k_{j1}^{(1)} R_2^{(1)} + k_{j2}^{(1)} R_3^{(1)})\, H_1 - (k_{j1}^{(2)} R_2^{(2)} + k_{j2}^{(2)} R_3^{(2)})\, H_2 \,. \tag{4.252}$$

Each component Y_j contains twelve unknowns: C_ν ($\nu = 1, 2, \ldots, 8$), $R_2^{(1)}$, $R_3^{(1)}$, $R_2^{(2)}$, and $R_3^{(2)}$. These unknowns are to be determined from the boundary conditions at $\eta = 0$ (four equations) and at $\eta = 1$ (four equations) and from the conditions at $\eta = \eta_1$ (two equations) and at $\eta = \eta_2 =$ (two equations).

At $\eta = \eta_1$ and $\eta = \eta_2$, we get

$$\begin{aligned} u_2(\eta_1) &= Y_7(\eta_1) = 0\,; \qquad u_3(\eta_1) = Y_8(\eta_1) = 0\,; \\ u_2(\eta_2) &= Y_7(\eta_2) = 0\,; \qquad u_3(\eta_2) = Y_8(\eta_2) = 0\,; \end{aligned} \tag{4.253}$$

In expanded notation, we have

$$\begin{aligned} Y_7(\eta_1) &= \sum_{j=1}^{4} k_{7j}(\eta_1)\, C_j = 0\,; \\ Y_8(\eta_1) &= \sum_{j=1}^{4} k_{8j}(\eta_1)\, C_j = 0\,; \\ Y_7(\eta_2) &= \sum_{j=1}^{4} k_{7j}(\eta_2)\, C_j - k_{71}^{(1)}(\eta_2 - \eta_1)\, R_2^{(1)} - k_{72}^{(1)} R_3^{(1)} = 0\,; \\ Y_8(\eta_2) &= \sum_{j=1}^{4} k_{8j}(\eta_2)\, C_j - k_{81}^{(1)}(\eta_2 - \eta_1)\, R_2^{(1)} - k_{82}^{(1)} R_3^{(1)} = 0\,. \end{aligned} \tag{4.254}$$

For a rod subjected to a dead load (Fig. 4.29), we have the following boundary conditions at $\eta = 0$: $\vartheta_2 = Y_5 = 0$, $\vartheta_3 = Y_6 = 0$, $u_2 = Y_7 = 0$, and $u_3 = Y_8 = 0$; hence, $C_5 = C_6 = C_7 = C_8 = 0$.

The boundary conditions at $\eta = 1$ have been already obtained (see (4.235)). Using (4.251), we can represent (4.235) as follows:

$$\begin{aligned} &\sum_{j=1}^{4} (k_{1j} - P_* k_{6j})\, C_j - (k_{11}^{(1)} - P_* k_{61}^{(1)})\, R_2^{(1)} - (k_{12}^{(1)} - P_* k_{62}^{(1)})\, R_3^{(1)} \\ &\qquad - (k_{11}^{(2)} - P_* k_{61}^{(2)})\, R_2^{(2)} - (k_{12}^{(2)} - P_* k_{62}^{(2)})\, R_3^{(2)} = 0\,; \\ &\sum_{j=1}^{4} (k_{2j} + P_* k_{5j})\, C_j - (k_{21}^{(1)} + P_* k_{51}^{(1)})\, R_2^{(1)} - (k_{22}^{(1)} + P_* k_{52}^{(1)})\, R_3^{(1)} \\ &\qquad - (k_{21}^{(2)} + P_* k_{51}^{(2)})\, R_2^{(2)} - (k_{22}^{(2)} + P_* k_{52}^{(2)})\, R_3^{(2)} = 0\,; \end{aligned}$$

$$\sum_{j=1}^{4}(k_{3j}+T_*k_{6j})\,C_j-(k_{31}^{(1)}+T_*k_{61}^{(1)})\,R_2^{(1)}-(k_{32}^{(1)}+T_*k_{62}^{(1)})\,R_3^{(1)}$$

$$-\,(k_{31}^{(2)}+T_*k_{61}^{(2)})\,R_2^{(2)}-(k_{32}^{(2)}+T_*k_{62}^{(2)})\,R_3^{(2)}=0\,;$$

$$\sum_{j=1}^{4}(k_{4j}-T_*k_{5j})\,C_j-(k_{41}^{(1)}-T_*k_{51}^{(1)})\,R_2^{(1)}-(k_{42}^{(1)}-T_*k_{52}^{(1)})\,R_3^{(1)}$$

$$-\,(k_{41}^{(2)}-T_*k_{51}^{(2)})\,R_2^{(2)}-(k_{42}^{(2)}-T_*k_{52}^{(2)})\,R_3^{(2)}=0\,. \tag{4.255}$$

Combining (4.254) and (4.255), we get

$$\mathbf{B}\mathbf{Z}=0\,, \tag{4.256}$$

where

$$\mathbf{Z}=(C_1\,,\,C_2\,,\,C_3\,,\,C_4\,,\,R_2^{(1)}\,,\,R_3^{(1)}\,,\,R_2^{(2)}\,,\,R_3^{(2)})^{\mathrm{T}}\,.$$

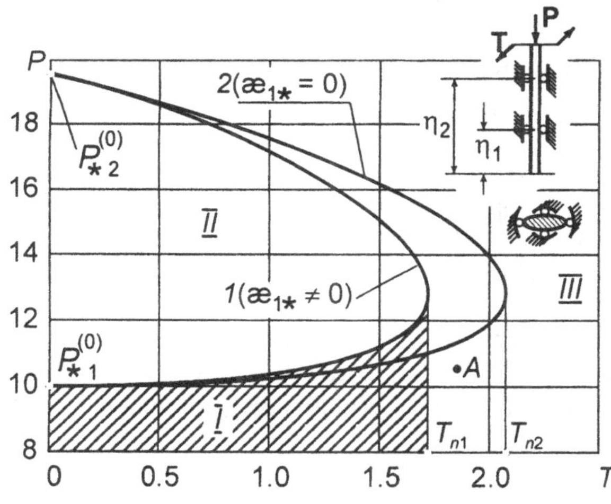

Fig. 4.30.

By use of the equation $\det \mathrm{B}=0$, we can determine the eigenvalues P_{*j} as functions of T, η_1, and η_2.

Equations (4.245) were solved for the numerical values of the physical parameters taken from the previous example.

The relationships between the first two eigenvalues and the moment T were obtained numerically for the following cases: (1) the dead force and the dead moment (Fig. 4.30); (2) the dead force and the follower moment (Fig. 4.31). The plots were obtained for $\eta_1 = 0.1$ and $\eta_2 = 0.9$ both with regard for the initial twist ($æ_{1*} \neq 0$) and with no regard for the initial twist ($æ_{1*} = 0$). From the plots it is clearly seen that the eigenvalues are insensitive to behavior of the torsional moment while the initial twist may seriously affect the eigenvalues. For example, the limit value T_{n2} is 20% higher than the limit value T_{n1} (Fig. 4.30).

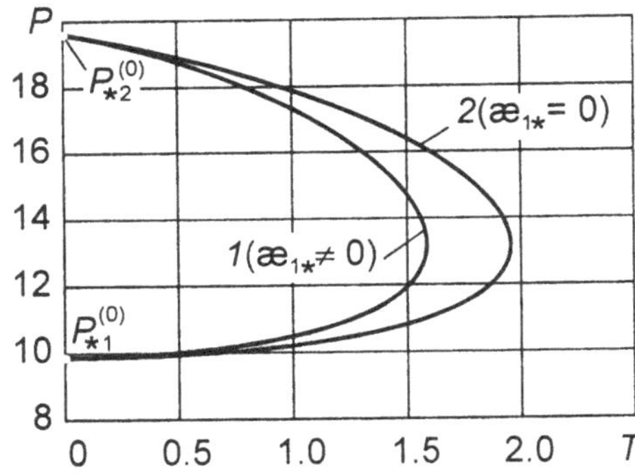

Fig. 4.31.

Let us consider the values of the parameters corresponding to the point A. Considering the plots corresponding to the case $æ_{1*} = 0$, we conclude that the rod is in a stable configuration, while exactly the opposite conclusion follows from the plots corresponding to the case $æ_{1*} \neq 0$.

Let us consider a rod with a local elastic support (see Fig. 4.29, variant *III*). Of course, this elastic support is a simplified model of a real support. Assume that the radial stiffness c of the support is the same in all directions perpendicular to the rod axis. This model may give a better insight into the role of local constraints.

Considering small displacements of the axial points, we can assume that the vector $\mathbf{u}$ lies in a plane perpendicular to the axial line. Hence, the reaction $\mathbf{R}$, which appears after the loss of stability, is proportional to the displacement vector of the point K, that is,

$$\mathbf{R} = \mathbf{R} = R_2\mathbf{e}_2 + R_3\mathbf{e}_3 = -c\mathbf{u}_K \,, \tag{4.257}$$

where c is the stiffness of the elastic support.

To derive the equilibrium equations, we should add the vector $\mathbf{b}^{(1)}$

$$\begin{aligned} \mathbf{b}^{(1)} &= (-R_2\,\delta\,,\,-R_3\,\delta\,,\,0\,,\,0\,,\,0\,,\,0\,,\,0\,,\,0)^{\mathrm{T}} \\ &= (cu_{k2}\,\delta\,,\,cu_{k3}\,\delta\,,\,0\,,\,0\,,\,0\,,\,0\,,\,0\,,\,0)^{\mathrm{T}} \end{aligned} \tag{4.258}$$

to the right-hand side of (4.245). The solution of (4.245) in which the vector $\mathbf{b}$ is replaced by the vector $\mathbf{b}^{(1)}$ is identical in form to (4.251), that is,

$$\mathbf{Y} = \mathrm{K}\,(\eta)\,\mathbf{C} + \mathrm{K}\,(\eta - \eta_k)\,\mathbf{b}^{(1)}H\,. \tag{4.259}$$

For a cantilever beam subjected to a dead load, the boundary conditions at $\eta = 0$ and $\eta = 1$ are the same as those from the preceding example (see (4.235)).

Taking into account (4.258), we obtain the following boundary-condition equations (see (4.255)):

$$\begin{aligned}
&\sum_{j=1}^{4}(k_{1j} - P_* k_{6j})\,C_j - (k_{11}^{(1)} - P_* k_{61}^{(1)})\,cu_{k2} + (k_{12}^{(1)} - P_* k_{62}^{(1)})\,cu_{k3} = 0\,;\\
&\sum_{j=1}^{4}(k_{2j} + P_* k_{5j})\,C_j + (k_{21}^{(1)} + P_* k_{51}^{(1)})\,cu_{k2} + (k_{22}^{(1)} + P_* k_{52}^{(1)})\,cu_{k3} = 0\,;\\
&\sum_{j=1}^{4}(k_{3j} + T_* k_{6j})\,C_j - (k_{31}^{(1)} + T_* k_{61}^{(1)})\,cu_{k2} + (k_{32}^{(1)} + T_* k_{62}^{(1)})\,cu_{k3} = 0\,;\\
&\sum_{j=1}^{4}(k_{4j} - T_* k_{5j})\,C_j + (k_{41}^{(1)} - T_* k_{51}^{(1)})\,cu_{k2} + (k_{42}^{(1)} - T_* k_{52}^{(1)})\,cu_{k3} = 0\,.
\end{aligned} \tag{4.260}$$

In the case under consideration, the conditions at $\eta = \eta_K$ differ from those in the previous example of hinge supports. It is necessary to consider these conditions in greater detail. For the case of hinge supports, after the loss of stability, the displacements of the points $\eta = \eta_1$ and $\eta = \eta_2$ equal zero, hence, the displacements are known. In the case of elastic support, the displacements u_{K_2} and u_{K_3} are unknown. Consider the expressions for u_2 and u_3 (they coincide with the components Y_7 and Y_8 of the vector $\mathbf{Y}$, respectively). We have

$$Y_7(\eta_k) = u_{k2} = \sum_{j=1}^{4} k_{7j}(\eta_k)\,C_j\,; \qquad Y_8(\eta_k) = u_{k3} = \sum_{j=1}^{4} k_{8j}(\eta_k)\,C_j$$

or

$$\sum_{j=1}^{4} k_{7j}(\eta_k)\,C_j - u_{k2} = 0\,; \qquad \sum_{j=1}^{4} k_{8j}(\eta_k)\,C_j - u_{k3} = 0\,. \tag{4.261}$$

As a result, we have obtained the system of six homogeneous equations (4.260) and (4.261) in the six unknowns C_1, C_2, C_3, C_4, u_{K_2}, and u_{K_3}. For a nontrivial solution to exist, the determinant of this system must equal zero. This allows us to find the eigenvalues P_{*j} as functions of the moment T and of the stiffness of elastic support c.

The methods presented in this section enable us to examine the stability of rods for an arbitrary set of local constraints.

4.8 Problems

4.1. A rod of constant cross section is subjected to a follower force $\mathbf{P}_0$ permanently directed at a fixed point O as illustrated in Fig. 4.32. Find a one-term approximation for the displacement of the point K using the principle of virtual displacements. The magnitude of $\mathbf{P}_0$ remains unchanged. Assume that the displacements of the axial points are small.

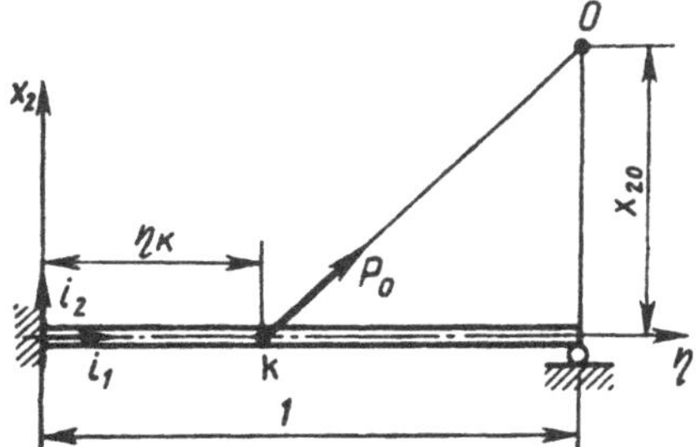

Fig. 4.32.

4.2. By use of the principle of minimum of potential energy, determine the displacements of the point K in Problem 4.1.

4.3. A rod of constant cross section is subjected to a distributed load of constant magnitude as shown in Fig. 4.33. The load vectors are permanently directed at a fixed point O with coordinates $\eta_0 = 0.5$ and $x_{20} = 0.5$. Determine the displacements of the point K. Use the principle of virtual displacements.

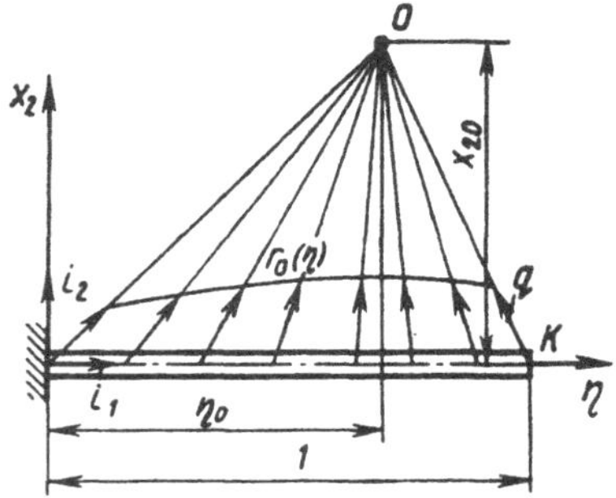

Fig. 4.33.

4.4. Determine the displacement of the point K in Problem 4.3. Use the principle of minimum of potential energy.

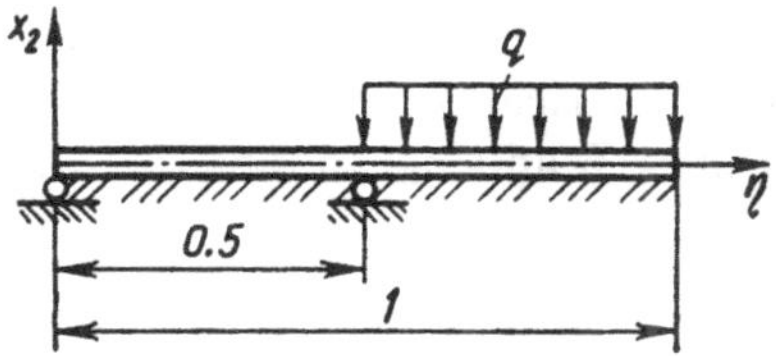

Fig. 4.34.

4.5. A rod of constant cross section resting on a linearly elastic foundation is shown in Fig. 4.34. The rod is subjected to a uniformly distributed load. Determine the displacements of the axial points.

4.6. A rod of constant cross section and a linearly elastic foundation are in contact at $0 \leq \eta \leq 0.5$ as shown in Fig. 4.35. Determine the displacements of the axial points. Use the principle of virtual displacements.

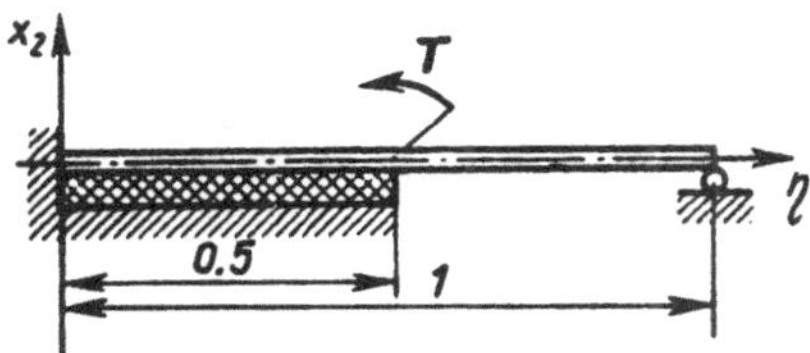

Fig. 4.35.

4.7. A rod of constant cross section (Fig.4.36) subjected to a distributed load rests on an elastic foundation. The nonlinear characteristic of the foundation is as follows: $q_{\mathrm{f}} = 4\alpha_1^4 u_{x_2} + \gamma u_{x_2}^3$. Obtain a one-term approximation for the rod deflections.

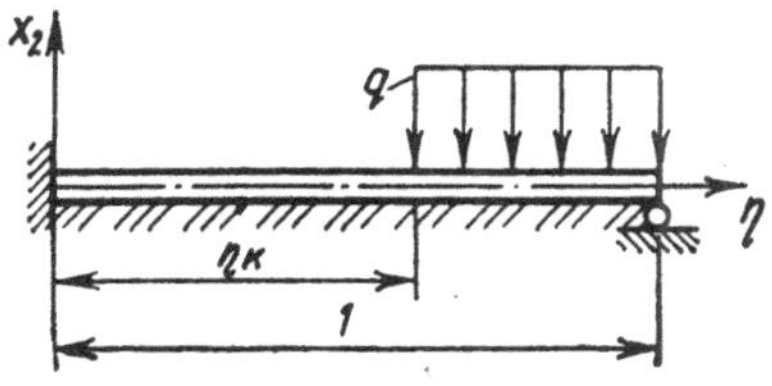

Fig. 4.36.

4.8. Assume that displacements of the axial points are small (see Fig. 4.32). Determine numerically the stress-strain state of the rod on the basis of the equilibrium equations. Put $|\mathbf{P}_0| = 0.5$. Use the equations of the zeroth approximation.

4.9. Derive the equilibrium equations of the zeroth and the first approximation for a rod shown in Fig. 4.32. Obtain the numerical solution for $|\mathbf{P}_0| = 0.5$ and $\eta_K = 0.5$. Estimate the error.

4.10. Derive the equilibrium equations for a rod shown in Fig. 4.32 for large displacements of the axial points. Obtain the numerical solution of the non-linear equilibrium equations for the case $|\mathbf{P}_0| = 2$ and $\eta_K = 0.5$. Use the method of step-by-step loading.

4.11. Obtain the numerical solution of the equilibrium equations of the zeroth approximation for a rod shown in Fig. 4.33. Put $q_0 = 1$.

5. Curvilinear Rods

In Chap. 1, the equilibrium equations of the most general form were derived. A rod axial line in the natural configuration was assumed to be a space-curved line. We considered special cases of the equations when investigating problems of statics of a rod whose axial line in the natural configuration is a straight line (see Chap. 4) or a plane curve.

The equations governing behavior of rods of helical axial line (spirals) can also be obtained by direct specialization of the general equilibrium equations. As illustrated in the Introduction, helical rods are widely used in engineering.

5.1 Plane Rods

5.1.1 Equilibrium Equations for a Rod Whose Axial Line Remains a Plane Curve During Deformation

These equations can be obtained from the general equilibrium equations derived in Chap. 1. To do this, we must put

(1) in the basis $\{\mathbf{e}_j\}$:	(2) in the basis $\{\mathbf{i}_j\}$:
$\mathbf{Q} = Q_{x_1}\mathbf{i}_1 + Q_{x_2}\mathbf{i}_2\,;$	$\mathbf{Q} = Q_1\mathbf{e}_1 + Q_2\mathbf{e}_2\,;$
$\mathbf{M} = M_{x_3}\mathbf{i}_3\,;$	$\mathbf{M} = M_3\mathbf{e}_3\,;$
$\mathbf{q} = q_{x_1}\mathbf{i}_1 + q_{x_2}\mathbf{i}_2\,;$	$\mathbf{q} = q_1\mathbf{e}_1 + q_2\mathbf{e}_2\,;$
$\boldsymbol{\mu} = \mu_{x_3}\mathbf{i}_3\,;$	$\boldsymbol{\mu} = \mu_3\mathbf{e}_3\,;$
$\mathbf{P}^{(i)} = P^{(i)}_{x_1}\mathbf{i}_1 + P^{(i)}_{x_2}\mathbf{i}_2\,;$	$\mathbf{P}^{(i)} = P^{(i)}_1\mathbf{e}_1 + P^{(i)}_2\mathbf{e}_2\,;$
$\mathbf{T}^{(\nu)} = T^{(\nu)}_{x_3}\mathbf{i}_3\,;$	$\mathbf{T}^{(\nu)} = T^{(\nu)}_3\mathbf{e}_3\,;$
$\mathbf{u} = u_{x_1}\mathbf{i}_1 + u_{x_2}\mathbf{i}_2\,;$	$\mathbf{u} = u_1\mathbf{e}_1 + u_2\mathbf{e}_2\,;$
$\boldsymbol{\vartheta} = \vartheta_{x_3}\mathbf{i}_3\,;$	$\boldsymbol{\vartheta} = \vartheta_3\mathbf{e}_3\,;$
$\boldsymbol{æ} = æ_{x_3}\mathbf{i}_3\,, \qquad æ_1 = æ_2 = 0\,;$	$\boldsymbol{æ} = æ_3\mathbf{e}_3\,, \qquad æ_1 = æ_2 = 0\,;$
$æ_{10} = æ_{20} = 0\,;$	$æ_{10} = æ_{20} = 0\,.$

For the case of in-plane bending, we have $\vartheta_1 = \vartheta_2 = 0$ and $\vartheta_{10} = \vartheta_{20} = 0$. Then, the matrices L (A.44) and L^0 (A.55) take the form

$$\mathrm{L} = [\, l_{ij} \,] = \begin{bmatrix} \cos\vartheta_3 & \sin\vartheta_3 & 0 \\ -\sin\vartheta_3 & \cos\vartheta_3 & 0 \\ 0 & 0 & 1 \end{bmatrix} ;$$

$$\mathrm{L}^0 = [\, l_{ij}^0 \,] = \begin{bmatrix} \cos\vartheta_{30} & \sin\vartheta_{30} & 0 \\ -\sin\vartheta_{30} & \cos\vartheta_{30} & 0 \\ 0 & 0 & 1 \end{bmatrix} .$$

Recall that L is the matrix of transformation form the basis $\{\mathbf{e}_{i0}\}$ to the basis $\{\mathbf{e}_i\}$ and L^0 is the matrix of transformation from the basis $\{\mathbf{i}_j\}$ to the basis $\{\mathbf{e}_{i0}\}$.

Nonlinear Equilibrium Equations in the Attached Coordinate System Using (1.64)–(1.68), we get the following system:

$$\frac{dQ_1}{d\eta} - Q_2 æ_3 + q_1 + \sum_{i=1}^{n} P_1^{(i)}\, \delta(\eta - \eta_i) = 0 ; \tag{5.1}$$

$$\frac{dQ_2}{d\eta} + Q_1 æ_3 + q_2 + \sum_{i=1}^{n} P_2^{(i)}\, \delta(\eta - \eta_i) = 0 ; \tag{5.2}$$

$$\frac{dM_3}{d\eta} + Q_2 + \mu_3 + \sum_{\nu=1}^{\rho} T_3^{(\nu)}\, \delta(\eta - \eta_\nu) = 0 ; \tag{5.3}$$

$$\frac{d\vartheta_3}{d\eta} - \frac{1}{A_{33}} M_3 = æ_{30} ; \tag{5.4}$$

$$\frac{du_1}{d\eta} - u_2 æ_3 + l_{11} - 1 = 0 , \qquad l_{11} = \cos\vartheta_3 ; \tag{5.5}$$

$$\frac{du_2}{d\eta} + u_1 æ_3 + l_{21} = 0 , \qquad l_{21} = \sin\vartheta_3 ; \tag{5.6}$$

$$M_3 = A_{33}(æ_3 - æ_{30}) , \qquad æ_3 = \frac{d\vartheta_3}{d\eta} , \qquad æ_{30} = \frac{d\vartheta_{30}}{d\eta} ; \tag{5.7}$$

here ϑ_3 is the angle between the tangents to the axial line before and after deformation as indicated in Fig. 5.1. Equations (5.1)–(5.7) present a system in the seven unknowns Q_1, Q_2, M_3, $æ_3$, ϑ_3, u_1, and u_2.

5.1.2 Nonlinear Equilibrium Equations in the Cartesian Coordinate System

If the components of external loads are given only in a Cartesian coordinate system, then it is more convenient to represent (1.89)–(1.93) in terms of projections on the Cartesian axes. As a result, we have

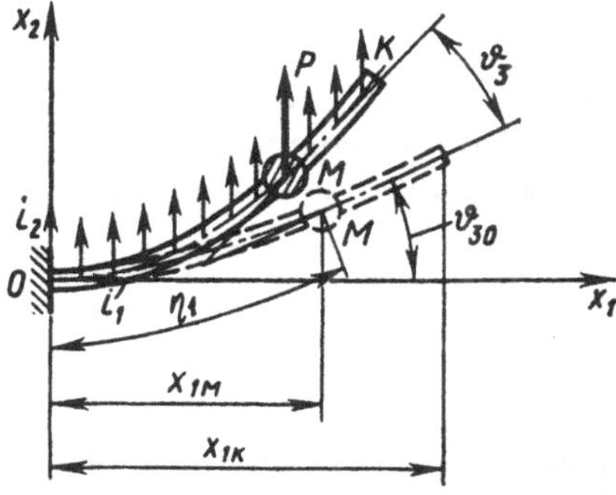

Fig. 5.1.

$$\frac{dQ_{x_1}}{d\eta} + q_{x_1} + \sum_{i=1}^{n} P_{x_1}^{(i)}\,\delta(\eta - \eta_i) = 0\,; \tag{5.8}$$

$$\frac{dQ_{x_2}}{d\eta} + q_{x_2} + \sum_{i=1}^{n} P_{x_2}^{(i)}\,\delta(\eta - \eta_i) = 0\,; \tag{5.9}$$

$$\frac{dM_{x_3}}{d\eta} + (l_{11}l_{11}^0 + l_{12}l_{21}^0)\,Q_{x_1} - (l_{11}l_{12}^0 + l_{12}l_{22}^0)\,Q_{x_2} + \mu_{x_3}\sum_{\nu=1}^{\rho} T_3^{(\nu)}\,\delta(\eta - \eta_\nu) = 0\,; \tag{5.10}$$

$$\frac{d\vartheta_3}{d\eta} - \frac{M_{x_3}}{A_{33}} - \frac{d\vartheta_{x30}}{d\eta} = 0\,; \tag{5.11}$$

$$\frac{du_{x_1}}{d\eta} - (l_{11}l_{11}^0 + l_{12}l_{21}^0) + l_{11}^0 = 0\,; \tag{5.12}$$

$$\frac{du_{x_2}}{d\eta} - (l_{11}l_{12}^0 + l_{12}l_{22}^0) + l_{12}^0 = 0\,, \tag{5.13}$$

where

$$(l_{11}l_{11}^0 + l_{12}l_{21}^0) = \cos(\vartheta_3 + \vartheta_{30})\,; \qquad l_{11}^0 = \cos\vartheta_{30}\,; \tag{5.14}$$

$$(l_{11}l_{12}^0 + l_{12}l_{22}^0) = \sin(\vartheta_3 + \vartheta_{30})\,; \qquad l_{12}^0 = \sin\vartheta_{30}\,. \tag{5.15}$$

Using (5.14) and (5.15), we can rewrite (5.10), (5.12), and (5.13) as follows:

$$\frac{dM_{x_3}}{d\eta} + \cos(\vartheta_3 + \vartheta_{30})\,Q_{x_2} - \sin(\vartheta_3 + \vartheta_{30})\,Q_{x_1} + \mu_{x_3} + \sum_{\nu=1}^{\rho} T_{x_3}^{(\nu)}\,\delta(\eta - \eta_\nu) = 0\,; \tag{5.16}$$

$$\frac{du_{x_1}}{d\eta} - \cos(\vartheta_3 + \vartheta_{30}) + \cos\vartheta_{30} = 0\,; \tag{5.17}$$

$$\frac{du_{x_2}}{d\eta} - \sin(\vartheta_3 + \vartheta_{30}) + \sin\vartheta_{30} = 0 . \tag{5.18}$$

Consider an example of distributed forces. Let the Cartesian components of these forces be given. Suppose that a helical rod is mounted on an object moving with a nonzero acceleration **a** as illustrated in Fig. 0.1. In this case, the rod is subjected to a distributed load

$$\mathbf{q} = m_0\mathbf{a} = \sum_{j=1}^{2} q_{x_j}\mathbf{i}_j = m_0|\mathbf{a}|\cos\alpha\cdot\mathbf{i}_1 + m_0|\mathbf{a}|\sin\alpha\cdot\mathbf{i}_2 .$$

The representation of equilibrium equations in the form of a system of first-order differential equations is suitable for application of numerical methods. The equilibrium equations can be written in terms of the projections of the vector **Q** on the attached axes as follows:

$$Q_{x_1} = Q_1\frac{dx_1}{d\eta} - Q_2\frac{dx_2}{d\eta}; \qquad Q_{x_2} = Q_1\frac{dx_2}{d\eta} + Q_2\frac{dx_1}{d\eta}; \tag{5.19}$$

then, we have

$$\frac{d}{d\eta}(Q_1x_1' - Q_2x_2') + q_{x_1} + \sum_{i=1}^{n} P_{x_1}^{(i)}\,\delta(\eta - \eta_i) = 0; \tag{5.20}$$

$$\frac{d}{d\eta}(Q_1x_2' + Q_2x_1') + q_{x_2} + \sum_{i=1}^{n} P_{x_2}^{(i)}\,\delta(\eta - \eta_i) = 0; \tag{5.21}$$

$$\frac{dM_{x_3}}{d\eta} + Q_2 + \mu_{x_3} + \sum_{\nu=1}^{\rho} T_{x_3}^{(\nu)}\,\delta(\eta - \eta_\nu) = 0; \tag{5.22}$$

$$M_{x_3} = A_{33}\left(\frac{d\vartheta_3}{d\eta} - \frac{d\vartheta_{30}}{d\eta}\right);$$

$$x_1' = \cos\vartheta_3;$$

$$x_2' = \sin\vartheta_3; \tag{5.23}$$

$$\frac{du_{x_1}}{d\eta} = x_1' - x_{10}';$$

$$\frac{du_{x_2}}{d\eta} = x_2' - x_{20}' . \tag{5.24}$$

Solving for the unknowns Q_1, Q_2, M_{x_3}, ϑ_3, x_1, and x_2 from (5.20)–(5.23), we determine the displacements of the axial points as follows:

$$u_{x_1} = x_1 - x_{10} + u_{x_10}; \qquad u_{x_2} = x_2 - x_{20} + u_{x_20};$$

here u_{x_10} and u_{x_20} are the initial displacements. If the rod is clamped at $\eta = 0$ (i.e. $u_{x_i}(0) = 0$), then $u_{x_10} = u_{x_20} = 0$.

As an example, consider a cantilever beam of constant cross section shown in Fig. 5.1. A mass point is joined to the beam. The natural configuration of the beam is shown as a dashed line. The functions $x_{10}(\eta)$, $x_{20}(\eta)$, and ϑ_η, which describe the axial line in the natural configuration, are assumed to be known. Suppose that the beam moves with constant acceleration a. Hence, the beam is subjected to a distributed load $\mathbf{q} = m_0 a \mathbf{i}_2$ and a concentrated load $\mathbf{P} = Ma\mathbf{i}_2$. Let us determine the new equilibrium configuration of the beam and the internal forces Q_1, Q_2, and M_{x_3}.

The nondimensional values of $\mathbf{q}$ and $\mathbf{P}$ are

$$\mathbf{q} = \frac{m_0 a l^3}{A_{33}(0)} \mathbf{i}_2 = q_{x_2}\mathbf{i}_2 ; \qquad \mathbf{P} = \frac{Mal^2}{A_{33}(0)} \mathbf{i}_2 = P_{x_2}\mathbf{i}_2 ,$$

where

$$q_{x_1} = \frac{m_0 a l^3}{A_{33}(0)} ; \qquad P_{x_2} = \frac{Mal^2}{A_{33}(0)} .$$

For rods of constant cross section, the nondimensional bending stiffness is equal to unity, $A_{33} = 1$.

Integration of (5.8) and (5.9) for $q_{x_1} = P_{x_1} = 0$ yields

$$Q_{x_1} = C_1 ; \qquad Q_{x_2} = -q_{x_2}\eta - P_{x_2} H(\eta - \eta_1) + C_2 . \tag{5.25}$$

At the free end of the beam, $\eta = 1$, we have $Q_{x_1} = Q_{x_2} = 0$, hence, $C_1 = 0$ and $C_2 = q_{x_2} + P_{x_2}$; consequently,

$$Q_{x_2} = -q_{x_2}(1-\eta) + P_{x_2} - P_{x_2} H(\eta - \eta_1) . \tag{5.26}$$

Introducing a new variable $\vartheta_{31} = \vartheta_3 + \vartheta_{30}$, we can rewrite (5.11) and (5.16) as follows:

$$\begin{aligned} &\frac{\mathrm{d}M_{x_3}}{\mathrm{d}\eta} + \cos\vartheta_{31} \left[q_{x_2}(1-\eta) + P_{x_2} - P_{x_2} H(\eta - \eta_1) \right] = 0 ; \\ &\frac{\mathrm{d}\vartheta_3}{\mathrm{d}\eta} - M_{x_3} - 2\,\frac{\mathrm{d}\vartheta_{30}}{\mathrm{d}\eta} = 0 . \end{aligned} \tag{5.27}$$

An analysis of (5.27) is complicated since in mechanics of rods we usually deal with two-point boundary-value problems, that is, the solution of (5.27) is to satisfy the boundary-condition equations at $\eta = 0$ and $\eta = 1$. In the example under consideration, $\vartheta_3(0) = 0$. At $\eta = 1$ we have $M_{x_3}(1) = 0$. Hence, $M_{x_3}(0)$ should be so chosen that the condition $M_{x_3}(1) = 0$ holds. The sought value $M_{x_3}(0)$ can be obtained by the method of iterations. The more accurately the initial approximate value of $M_{x_3}(0)$ selected, the fewer iterations needed. In the problem under consideration, it is reasonable to put

$$M_{x_3}(0) = \beta \left(\frac{1}{2} q_{x_2} x_{1K}^2 + P_{x_2} x_{1M} \right) ,$$

where $\beta \leq 1$ and x_{1K} and x_{1M} are nondimensional coordinates.

Knowing $\vartheta_3(\eta)$, we can use the relations

$$\frac{dx_1}{d\eta} = \cos\vartheta_3 ; \qquad \frac{dx_2}{d\eta} = \sin\vartheta_3$$

and, thus, obtain the coordinates of the axial points.

Integration of these relations yields

$$x_1 = \int_0^\eta \cos\vartheta_3 \, d\zeta + C_3 ; \qquad x_2 = \int_0^\eta \sin\vartheta_3 \, d\zeta + C_4 .$$

At $\eta = 0$ we have $x_1 = x_2 = 0$, hence, $C_3 = C_4 = 0$.

The displacements of the axial points are

$$u_{x_1} = x_1 - x_{10} ; \qquad u_{x_2} = x_2 - x_{20} ,$$

where

$$x_{10} = \int_0^\eta \cos\vartheta_{30} \, d\zeta ; \qquad x_{20} = \int_0^\eta \sin\vartheta_{30} \, d\zeta .$$

To determine the axial and shear forces Q_1 and Q_2, we use (5.19)

$$Q_{x_1} = Q_1 x_1' - Q_2 x_2' ; \qquad Q_{x_2} = Q_1 x_2' + Q_2 x_1' ;$$

solving these relations for Q_1 and Q_2, we get

$$Q_1 = Q_{x_1} x_1' + Q_{x_2} x_2' ; \qquad Q_2 = -Q_{x_1} x_2' + Q_{x_2} x_1' .$$

Since in our example $Q_{x_1} = 0$, we finally arrive at

$$\begin{aligned} Q_1 &= Q_{x_2} x_2' = [\, q_{x_2}(1-\eta) + P_{x_2} - P_{x_2} H(\eta - \eta_1) \,] \sin\vartheta_3 ; \\ Q_2 &= Q_{x_2} x_1' = [\, q_{x_2}(1-\eta) + P_{x_2} - P_{x_2} H(\eta - \eta_1) \,] \cos\vartheta_3 . \end{aligned} \tag{5.28}$$

The plots of Q_1, Q_2, and M_{x_3} versus η for different values of the nondimensional parameters q_{x_2} and P_1 are illustrated in Fig. 5.2. It is assumed that $M = m_0 l$. When solving the equations numerically, we put $\eta_1 = 0.5$. The plots of Q_1 and Q_2 have discontinuities at $\eta = \eta_1$. Recall that the concentrated force P_{x_2} is applied at the point $\eta = \eta_1$. Plots *1*, *2*, *3*, and *4* correspond to the nondimensional load values: $P_{x_2} = q_{x_2} = 0.5$, $P_{x_2} = q_{x_2} = 1$, $P_{x_2} = q_{x_2} = 1.5$, and $P_{x_2} = q_{x_2} = 2$, respectively. Figure 5.2*d* illustrates the shape of the rod axis in a deformed configuration. The initial shape of the axial line is described by the equation $\vartheta_{30} = \pi\eta^2/6$.

As another example, consider an elastic element that is a part of a controller of velocity of rotation (see Fig. 5.3*a*). Owing to the symmetry of the element, it is sufficient to investigate just one of its halves, say, from the clamped point A to the point B. A mass point M is fixed at the point B (Fig. 5.3*b*). The cross section B of the element does not rotate. The element is subjected to a distributed load

$$\mathbf{q} = m_0 \omega^2 (x_2 + R_0) \mathbf{i}_2$$

Q1
1.75
1.5
1.25
1
0.75
0.5
0.25
0
0.25
0.5
0.75
(a)
Q2
3.5
3
2.5
2
1.5
1
0.5
0
(b)
Mx3
(c)
x2
x1
(d)
1
2
3
4

Fig. 5.2.

P
M
B
A
l
C
C
R0
ω
b
h
(a)
P/2
B
M/2
q
x2
i2
A
0
i1
x1
R0
(b)

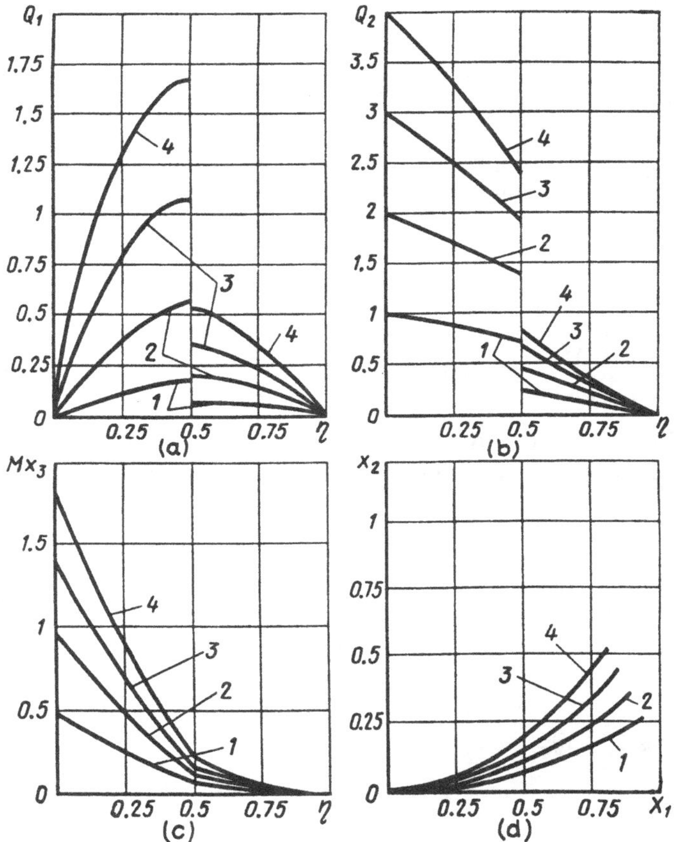

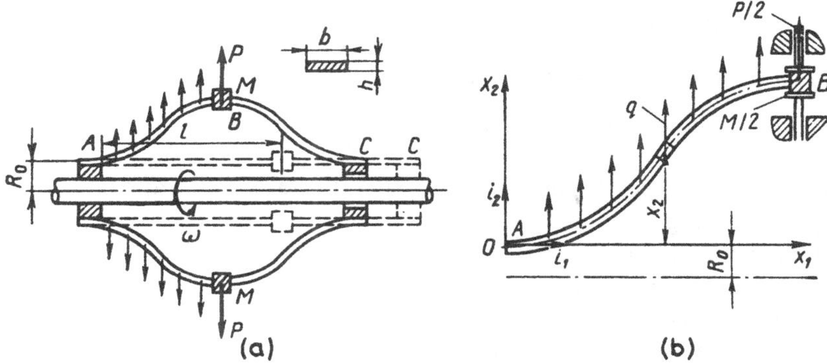

Fig. 5.3.

and a concentrated force

$$\mathbf{P} = M\omega^2 x_{2K}\mathbf{i}_2 \,.$$

To operate with nondimensional quantities, we put

$$\omega = \tilde{\omega} P_0 \,, \qquad P_0 = \frac{1}{l^2}\sqrt{\frac{A_{33}}{m_0}} \,.$$

As a result, we have

$$\begin{aligned} \tilde{\mathbf{q}} &= \mathbf{q}\,\frac{l^3}{A_{33}} = \tilde{\omega}^2(\tilde{x}_2 + \tilde{R}_0)\,\mathbf{i}_2 \,; \\ \frac{\tilde{\mathbf{P}}}{2} &= \frac{\mathbf{P}}{2}\,\frac{l^2}{A_{33}} = \frac{1}{2}\left(\frac{M}{m_0 l}\right)\tilde{\omega}^2 \tilde{x}_{2K}\mathbf{i}_2 \,, \end{aligned} \tag{5.29}$$

where nondimensional variables are marked with a tilde.

Taking into account (5.29) and using (5.8)–(5.18), we arrive at

$$\frac{\mathrm{d}Q_{x_1}}{\mathrm{d}\eta} = 0 \,; \tag{5.30}$$

$$\frac{\mathrm{d}Q_{x_2}}{\mathrm{d}\eta} + \omega^2(x_2 + R_0) = 0 \,; \tag{5.31}$$

$$\frac{\mathrm{d}\vartheta_3}{\mathrm{d}\eta} - M_{x_3} = 0 \,; \tag{5.32}$$

$$\frac{\mathrm{d}M_{x_3}}{\mathrm{d}\eta} + Q_{x_2}\cos\vartheta_3 - Q_{x_1}\sin\vartheta_3 = 0 \,; \tag{5.33}$$

$$\frac{\mathrm{d}x_1}{\mathrm{d}\eta} - \cos\vartheta_3 = 0 \,; \tag{5.34}$$

$$\frac{\mathrm{d}x_2}{\mathrm{d}\eta} - \sin\vartheta_3 = 0 \,; \tag{5.35}$$

here tildes are dropped out. Solutions of (5.30)–(5.35) must satisfy the following boundary conditions:

$$\begin{aligned} &(1) \quad \text{at} \quad \eta = 0 \,, \quad x_1 = x_2 = \vartheta_3 = 0 \,; \\ &(2) \quad \text{at} \quad \eta = 1 \,, \quad \vartheta_3 = 0 \quad \text{and} \quad Q_{x_2} = \frac{P}{2} \,. \end{aligned}$$

Consider an elastic element such that $l = 10\,\mathrm{cm}$, $b = 1\,\mathrm{cm}$, $h = 0.2\,\mathrm{cm}$, and $R_0 = 5\,\mathrm{cm}$ (Fig. 5.3*a*).

Applying the method of step-by-step loading to (5.30)–(5.35), we get the results presented in Figs. 5.4 and 5.5. The equilibrium configurations of the half of the element for two values of the velocity of rotation $\omega = 60$ and $\omega = 80$ are shown in Fig. 5.4 (curves *1* and *2*, respectively). For these values of the velocity of rotation, the plots of Q_{x_2} and M_{x_3} versus η are shown in Figs. 5.5*a*,*b* (curves *1* and *2*, respectively).

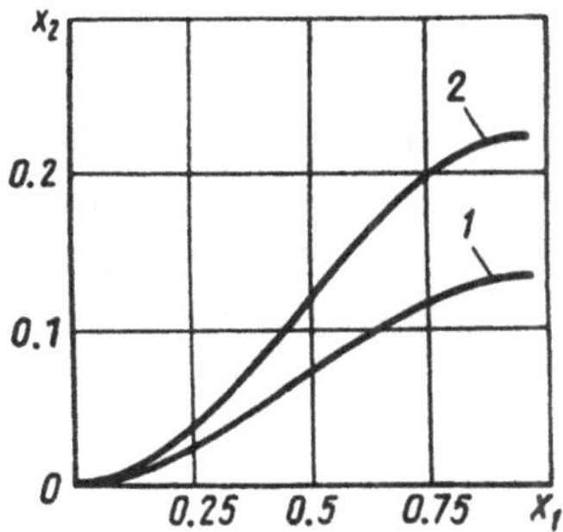

Fig. 5.4.

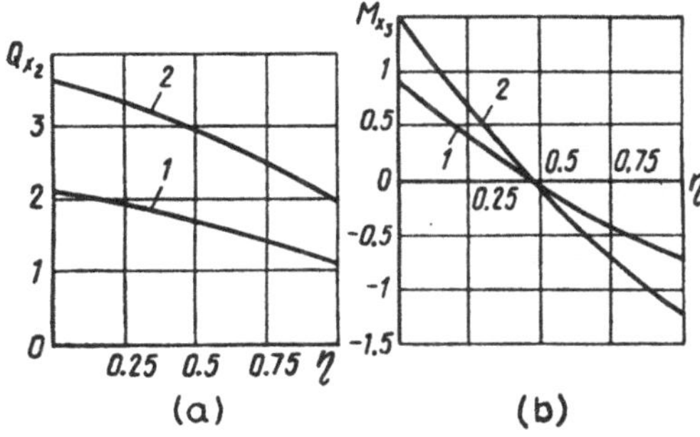

Fig. 5.5.

5.1.3 Equilibrium Equations for a Rod Whose Axial Line is a Spatial Curve in a Deformed Configuration

For this case, the equilibrium equations can be obtained from the general equilibrium equations (see Chap. 1). Considering these equations, we should put

$$æ_0^{(1)} = æ_{30}\mathbf{e}_3 ; \qquad \mathrm{L}^0 = \begin{bmatrix} \cos\vartheta_{30} & \sin\vartheta_{30} & 0 \\ -\sin\vartheta_{30} & \cos\vartheta_{30} & 0 \\ 0 & 0 & 1 \end{bmatrix} ;$$

as a result, we have

$$M_1 = A_{11}æ_1 ; \qquad M_2 = A_{22}æ_2 ; \qquad M_3 = A_{33}(æ_3 - æ_{30}) ,$$

where

$$æ_1 = \Omega_1 + \frac{\mathrm{d}\vartheta_1}{\mathrm{d}\eta} , \qquad æ_2 = \Omega_3 \sin\vartheta_1 , \qquad æ_3 = \Omega_3 \cos\vartheta_1 .$$

We do not write out the equilibrium equations as they differ little from (1.64)–(1.68).

5.1.4 Equilibrium Equations in the Case of Small Displacement of Axial Points

In this subsection, we derive equilibrium equations of the zeroth and first approximation. The equations will be represented in terms of projections on the attached axes. We assume that the displacements and the angles of rotation are small and that during deformation the rod axis remains a plane curve.

The equilibrium equations of the zeroth approximation are as follows:

$$\frac{dQ_1^{(0)}}{d\eta} - æ_{30} Q_2^{(0)} + P_{10} = 0\,; \tag{5.36}$$

$$\frac{dQ_2^{(0)}}{d\eta} + æ_{30} Q_1^{(0)} + P_{20} = 0\,; \tag{5.37}$$

$$\frac{dM_3^{(0)}}{d\eta} + Q_2^{(0)} + T_{30} = 0\,; \tag{5.38}$$

$$\frac{d\vartheta_3^{(0)}}{d\eta} - \frac{1}{A_{33}} M_3^{(0)} = 0\,, \qquad M_3^{(0)} = A_{33}\, \Delta æ_3^{(0)}\,; \tag{5.39}$$

$$\frac{du_1^{(0)}}{d\eta} - æ_{30} u_2^{(0)} = 0\,; \tag{5.40}$$

$$\frac{du_2^{(0)}}{d\eta} + æ_{30} u_1^{(0)} - \vartheta_3^{(0)} = 0\,. \tag{5.14}$$

The equilibrium equations of the first approximation are

$$\frac{dQ_1^{(1)}}{d\eta} - æ_{30} Q_2^{(1)} - Q_2^{(0)}\, \Delta æ_3^{(1)} = -\Delta P_1 + Q_2^{(0)}\, \Delta æ_3^{(0)}\,; \tag{5.42}$$

$$\frac{dQ_2^{(1)}}{d\eta} + æ_{30} Q_1^{(1)} + Q_2^{(1)}\, \Delta æ_3^{(1)} = -\Delta P_2 - Q_1^{(0)}\, \Delta æ_3^{(0)}\,; \tag{5.43}$$

$$\frac{dM_3^{(1)}}{d\eta} + Q_2^{(1)} = -\Delta T_3\,; \tag{5.44}$$

$$\frac{d\vartheta_3^{(1)}}{d\eta} - \frac{1}{A_{33}} M_3^{(1)} = 0\,, \qquad M_3^{(1)} = A_{33}\, \Delta æ^{(1)}\,; \tag{5.45}$$

$$\frac{du_1^{(1)}}{d\eta} - æ_{30} u_2^{(1)} = 0\,; \tag{5.46}$$

$$\frac{du_2^{(1)}}{d\eta} + æ_{30} u_1^{(1)} - \vartheta_3^{(1)} = 0\,. \tag{5.47}$$

Using vector notation, we can represent the equations of the zeroth approximation as follows:

$$\mathbf{Y}^{(0)\prime} + \mathrm{A}^{(0)}(\eta)\, \mathbf{Y}^{(0)} = \mathbf{f}^{(0)}\,; \tag{5.48}$$

here

$$\mathbf{Y}^{(0)} = (Q_1^{(0)}, Q_2^{(0)}, M_3^{(0)}, \vartheta_3^{(0)}, u_1^{(0)}, u_2^{(0)})^{\mathrm{T}};$$
$$\mathbf{f}^{(0)} = (-P_{10}, -P_{20}, -T_{30}, 0, 0, 0)^{\mathrm{T}};$$
$$\mathrm{A}^{(0)}(\eta) = \begin{bmatrix} 0 & -æ_{30} & 0 & 0 & 0 & 0 \\ æ_{30} & 0 & 0 & 0 & 0 & 0 \\ 0 & 1 & 0 & 0 & 0 & 0 \\ 0 & 0 & -\dfrac{1}{A_{33}} & 0 & 0 & 0 \\ 0 & 0 & 0 & 0 & 0 & -æ_{30} \\ 0 & 0 & 0 & 1 & æ_{30} & 0 \end{bmatrix}.$$

Vector form of the equations of the first approximation is

$$\mathbf{Y}^{(1)\prime} + \mathrm{A}^{(1)}(\eta)\,\mathbf{Y}^{(1)} = \mathbf{f}^{(1)}, \tag{5.49}$$

where

$$\mathbf{Y}^{(1)} = (Q_1^{(1)}, Q_2^{(1)}, M_3^{(1)}, \vartheta_3^{(1)}, u_1^{(1)}, u_2^{(1)})^{\mathrm{T}};$$
$$\mathbf{f}^{(1)} = \begin{bmatrix} -\Delta P_1 + Q_2^{(0)}\,\Delta æ_3^{(0)} \\ -\Delta P_2 - Q_1^{(0)}\,\Delta æ_3^{(0)} \\ -\Delta T_3 \\ 0 \\ 0 \\ 0 \end{bmatrix};$$
$$\mathrm{A}^{(1)}(\eta) = \begin{bmatrix} 0 & -æ_{30} & -\dfrac{Q_2^{(0)}}{A_{33}} & q_{20} & 0 & 0 \\ æ_{30} & 0 & \dfrac{Q_1^{(0)}}{A_{33}} & -q_{10} & 0 & 0 \\ 0 & 1 & 0 & 0 & 0 & 0 \\ 0 & 0 & -\dfrac{1}{A_{33}} & 0 & 0 & 0 \\ 0 & 0 & 0 & 0 & 0 & -æ_{30} \\ 0 & 0 & 0 & -1 & æ_{30} & 0 \end{bmatrix};$$
$$\Delta P_1 = \left(\sum_{i=1}^{n} P_{20}^{(i)}\,\delta(\eta-\eta_i)\right)\vartheta_3^{(1)} - q_{20}\vartheta_3^{(0)} + \left(\sum_{i=1}^{n} P_{20}^{(i)}\,\delta(\eta-\eta_i)\right)\vartheta_3^{(0)};$$
$$\Delta P_2 = -\left(\sum_{i=1}^{n} P_{10}^{(i)}\,\delta(\eta-\eta_i)\right)\vartheta_3^{(1)} + q_{10}\vartheta_3^{(0)} - \left(\sum_{i=1}^{n} P_{10}^{(i)}\,\delta(\eta-\eta_i)\right)\vartheta_3^{(0)};$$
$$\Delta T_3 = 0.$$

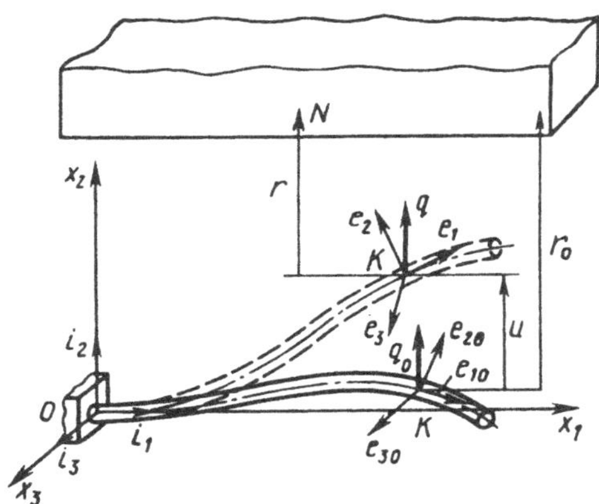

Fig. 5.6.

As an example, consider a steel rod subjected to a magnetic field (Fig. 5.6). A distributed attracting force $\mathbf{q}_0 = q_{x_{20}}(\mathbf{r}_0)\mathbf{i}_2$ acts on the rod. The absolute value of the force depends on the distance from the points of the rod axis to the magnet surface. Under the action of the force, the rod deforms into a configuration shown in Fig. 5.6 as a dashed line. This deformation causes a change of the magnitude of the attracting force $\mathbf{q}$ but the force direction remains the same with respect to the Cartesian axes because $\mathbf{q} = q_{x_2}\mathbf{i}_2$. Hence, the force $\mathbf{q}$ may be called a *dead* force. Let us obtain the expression for $\Delta\mathbf{q}$ assuming that displacements of the axial points are small. The magnet surface is a plane which is parallel to the plane x_1Ox_3, hence, the magnitude of the force is a function of $|\mathbf{r}|$, where $|\mathbf{r}| = |\mathbf{r}_0| - u_{x_2}$. Expanding $q_{x_2}(r_0 - u_{x_2})$ into a series and omitting high-order terms, we get

$$q_{x_2} = q_{x_20} + \left.\frac{\partial q_{x_2}}{\partial r}\right|_{r=r_0} u_{x_2} \,. \tag{5.50}$$

Therefore,

$$\Delta\mathbf{q} = (q_{x_2} - q_{x_20})\,\mathbf{i}_2 = a u_{x_2}\mathbf{i}_2\,, \qquad a = \left.\frac{\partial q_{x_2}}{\partial r}\right|_{r=r_0} . \tag{5.51}$$

To deal with the equilibrium equations in the attached coordinate system, it is convenient to represent the increment $\Delta\mathbf{q}$ as follows:

$$\Delta\mathbf{q} = \mathbf{q} - \mathbf{q}_0^{(1)} . \tag{5.52}$$

To obtain the increment $\Delta\mathbf{q}$ in the basis $\{\mathbf{e}_j\}$, we are to represent the vectors $\mathbf{q}$ and $\mathbf{q}_0^{(1)}$ in the attached basis as follows:

$$\mathbf{q} = \sum_{i=1}^{3} q_i\mathbf{e}_i = \sum_{i=1}^{3} q_{x_2}(\mathbf{i}_2 \cdot \mathbf{e}_i)\,\mathbf{e}_i\,;$$

the components of the vector $\mathbf{q}_0^{(1)}$ in the bases $\{\mathbf{e}_j\}$ and $\{\mathbf{e}_{j0}\}$ coincide, i.e.

$$\mathbf{q}_0^{(1)} = \sum_{i=1}^{3} q_{i0}\mathbf{e}_i = \sum_{i=1}^{3} q_{x_20}(\mathbf{i}_2 \cdot \mathbf{e}_{i0})\,\mathbf{e}_i\,.$$

To obtain the vector $\mathbf{i}_2$ in the basis $\{\mathbf{e}_j\}$, we shall use the matrix of transformation from the basis $\{\mathbf{i}_j\}$ to the basis $\{\mathbf{e}_j\}$; we have $\mathrm{L}^{(1)} = \mathrm{L}\mathrm{L}^0$, where L^0 is the matrix of transformation from the basis $\{\mathbf{i}_j\}$ to the basis $\{\mathbf{e}_{j0}\}$ and L is the matrix of transformation from the basis $\{\mathbf{e}_{j0}\}$ to the basis $\{\mathbf{e}_j\}$. The elements l_{ij}^0 of the matrix L^0 are assumed to be known. For small angles of rotation of the attached axes, we have (see Appendix 1)

$$\mathrm{L} = \begin{bmatrix} 1 & \vartheta_3 & -\vartheta_2 \\ -\vartheta_3 & 1 & \vartheta_1 \\ \vartheta_2 & -\vartheta_1 & 1 \end{bmatrix},$$

consequently, in the basis $\{\mathbf{e}_j\}$, the vector $\mathbf{i}_2$ takes the form

$$\mathbf{i}_2 = \sum_{j=1}^{3} l_{j2}^{(1)}\mathbf{e}_j = \sum_{j=1}^{3} l_{j2}^{(0)}\mathbf{e}_j + (l_{22}^0\vartheta_3 - l_{32}^0\vartheta_2)\,\mathbf{e}_1$$
$$+ (-l_{12}^0\vartheta_3 + l_{32}^0\vartheta_1)\,\mathbf{e}_2 + (l_{12}^0\vartheta_2 - l_{22}^0\vartheta_1)\,\mathbf{e}_3\,. \tag{5.53}$$

Similarly, the vector $\mathbf{i}_2$ in the basis $\{\mathbf{e}_{j0}\}$ reads

$$\mathbf{i}_2 = \sum_{j=1}^{3} l_{j2}^{(0)}\mathbf{e}_{j0}\,.$$

Using (5.50) and (5.52), we get

$$\Delta\mathbf{q} = (q_{x_20} + au_{x_2})\sum_{j=1}^{3} l_{j2}^{(1)}\mathbf{e}_j - q_{x_20}\sum_{j=1}^{3} l_{j2}^{(0)}\mathbf{e}_j\,,$$

Using (5.53), we have

$$\Delta\mathbf{q} = au_{x_2}\left(\sum_{j=1}^{3} l_{j2}^{(0)}\mathbf{e}_j\right) + q_{x_20}(l_{22}^0\vartheta_3 - l_{32}^0\vartheta_2)\,\mathbf{e}_1$$
$$+ q_{x_20}(-l_{12}^0\vartheta_3 + l_{32}^0\vartheta_1)\,\mathbf{e}_2 + q_{x_20}(l_{12}^0\vartheta_2 - l_{22}^0\vartheta_1)\,\mathbf{e}_3 \tag{5.54}$$

or, in vector form,

$$\Delta\mathbf{q} = \mathrm{C}^{(1)}\boldsymbol{\vartheta} + \mathrm{C}^{(2)}\mathbf{u}\,, \tag{5.55}$$

where

$$\mathrm{C}_1^{(1)} = \begin{bmatrix} 0 & -q_{x_20}l_{32}^0 & q_{x_20}l_{22}^0 \\ q_{x_20}l_{32}^0 & 0 & -q_{x_20}l_{12}^0 \\ -q_{x_20}l_{22}^0 & q_{x_20}l_{12}^0 & 0 \end{bmatrix};$$

$$\mathrm{C}^{(2)} = \begin{bmatrix} 0 & al_{12}^0 & 0 \\ 0 & al_{22}^0 & 0 \\ 0 & al_{32}^0 & 0 \end{bmatrix}.$$

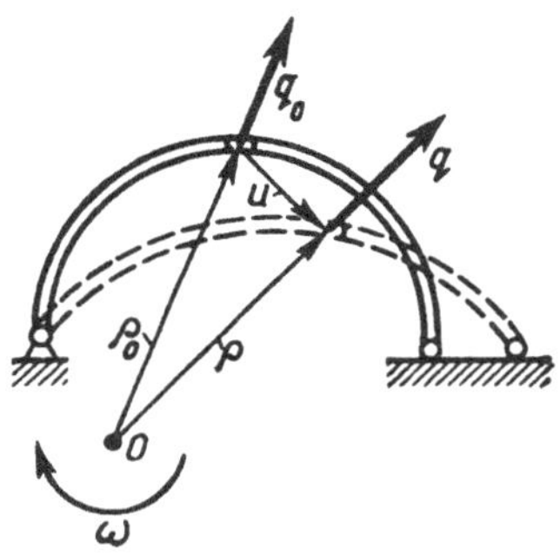

Fig. 5.7.

If the loads depend on displacements of the axial points, the determination of the increments of external loads can be illustrated by the following example. A curvilinear plane rod fixed on a rotating foundation is shown in Fig. 5.7. A deformed configuration of the rod is shown as a dashed line. Since

$$\mathbf{q}_0 = m_0\omega^2\rho_0\,; \qquad \mathbf{q} = m_0\omega^2\rho\,; \qquad \rho = \rho_0 + \mathbf{u}\,,$$

we get

$$\Delta\mathbf{q} = \mathbf{q} - \mathbf{q}_0 = m_0\omega^2\mathbf{u}\,, \tag{5.56}$$

where m_0 is the mass of a rod element of unit length and ω is the velocity of rotation of the foundation.

Now let us derive the equilibrium equations for a rod whose axial line becomes a space-curved line during deformation. The equations of the zeroth approximation can be obtained from (1.116)–(1.119) as follows:

$$\frac{\mathrm{d}Q_1^{(0)}}{\mathrm{d}\eta} - æ_{30}Q_2^{(0)} + P_{10} = 0\,;$$

$$\frac{\mathrm{d}Q_2^{(0)}}{\mathrm{d}\eta} + æ_{30}Q_1^{(0)} + P_{20} = 0\,;$$

$$\frac{\mathrm{d}Q_3^{(0)}}{\mathrm{d}\eta} + P_{30} = 0\,; \tag{5.57}$$

$$\frac{\mathrm{d}M_1^{(0)}}{\mathrm{d}\eta} - æ_{30} M_2^{(0)} + T_{10} = 0\,;$$
$$\frac{\mathrm{d}M_2^{(0)}}{\mathrm{d}\eta} + æ_{30} M_1^{(0)} - Q_3^{(0)} + T_{20} = 0\,;$$
$$\frac{\mathrm{d}M_3^{(0)}}{\mathrm{d}\eta} + Q_2^{(0)} + T_{30} = 0\,; \tag{5.58}$$
$$\frac{\mathrm{d}\vartheta_1^{(0)}}{\mathrm{d}\eta} - æ_{30} \vartheta_2^{(0)} - \Delta æ_1^{(0)} = 0\,;$$
$$\frac{\mathrm{d}\vartheta_2^{(0)}}{\mathrm{d}\eta} + æ_{30} \vartheta_1^{(0)} - \Delta æ_2^{(0)} = 0\,;$$
$$\frac{\mathrm{d}\vartheta_3^{(0)}}{\mathrm{d}\eta} - \Delta æ_3^{(0)} = 0\,; \tag{5.59}$$
$$\frac{\mathrm{d}u_1^{(0)}}{\mathrm{d}\eta} - æ_{30} u_2^{(0)} = 0\,;$$
$$\frac{\mathrm{d}u_2^{(0)}}{\mathrm{d}\eta} + æ_{30} u_1^{(0)} - \vartheta_3^{(0)} = 0\,;$$
$$\frac{\mathrm{d}u_3^{(0)}}{\mathrm{d}\eta} + \vartheta_2^{(0)} = 0\,; \tag{5.60}$$
$$M_1^{(0)} = A_{11}\, \Delta æ_1^{(0)}\,;$$
$$M_2^{(0)} = A_{22}\, \Delta æ_2^{(0)}\,;$$
$$M_3^{(0)} = A_{33}\, \Delta æ_3^{(0)}\,; \tag{5.61}$$

here

$$P_{j0} = q_{j0} + \sum_{i=1}^{n} P_{jo}^{(i)}\, \delta\left(\eta - \eta_i\right);$$
$$T_{jo} = \mu_{jo} + \sum_{\nu=1}^{\rho} T_{j0}^{(\nu)}\, \delta\left(\eta - \eta_\nu\right), \qquad j = 1\,, 2\,, 3\,. \tag{5.62}$$

If the forces are not follower, the equations of the first approximation can be derived on the basis of (1.168) and (1.169) (see Appendix 6) as follows:

$$\frac{\mathrm{d}Q_1^{(1)}}{\mathrm{d}\eta} - æ_{30}^{(1)} Q_2^{(1)} - \frac{Q_3^{(0)}}{A_{22}} M_2^{(1)} + \frac{Q_2^{(0)}}{A_{33}} M_3^{(1)} - q_{30}\vartheta_2^{(1)} + q_{20}\vartheta_3^{(1)}$$
$$= \left(\sum_{i=1}^{n} P_{30}^{(i)}\, \delta\left(\eta - \eta_i\right)\right) \vartheta_2^{(1)} - \left(\sum_{i=1}^{n} P_{20}^{(i)}\, \delta\left(\eta - \eta_i\right)\right) \vartheta_3^{(1)}$$
$$+ P_{30}\vartheta_2^{(0)} - P_{20}\vartheta_3^{(0)} - Q_3^{(0)}\, \Delta æ_2^{(0)} + Q_2^{(0)}\, \Delta æ_3^{(0)}\,, \qquad æ_{30}^{(1)} = æ_{30} + \Delta æ_3^{(0)}\,;$$

$$\frac{\mathrm{d}Q_2^{(1)}}{\mathrm{d}\eta} + æ_{30}^{(1)} Q_1^{(1)} - \frac{Q_3^{(0)}}{A_{11}} M_1^{(1)} + \frac{Q_1^{(0)}}{A_{33}} M_3^{(1)} - q_{10}\vartheta_3^{(1)} + q_{30}\vartheta_1^{(1)}$$
$$= -\left(\sum_{i=1}^{n} P_{30}^{(i)}\,\delta(\eta-\eta_i)\right)\vartheta_1^{(1)} + \left(\sum_{i=1}^{n} P_{10}^{(i)}\,\delta(\eta-\eta_i)\right)\vartheta_3^{(1)}$$
$$- P_{30}\vartheta_1^{(0)} + P_{10}\vartheta_3^{(0)} + Q_3^{(0)}\,\Delta æ_1^{(0)} - Q_1^{(0)}\,\Delta æ_3^{(0)}\,;$$

$$\frac{\mathrm{d}Q_3^{(1)}}{\mathrm{d}\eta} + \frac{Q_2^{(0)}}{A_{11}} M_1^{(1)} - \frac{Q_1^{(0)}}{A_{22}} M_2^{(1)} - q_{20}\vartheta_1^{(1)} + q_{10}\vartheta_2^{(1)}$$
$$= -\left(\sum_{i=1}^{n} P_{20}^{(i)}\,\delta(\eta-\eta_i)\right)\vartheta_1^{(1)} + \left(\sum_{i=1}^{n} P_{10}^{(i)}\,\delta(\eta-\eta_i)\right)\vartheta_2^{(1)}$$
$$+ P_{20}\vartheta_1^{(0)} - P_{10}\vartheta_2^{(0)} + Q_2^{(0)}\,\Delta æ_1^{(0)} - Q_1^{(0)}\,\Delta æ_2^{(0)}\,; \tag{5.63}$$

$$\frac{\mathrm{d}M_1^{(1)}}{\mathrm{d}\eta} - æ_{30}^{(1)} M_2^{(1)} - M_3^{(0)}\,\Delta æ_2^{(1)} + M_2^{(0)}\,\Delta æ_3^{(1)} - \mu_{30}\vartheta_2^{(1)} + \mu_{20}\vartheta_3^{(1)}$$
$$= \left(\sum_{\nu=1}^{\rho} T_{30}^{(\nu)}\,\delta(\eta-\eta_\nu)\right)\vartheta_2^{(1)} - \left(\sum_{\nu=1}^{\rho} T_{20}^{(\nu)}\,\delta(\eta-\eta_\nu)\right)\vartheta_3^{(1)}$$
$$+ T_{30}\vartheta_2^{(0)} - T_{20}\vartheta_3^{(0)} - M_3^{(0)}\,\Delta æ_2^{(0)} + M_2^{(0)}\,\Delta æ_3^{(0)}\,;$$

$$\frac{\mathrm{d}M_2^{(1)}}{\mathrm{d}\eta} - æ_{30}^{(1)} M_1^{(1)} - M_3^{(0)}\,\Delta æ_1^{(1)} + M_1^{(0)}\,\Delta æ_3^{(1)} - Q_3^{(1)} + \mu_{30}\vartheta_1^{(1)} - \mu_{10}\vartheta_3^{(1)}$$
$$= \left(\sum_{\nu=1}^{\rho} T_{30}^{(\nu)}\,\delta(\eta-\eta_\nu)\right)\vartheta_1^{(1)} - \left(\sum_{\nu=1}^{\rho} T_{10}^{(\nu)}\,\delta(\eta-\eta_\nu)\right)\vartheta_3^{(1)}$$
$$- T_{30}\vartheta_1^{(0)} + T_{10}\vartheta_3^{(0)} + M_3^{(0)}\,\Delta æ_1^{(0)} - M_1^{(0)}\,\Delta æ_3^{(0)}\,;$$

$$\frac{\mathrm{d}M_3^{(1)}}{\mathrm{d}\eta} + M_2^{(0)}\,\Delta æ_1^{(1)} - M_1^{(0)}\,\Delta æ_2^{(1)} - \mu_{20}\vartheta_1^{(1)} + \mu_{10}\vartheta_2^{(1)} + Q_2^{(1)}$$
$$= -\left(\sum_{\nu=1}^{\rho} T_{20}^{(\nu)}\,\delta(\eta-\eta_\nu)\right)\vartheta_1^{(1)} + \left(\sum_{\nu=1}^{\rho} T_{10}^{(\nu)}\,\delta(\eta-\eta_\nu)\right)\vartheta_2^{(1)}$$
$$+ T_{20}\vartheta_1^{(0)} - T_{10}\vartheta_2^{(0)} + M_2^{(0)}\,\Delta æ_1^{(0)} - M_1^{(0)}\,\Delta æ_2^{(0)}\,; \tag{5.64}$$

$$\frac{\mathrm{d}\vartheta_1^{(1)}}{\mathrm{d}\eta} - æ_{30}^{(1)}\vartheta_2^{(1)} - \frac{1}{A_{11}} M_1^{(1)} = 0\,;$$

$$\frac{\mathrm{d}\vartheta_2^{(1)}}{\mathrm{d}\eta} + \ae_{30}^{(1)}\vartheta_1^{(1)} - \frac{1}{A_{22}}\,M_2^{(1)} = 0\,;$$

$$\frac{\mathrm{d}\vartheta_3^{(1)}}{\mathrm{d}\eta} - \frac{1}{A_{33}}\,M_3^{(1)} = 0\,; \tag{5.65}$$

$$\frac{\mathrm{d}u_1^{(1)}}{\mathrm{d}\eta} - \ae_{30}^{(1)}u_2^{(1)} = 0\,;$$

$$\frac{\mathrm{d}u_2^{(1)}}{\mathrm{d}\eta} + \ae_{30}^{(1)}u_1^{(1)} - \vartheta_3^{(1)} = 0\,;$$

$$\frac{\mathrm{d}u_3^{(1)}}{\mathrm{d}\eta} + \vartheta_2^{(1)} = 0\,; \tag{5.66}$$

$$\Delta\ae_1^{(1)} = \frac{1}{A_{11}}\,M_1^{(1)}\,;$$

$$\Delta\ae_2^{(1)} = \frac{1}{A_{22}}\,M_2^{(1)}\,;$$

$$\Delta\ae_3^{(1)} = \frac{1}{A_{33}}\,M_3^{(1)}\,. \tag{5.67}$$

The components of the vectors of the first approximation, $Q_i^{(1)}$, $M_j^{(1)}$, etc., and of the zeroth approximation, $Q_i^{(0)}$, $M_j^{(0)}$, etc., must satisfy the same boundary conditions. The equations of the zeroth approximation (5.57)–(5.61) can be represented in vector form as follows:

$$\mathbf{Y}^{(0)\prime} + \mathrm{A}^{(0)}(\eta)\,\mathbf{Y}^{(0)} = \mathbf{f}^{(0)}\,; \tag{5.68}$$

here

$$\mathbf{Y}^{(0)} = (Q_1^{(0)}, Q_2^{(0)}, Q_3^{(0)}, M_1^{(0)}, M_2^{(0)}, M_3^{(0)}, \vartheta_1^{(0)}, \vartheta_2^{(0)}, \vartheta_3^{(0)}, u_1^{(0)}, u_2^{(0)}, u_3^{(0)})^{\mathrm{T}}\,;$$

$$\mathbf{f}^{(0)} = (-P_{10}\,, -P_{20}\,, -P_{30}\,, -T_{10}\,, -T_{20}\,, -T_{30}\,, 0\,, 0\,, 0\,, 0\,, 0\,, 0)^{\mathrm{T}}\,;$$

$$\mathrm{A}^{(0)}(\eta) =$$

$$\begin{bmatrix}
0 & -\ae_{30} & 0 & 0 & 0 & 0 & 0 & 0 & 0 & 0 & 0 & 0 \\
\ae_{30} & 0 & 0 & 0 & 0 & 0 & 0 & 0 & 0 & 0 & 0 & 0 \\
0 & 0 & 0 & 0 & 0 & 0 & 0 & 0 & 0 & 0 & 0 & 0 \\
0 & 0 & 0 & 0 & -\ae_{30} & 0 & 0 & 0 & 0 & 0 & 0 & 0 \\
0 & 0 & -1 & \ae_{30} & 0 & 0 & 0 & 0 & 0 & 0 & 0 & 0 \\
0 & 1 & 0 & 0 & 0 & 0 & 0 & 0 & 0 & 0 & 0 & 0 \\
0 & 0 & 0 & -\frac{1}{A_{11}} & 0 & 0 & 0 & -\ae_{30} & 0 & 0 & 0 & 0 \\
0 & 0 & 0 & 0 & -\frac{1}{A_{22}} & 0 & \ae_{30} & 0 & 0 & 0 & 0 & 0 \\
0 & 0 & 0 & 0 & 0 & -\frac{1}{A_{11}} & 0 & 0 & 0 & 0 & 0 & 0 \\
0 & 0 & 0 & 0 & 0 & 0 & 0 & 0 & 0 & 0 & -\ae_{30} & 0 \\
0 & 0 & 0 & 0 & 0 & 0 & 0 & 0 & 1 & \ae_{30} & 0 & 0 \\
0 & 0 & 0 & 0 & 0 & 0 & 0 & -1 & 0 & 0 & 0 & 0
\end{bmatrix}.$$

Similarly, equations of the first approximation (5.63)–(5.67) are

$$\mathbf{Y}^{(1)'} + \mathrm{A}^{(1)}(\eta)\,\mathbf{Y}^{(1)} = \mathbf{f}^{(1)}\,. \tag{5.69}$$

The vector $\mathbf{f}^{(1)}$ depends on the solutions of (5.68) and on the angles $\vartheta_j^{(1)}$, where $\vartheta_j^{(1)}$ are the angles of rotation of the attached coordinate axes.

5.2 Elementary Theory of Cylindrical Springs

5.2.1 Helical Rods

Curvilinear elastic rods are widely used in engineering. They are used as accumulators of mechanical energy, sensitive elements in various devices, frequency indicators, etc. In what follows, we will examine in greater detail helical rods or cylindrical springs (see Fig. 0.7). Other (not cylindrical) springs could be obtained if instead of a cylinder we use another guide surface to coil wire around, e.g. a cone (Fig. 0.8) or a surface of revolution (dashed lines in Fig. 0.8).

There are helical rods of constant angle of helix ($\alpha_0 = \text{const}$) or of variable angle of helix ($\alpha_0 \neq \text{const}$). In the first case, the curvature Ω_{30} and the twist Ω_{10} of a rod axis are constants (see (A.104) and (A.105)) given by the relations

$$\Omega_{10} = \frac{\sin\alpha_0\cos\alpha_0}{R_0}\,; \qquad \Omega_{30} = \frac{\cos^2\alpha_0}{R_0}\,, \qquad R_0 = \frac{R_{00}}{l}\,, \tag{5.70}$$

where l is the length of the axial line. In the second case, $\Omega_1(\eta)$ and $\Omega_3(\eta)$ can be determined on the basis of the equation of the rod axis $x_{i0}(\eta)$ (see (A.91) and (A.100)). The curvature Ω_1 and the twist Ω_3 are characteristics of a rod axis in the unloaded configuration.

In the theory of cylindrical springs (both for $\alpha_0 = \text{const}$ and $\alpha_0 \neq \text{const}$) there exist two types of problems:

(1) Problems of statics of cylindrical springs. In these problems, the increments of the geometrical characteristics $\Delta\Omega_i$, $\Delta\alpha$, ΔR_0, and ΔH of a spring are assumed to be small. Problems of this type are referred to the linear theory of cylindrical springs.

(2) Nonlinear problems. Considering these problems, we cannot assume the increments of the parameters $\Delta\Omega_i$, $\Delta\alpha$, ΔR_0, and ΔH to be small.

In the case of linear theory, for all types of loading (symmetrical (see Fig. 0.7*a*) or nonsymmetrical (see Fig. 0.7*b*)), we can use the equations of the zeroth approximation (1.101)–(1.111) derived in Sect. 1.4. In the second case, we should use the general nonlinear equilibrium equations obtained in Sect. 1.3.

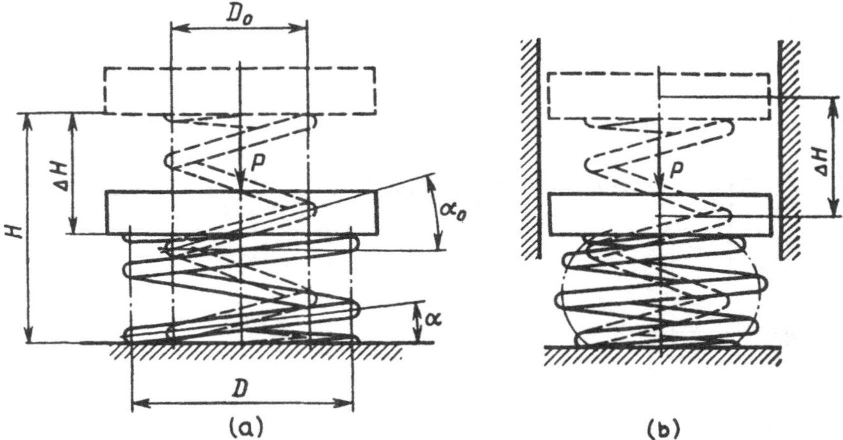

Fig. 5.8.

In what follows, we assume that the problems of statics of helical rods are linear in physical sense, i.e. strains depend linearly on stresses. A cylindrical spring loaded symmetrically at its ends by concentrated forces is discussed in detail in literature (e.g. Ponomarev, Andreeva (1980); in this book, deformations of springs are assumed to be small).

It should be noted that the complicity of a problem strictly depends on the type of loading. In the case of symmetrical loading, small increments of the curvature $\Delta\Omega_1$ and the twist $\Delta\Omega_3$ may be assumed to be the same at any point of the axial line. In the case of nonsymmetrical loading, these increments depend on the arc coordinate s. A helical rod subjected to an external symmetrical loading remains a helical rod but with other geometrical characteristics R, α, Ω_1, and Ω_3. If the length of a spring is large as compared to its diameter D or if rod ends are not constrained against moving in a plane perpendicular to the axial line, then the influence of the end fixity conditions is negligibly small. In these cases, the following unknown constants appear in the equilibrium equations: the curvature Ω_1, the twist Ω_3, the diameter D, and the angle α (see Fig. 5.8*a*). Increments ΔH, $\Delta\alpha$, and ΔD (no matter small or large) can be obtained without use of differential equilibrium equations.

If a cylindrical spring with clamped ends is subjected to symmetrical forces such that ΔH becomes large (see Fig. 5.8*b*), then the spring axis is no longer a circular helix. Hence, $æ_1$ and $æ_3$ depend on the coordinate s.

5.2.2 Linear Theory of Cylindrical Springs

A symmetrically loaded helical rod is shown in Fig. 5.9*a*. A force $\mathbf{P}$ and a moment $\mathbf{T}$ lie in a plane perpendicular to the rod axis. As a result, an internal force $\mathbf{Q}$ and an internal moment $\mathbf{M}$ appear. The components of the vectors $\mathbf{Q}$

and **M** in the attached basis depend both on the geometrical characteristics of the spring and on the end fixity conditions.

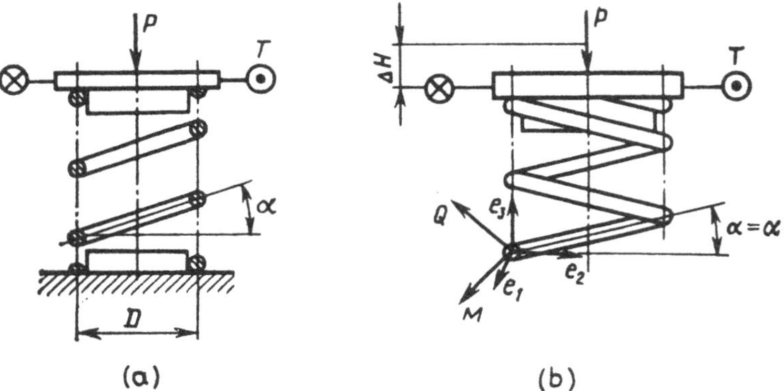

Fig. 5.9.

There exist end fixity conditions such that the internal loads **M** and **Q** can be found without use of differential equilibrium equations irrespective of the type of external loading. Let us put $\alpha \approx \alpha_0 = \text{const}$ and $D = D_0 = \text{const}$, i.e. we neglect the spring deformation in the equilibrium equations. From summation of forces and moments in the directions of the attached axes (see Fig. 5.9*b*), we get six algebraic equations in the six unknowns Q_j and M_j, $j = 1\,, 2\,, 3$. These equations are valid both for constant and variable angle α. In this case, the deflection of the spring ΔH and the angle of rotation of one end relative to the other $\Delta\psi$ may be found according to Mohr's method (Feodosiev (1996)). This method does not imply any use of differential equations. Thus, we see that under certain conditions, problems of statics of helical springs may be solved without use of differential equilibrium equations. These conditions are as follows: the top end of a spring is not constrained against rotation and displacement in the axial direction. If this is not the case, the problem becomes a statically indeterminate one and, thus, the differential equilibrium equations are to be used.

Let us consider in greater detail a symmetrically loaded spring such that $\alpha_0 = \text{const}$ (see Fig. 5.9*a*). We assume that displacements of the axial points are small. If the top end is not constrained against axial displacement and rotation, then this problem is a statically determinate one. A helical curve and its involute curve are shown in Fig. 5.10. If the angle of helix α_0 is constant, then the involute curve is a straight line. In this case, the angle α_0 and the nondimensional height of the cylindrical spring H_0 are related as follows:

$$H_0 = \sin\alpha_0\,, \qquad H_0 = \frac{H_{00}}{l}\,; \tag{5.71}$$

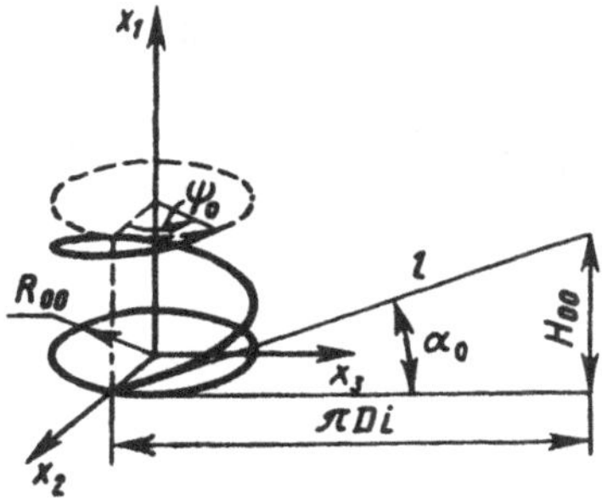

Fig. 5.10.

here l is the length of the rod. Moreover, the geometrical parameters of the spring (see Fig. 5.10) satisfy the following relation:

$$\tan \alpha_0 = \frac{H_{00}}{\pi D_{00} i} = \frac{H_0}{\pi D_0 i}, \qquad D_0 = \frac{D_{00}}{l}; \tag{5.72}$$

here i is the number of coils. The central angle can be found from the equation

$$\psi_0 = 2\pi = \frac{\cos \alpha_0}{R_0}, \qquad R_0 = \frac{R_{00}}{l}. \tag{5.73}$$

If a rod has a circular or square cross section, then the principal axes and the natural axes of a cross section coincide. Then, $æ_{20} = \Omega_{20} = 0$. For small displacements of the axial points, the increments ΔH, $\Delta \alpha$, and ΔR as well as $\Delta \Omega_1$ and $\Delta \Omega_3$ can be considered small. Therefore, in the linear approximation for inextensible rods ($l = \text{const}$), from (5.70)–(5.73) it follows that

$$\Delta H_0 = \cos \alpha_0 \, \Delta \alpha; \tag{5.74}$$

$$\Delta \psi = -\frac{1}{R_0} \sin \alpha_0 \, \Delta \alpha - \frac{1}{R_0^2} \cos \alpha_0 \, \Delta R_0, \qquad \Delta R_0 = \frac{\Delta R_{00}}{l}; \tag{5.75}$$

$$\Delta \Omega_1 = \frac{1}{R_0} \cos 2\alpha_0 \, \Delta \alpha - \frac{\sin 2\alpha_0}{2R_0^2} \Delta R_0; \tag{5.76}$$

$$\Delta \Omega_3 = -\frac{2 \sin \alpha_0 \cos \alpha_0 \, \Delta \alpha}{R_0} - \frac{\cos^2 \alpha_0}{R_0^2} \Delta R_0. \tag{5.77}$$

Using (5.76) and (5.77), we get

$$\Delta \alpha = R_0 \left(\Delta \Omega_1 - \Delta \Omega_3 \tan \alpha_0 \right);$$
$$\Delta R_0 = -R_0^2 \, \Delta \Omega_3 (1 - \tan^2 \alpha_0) - 2R_0^2 \, \Delta \Omega_1 \tan \alpha_0 .$$

When $\Delta \alpha$ and ΔR are found, we have

$$\Delta H_0 = \cos \alpha_0 R_0 \left(\Delta \Omega_1 - \Delta \Omega_3 \tan \alpha_0 \right);$$
$$\Delta \psi = \sin \alpha_0 \, \Delta \Omega_1 + \frac{1}{2 \cos \alpha_0} \Delta \Omega_3 .$$

Solving for $\Delta æ_1$ and $\Delta æ_3$ from (1.107)–(1.111), we get ΔH_0 and $\Delta\psi$ for the case indicated in Fig. 5.9*a*, that is, when one end of the spring rotates freely (this is always so in statically determinate problems). The increments of the twist $\Delta\Omega_1$ and the curvature $\Delta\Omega_3$ are related to the bending moments as follows:

$$\Delta\Omega_1 = \frac{M_1}{A_{11}}; \qquad \Delta\Omega_3 = \frac{M_3}{A_{33}}.$$

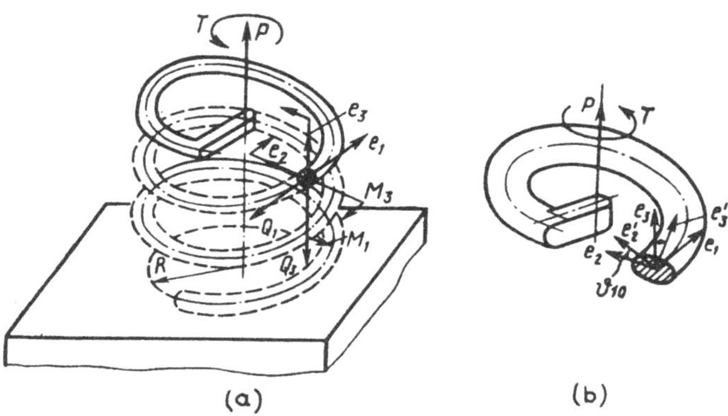

Fig. 5.11.

For statically determinate problems, the internal forces may be found from the equations that govern equilibrium of a part of the spring. For the spring shown in Fig. 5.11*a*, we have

$$\begin{aligned} M_1 &= PR\cos\alpha + T\sin\alpha\,; \\ M_3 &= T\cos\alpha - PR\sin\alpha\,; \\ Q_1 &= P\sin\alpha\,; \\ Q_3 &= P\cos\alpha\,, \end{aligned}$$

where α and R depend on the external loads. For this reason, in the exact statement, the problem of determination of the stress-strain state of cylindrical springs is nonlinear. Meanwhile, for small displacements of the axial points, we have $\alpha \approx \alpha_0$ and $R \approx R_0$, hence,

$$\begin{aligned} M_1 &= PR_0\cos\alpha_0 + T\sin\alpha_0\,; \\ M_3 &= T\cos\alpha_0 - PR_0\sin\alpha_0\,; \\ Q_1 &= P\sin\alpha_0\,; \\ Q_3 &= P\cos\alpha_0\,. \end{aligned} \tag{5.78}$$

Knowing M_1, M_3, and Q_1, we can sequentially obtain the values of $\Delta\Omega_1$, $\Delta\Omega_3$, ΔH_0, and $\Delta\psi$. If one end of the spring cannot rotate relative to the other, we get an additional relation

$$\Delta\psi = \sin\alpha_0\,\Delta\Omega_1 + \frac{1}{2\cos\alpha_0}\,\Delta\Omega_3 = 0\,.$$

The moment $\mathbf{T}$ can be obtained from this relation.

Let us consider a case when the principal axes of cross sections do not coincide with the natural axes. An element of a cylindrical spring made of a rod with different moments of inertia of cross sections ($A_{22} \neq A_{33}$) is shown in Fig. 5.11*b*.

Consider a helical rod such that $A_{22} \neq A_{33}$ and $\vartheta_{10} = \text{const}$. In the basis of the principal axes, the vector æ can be written as follows:

$$æ = \frac{\sin 2\alpha}{2R}\,\mathbf{e}_1^{(1)} + \frac{\cos^2\alpha}{R}\,\sin\vartheta_{10}\mathbf{e}_2^{(1)} + \frac{\cos^2\alpha}{R}\,\cos\vartheta_{10}\mathbf{e}_3^{(1)}\,.$$

In the general case (Fig. 5.11*b*), the angle ϑ_{10} changes and $\Delta\vartheta_{10} \neq 0$. Then, we get

$$\Delta æ_1 = \frac{\mathrm{d}\Delta\vartheta_{10}}{\mathrm{d}\eta} + \frac{1}{R_0}\cos 2\alpha_0\,\Delta\alpha - \frac{\sin 2\alpha_0}{2R_0^2}\,\Delta R_0\,; \tag{5.79}$$

$$\Delta æ_2 = -\left(\frac{\sin 2\alpha_0}{R_0}\,\Delta\alpha + \frac{\cos^2\alpha_0}{R_0^2}\,\Delta R_0\right)\sin\vartheta_{10} + \frac{\cos^2\alpha_0}{R_0}\,\cos\vartheta_{10}\vartheta_{10}\,; \tag{5.80}$$

$$\Delta æ_3 = -\left(\frac{\sin 2\alpha_0}{R_0}\,\Delta\alpha + \frac{\cos^2\alpha_0}{R_0^2}\,\Delta R_0\right)\cos\vartheta_{10} - \frac{\cos^2\alpha_0}{R_0}\,\sin\vartheta_{10}\,\Delta\vartheta_{10}\,. \tag{5.81}$$

In the principal axes, the nondimensional equilibrium equations can be written as follows:

$$M = \frac{\Delta æ_1}{A_{11}} = (PR_0\cos\alpha_0 + T\sin\alpha_0)\,; \tag{5.82}$$

$$M_2 = \frac{\Delta æ_2}{A_{22}} = (T\cos\alpha_0 - PR_0\sin\alpha_0)\,\sin\vartheta_{10}\,; \tag{5.83}$$

$$M_3 = \frac{\Delta æ_3}{A_{33}} = (T\cos\alpha_0 - PR_0\sin\alpha_0)\,\cos\vartheta_{10}\,; \tag{5.84}$$

$$\begin{aligned} Q_1 &= P\sin\alpha_0\,;\\ Q_2 &= -P\cos\alpha_0\sin\vartheta_{10}\,;\\ Q_3 &= P\cos\alpha_0\cos\vartheta_{10}\,. \end{aligned} \tag{5.85}$$

From (5.80) and (5.81) it follows that

$$\Delta æ_2 \cos\vartheta_{10} - \Delta æ_3 \sin\vartheta_{10} = \frac{\cos^2\alpha_0}{R_0}\,\Delta\vartheta_{10}\,. \tag{5.86}$$

Substituting (5.83) and (5.84) into (5.86), we get

$$(T\cos\alpha_0 - PR_0\sin\alpha_0)\,(A_{22} - A_{33})\,\cos\vartheta_{10}\sin\vartheta_{10} = \frac{\cos^2\alpha_0}{R_0}\,\Delta\vartheta_{10}\,. \tag{5.87}$$

From (5.87) it is seen that if the angle ϑ_{10} is constant, then the increment $\Delta\vartheta_{10}$ is constant too. The derivative of $\Delta\vartheta_{10}$ with respect to η equals zero. The increments $\Delta æ_1$ can be determined from the relation

$$\Delta æ_1 = \frac{\cos 2\alpha_0}{R_0}\,\Delta\alpha - \frac{\sin 2\alpha_0}{2R_0^2}\,\Delta R_0\,. \tag{5.88}$$

Using (5.80) and (5.81), we get

$$\Delta æ_2 \sin\vartheta_{10} + \Delta æ_3 \cos\vartheta_{10} = -\frac{\sin 2\alpha_0}{R_0}\,\Delta\alpha - \frac{\cos^2\alpha_0}{R_0^2}\,\Delta R_0\,. \tag{5.89}$$

Combining (5.89) and (5.88), we arrive at

$$\Delta\alpha = R_0[\,\Delta æ_1 - (\Delta æ_3\cos\vartheta_{10} + \Delta æ_2\sin\vartheta_{10})\,\tan\alpha_0\,]\,;$$
$$\Delta R_0 = -R_0^2(\Delta æ_3\cos\vartheta_{10} + \Delta æ_2\sin\vartheta_{10})\,(1-\tan^2\alpha_0) - 2R_0^2\tan\alpha_0\,\Delta æ_1\,.$$

Then,

$$\Delta H = R_0\cos\alpha_0[\,\Delta æ_1 - (\Delta æ_3\cos\vartheta_{10} + \Delta æ_2\sin\vartheta_{10})\,\tan\alpha_0\,]\,;$$
$$\Delta\psi = \sin\alpha_0\,\Delta æ_1 + \frac{1}{2\cos\alpha_0}\,(\Delta æ_3\cos\vartheta_{10} + \Delta æ_2\sin\vartheta_{10})\,.$$

Suppose now that ϑ_{10} depends on η. We have

$$\Delta æ_1 = \frac{\mathrm{d}\Delta\vartheta_{10}}{\mathrm{d}\eta} + \frac{\cos 2\alpha_0}{R_0}\,\Delta\alpha - \frac{\sin 2\alpha_0}{2R_0^2}\,\Delta R_0\,. \tag{5.90}$$

From (5.87) it follows that

$$\Delta\vartheta_{10}(\eta) = c\sin 2\vartheta_{10}(\eta)\,,$$

where

$$C = \frac{R_0}{2\cos^2\alpha_0}\,(A_{22} - A_{33})\,(T\cos\alpha_0 - PR_0\sin\alpha_0)\,.$$

Substituting $\Delta\vartheta_{10}$ into (5.90) and using (5.89), we get

$$\Delta æ_1 = 2c\vartheta_{10}{}'\cos 2\vartheta_{10} + \frac{1}{R_0}\,\cos 2\alpha_0\,\Delta\alpha - \frac{\sin 2\alpha_0}{2R_0^2}\,\Delta R_0\,;$$

$$\Delta æ_2\sin\vartheta_{10} + \Delta æ_3\cos\vartheta_{10} = -\frac{\sin 2\alpha_0}{R_0}\,\Delta\alpha - \frac{\cos^2\alpha_0}{R_0^2}\,\Delta R_0\,,$$

hence,

$$\Delta\alpha = R_0[\,\Delta æ_1 - 2c\vartheta'_{10}\cos 2\vartheta_{10} - (\Delta æ_2 \sin\vartheta_{10} + \Delta æ_3 \cos\vartheta_{10})\,\tan\alpha_0\,];$$
$$\Delta R_0 = -R_0^2(\Delta æ_3 \sin\vartheta_{10} + \Delta æ_3 \cos\vartheta_{10})\,(1 - \tan^2\alpha_0) - 2R_0^2(\Delta æ_1 - 2c\vartheta'_{10}\cos 2\vartheta_{10})\,\tan\alpha_0\,.$$

Since $\Delta\alpha$ and ΔR_0 depend on η, we finally arrive at

$$\Delta H_0 = \cos\alpha_0 \int_0^1 \Delta\alpha\,\mathrm{d}\eta\,;$$
$$\Delta\psi = -\frac{\cos\alpha_0}{R_0^2}\int_0^1 \Delta R_0\,\mathrm{d}(\eta) - \frac{\sin\alpha_0}{R_0}\int_0^1 \Delta\alpha\,\mathrm{d}\eta\,.$$

5.2.3 Basics of Nonlinear Theory of Cylindrical Springs

Consider a spring subjected to an axial force **P** and a moment **T** as indicated in Fig. 5.1. For some values of **P** and **T**, the increments $\Delta\alpha$, ΔR, etc. cannot be considered small. The finite increments of the nondimensional geometrical characteristics of the spring are as follows:

$$\Delta H_0 = \sin\alpha - \sin\alpha_0\,; \tag{5.91}$$
$$\Delta\psi = \psi - \psi_0 = \frac{\cos\alpha}{R} - \frac{\cos\alpha_0}{R_0}\,; \tag{5.92}$$
$$\Delta\Omega_1 = \Omega_1 - \Omega_{10} = \frac{\sin 2\alpha}{R} - \frac{\cos 2\alpha_0}{R_0}\,; \tag{5.93}$$
$$\Delta\Omega_3 = \Omega_3 - \Omega_{30} = \frac{\cos^2\alpha}{R} - \frac{\cos^2\alpha_0}{R_0}\,. \tag{5.94}$$

The problem under consideration is nonlinear in geometrical sense and linear in physical sense. Hence, the quantities M_1, M_3, $\Delta\Omega_1$, and $\Delta\Omega_3$ satisfy the relations derived above for the linear problem,

$$\Delta\Omega_1 = \frac{M_1}{A_{11}}\,; \qquad \Delta\Omega_3 = \frac{M_3}{A_{33}}\,. \tag{5.95}$$

The relations (5.91)–(5.95) are valid provided the geometrical characteristics α, D, Ω_1, and Ω_3 of the spring in a deformed configuration are the same for any cross section. This condition is satisfied provided the rod is of constant cross section, $A_{33} = 1$, and the end coils can move freely in the planes perpendicular to the spring axis. Combining (5.95), (5.78), (5.93), and (5.73), we get the following equations in the unknowns α and R:

$$\frac{\sin 2\alpha}{2R} - \frac{\sin 2\alpha_0}{2R_0} = \frac{PR}{A_{11}}\cos\alpha_0 + \frac{T}{A_{11}}\sin\alpha\,; \tag{5.96}$$
$$\frac{\cos^2\alpha}{R} - \frac{\cos^2\alpha_0}{R_0} = T\cos\alpha - PR\cos\alpha\,. \tag{5.97}$$

Once α and R are found, we can determine the stress-strain state of the cylindrical spring for large deformations.

Eliminating R from (5.96) and (5.97), we can establish a relationship between P and α. For example, setting $T = 0$ in (5.96) and (5.97), we get

$$P = \frac{1}{R_0} \sin(\alpha - \alpha_0) \frac{\frac{1}{A_{11}} \cos^3 \alpha \cos \alpha_0 + \sin \alpha \sin \alpha_0}{\cos \alpha \left(\frac{1}{A_{11}} \cos^2 \alpha + \sin^2 \alpha \right)^2} . \tag{5.98}$$

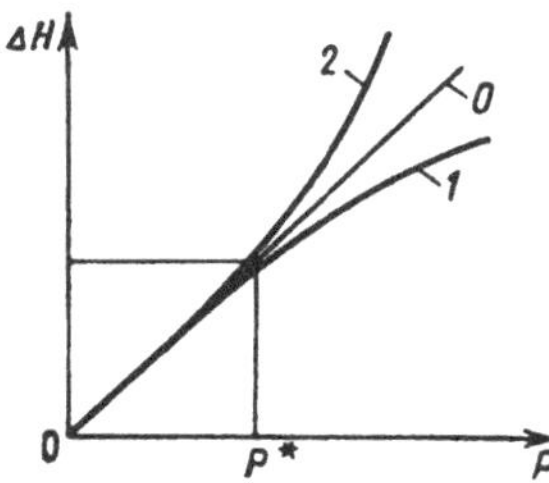

Fig. 5.12.

Using (5.98) and (5.91), we can obtain the relation $\Delta H = \Delta H(P)$. The formulas (5.96) and (5.97) are valid provided the angle α does not exceed its limit value at which the coils interlock. For large deformations, for $T = 0$, the relationship between ΔH and P for a rod of circular cross section is qualitatively illustrated in Fig. 5.12. Line *1* corresponds to compression of the rod, line *2* corresponds to its tension.

The formulas presented above allows us to obtain explicit analytical relations which are valid only for a rather narrow set of external loads, say, for an axial load $\mathbf{P}$ and a moment $\mathbf{T}$ indicated in Fig. 5.11. In more complicated cases, when a force is not directed exactly along the axial line or a spring is subjected to a set of force fields, the strain-stress state of a spring can be determined only by use of the differential equilibrium equations. For this reason, let us turn our attention to more general methods.

5.3 General Theory of Cylindrical Springs

5.3.1 Linear Equilibrium Equations

Helical rods are used as sensitive elements of devices (e.g. accelerometers). During operation, such an element is subjected to distributed forces. The vector $\mathbf{q}$ of these forces may be of arbitrary direction. In this case, the strain-stress state of the rod can be determined from the differential equilibrium equations. For small displacements of the axial points, we can use the equilibrium equations of the zeroth approximation (1.107)–(1.111). For $\mu_0 = 0$, we have

$$\frac{d\mathbf{Q}^{(0)}}{d\eta} + \mathrm{A}_{æ}\mathbf{Q}^{(0)} + \mathbf{q}_0 = \sum_{i=1}^{n} \mathbf{P}^{(i)}\,\delta(\eta - \eta_i)\,; \tag{5.99}$$

$$\frac{d\mathbf{M}^{(0)}}{d\eta} + \mathrm{A}_{æ}\mathbf{M}^{(0)} + \mathrm{A}_1\mathbf{Q}^{(0)} = \sum_{\nu=1}^{\rho} T^{(\nu)}\,\delta(\eta - \eta_\nu)\,; \tag{5.100}$$

$$\frac{d\boldsymbol{\vartheta}^{(0)}}{d\eta} + \mathrm{A}_{æ}\boldsymbol{\vartheta}^{(0)} - \mathrm{A}^{-1}\mathbf{M}^{(0)} = 0\,, \qquad \mathbf{M}^{(0)} = \mathrm{A}\,\Delta æ^{(0)}\,; \tag{5.101}$$

$$\frac{d\mathbf{u}^{(0)}}{d\eta} + \mathrm{A}_{æ}\mathbf{u}^{(0)} + \mathrm{A}_1\boldsymbol{\vartheta}^{(0)} = 0\,. \tag{5.102}$$

If $\vartheta_{10} = 0$ or $\vartheta_{10} = \text{const}$ and $A_{ii} = \text{const}$, then the elements $æ_{i0}$ of the matrix $\mathrm{A}_{æ}$ are constants and (5.99)–(5.102) become linear differential equations with constant coefficients. For $\vartheta_{10} = \text{const}$, we have

$$æ_{10} = \frac{\sin\alpha_0\cos\alpha_0}{R_0}\,;$$

$$æ_{20} = \frac{\cos^2\alpha_0\sin\vartheta_{10}}{R_0}\,;$$

$$æ_{30} = \frac{\cos^2\alpha_0\cos\vartheta_{10}}{R_0}\,.$$

The method of successive integration will be applied. First, let us solve the homogeneous equations corresponding to (5.99)–(5.102). If we put

$$\mathbf{Q}^{(0)} = \mathbf{C}e^{\lambda\eta}\,,$$

then from the homogeneousn equation (5.99) we get the equality

$$\det[\mathrm{A}_{æ} + \lambda\mathrm{E}] = 0$$

or, in an expanded form,

$$\lambda^3 + (æ_{10}^2 + æ_{20}^2 + æ_{30}^2)\,\lambda = 0\,.$$

The equation has the following roots:

$$\lambda_1 = 0\,; \qquad \lambda_{2,3} = \pm i\sqrt{æ_{10}^2 + æ_{20}^2 + æ_{30}^2} = \pm i\lambda_0\,.$$

Substituting $æ_{i0}$ into the expression for $\lambda_{2,3}$, we get

$$\lambda_{2,3} = \pm i\lambda_0 = 0 \pm i\,\frac{\cos\alpha_0}{R_0}\,.$$

Let us derive the partial solutions of (5.99) that correspond to the roots λ_i, $i = 1, 2, 3$. For $\lambda_1 = 0$, we have the following homogeneous equations in the unknowns c_{1i} which are the components of the vector $\mathbf{C}$:

$$\begin{aligned} &-æ_{30}c_{12} + æ_{20}c_{13} = 0\,; \\ &æ_{30}c_{11} - æ_{10}c_{13} = 0\,; \\ &-æ_{20}c_{11} + æ_{10}c_{12} = 0\,. \end{aligned} \tag{5.103}$$

Substituting $c_{11} = 1$ into (5.103), we get

$$c_{12} = \frac{æ_{20}}{æ_{10}}\,; \qquad c_{13} = \frac{æ_{30}}{æ_{10}}\,, \tag{5.104}$$

i.e. the partial solution that corresponds to $\lambda_1 = 0$ is a vector

$$\mathbf{C}_1 = \left(1\,,\, \frac{æ_{20}}{æ_{10}}\,,\, \frac{æ_{30}}{æ_{10}}\right)^{\mathrm{T}}.$$

The eigenvectors $\mathbf{C}_2$ and $\mathbf{C}_3$ correspond to λ_2 and λ_3, respectively. We have

$$\begin{aligned} &\pm i\lambda_0 c_{21} - æ_{30}c_{22} + æ_{20}c_{23} = 0\,; \\ &æ_{30}c_{21} \pm i\lambda_0 c_{22} - æ_{10}c_{23} = 0\,; \\ &-æ_{20}c_{21} + æ_{10}c_{22} \pm i\lambda_0 c_{23} = 0\,. \end{aligned} \tag{5.105}$$

For the root $\lambda_1 = +i\lambda_0$, we put $c_{21} = 1$ and, from the first two equations of (5.105), we get

$$\begin{aligned} c_{22} &= \frac{(i\lambda_0 æ_{10} + æ_{20}æ_{30})\,(i\lambda_0 æ_{20} + æ_{10}æ_{30})}{(æ_{10}^2æ_{30}^2 + \lambda_0^2æ_{20}^2)}\,; \\ c_{23} &= \frac{æ_{30}}{æ_{10}} + i\,\frac{\lambda_0}{æ_{10}}\;\frac{(i\lambda_0 æ_{10} + æ_{20}æ_{30})\,(i\lambda_0 æ_{20} + æ_{10}æ_{30})}{(æ_{10}^2æ_{30}^2 + \lambda_0^2æ_{20}^2)} \end{aligned}$$

or

$$c_{22} = c_{22}^{(1)} + ic_{22}^{(2)}\,; \qquad c_{23}^{(1)} + ic_{23}^{(2)} = c_{23}\,,$$

where

$$\begin{aligned} c_{22}^{(1)} &= -\frac{æ_{10}æ_{20}}{æ_{20}^2 + æ_{30}^2}\,; \qquad & c_{22}^{(2)} &= \frac{\lambda_0 æ_{30}}{æ_{20}^2 + æ_{30}^2}\,; \\ c_{23}^{(1)} &= -\frac{æ_{10}æ_{30}}{æ_{20}^2 + æ_{30}^2}\,; \qquad & c_{23}^{(2)} &= -\frac{\lambda_0 æ_{20}}{æ_{20}^2 + æ_{30}^2}\,. \end{aligned}$$

For the root $\lambda_3 = -i\lambda_0$, we put $c_{31} = 1$ and thus obtain the following expressions for c_{3i}:

$$c_{32} = c_{32}^{(1)} + ic_{32}^{(2)}\,; \qquad c_{33} = c_{33}^{(1)} + ic_{33}^{(2)}\,;$$

here

$$c_{32}^{(1)} = c_{22}^{(1)}\,; \qquad c_{32}^{(2)} = -c_{22}^{(2)}\,; \qquad c_{33}^{(1)} = c_{23}^{(1)}\,; \qquad c_{33}^{(2)} = -c_{23}^{(2)}\,.$$

As a result, we arrive at

$$\mathbf{C}_3 = \begin{bmatrix} 1 \\ c_{32}^{(1)} \\ c_{33}^{(1)} \end{bmatrix} + i \begin{bmatrix} 1 \\ c_{32}^{(2)} \\ c_{33}^{(2)} \end{bmatrix} = \begin{bmatrix} 1 \\ c_{22}^{(1)} \\ c_{23}^{(1)} \end{bmatrix} + i \begin{bmatrix} 1 \\ -c_{22}^{(2)} \\ -c_{23}^{(2)} \end{bmatrix} .$$

On rearrangement, we get the general solution of the homogeneous equation

$$\begin{aligned} \mathbf{Q}_0 = c_1 \mathbf{C}_1 &+ c_2 \left(\mathbf{C}_2^{(1)} \cos \lambda_0 \eta - \mathbf{C}_2^{(2)} \sin \lambda_0 \eta \right) \\ &+ c_3 \left(\mathbf{C}_3^{(1)} \sin \lambda_0 \eta + \mathbf{C}_3^{(2)} \cos \lambda_0 \eta \right) , \end{aligned} \qquad (5.106)$$

where

$$\mathbf{C}_2^{(1)} = \begin{bmatrix} 1 \\ c_{22}^{(1)} \\ c_{23}^{(1)} \end{bmatrix} ; \qquad \mathbf{C}_2^{(2)} = \begin{bmatrix} 0 \\ c_{22}^{(2)} \\ c_{23}^{(2)} \end{bmatrix} ;$$

$$\mathbf{C}_3^{(1)} = \begin{bmatrix} 1 \\ c_{22}^{(1)} \\ c_{23}^{(1)} \end{bmatrix} ; \qquad \mathbf{C}_3^{(2)} = \begin{bmatrix} 0 \\ -c_{22}^{(2)} \\ -c_{23}^{(2)} \end{bmatrix} ;$$

here c_i are arbitrary constants.

The solution of the homogeneous equation (5.99) can be represented as follows:

$$\mathbf{Q}^{(0)} = \mathrm{K}_1(\eta)\, \mathbf{C}^{(1)}\,; \qquad (5.107)$$

here

$$\mathrm{K}_1(\eta) = \begin{bmatrix} 1 & \cos \lambda_0 \eta & \sin \lambda_0 \eta \\ \dfrac{æ_{20}^{(1)}}{æ_{10}^{(1)}} & c_{22}^{(1)} \cos \lambda_0 \eta - c_{22}^{(2)} \sin \lambda_0 \eta & c_{22}^{(1)} \sin \lambda_0 \eta + c_{22}^{(2)} \cos \lambda_0 \eta \\ \dfrac{æ_{30}^{(1)}}{æ_{10}^{(1)}} & c_{23}^{(1)} \cos \lambda_0 \eta - c_{23}^{(2)} \sin \lambda_0 \eta & c_{23}^{(1)} \sin \lambda_0 \eta + c_{23}^{(2)} \cos \lambda_0 \eta \end{bmatrix} ;$$

$$\mathbf{C}^{(1)} = (c_1\,,\, c_2\,,\, c_3)^{\mathrm{T}}\,.$$

In scalar form,

$$Q_1^{(0)} = c_1 + c_2 \cos\lambda_0\eta + c_3 \sin\lambda_0\eta\,; \tag{5.108}$$

$$Q_2^{(0)} = c_1 \frac{æ_{20}}{æ_{10}} + c_2 \left(c_{22}^{(1)} \cos\lambda_0\eta - c_{22}^{(2)} \sin\lambda_0\eta\right) + c_3 \left(c_{22}^{(1)} \sin\lambda_0\eta + c_{22}^{(2)} \cos\lambda_0\eta\right); \tag{5.109}$$

$$Q_3^{(0)} = c_1 \frac{æ_{30}}{æ_{10}} + c_2 \left(c_{23}^{(1)} \cos\lambda_0\eta - c_{23}^{(2)} \sin\lambda_0\eta\right) + c_3 \left(c_{23}^{(1)} \sin\lambda_0\eta + c_{23}^{(2)} \cos\lambda_0\eta\right). \tag{5.110}$$

The expressions for the internal forces $Q_i^{(0)}$ are valid for any values of the angle of helix α_0.

The fundamental matrix $\mathrm{K}(\eta)$ such that $\mathrm{K}(0) = \mathrm{E}$ is as follows:

$$\mathrm{K}(\eta) = \mathrm{K}_1(\eta)\,\mathrm{K}_1^{-1}(0) = A + \mathrm{B}\cos\lambda_0\eta + \mathrm{C}\sin\lambda_0\eta\,. \tag{5.111}$$

On rearrangement, we get the following expressions for the matrices A, B, and C:

$$\mathrm{A} = \begin{bmatrix} \dfrac{æ_{10}^2}{\lambda_0^2} & \dfrac{æ_{10}æ_{20}}{\lambda_0^2} & \dfrac{æ_{10}æ_{30}}{\lambda_0^2} \\ \dfrac{æ_{10}æ_{20}}{\lambda_0^2} & \dfrac{æ_{20}^2}{\lambda_0^2} & \dfrac{æ_{20}æ_{30}}{\lambda_0^2} \\ \dfrac{æ_{30}æ_{10}}{\lambda_0^2} & \dfrac{æ_{30}æ_{20}}{\lambda_0^2} & \dfrac{æ_{30}^2}{\lambda_0^2} \end{bmatrix};$$

$$\mathrm{B} = \begin{bmatrix} æ_{20}^2 + \dfrac{æ_{30}^2}{\lambda_0^2} & -\dfrac{æ_{10}æ_{20}}{\lambda_0^2} & -\dfrac{æ_{10}æ_{30}}{\lambda_0^2} \\ -\dfrac{æ_{20}æ_{10}}{\lambda_0^2} & æ_{10}^2 + \dfrac{æ_{30}^2}{\lambda_0^2} & -\dfrac{æ_{20}æ_{30}}{\lambda_0^2} \\ -\dfrac{æ_{10}æ_{30}}{\lambda_0^2} & -\dfrac{æ_{20}æ_{30}}{\lambda_0^2} & æ_{10}^2 + \dfrac{æ_{20}^2}{\lambda_0^2} \end{bmatrix};$$

$$\mathrm{C} = \begin{bmatrix} 0 & \dfrac{æ_{30}}{\lambda_0} & -\dfrac{æ_{20}}{\lambda_0} \\ -\dfrac{æ_{30}}{\lambda_0} & 0 & \dfrac{æ_{10}}{\lambda_0} \\ \dfrac{æ_{20}}{\lambda_0} & -\dfrac{æ_{10}}{\lambda_0} & 0 \end{bmatrix} = \frac{1}{\lambda_0}\mathrm{A}_{æ}$$

or

$$\mathrm{A} = \begin{bmatrix} \sin^2\alpha_0 & \sin\alpha_0\cos\alpha_0\sin\vartheta_{10} & \sin\alpha_0\cos\alpha_0 \times\cos\vartheta_{10} \\ \sin\alpha_0\cos\alpha_0\sin\vartheta_{10} & \cos^2\alpha_0\sin^2\vartheta_{10} & \cos^2\alpha_0\sin\vartheta_{10} \times\cos\vartheta_{10} \\ \sin\alpha_0\cos\alpha_0 \times\cos\vartheta_{10} & \cos^2\alpha_0\sin\vartheta_{10}\cos\vartheta_{10} & \cos^2\alpha_0\cos^2\vartheta_{10} \end{bmatrix};$$

$$\mathrm{B} = \begin{bmatrix} \cos^2\alpha_0 & -\sin\alpha_0\cos\alpha_0\sin\vartheta_{10} & -\sin\alpha_0\cos\alpha_0 \\ & & \times\cos\vartheta_{10} \\ -\sin\alpha_0\cos\alpha_0 & \sin^2\alpha_0+\cos^2\alpha_0\cos^2\vartheta_{10} & -\cos^2\alpha_0\sin\vartheta_{10} \\ \times\sin\vartheta_{10} & & \times\cos\vartheta_{10} \\ -\sin\alpha_0\cos\alpha_0 & -\cos^2\alpha_0\sin\vartheta_{10}\cos\vartheta_{10} & \sin^2\alpha_0+\cos^2\alpha_0 \\ \times\cos\vartheta_{10} & & \times\sin^2\vartheta_{10} \end{bmatrix};$$

$$\mathrm{C} = \begin{bmatrix} 0 & \cos\alpha_0\cos\vartheta_{10} & -\cos\alpha_0\sin\vartheta_{10} \\ -\cos\alpha_0\cos\vartheta_{10} & 0 & \sin\alpha_0 \\ \cos\alpha_0\sin\vartheta_{10} & -\sin\alpha_0 & 0 \end{bmatrix}.$$

If the principal axes coincide with the natural axes, we have $\vartheta_{10} = 0$ and $\Omega_{20} = 0$ and, therefore, the matrix $\mathrm{K}_1(\eta)$ reads

$$\mathrm{K}_1(\eta) = \begin{bmatrix} 1 & \cos\lambda_0\eta & \sin\lambda_0\eta \\ 0 & 0 & \dfrac{\lambda_0\cos\lambda_0\eta}{\Omega_{30}} \\ \dfrac{\Omega_{30}}{\Omega_{10}} & \dfrac{-\cos\lambda_0\eta}{\Omega_{10}\Omega_{30}} & 0 \end{bmatrix}.$$

In this case, the fundamental matrix $\mathrm{K}(\eta)$ such that $\mathrm{K}(0) = \mathrm{E}$ takes the form

$$\mathrm{K}_0(\eta) = \begin{bmatrix} \dfrac{1}{\lambda_0^2}\left(\Omega_{10}^2+\Omega_{30}^2\cos\lambda_0\eta\right) & \dfrac{\Omega_{30}}{\lambda_0}\sin\lambda_0\eta & \dfrac{\Omega_{10}\Omega_{30}}{\lambda_0^2}(1-\cos\lambda_0\eta) \\ -\dfrac{\Omega_{20}}{\lambda_0}\sin\lambda_0\eta & \cos\lambda_0\eta & \dfrac{\Omega_{10}}{\lambda_0}\sin\lambda_0\eta \\ \dfrac{\Omega_{10}\Omega_{30}}{\lambda_0^2}(1-\cos\lambda_0\eta) & -\dfrac{\Omega_{10}}{\lambda_0}\sin\lambda_0\eta & \dfrac{1}{\lambda_0^2}\left(\Omega_{30}^2+\Omega_{10}^2\cos\lambda_0\eta\right) \end{bmatrix},$$

Once the fundamental matrix (5.111) is found, we can represent the general solution of the inhomogeneous equation (5.99) as follows:

$$\mathbf{Q}^{(0)} = \mathrm{K}(\eta)\,\mathbf{C}^{(1)} + \int_0^\eta \mathrm{K}(\eta-\zeta)\left[-\mathbf{q}_0 + \sum_{i=1}^n \mathbf{P}_0^{(i)}\,\delta(\zeta-\eta_i)\right] d\zeta\,. \quad (5.112)$$

Using (5.111), we get

$$\begin{aligned}\mathbf{Q}^{(0)} = {} & \mathrm{K}(\eta)\,\mathbf{C}^{(1)} + \mathrm{A}\int_0^\eta \left[-\mathbf{q}_0 + \sum_{i=1}^n \mathbf{P}_0^{(i)}\,\delta(\zeta-\eta_i)\right] d\zeta \\ & - \mathrm{B}\int_0^\eta \cos\lambda_0(\eta-\zeta)\left[-\mathbf{q}_0 + \sum_{i=1}^n \mathbf{P}_0^{(i)}\,\delta(\zeta-\eta_i)\right] d\zeta \\ & - \mathrm{C}\int_0^\eta \sin\lambda_0(\eta-\zeta)\left[-\mathbf{q}_0 + \sum_{i=1}^n \mathbf{P}_0^{(i)}\,\delta(\zeta-\eta_i)\right] d\zeta\,. \end{aligned} \quad (5.113)$$

Similarly, the solution of (5.100) is as follows:

$$\mathbf{M}^{(0)} = \mathrm{K}(\eta)\,\mathbf{C}^{(2)} + \int_0^{\eta} \mathrm{K}(\eta-\zeta)\left[-\mathrm{A}_1\mathbf{Q}^{(0)} + \sum_{\nu=1}^{\rho} \mathbf{T}_0^{(\nu)}\,\delta(\zeta-\eta_\nu)\right] \mathrm{d}\zeta\,. \tag{5.114}$$

Under certain conditions, the explicit expressions for solutions (the expressions for the components of the vectors $\mathbf{Q}^{(0)}$ and $\mathbf{M}^{(0)}$) can be obtained on the basis of (5.113) and (5.114). When solving (5.99) and (5.100) numerically, it is convenient to evaluate the partial solutions for the zero initial conditions.

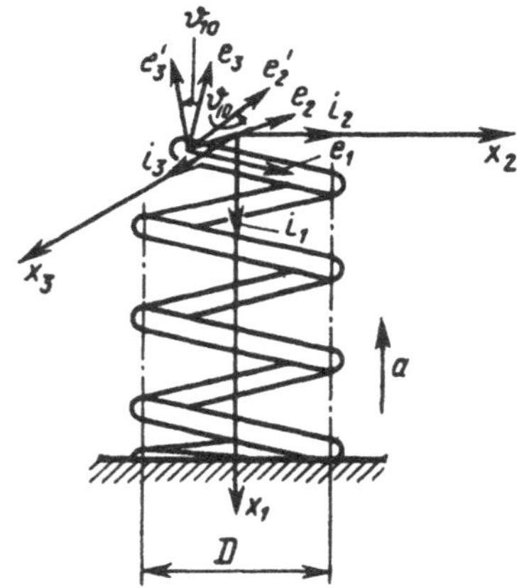

Fig. 5.13.

Let us consider some illustrative examples of determination of internal forces $Q_i^{(0)}$ and moments $M_i^{(0)}$. A cylindrical spring mounted on a moving object is shown in Fig. 5.13. The vector of acceleration of the object $\mathbf{a}$ is directed along the axis of the spring. We assume that the principal axes of a rod cross section do not coincide with the natural axes. The spring is subjected to the distributed inertial load. In what follows, we deal with the nondimensional load $\mathbf{q}_0$ such that

$$|\mathbf{q}_0| = \frac{m_0 a l^3}{A_{33}}\,,$$

where m_0 is the mass of unit length of the rod. In terms of projections on the natural axes, we have

$$\mathbf{q}_0 = q\sin\alpha_0\mathbf{e}_1 + q\cos\alpha_0\mathbf{e}_3\,, \qquad q = |\mathbf{q}_0|\,.$$

Using the principal axes, we get

$$\mathbf{q}_0 = q \sin\alpha_0 \mathbf{e}_1' + q\cos\alpha_0 \sin\vartheta_{10}\mathbf{e}_2' + q\cos\alpha_0\cos\vartheta_{10}\mathbf{e}_3' \,.$$

At $\eta = 0$, we have $\mathbf{Q}^{(0)}(0) = 0$. This leads to $\mathbf{C}^{(1)} = 0$ and, in view of (5.113), we get ($\mathbf{P}_0^{(i)} = 0$)

$$\mathbf{Q}^{(0)} = -\int_0^\eta \mathrm{K}\,(\eta - \zeta)\,\mathbf{q}\,\mathrm{d}\zeta \,,$$

or, by use of (5.111),

$$\mathbf{Q}^{(0)} = -\mathrm{A}\int_0^\eta \mathbf{q}_0\,\mathrm{d}\zeta - \mathrm{B}\int_0^\eta \cos\lambda_0\,(\eta-\zeta)\,\mathbf{q}_0\,\mathrm{d}\zeta - \mathrm{C}\int_0^\eta \sin\lambda_0\,(\eta-\zeta)\,\mathbf{q}_0\,\mathrm{d}\zeta \,.$$

Integration of this relation yields

$$\mathbf{Q}^{(0)} = -\mathbf{A}\mathbf{q}_1 - \mathbf{B}\mathbf{q}_2 - \mathbf{C}\mathbf{q}_3 \,, \tag{5.115}$$

where

$$\begin{aligned}
\mathbf{q}_1 &= \sum_{i=1}^{3}(q_{1i}\mathbf{e}_i') \\
&= (q\eta\sin\alpha_0)\,\mathbf{e}_1' + (q\eta\cos\alpha_0\sin\vartheta_{10})\,\mathbf{e}_2' + (q\eta\cos\alpha_0\cos\vartheta_{10})\,\mathbf{e}_3' \,; \\
\mathbf{q}_2 &= \sum_{i=1}^{3}(q_{2i}\mathbf{e}_i') = \left(q\,\frac{\eta}{\lambda_0}\,\sin\alpha_0\sin\lambda_0\eta\right)\mathbf{e}_1' \\
&\quad + q\left(\frac{1}{\lambda_0}\,\cos\alpha_0\sin\vartheta_{10}\sin\lambda_0\eta\right)\mathbf{e}_2' \\
&\quad + \left(\frac{q\cos\alpha_0\cos\vartheta_{10}}{\lambda_0}\,\sin\lambda_0\eta\right)\mathbf{e}_3' \,; \\
\mathbf{q}_3 &= \sum_{i=1}^{3}(q_{3i}\mathbf{e}_i') = \left[\frac{q\sin\alpha_0}{\lambda_0}\,(1-\cos\lambda_0\eta)\right]\mathbf{e}_1' \\
&\quad + \left[\frac{q\cos\alpha_0\sin\vartheta_{10}}{\lambda_0}\,(1-\cos\lambda_0\eta)\right]\mathbf{e}_2' \\
&\quad + \left[q\,\frac{q\cos\alpha_0\cos\vartheta_{10}}{\lambda_0}\,(1-\cos\lambda_0\eta\right]\mathbf{e}_3' \,.
\end{aligned}$$

When the principal axes coincide with the natural axes, we have

$$\mathbf{Q}^0 = K_0(\eta)\,\mathbf{C}^{(1)} - \int_0^\eta \mathrm{K}_0(\eta-\zeta)\,\mathbf{q}\,\mathrm{d}\zeta \,, \tag{5.116}$$

where the components of the matrix $\mathrm{K}_0(\eta - \zeta)$ are as follows:

$$(K_0)_{11} = \frac{1}{\lambda_0^2}\,[\,\Omega_{10}^2 + \Omega_{30}^2 \cos\lambda_0(\eta-\zeta)\,]\,;$$
$$(K_0)_{12} = \frac{\Omega_{30}}{\lambda_0}\,\sin\lambda_0(\eta-\zeta)\,;$$
$$(K_0)_{13} = \frac{\Omega_{10}\Omega_{30}}{\lambda_0^2}\,[\,1-\cos\lambda_0(\eta-\zeta)\,]\,;$$
$$(K_0)_{21} = -\frac{\Omega_{30}}{\lambda_0}\,\sin\lambda_0(\eta-\zeta)\,;$$
$$(K_0)_{22} = \cos\lambda_0(\eta-\zeta)\,;$$
$$(K_0)_{23} = \frac{\Omega_{10}}{\lambda_0}\,\sin\lambda_0(\eta-\zeta)\,;$$
$$(K_0)_{31} = \frac{\Omega_{10}\Omega_{30}}{\lambda_0^2}\,[\,1-\cos\lambda_0(\eta-\zeta)\,]\,;$$
$$(K_0)_{32} = -\frac{\Omega_{10}}{\lambda_0}\,\sin\lambda_0(\eta-\zeta)\,;$$
$$(K_0)_{33} = \frac{1}{\lambda_0^2}\,[\,\Omega_{30}^2 + \Omega_{10}^2 \cos\lambda_0(\eta-\zeta)\,]\,.$$

Using the expressions for $æ_{10}$ and $æ_{30}$ (5.70), we get

$$\begin{aligned}
(K_0)_{11} &= \sin^2\alpha_0 + \cos^2\alpha_0 \cos\lambda_0(\eta-\zeta)\,;\\
(K_0)_{12} &= \cos\alpha_0 \sin\lambda_0(\eta-\zeta)\,;\\
(K_0)_{13} &= \frac{1}{2}\sin^2\alpha_0[\,1-\cos\lambda_0(\eta-\zeta)\,]\,;\\
(K_0)_{21} &= -\cos\alpha_0 \sin\lambda_0(\eta-\zeta)\,;\\
(K_0)_{22} &= \cos\lambda_0(\eta-\zeta)\,;\\
(K_0)_{23} &= \sin\alpha_0 \sin\lambda_0(\eta-\zeta)\,;\\
(K_0)_{31} &= \frac{1}{2}\sin^2\alpha_0[\,1-\cos\lambda_0(\eta-\zeta)\,]\,;\\
(K_0)_{32} &= -\sin\alpha_0 \sin\alpha_0(\eta-\zeta)\,;\\
(K_0)_{33} &= \cos^2\alpha_0 + \sin^2\alpha_0 \cos\lambda_0(\eta-\zeta)\,.
\end{aligned} \tag{5.117}$$

On rearrangement, from (5.116) (for $\mathbf{C}_1 = 0$), it follows that

$$\mathbf{Q}_1^{(0)} = -q\eta\sin\alpha_0\,; \qquad \mathbf{Q}_2^{(0)} = 0\,; \qquad \mathbf{Q}_3^{(0)} = -q\eta\cos\alpha_0\,.$$

As another example, let us consider a rotating cylindrical spring shown in Fig. 5.14. The spring is subjected to a distributed load. The nondimensional value of the load is as follows:

$$q_0 = \frac{Dm_0\omega^2 l^3}{2A_{33}}\,.$$

The vector $\mathbf{q}_0$ is opposite in direction to the principal normal to the rod axis, consequently,

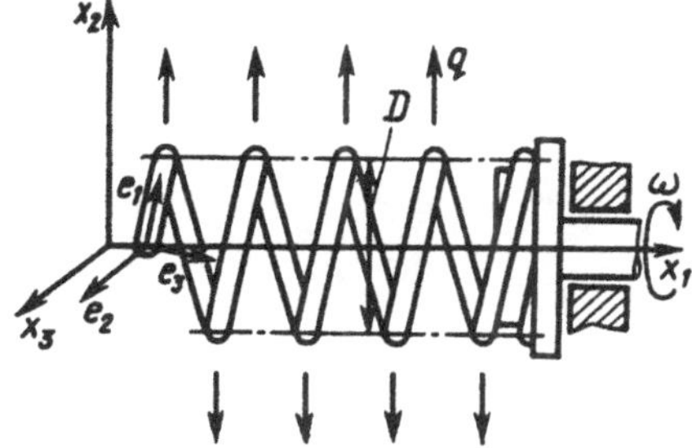

Fig. 5.14.

$$\mathbf{q}_0 = -q\mathbf{e}_2\,, \qquad q = |\mathbf{q}_0| = \frac{m_0\omega^2 D l^3}{2A_{33}}\,.$$

In the natural axes, the vector $\mathbf{q}_0$ can be written as follows:

$$\mathbf{q}_0 = q\cos\vartheta_{10}\mathbf{e}_2' - q\sin\vartheta_{10}\mathbf{e}_3'\,.$$

The vectors $\mathbf{q}_i$ in (5.115) are

$$\begin{aligned}
\mathbf{q}_1 &= q\eta\cos\vartheta_{10}\mathbf{e}_2' - q\eta\sin\vartheta_{10}\mathbf{e}_3'\,;\\
\mathbf{q}_2 &= \frac{q\cos\vartheta_{10}}{\lambda_0}\sin\lambda_0\eta\mathbf{e}_2' - \frac{q\sin\vartheta_{10}}{\lambda_0}\sin\lambda_0\eta\mathbf{e}_3'\,;\\
\mathbf{q}_3 &= \frac{q\cos\vartheta_{10}}{\lambda_0}(1-\cos\lambda_0\eta)\,\mathbf{e}_2' - \frac{q\sin\vartheta_{10}}{\lambda_0}(1-\cos\lambda_0\eta)\,\mathbf{e}_3'\,.
\end{aligned}$$

Let $\vartheta_{10} = 0$. Suppose that the principal axes coincide with the natural axes, that is, $æ_{20} = 0$; then, using (5.115), we get

$$Q_1^{(0)} = -\frac{\varOmega_{30}}{\lambda_0^2}\,q\,(1-\cos\lambda_0\eta) = -qR_0(1-\cos\lambda_0\eta)\,; \tag{5.118}$$

$$Q_2^{(0)} = -\frac{q}{\lambda_0}\sin\lambda_0\eta = -\frac{qR_0}{\cos\alpha_0}\sin\lambda_0\eta\,; \tag{5.119}$$

$$Q_3^{(0)} = -\frac{\varOmega_{10}}{\lambda_0^2}\,q\,(1-\cos\lambda_0\eta) = qR_0\tan\alpha_0(1-\cos\lambda_0\eta)\,. \tag{5.120}$$

In the example under consideration, the projection of the vector $\mathbf{Q}^{(0)}$ on the axis x_1 must equal zero, that is

$$Q_{x_1}^{(0)} = (\mathbf{Q}^{(0)}\cdot\mathbf{i}_1) = Q_1^{(0)}(\mathbf{e}_1\cdot\mathbf{i}_1) + Q_3^{(0)}(\mathbf{e}_3\cdot\mathbf{i}_1) = 0\,.$$

Let us prove this equality. We have $\varOmega_{10}\cos\alpha_0 - \varOmega_{30}\sin\alpha_0 = 0$, hence,

$$\begin{aligned}
Q_{x_1}^{(0)} &= -\frac{\varOmega_{30}}{\lambda_0^2}\,q\sin\alpha_0(1-\cos\lambda_0\eta) + \frac{\varOmega_{10}}{\lambda_0^2}\,q\cos\alpha_0(1-\cos\lambda_0\eta)\\
&= (\varOmega_{10}\cos\alpha_0 - \varOmega_{30}\sin\alpha_0)\,\frac{q}{\lambda_0^2}\,(1-\cos\lambda_0\eta) = 0\,.
\end{aligned}$$

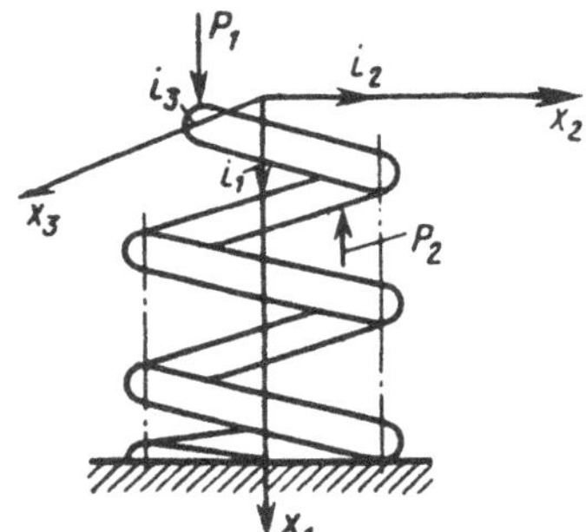

Fig. 5.15.

A cylindrical spring subjected to concentrated forces $\mathbf{P}^{(1)}$ and $\mathbf{P}^{(2)}$ is shown in Fig. 5.15. The forces are applied to the cross sections with coordinates $\eta = \eta_1 = 0$ and $\eta = \eta_2$. We assume that the principal axes coincide with the natural axes. Our purpose is to determine $Q_i^{(0)}(\eta)$. In terms of projections on the natural axes, the applied forces take the form

$$\mathbf{P}^{(1)} = P^{(1)} \sin\alpha_0 \mathbf{e}_1 + P^{(1)} \cos\alpha_0 \mathbf{e}_3 , \qquad P^{(1)} = |\mathbf{P}^{(1)}| ;$$
$$\mathbf{P}^{(2)} = -\mathbf{P}^{(2)} \sin\alpha_0 \mathbf{e}_1 - P^{(2)} \cos\alpha_0 \mathbf{e}_3 , \qquad P^{(2)} = |\mathbf{P}^{(2)}| .$$

From (5.112) it follows that

$$\mathbf{Q}^{(0)} = -\int_0^\eta \mathrm{K}_0(\eta - \zeta)\,[\,\mathbf{P}^{(1)}\,\delta(\zeta) + \mathbf{P}^{(2)}\,\delta(\zeta - \eta_2)\,]\,\mathrm{d}\zeta ;$$

in scalar form, we have

$$\begin{aligned} Q_j^{(0)} &= -P^{(1)} \sin\alpha_0 \int_0^\eta k_{j1}(\eta - \zeta)\,\delta(\zeta)\,\mathrm{d}\zeta \\ &\quad - P^{(1)} \cos\alpha_0 \int_0^\eta k_{j3}(\eta - \zeta)\,\delta(\zeta)\,\mathrm{d}\zeta \\ &\quad + P^{(2)} \sin\alpha_0 \int_0^\eta k_{j1}(\eta - \zeta)\,\delta(h - \eta_2)\,\mathrm{d}\zeta \\ &\quad + P^{(2)} \cos\alpha_0 \int_0^\eta k_{j3}(\eta - \zeta)\,\delta(\zeta - \eta_2)\,\mathrm{d}\zeta . \end{aligned}$$

Since (see Appendix 4)

$$\int_a^b f(\eta)\,\delta(\eta - \eta_k)\,\mathrm{d}\eta = f(\eta_k) , \qquad a \le \eta_k \le b ,$$

we have

$$\begin{aligned} Q_j^{(0)} &= -P^{(1)} \sin\alpha_0 k_{j1}(\eta) - P^{(1)} \cos\alpha_0 k_{j3}(\eta) \\ &\quad + P^{(2)} \sin\alpha_0 k_{j1}(\eta - \eta_2)\,H(\eta - \eta_2) + P^{(2)} \cos\alpha_0 k_{j3}(\eta - \eta_2)\,H(\eta - \eta_2) , \end{aligned}$$

where $H(\eta - \eta_k)$ is the Heaviside function. It is possible to dispense with the Heaviside function if we adopt a usual assumption from structural mechanics that the functions $k_{ji}(\eta - \eta_k)$ do not equal zero only when $\eta - \eta_k > 0$ and $k_{ji} = 0$ for $\eta - \eta_k < 0$.

Now our goal is to determine the moment $\mathbf{M}^{(0)}$ and the vectors $\mathbf{u}^{(0)}$ and $\boldsymbol{\vartheta}^{(0)}$. The general solution of (5.100) can be represented as follows:

$$\mathbf{M}^{(0)} = \mathrm{K}(\eta)\,\mathbf{C}^{(2)} + \int_0^{\eta} \mathrm{K}(\eta - \zeta)\left[-\mathrm{A}_1\mathbf{Q}^{(0)} + \sum_{\nu=1}^{\rho} \mathbf{T}_0^{(\nu)}\,\delta(\zeta - \eta_\nu)\right] \mathrm{d}\zeta\,; \tag{5.121}$$

in view of (5.112), we get

$$\begin{aligned}\mathbf{M}^{(0)} = \mathrm{K}(\eta)\,\mathbf{C}^{(2)} - \int_0^{\eta} \mathrm{K}(\eta - \zeta)\mathrm{A}_1 \Bigg\{ \mathrm{K}(\zeta)\,\mathbf{C}^{(1)} \\ + \int_0^{\zeta} \mathrm{K}(\zeta - x)\left[-\mathbf{q}_0 + \sum_{i=1}^{n} \mathbf{P}_0^{(i)}\,\delta(x - \zeta)\right] \mathrm{d}x \\ - \int_0^{\eta} \mathrm{K}(\eta - \zeta)\left[\sum_{\nu=1}^{\rho} \mathbf{T}_0^{(\nu)}\,\delta(\zeta - \eta_\nu)\right]\Bigg\} \mathrm{d}\zeta\,.\end{aligned} \tag{5.122}$$

Let us obtain $\mathbf{M}^{(0)}$. In the case shown in Fig. 5.13, we have $\mathbf{C}^{(2)} = 0$, hence, $\mathbf{T}_0^{(\nu)} = 0$ and $\mathbf{M}^{(0)}(0) = 0$. Consequently, from (5.121) it follows that

$$\mathbf{M}^{(0)} = -\int_0^{\eta} \mathrm{K}(\eta - \zeta)\,\mathrm{A}_1\mathbf{Q}^{(0)}\,\mathrm{d}\zeta\,. \tag{5.123}$$

For $\vartheta_{10} = 0$ and in view of (5.117), we have $\mathrm{A}_1\mathbf{Q}^{(0)} = -Q_3^{(0)}\mathbf{e}_3$. Hence, from (5.123) it follows that

$$\mathbf{M}^{(0)} = \int_0^{\eta} \mathrm{K}(\eta - \zeta)\,Q_3^{(0)}\mathbf{e}_3\,\mathrm{d}\zeta\,,$$

or, in scalar form,

$$\begin{aligned} M_1^{(0)} &= \int_0^{\eta} k_{13}(\eta - \zeta)\,Q_3^{(0)}\,\mathrm{d}\zeta \\ &= -\frac{q\sin^2\alpha_0\cos\alpha_0}{2}\eta^2 + qR_0^2\frac{\sin^2\alpha_0}{\cos\alpha_0}(1 - \cos\alpha_0\eta)\,; \\ M_2^{(0)} &= \int_0^{\eta} k_{23}(\eta - \zeta)\,Q_3^{(0)}\,\mathrm{d}\zeta = -qR_0\frac{\sin^2\alpha_0}{\cos\alpha_0}\left(\eta - \frac{1}{\lambda_0}\sin\lambda_0\eta\right)\,; \\ M_3^{(0)} &= \int_0^{\eta} k_{33}(\eta - \zeta)\,Q_3^{(0)}\,\mathrm{d}\zeta \\ &= -\frac{qR_0^2}{2}\cos\alpha_0\sin\alpha_0\eta^2 - \frac{q\sin^2\alpha_0 R_0^4}{\cos^2\alpha_0}(1 - \cos\lambda_0\eta)\,. \end{aligned}$$

Consider the rotating spring shown in Fig. 5.14. Let $æ_{20} = \vartheta_{10} = 0$. Then, we have

$$M_i^{(0)} = -\int_0^\eta k_{i2} Q_2^{(0)}\,\mathrm{d}\zeta + \int_0^\eta k_{i3} Q_3^{(0)}\,\mathrm{d}\zeta\,, \qquad i = 1\,,\,2\,,\,3\,, \tag{5.124}$$

where $Q_2^{(0)}$ and $Q_3^{(0)}$ are defined by (5.119) and (5.120), respectively, and k_{ij} are the elements of the matrix (5.117).

When $\mathbf{M}^{(0)}$ is found, we get

$$\Delta æ^{(0)} = \mathrm{A}^{-1} M^{(0)}\,.$$

From (5.101) it follows that

$$\boldsymbol{\vartheta}^{(0)} = \mathrm{K}\,(\eta)\,\mathbf{C}^{(3)} + \int_0^\eta \mathrm{K}\,(\eta - \zeta)\,\mathrm{A}^{-1}\mathbf{M}^{(0)}\,\mathrm{d}\zeta\,. \tag{5.125}$$

The displacement vector $\mathbf{u}$ can be found from the equation

$$\mathbf{u}^{(0)} = \mathrm{K}\,(\eta)\,\mathbf{C}^{(4)} - \int_0^\eta \mathrm{K}(\eta - \zeta)\,\mathrm{A}_1 \boldsymbol{\vartheta}^{(0)}\,\mathrm{d}\zeta\,. \tag{5.126}$$

The relations (5.112), (5.121), (5.125), and (5.126) include the four vectors $\mathbf{C}^{(1)}$, $\mathbf{C}^{(2)}$, $\mathbf{C}^{(3)}$, and $\mathbf{C}^{(4)}$ with arbitrary components (twelve arbitrary constants). Note that we have the twelve boundary-condition equations (six at each end). For a spring whose one end, $\eta = 1$, is clamped (Figs. 5.14 and 5.15), we have $\mathbf{u}^{(0)}(1) = 0$ and $\boldsymbol{\vartheta}^{(0)}(1) = 0$. The vectors $\mathbf{C}^{(3)}$ and $\mathbf{C}^{(4)}$ can be obtained from the two algebraic vector equations (see (5.125) and (5.126)):

$$\mathrm{K}\,(1)\,\mathbf{C}^{(3)} = -\int_0^1 \mathrm{K}\,(1 - h)\,\mathrm{A}^{-1}\mathbf{M}^{(0)}\,\mathrm{d}h\,;$$

$$\mathrm{K}\,(1)\,\mathbf{C}^{(4)} = \int_0^1 \mathrm{K}\,(1 - h)\,\mathrm{A}_1 \boldsymbol{\vartheta}^{(0)}\,\mathrm{d}h\,.$$

To determine displacements of the axial points in the direction x_1, it is sufficient to consider the scalar product

$$u_{x_1}^{(0)} = \sum_{i=1}^{3} u_i^{(0)} (\mathbf{e}_i^{(1)} \mathbf{i}_1) = \sum_{i=1}^{3} u_i^{(0)} k_{i1}^{(0)}\,.$$

The value of extension (shortening) of the spring is

$$\Delta H = \sum_{i=1}^{3} u_i^{(0)}(0)\,(\mathbf{e}_i' \mathbf{i}_1)\,.$$

Finally, vector form of the general solution of the system (5.99)–(5.102) with $\vartheta_{10} = \mathrm{const}$ is as follows:

$$\mathbf{Q}^{(0)} = \mathrm{K}(\eta)\,\mathbf{C}^{(1)} + \int_0^{\eta} \mathrm{K}(\eta-\zeta)\left[-\mathbf{q}_0 + \sum_{i=1}^{n} \mathbf{P}_0^{(i)}\,\delta(\zeta-\eta_i)\right] \mathrm{d}\zeta\,; \tag{5.127}$$

$$\mathbf{M}^{(0)} = \mathrm{K}(\eta)\,\mathbf{C}^{(2)} + \int_0^{\eta} \mathrm{K}(\eta-\zeta)\left[-\mathrm{A}_1\mathbf{Q}^{(0)} + \sum_{\nu=1}^{\rho} \mathbf{T}_0^{(\nu)}\,\delta(\zeta-\eta_\nu)\right] \mathrm{d}\zeta\,; \tag{5.128}$$

$$\Delta æ^{(0)} = \mathrm{A}^{-1} M^{(0)}\,; \tag{5.129}$$

$$\boldsymbol{\vartheta}^{(0)} = \mathrm{K}(\eta)\,\mathbf{C}^{(3)} + \int_0^{\eta} \mathrm{K}(\eta-\zeta)\,\mathrm{A}^{-1}\mathbf{M}^{(0)}\,\mathrm{d}\zeta\,; \tag{5.130}$$

$$\mathbf{u}^{(0)} = \mathrm{K}(\eta)\,\mathbf{C}^{(4)} - \int_0^{\eta} \mathrm{K}(\eta-\zeta)\,\mathrm{A}_1\boldsymbol{\vartheta}^{(0)}\,\mathrm{d}\zeta\,. \tag{5.131}$$

The relations (5.127)–(5.131) present the general solution to the problem when the displacements u_j and the angles of rotation ϑ_j are small and the principal axes do not coincide with the natural axes (and under the condition $\vartheta_{10} = \mathrm{const}$).

5.3.2 Cylindrical Springs of Variable Angle of Helix

First, let us consider the geometrical parameters of a helical line with variable angle of helix α_0. The nondimensional Cartesian coordinates of a point B lying on the axial line are (see Fig. 5.16)

$$x_1 = f(\varphi)\,;$$
$$x_2 = R_0 \cos\varphi\,;$$
$$x_3 = R_0 \sin\varphi\,, \qquad R_0 = \frac{R_{00}}{l}\,. \tag{5.132}$$

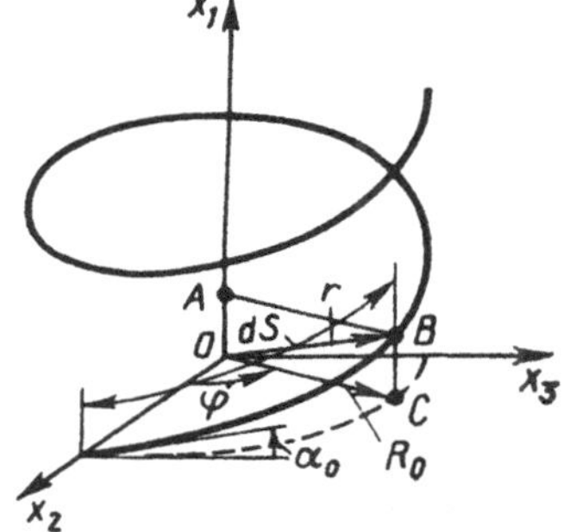

Fig. 5.16.

For a helical line with constant angle of helix, the nondimensional coordinate x_1 depends linearly on the parameter φ, i.e. $x_1 = b\varphi$. In the case of nonlinear relationship between x_1 and φ, the angle of helix $\alpha_0(\eta)$ is not constant. For example,

$$x_1 = b\varphi + b_1\varphi^2 , \tag{5.133}$$

where b and b_1 are constants and the angle φ is a function of the nondimensional axial coordinate η.

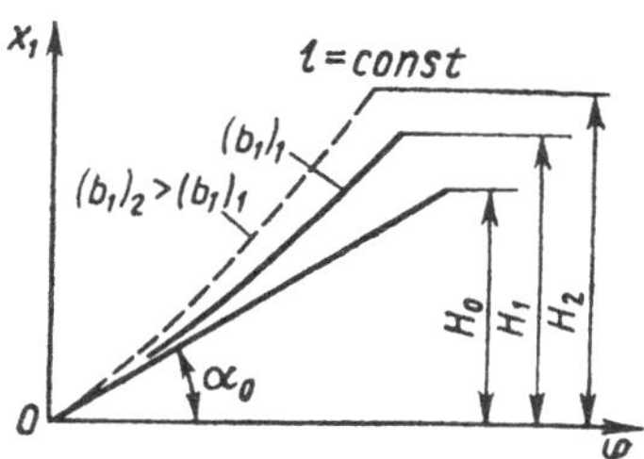

Fig. 5.17.

The involute curve of a helical line with constant angle of helix is a straight line $x_1 = b\varphi$ as indicated in Fig. 5.17. If $x_1 = b\varphi + b_1\varphi^2$, then the involute curve is not a straight line. In this case, the shape of the involute curve depends on the coefficient b_1 (see Fig. 5.17). The angle φ is related to the nondimensional coordinate η as follows:

$$\eta = \int_0^{\varphi} \sqrt{x_1'^{\,2} + x_2'^{\,2} + x_3'^{\,2}}\, \mathrm{d}\varphi . \tag{5.134}$$

Substituting (5.132) and expression (5.133) into (5.134), we get

$$\eta = \int_0^{\varphi} \sqrt{R_0^2 + (b + 2b_1\varphi)^2}\, \mathrm{d}\varphi ,$$

After integration, we obtain

$$\begin{aligned}\eta = \frac{1}{4b_1} \Big\{ &(b + 2b_1\varphi)\sqrt{R_0^2 + (b + 2b_1\varphi)^2} \\ &+ R_0^2 \ln \left[b + 2b_1\varphi + \sqrt{R_0^2 + (b + 2b_1\varphi)^2} \right] \\ &- b\sqrt{R_0^2 + b^2} - R_0^2 \ln \left[b + \sqrt{R_0^2 + b^2} \right] \Big\} .\end{aligned}$$

If the coordinate x_1 is linear in φ (when $\alpha = \text{const}$), then the coordinate η is also linear in φ,

$$\eta = \sqrt{R_0^2 + b^2}\, \varphi .$$

For determination of the curvature Ω_{30} and the twist Ω_{10} of the axial line, the following expressions may be used:

$$\Omega_{30}(\varphi) = \\ = \frac{\sqrt{({x'_1}^2 + {x'_2}^2 + {x'_3}^2)({x''_1}^2 + {x''_2}^2 + {x''_3}^2) - (x'_1 x''_1 + x'_2 x''_2 + x'_3 x''_3)^2}}{({x'_1}^2 + {x'_2}^2 + {x'_3}^2)^{3/2}};$$

$$\Omega_{10}(\varphi) = \frac{1}{\Omega_{30}^2({x'_1}^2 + {x'_2}^2 + {x'_3}^2)^3} \begin{vmatrix} x'_1 & x'_2 & x'_3 \\ x''_1 & x''_2 & x''_3 \\ x'''_1 & x'''_2 & x'''_3 \end{vmatrix}.$$

Substituting the derivatives of x_i, which can be found from (5.132), into these expressions, we get

$$\Omega_{30} = \frac{R_0\sqrt{R_0^2 + f'^2 + f''^2}}{(R_0^2 + f'^2)^{3/2}}; \tag{5.135}$$

$$\Omega_{10} = \frac{f'' + f'''}{R_0^2 + f'^2 + f''^2}. \tag{5.136}$$

If the principal moments of inertia of a cross section are different, then the components $æ_i$ are as follows:

$$æ_1 = \frac{\mathrm{d}\vartheta_{10}}{\mathrm{d}\eta} + \Omega_{10}; \qquad æ_2 = \Omega_{30}\sin\vartheta_{10}; \qquad æ_3 = \Omega_{30}\cos\vartheta_{10}.$$

Derivatives of functions with respect to the parameter φ appear in (5.135) and (5.136). For example, if x_1 is given by (5.133), we have

$$\Omega_{30} = \frac{R_0\sqrt{R_0^2 + (b + 2b_1\varphi)^2 + 4b_1^2}}{[\,R_0^2 + (b + 2b_1\varphi)^2\,]^{3/2}};$$

$$\Omega_{10} = \frac{b + 2b_1\varphi}{R_0^2 + (b + 2b_1\varphi)^2 + 4b_1^2}.$$

Using the expression that relates η to φ, we can establish numerically the relationship between Ω_{10}, Ω_{30} and η. These relations are necessary for analysis of equilibrium equations. If the angle of helix α is variable, then the elements of the matrix $\mathrm{A}_æ$ depend on η. Hence, (5.99) and (5.100) can be solved only numerically.

Let us consider a particular case. In vector notation, (5.99) and (5.100) are

$$\frac{\mathrm{d}\mathbf{Y}^{(0)}}{\mathrm{d}\eta} + \mathrm{A}^{(0)}(\eta)\,\mathbf{Y}^{(0)} = \mathbf{f}^{(0)}(\eta), \tag{5.137}$$

where

$$\mathbf{Y}^{(0)} = \begin{bmatrix} \mathbf{Q}^{(0)} \\ \mathbf{M}^{(0)} \\ \boldsymbol{\vartheta}^{(0)} \\ \mathbf{u}^{(0)} \end{bmatrix}; \qquad \mathbf{f}^{(0)} = \begin{bmatrix} -\mathbf{q}_0 + \sum_{i=1}^{n} \mathbf{P}_0^{(i)} \delta(\eta - \eta_i) \\ \sum_{\nu=1}^{\rho} \mathbf{T}_0^{(\nu)} \delta(\eta - \eta_\nu) \\ \mathbf{0} \\ \mathbf{0} \end{bmatrix};$$

$$\mathrm{A}^{(0)}(\eta) = \begin{bmatrix} A_æ & 0 & 0 & 0 \\ A_0 & A_æ & 0 & 0 \\ 0 & -A^{-1} & A_æ & 0 \\ 0 & 0 & A_1 & A_æ \end{bmatrix}.$$

The general solution of (5.137) can be represented as follows:

$$\mathbf{Y}^{(0)}(\eta) = \mathbf{Y}_1 + \mathbf{Y}^{(1)};$$

here $\mathbf{Y}_1 = \mathrm{K}(\eta)\mathbf{C}$ is a solution of the homogeneous equation (5.137) and $\mathbf{Y}^{(1)}$ is a partial solution of the inhomogeneous equation (5.137). Computer-oriented methods for determination of the vectors $\mathbf{Y}_1$, $\mathbf{Y}^{(1)}$, and $\mathbf{C}$ are presented in Sect. 2.1.

5.4 Flexible Rods in a Rigid Conduit

5.4.1 Statement of the Problem

Flexible rods (flexible shafts) can be used to transfer a rotation or a torque from a cross section A (input) to a cross section B (output) (see Fig. 5.18*a*). We assume that there is no clearance between the rod and the walls of the conduit. For the device to work properly, smooth rotations of the cross section A must result in smooth rotations of the cross section B. In practice, due to the loss of stability, the cross section B may start vibrating while the cross section A rotates smoothly (Panovko, Gubanova (1979)).

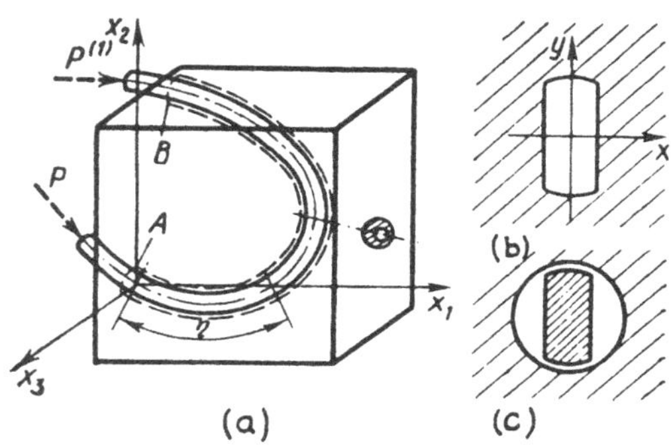

Fig. 5.18.

In what follows, we assume that the rod may be of variable or constant cross section and of equal ($A_{22} = A_{33}$) or different ($A_{22} \neq A_{33}$) bending stiffnesses. We also assume that the rod rotates rather slowly so that dynamic effects can be neglected.

The following characteristics are assumed to be given a priori:

(1) Ω^{c}_{10}, Ω^{c}_{30}, and $\vartheta^{\mathrm{c}}_{10}$, which describe the axial line of the conduit and the location of the principal axes of a conduit cross section (the moments of inertia of a conduit cross section may be different as indicated in Fig. 5.18*b*,*c*);

(2) Ω^{r}_{10}, Ω^{r}_{30}, and $\vartheta^{\mathrm{r}}_{10}$ are characteristics of the axial line of the rod and the location of the principal axes in the natural configuration. Quantities associated with the rod are marked by the superscript 'r', and those associated with the conduit are marked by the superscript 'c'.

5.4.2 Equilibrium Equations

Let us derive the general equilibrium equations for a rod in a rigid conduit. The rod axis in its natural configuration as well as the axial line of the conduit are not necessarily straight. In the absence of concentrated forces and moments, the system of equations (1.64)–(1.66) takes the form

$$\frac{\mathrm{d}Q_1}{\mathrm{d}\eta} + Q_3 æ_2 - Q_2 æ_3 + q_1 = 0\,; \tag{5.138}$$

$$\frac{\mathrm{d}Q_2}{\mathrm{d}\eta} + Q_1 æ_3 - Q_3 æ_1 + q_2 = 0\,; \tag{5.139}$$

$$\frac{\mathrm{d}Q_3}{\mathrm{d}\eta} + Q_2 æ_1 - Q_1 æ_2 + q_3 = 0\,; \tag{5.140}$$

$$\frac{\mathrm{d}M_1}{\mathrm{d}\eta} + M_3 æ_2 - M_2 æ_3 + \mu_1 = 0\,; \tag{5.141}$$

$$\frac{\mathrm{d}M_2}{\mathrm{d}\eta} + M_1 æ_3 - M_3 æ_1 - Q_3 + \mu_2 = 0\,; \tag{5.142}$$

$$\frac{\mathrm{d}M_3}{\mathrm{d}\eta} + M_2 æ_1 - M_1 æ_2 + Q_2 + \mu_3 = 0\,, \tag{5.143}$$

where

$$\begin{aligned} M_1 &= A_{11}(æ_1 - æ_{10})\,; \\ M_2 &= A_{22}(æ_2 - æ_{20})\,; \\ M_3 &= A_{33}(æ_3 - æ_{30})\,. \end{aligned} \tag{5.144}$$

These equations are nonlinear because M_j and $æ_j$ depend on the unknown angle ϑ_{10}.

The components $æ_i$, which appear in (1.64)–(1.66), satisfy the relations

$$\begin{aligned} æ_1 &= \Omega^{c}_{10} + \frac{d\vartheta^{c}_{10}}{d\eta} ; \\ æ_2 &= \Omega^{c}_{30} \sin \vartheta^{c}_{10} ; \\ æ_3 &= \Omega^{c}_{30} \cos \vartheta^{c}_{10} , \end{aligned} \tag{5.145}$$

where Ω^{c}_{j0} are the components of the Darboux vector for the axial line of the conduit. For the rod, we have similar relations,

$$\begin{aligned} æ_{10} &= \Omega^{r}_{10} + \frac{d\vartheta^{r}_{10}}{d\eta} ; \\ æ_{20} &= \Omega^{r}_{30} \sin \vartheta^{r}_{10} ; \\ æ_{30} &= \Omega^{r}_{30} \cos \vartheta^{r}_{10} , \end{aligned} \tag{5.146}$$

where Ω^{r}_{j0} are components of the Darboux vector for the rod axis in the natural configuration.

The internal moments M_j are as follows:

$$M_1 = A_{11} \left(\Omega^{c}_{10} + \frac{d\vartheta^{c}_{10}}{d\eta} - \Omega^{r}_{10} - \frac{d\vartheta^{r}_{10}}{d\eta} \right) ; \tag{5.147}$$

$$M_2 = A_{22} \left(\Omega^{c}_{30} \sin \vartheta^{c}_{10} - \Omega^{r}_{30} \sin \vartheta^{r}_{10} \right) ; \tag{5.148}$$

$$M_3 = A_{33} \left(\Omega^{c}_{30} \cos \vartheta^{c}_{10} - \Omega^{r}_{30} \cos \vartheta^{r}_{10} \right) . \tag{5.149}$$

For a rod of constant cross section, we have $A_{33} = 1$.

If a conduit cross section is the one shown in Fig. 5.18*b*, then the three moments M_j are known. If the conduit is of circular cross section (Fig. 5.18*c*), then the moments M_j are unknown because the angle ϑ^{c}_{10} is unknown. In this case, we have $\vartheta^{c}_{10} = \vartheta_{10}$, where ϑ_{10} is an unknown angle describing the location of the rod principal axes with respect to the natural axes associated with the axial line of the conduit. A rod inserted into a conduit of circular cross section may rotate freely about its axial line. In what follows, only conduits of circular cross section will be considered.

Now, let us examine in greater detail loads acting on a rod as it slowly moves inside a conduit. An important feature of the problem is that the interaction forces q_j and μ_j between the rod and the conduit walls are unknown a priori. If the rod rotates about its axial line and moves along the conduit, then all the three components of $\mathbf{q}$ and $\boldsymbol{\mu}$ are different from zero (provided the friction force is taken into consideration). In the case of pure rotation of the rod, the components q_1, μ_2, and μ_3 equal zero. The distributed rotating moment μ_1 depends on the coefficient of friction. If the friction forces are not taken into consideration, then $\mu_1 = 0$. Taking into account the friction forces, we have

$$\mu_1 = f \frac{d}{2} \sqrt{q_2^2 + q_3^2} , \tag{5.150}$$

where f is the coefficient of friction.

Presence of friction forces substantially complicates the analysis of the equations. Indeed, due to the term μ_1, which depends on q_2 and q_3, the systems of equations (5.138)–(5.140) and (5.141)–(5.143) become coupled. If friction forces are neglected, then these systems are uncoupled and thus can be solved sequentially. This fact simplifies the solution of the problem. Friction may be taken into consideration by the method of successive approximation as follows: assuming that $\mu_1 = 0$, we determine q_2 and q_3, then, substituting M_i into (5.141), we evaluate the angle ϑ_{10}.

5.4.3 Equilibrium Equations for Friction-Free Case

Let $\mu_1 = \mu_2 = \mu_3 = 0$. Substituting (5.146)–(5.148) into (5.141), we obtain the following equation in the unknown ϑ_{10}:

$$\vartheta_{10}'' + \frac{(A_{33} - A_{22})}{3A_{11}} (\Omega_{30}^{\mathrm{c}})^2 \sin 2\vartheta_{10} - \frac{A_{33}}{A_{11}} \Omega_{30}^{\mathrm{c}} \Omega_{30}^{\mathrm{r}} \cos \vartheta_{10}^{\mathrm{c}} \sin \vartheta_{10} + \frac{A_{22}}{A_{11}} \Omega_{30}^{\mathrm{c}} \Omega_{30}^{\mathrm{r}} \sin \vartheta_{10}^{\mathrm{r}} \cos \vartheta_{10} = -\Omega_{10}^{\mathrm{c}\,\prime} + \Omega_{10}^{\mathrm{r}\,\prime} + \vartheta_{10}^{\mathrm{r}\,\prime\prime}. \tag{5.151}$$

Solving for ϑ_{10} from (5.151) and using the boundary-condition equations, we can find $æ_j$ and M_j. To do this, we replace $\vartheta_{10}^{\mathrm{c}}$ by the angle ϑ_{10} in (5.145), (5.146), and (5.149). After that, we obtain Q_2 and Q_3 from (5.142) and (5.143) and, then, we get Q_1 from (5.138). Finally, $q_2(\eta)$ and $q_3(\eta)$ can be determined from (5.139) and (5.140).

To estimate the influence of the friction, we should evaluate $\mu_1(\eta)$ from (5.150) using the values of q_2 and q_3 corresponding to the friction-free case. Then, we must solve (5.138)–(5.148) again but now the right-hand side of (5.151) contains the new value of the function μ_1.

5.4.4 Specialization of Equilibrium Equations (5.151) for Rods of Different Bending Stiffnesses ($A_{22} \neq A_{33}$)

Let us consider some particular cases.

(1) The axial line of the conduit is a spatial curve.

(1a) The rod axis in the natural configuration is a plane curve. In this case, only geometrical characteristics of the rod axis $æ_{j0}$ (5.146) will differ from those in the previous subsection, namely $æ_{10} = æ_{20} = 0$, $æ_{30} = \Omega_{30}^{(\mathrm{c})}$, and $\vartheta_{10}^{(\mathrm{c})} = 0$. Equation (5.151) takes the form

$$\vartheta_{10}'' + \frac{(A_{33} - A_{22})}{2A_{11}} (\Omega_{30}^{\mathrm{c}})^2 \sin 2\vartheta_{10} - \frac{A_{33}}{A_{11}} \Omega_{30}^{\mathrm{c}} \Omega_{30}^{\mathrm{r}} \sin \vartheta_{10} = -\Omega_{10}^{\mathrm{c}\,\prime}. \tag{5.152}$$

The moments M_i are as follows:

$$M_1 = A_{11}\left(\Omega_{10}^{\mathrm{c}} + \frac{\mathrm{d}\vartheta_{10}}{\mathrm{d}\eta}\right);$$
$$M_2 = A_{22}\,\Omega_{30}^{\mathrm{c}} \sin\vartheta_{10};$$
$$M_3 = A_{33}\,(\Omega_{30}^{\mathrm{c}} \cos\vartheta_{10} - \Omega_{30}^{\mathrm{r}}).$$

(1b) The rod axis in the natural configuration is a straight line ($\text{æ}_{j0} = 0$). In this case, (5.151) and (5.152) take the form

$$\vartheta_{10}'' + \frac{A_{33} - A_{22}}{2A_{11}}\,(\Omega_{30}^{\mathrm{c}})^2 \sin 2\vartheta_{10} = -\Omega_{10}^{\mathrm{c}}{}'. \tag{5.153}$$

The moments M_i are as follows:

$$M_1 = A_{11}\left(\Omega_{10}^{\mathrm{c}} + \frac{\mathrm{d}\vartheta_{10}}{\mathrm{d}\eta}\right);$$
$$M_2 = A_{22}\Omega_{30}^{\mathrm{c}} \sin\vartheta_{10};$$
$$M_3 = \Omega_{30}^{\mathrm{c}} \cos\vartheta_{10}.$$

(2) The axial line of the conduit is a helical curve. In this case, $\Omega_0^{\mathrm{c}} = \text{const}$.

(3) The axial line of the conduit is a plane curve and the rod axis is a spatial curve. In this case, $\Omega_{10}^{\mathrm{c}} = 0$ and $\Omega_{30}^{\mathrm{c}} \neq 0$.

For space-curved rods, we have

$$M_1 = A_{11}\left(\frac{\mathrm{d}\vartheta_{10}}{\mathrm{d}\eta} - \Omega_{10}^{\mathrm{r}} - \frac{\mathrm{d}\vartheta_{10}^{\mathrm{r}}}{\mathrm{d}\eta}\right);$$
$$M_2 = A_{22}\,(\Omega_{30}^{\mathrm{c}} \sin\vartheta_{10} - \Omega_{30}^{\mathrm{r}} \sin\vartheta_{10}^{\mathrm{r}});$$
$$M_3 = A_{33}\,(\Omega_{30}^{\mathrm{c}} \cos\vartheta_{10} - \Omega_{30}^{\mathrm{r}} \cos\vartheta_{10}^{\mathrm{r}}).$$

(4) The axial line of the conduit and the axial line of the rod are plane curves with different radii of curvature. Hence, $\Omega_{10}^{\mathrm{c}} = \Omega_{10}^{\mathrm{r}} = \vartheta_{10}^{\mathrm{c}} = 0$. The equilibrium equation, a special case of (5.152), is as follows:

$$\vartheta_{10}'' + \frac{(A_{33} - A_{22})}{2A_{11}}\,(\Omega_{30}^{\mathrm{c}})^2 \sin 2\vartheta_{10} - \frac{A_{33}}{A_{11}}\,\Omega_{30}^{\mathrm{c}}\Omega_{30}^{\mathrm{r}} \sin\vartheta_{10} = 0. \tag{5.154}$$

Then,

$$M_1 = A_{11}\vartheta_{10}';$$
$$M_2 = A_{22}\Omega_{30}^{\mathrm{c}} \sin\vartheta_{10};$$
$$M_3 = A_{33}\,(\Omega_{30}^{\mathrm{c}} \cos\vartheta_{10} - \Omega_{30}^{\mathrm{r}}).$$

5.4.5 Specialization of Equilibrium Equations for Rods with Equal Bending Stiffnesses ($A_{22} = A_{33}$)

Let us consider some particular cases.

(1) A space-curved rod. From (5.151) it follows that

$$\vartheta''_{10} - \frac{A_{33}}{A_{11}}\, \Omega^{c}_{30}\Omega^{r}_{30} \sin\vartheta_{10} = -\Omega^{c}_{10}{}' + \Omega^{r}_{10}{}' . \tag{5.155}$$

(2) A plane rod. Putting $\vartheta^{c}_{10} = 0$ in (5.152), we get

$$\vartheta''_{10} - \frac{A_{33}}{A_{11}}\, \Omega^{c}_{30}\Omega^{r}_{30} \sin\vartheta_{10} = -\Omega^{c}_{10}{}' . \tag{5.156}$$

(3) A straight rod. From (5.153) it follows that

$$\vartheta''_{10} = -\Omega^{c}_{10}{}' . \tag{5.157}$$

(4) The axial line of the conduit is a helical curve. In this case, we must put $\Omega^{c}_{10}{}' = 0$ in (5.155)–(5.157).

5.4.6 Determination of Twisting Moments for Rods with Equal Bending Stiffnesses ($A_{22} = A_{33}$)

Let us establish conditions under which the internal twisting moment does not depend on the axial coordinate η. We have (see Fig. 5.19)

$$M_1 = A_{11}\left(\Omega^{c}_{10} + \frac{\mathrm{d}\vartheta_{10}}{\mathrm{d}\eta}\right) = \text{const} . \tag{5.158}$$

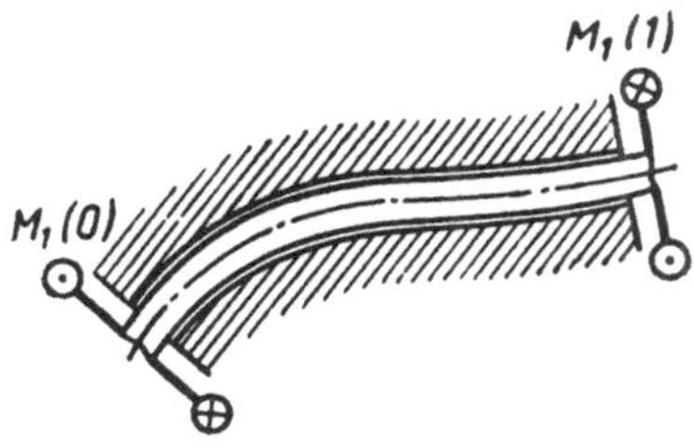

Fig. 5.19.

Equality (5.158) holds provided

$$\Omega^{c}_{10}(\eta) + \frac{\mathrm{d}\vartheta_{10}}{\mathrm{d}\eta} = \text{const} = c_1$$

or

$$\frac{d\vartheta_{10}}{d\eta} = c_1 - \Omega_{10}^c(\eta).$$

These equalities are satisfied provided rods are straight in the natural configuration and if $\Omega_{10}^c = \Omega_{30}^c = 0$. In this case, (5.155)–(5.157) reduce to the form

$$\vartheta_{10}'' = -\Omega_{10}^{c\,\prime}. \tag{5.159}$$

From (5.159) it follows that

$$\vartheta_{10}' = c_1 - \Omega_{10}^c$$

or

$$\Omega_{10}^c + \vartheta_{10}' = c_1 = \text{const}.$$

Suppose that the axial line of the conduit is a helical curve. From (5.159) it follows that $\vartheta_{10}'' = 0$ and $\vartheta_{10}' = \text{const}$. Consequently, $\Omega_{10}^c = \text{const}$ and, as a result, we obtain

$$M_1 = A_{11}(\Omega_{10}^c + \vartheta_{10}') = \text{const}, \tag{5.160}$$

Thus, for straight rods with $A_{22} = A_{33}$, the twisting moment M_1 is constant irrespective of the configuration (plane or spatial) of the axial line of the conduit.

The following examples illustrate equilibrium analysis of straight rods ($A_{22} = A_{33}$) subjected to axial forces only (e.g. see Fig. 5.18*a*). In this case, the twisting moment M_1 equals zero. Using (5.160), we obtain the following equation:

$$\frac{d\vartheta_{10}}{d\eta} = -\Omega_{10}^c(\eta);$$

hence,

$$\vartheta_{10} = -\int \Omega_{10}^c(\zeta)\, d\zeta + \vartheta_{100}. \tag{5.161}$$

If the axial line of the conduit is a helical curve, $\Omega_{10}^c = \text{const}$, then we have

$$\vartheta_{10} = -\Omega_{10}^c \eta + \vartheta_{100}. \tag{5.162}$$

If at $\eta = 0$ the equalities $\vartheta_{10} = 0$ and $\vartheta_{100} = 0$ hold, we conclude that $\vartheta_{10} = -\Omega_{10}^c \eta$.

In the general case, $\Omega_{10}^c \neq \text{const}$, using (5.162), we obtain the relationship between ϑ_{10} and η. Once the function $\vartheta_{10}(\eta)$ is known, we find the bending moments as follows:

$$M_2 = A_{33}\,\Omega_{30}^{c} \sin\vartheta_{10}\,; \qquad M_3 = A_{33}\,\Omega_{30}^{c} \cos\vartheta_{10}\,. \tag{5.163}$$

Hence, the concentrated moments $M_3(0)$, $M_2(1)$, and $M_3(1)$ applied to the end cross sections occur. These moments satisfy the relations

$$M_3(0) = A_{33}\,\Omega_{30}^{c}(0)\,;$$
$$M_2(1) = A_{33}\,\Omega_{30}^{c}(1) \sin\vartheta_{10}(1)\,;$$
$$M_3(1) = A_{33}\,\Omega_{30}^{c}(1) \cos\vartheta_{10}(1)\,.$$

The occurrence of the moments can be explained by the fact that the rod is in contact with the absolutely rigid walls of the conduit at the points A and B.

Combining (5.163), (5.142) and (5.143), we get

$$Q_3 = A_{11}\Omega_{30}^{c}{}' \sin\vartheta_{10} + A_{33}\Omega_{30}^{c} \cos\vartheta_{10}\frac{\mathrm{d}\vartheta_{10}}{\mathrm{d}\eta}\,; \tag{5.164}$$

$$Q_2 = -A_{33}\Omega_{30}^{c}{}' \cos\vartheta_{10} + A_{33}\Omega_{30}^{c} \sin\vartheta_{10}\frac{\mathrm{d}\vartheta_{10}}{\mathrm{d}\eta}\,. \tag{5.165}$$

From these relations it is seen that there exist concentrated forces applied to the rod at the end cross sections ($\eta = 0$ and $\eta = 1$). At the cross section A we have $\vartheta_{10}(0) = 0$ and, consequently,

$$Q_2(0) = -A_{33}\Omega_{30}^{c}{}'(0)\,; \qquad Q_3(0) = -A_{33}\Omega_{30}^{c}\Omega_{10}^{c}\,.$$

To determine $Q_1(\eta)$, we put $q_1 = 0$ in (5.138):

$$Q_1' = \Omega_{30}^{c}{}'\Omega_{30}^{c}A_{33}\,. \tag{5.166}$$

Integration of (5.166) yields

$$Q_1 = -\frac{A_{33}}{2}(\Omega_{30}^{c})^2 + \Omega_{10}\,. \tag{5.167}$$

From this relation it follows that if no external forces are applied at the cross section B, i.e. $Q_1(1) = 0$, then the rod will remain in equilibrium provided the axial force

$$P = -Q_1(0) = \frac{A_{33}}{2}[\,\Omega_{30}^{c}(0)\,]^2 - \frac{A_{33}}{2}[\,\Omega_{30}^{c}(1)\,]^2$$

is applied to the cross section A.

From (5.139) and (5.140) it follows that

$$q_2 = -Q_2' - \Omega_{30}^{c} \cos\vartheta_{10}\frac{A_{33}}{2}\{[\,\Omega_{30}^{c}(1)\,]^2 - [\,\Omega_{30}^{c}(\eta)\,]^2\}\,;$$
$$q_3 = -Q_3' + \Omega_{30}^{c} \sin\vartheta_{10}\frac{A_{33}}{2}\{[\,\Omega_{30}^{c}(1)\,]^2 - [\,\Omega_{30}^{c}(\eta)\,]^2\}\,.$$

Suppose that the axial line of the conduit is a helical curve. By use of (5.164) and (5.165), we get

$$Q_3 = A_{33}\Omega_{30}^{c} \cos\vartheta_{10}\vartheta_{10}' ;$$
$$Q_2 = A_{33}\Omega_{30}^{c} \sin\vartheta_{10}\vartheta_{10}' .$$

Using (5.162), we have

$$Q_3 = -A_{33}\Omega_{30}^{c}\Omega_{10}^{c} \cos\vartheta_{10}(\eta) ;$$
$$Q_2 = -A_{33}\Omega_{30}^{c}\Omega_{10}^{c} \sin\vartheta_{10}(\eta) .$$

From (5.166) and in view of ${\Omega_{30}^{c}}' = 0$ it follows that $Q_1 = \text{const}$.

The components of the contact pressure can be obtained from (5.139) and (5.140) as follows:

$$q_2 = -A_{33}\Omega_{30}^{c}(\Omega_{10}^{c})^2 \cos\Omega_{10}^{c}\eta ;$$
$$q_3 = -A_{33}\Omega_{30}^{c}(\Omega_{10}^{c})^2 \sin\Omega_{10}^{c}\eta .$$

The absolute value of the distributed contact pressure is

$$|\mathbf{q}| = \sqrt{q_2^2 + q_3^2} = A_{33}\Omega_{30}^{c}(\Omega_{10}^{c})^2 = \text{const} .$$

Recall that we analyze the equilibrium of naturally straight rods with $A_{22} = A_{33}$ subjected only to axial forces (see Fig. 5.18*a*). If the axial line of the conduit is a plane curve, then we have $\Omega_{10}^{c} = 0$, $\vartheta_{10} = 0$, and $\Omega_{30}^{c}(\eta) \neq 0$. From (5.160) and (5.163) it follows that $M_1 = 0$, $M_2 = 0$, and $M_3 = A_{33}\Omega_{30}^{c}$.

From the expression for M_3 it follows that there are concentrated moments of magnitudes $A_{33}\Omega_{30}^{c}(0)$ and $A_{33}\Omega_{30}^{c}(1)$ applied to the cross sections A and B, respectively. From (5.143) it follows that

$$Q_2 = -A_{33}{\Omega_{30}^{c}}'(\eta) . \tag{5.168}$$

The functions Q_1 and q_2 can be determined from (5.138) and (5.139) as follows:

$$Q_1(\eta) = -\frac{A_{33}}{2}(\Omega_{30}^{c})^2 + Q_{10} , \quad Q_{10} = \frac{1}{2}A_{33}(\Omega_{30}^{c}(1))^2 ; \tag{5.169}$$

$$q_2 = A_{33}{\Omega_{30}^{c}}'' + \frac{1}{2}A_{33}(\Omega_{30}^{c})^2 + Q_{10}\Omega_{30}^{c} ; \tag{5.170}$$

An arbitrary constant Q_{10} can be obtained from the condition $Q_1(1) = 0$.

From (5.168) it follows that there are concentrated forces of magnitudes $-A_{33}{\Omega_{30}^{c}}'(0)$ and $-A_{33}{\Omega_{30}^{c}}'(1)$ applied to the cross sections A and B, respectively.

Let us determine the axial force P that must be applied to maintain the rod in equilibrium. For $Q_1(1) = 0$, from (5.169) it follows that

$$P = \frac{A_{33}}{2}\{[\Omega_{30}^{c}(0)]^2 - [\Omega_{30}^{c}(1)]^2\} .$$

If $\Omega_{30}^{c}(0) = \Omega_{30}^{c}(1)$, then the force P vanishes. If $\Omega_{30}^{c}(0) < \Omega_{30}^{c}(1)$, then P is opposite in direction to the direction indicated in Fig. 5.18*a*.

If $\Omega_{30}^{c} = \text{const}$, i.e. the axial line of the conduit is a circle arc, then we have

$$M_3 = A_{30}\Omega_{30}^{c} = \text{const}\,; \qquad Q_2 = 0\,; \qquad Q_1 = 0\,; \qquad q_2 = 0\,.$$

The result $q_2 = 0$ can be explained by the fact that in this case we have pure bending caused by concentrated moments applied at the end cross sections A and B. Recall that the curvature of the axial line of the conduit Ω_{30}^{c} is equal to the curvature of the rod axis $æ_{30}$, hence, the contact pressure vanishes.

Consider a numerical example. Suppose that a rod is inserted into a rigid curvilinear conduit. Twisting moments are applied to the rod at its ends.

The axial line of the rod and the axial line of the conduit are plane curves. Their nondimensional curvatures are $æ_{30} = 5$ and $\Omega_3 = \Omega_{30} f(\eta)$, where $\Omega_{30} = 6$. The nondimensional stiffnesses are $A_{33} = A_{22} = 1$ and $A_{11} = 0.5$.

We have

$$\begin{aligned}
æ_1 &= \frac{\mathrm{d}\vartheta_{10}}{\mathrm{d}\eta}\,; \\
æ_2 &= \Omega_3 \sin\vartheta_{10}\,; \\
æ_3 &= \Omega_3 \cos\vartheta_{10}\,; \\
M_1 &= A_{11}\,\frac{\mathrm{d}\vartheta_{10}}{\mathrm{d}\eta}\,; \\
M_2 &= A_{22}\,\Omega_3 \sin\vartheta_{10}\,; \\
M_3 &= \Omega_3 \cos\vartheta_{10} - æ_{30}\,.
\end{aligned} \tag{5.171}$$

Using (5.141), we obtain the following equation in the unknown angle ϑ_{10} (for $A_{22} = A_{33}$):

$$\frac{\mathrm{d}^2\vartheta_{10}}{\mathrm{d}\eta^2} - \frac{A_{33}}{A_{11}} æ_{30}\Omega_{30} f(\eta) \sin\vartheta_{10} = 0\,. \tag{5.172}$$

Let us obtain the solution of (5.172) for the following functions $f(\eta)$: $f(\eta) = 1$ and $f(\eta) = \cos\pi\eta$, where $0 \le \eta \le 1$.

The sought solution must satisfy the following boundary conditions (see Fig. 5.18*a*):

$$\text{at} \quad \eta = 0\,, \qquad \vartheta_{10} = 0 \quad \text{and} \quad \frac{\mathrm{d}\vartheta_{10}}{\mathrm{d}\eta} A_{11} = M_1(0)\,.$$

For example, we put $M_1(0) = 0.1$. Inserting $\eta = 1$ into the solution of (5.172), we obtain the twisting moment that must be applied to maintain the rod in equilibrium; we also obtain the angle of mutual rotation of the end cross sections.

When $\vartheta_{10}(\eta)$ is determined, then, in view of (5.138)–(5.144), we can successively find the components M_j of the vector $\mathbf{M}$ (see (5.171)); then, we obtain the components Q_j of the vector $\mathbf{Q}$ and the components q_2 and q_3 of the vector $\mathbf{q}$ of the distributed contact pressure as follows:

$$Q_3 = M_2' + M_1 æ_3 - M_3 æ_1 ;$$
$$Q_2 = -M_3' - M_2 æ_1 + M_1 æ_2 ;$$
$$Q_1 = \int_0^{\eta} (Q_2 æ_3 - Q_3 æ_2)\, \mathrm{d}\zeta + Q_{10} ;$$
$$q_2 = -Q_2' - Q_1 æ_3 + Q_3 æ_1 ;$$
$$q_3 = -Q_3' - Q_2 æ_1 + Q_1 æ_2 ;$$
$$|\mathbf{q}| = \sqrt{q_2^2 + q_3^2} .$$

Assuming that $Q_1 = 0$ for $\eta = 0$, we get $Q_{10} = 0$. The stabilizing force that is to be applied to the rod at $\eta = 1$ satisfies the relation

$$P_1 = Q_1 = \int_0^1 (Q_2 æ_3 - Q_3 æ_2)\, \mathrm{d}\zeta .$$

The results of numerical analysis of (5.172) are illustrated in Fig. 5.20*a-f*. The plots in Fig. 5.20*a-c* correspond to the function $f(\eta) = 1$. The plots in Fig. 5.20*d-f* correspond to the function $f(\eta) = \cos \pi\eta$.

5.5 Stability of Plane Curvilinear Rods

Consider stability of a curvilinear rod whose axial line in the natural configuration is either a plane curve (Fig. 5.21, line *0*) or a straight line (Figs. 4.17 and 4.18). Assume that one of the principal axes lies in the plot of drawing.

Let the rod be subjected to forces and moments such that the force vectors lie in the plane x_1Ox_2 and the moment vectors are perpendicular to this plane. Under the action of forces and moments whose magnitudes are less than critical values, after the deformation, the rod axis remains a plane curve. When the loads exceed the critical values, an out-of-plane deformation of the rod takes place (Fig. 5.21, line *2*), i.e. the rod axial line becomes a spatial curve. The shape of the axial line before the loss of stability, line *1*, may differ considerably from the shape of the axial line in the natural configuration.

The equilibrium equations governing the behavior of the rod in a critical state can be derived from the general nonlinear equilibrium equations for space-curved rods (3.10)–(3.14). Scalar form of these equations for a plane rod is as follows (for $\mathbf{q} = 0$):

$$\frac{\mathrm{d}Q_{1*}}{\mathrm{d}\eta} + æ_{3*} Q_{2*} = 0 ;$$
$$\frac{\mathrm{d}Q_{2*}}{\mathrm{d}\eta} + æ_{3*} Q_{1*} = 0 ;$$

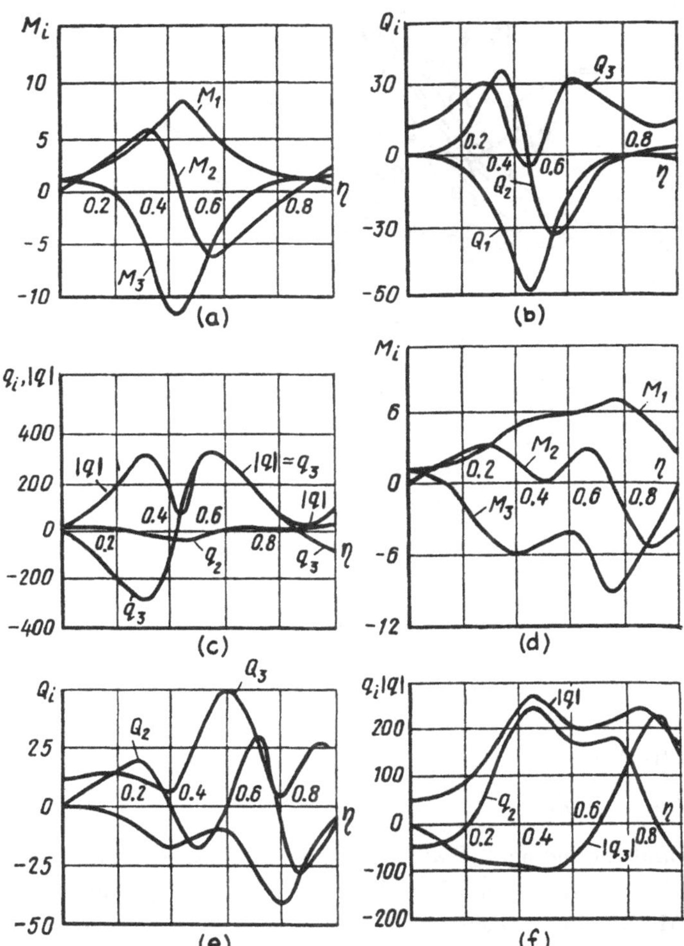

Fig. 5.20.

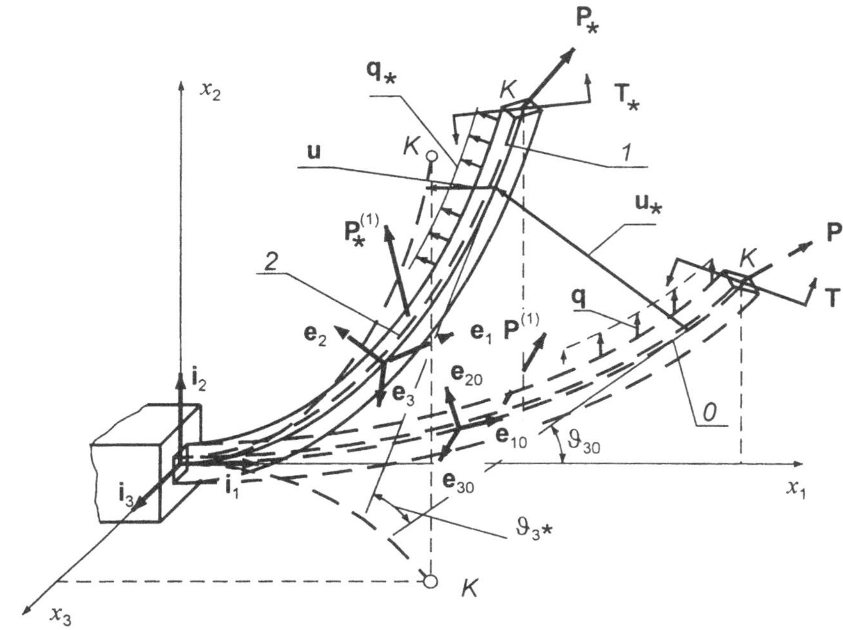

Fig. 5.21.

$$\frac{\mathrm{d}M_{3*}}{\mathrm{d}\eta} - Q_{2*} = 0;$$
$$\frac{\mathrm{d}\vartheta_{3*}}{\mathrm{d}\eta} - \frac{M_{3*}}{A_{33}} = 0;$$
$$\frac{\mathrm{d}u_{1*}}{\mathrm{d}\eta} - æ_{3*}u_{2*} + l_{11*} - 1 = 0;$$
$$\frac{\mathrm{d}u_{2*}}{\mathrm{d}\eta} + æ_{3*}u_{1*} + l_{21*} = 0;$$
$$M_{3*} = A_{33}(æ_{3*} - æ_{30}). \tag{5.173}$$

The force $\mathbf{P}$ and the moment $\mathbf{T}$ appear in the boundary-condition equations. For example, consider a dead force $\mathbf{P}^{(1)} = -P_{x_1}^{(1)}\mathbf{i}_1 - P_{x_2}^{(1)}\mathbf{i}_2$ and a dead moment $\mathbf{T} = T_3\mathbf{i}_3$; in this case, we have the following boundary-condition equations (for $\eta = 1$):

$$Q_{1*}(1) = -\sum_{j=1}^{2} P_{x_j*}(\mathbf{i}_j \cdot \mathbf{e}_1) = -P_{x_1*}\cos(\vartheta_{3*} + \vartheta_{30}) - P_{x_2*}\sin(\vartheta_{3*} + \vartheta_{30});$$
$$Q_{2*}(1) = -\sum_{j=1}^{2} P_{x_j*}(\mathbf{i}_j \cdot \mathbf{e}_2) = P_{x_1*}\sin(\vartheta_{3*} + \vartheta_{30}) - P_{x_2*}\cos(\vartheta_{3*} + \vartheta_{30});$$
$$M_{3*} = T_{3*}; \tag{5.174}$$

here ϑ_{3*} and ϑ_{30} are the values of the angles at $\eta = 1$.

The nonlinear equations (5.173) together with the boundary conditions (5.174) are to be solved by numerical methods, say, by the method of step-by-step loading or the method of discrete continuation with respect to a parameter (Grigolyuk, Shalashilin (1991)). At each step of loading, a new plane equilibrium configuration is determined numerically.

Let us consider a state vector corresponding to the kth step of loading,

$$\mathbf{Y}^{(k)} = (Q_1^{(k)}, Q_2^{(k)}, M_3^{(k)}, \vartheta_3^{(k)}, u_1^{(k)}, u_2^{(k)})^{\mathrm{T}}.$$

The components of $\mathbf{Y}^{(k)}$ satisfy the system of linear equations (3.33)–(3.36) in which we put $\Delta P_i = \Delta T_i = 0$:

$$\begin{aligned}
&\frac{\mathrm{d}Q_1}{\mathrm{d}\eta} - æ_3^{(k)} Q_2 - Q_2^{(k)} \frac{M_3}{A_{33}} = 0\,;\\
&\frac{\mathrm{d}Q_2}{\mathrm{d}\eta} + æ_3^{(k)} Q_1 + Q_1^{(k)} \frac{M_3}{A_{33}} = 0\,;\\
&\frac{\mathrm{d}Q_3}{\mathrm{d}\eta} + æ_3^{(k)} \frac{M_1}{A_{11}} - Q_1^{(k)} \frac{M_2}{A_{22}} = 0\,;\\
&\frac{\mathrm{d}M_1}{\mathrm{d}\eta} - æ_3^{(k)} M_2 + M_3^{(k)} \frac{M_2}{A_{22}} = 0\,;\\
&\frac{\mathrm{d}M_2}{\mathrm{d}\eta} + æ_3^{(k)} M_1 - M_3^{(k)} \frac{M_1}{A_{11}} - Q_3 = 0\,;\\
&\frac{\mathrm{d}M_3}{\mathrm{d}\eta} + Q_2 = 0\,;\\
&\frac{\mathrm{d}\vartheta_1}{\mathrm{d}\eta} - æ_3^{(k)} \vartheta_2 - \frac{M_1}{A_{11}} = 0\,;\\
&\frac{\mathrm{d}\vartheta_2}{\mathrm{d}\eta} + æ_3^{(k)} \vartheta_1 - \frac{M_2}{A_{22}} = 0\,;\\
&\frac{\mathrm{d}\vartheta_3}{\mathrm{d}\eta} - \frac{M_3}{A_{33}} = 0\,;\\
&\frac{\mathrm{d}u_1}{\mathrm{d}\eta} - æ_3^{(k)} u_2 = 0\,;\\
&\frac{\mathrm{d}u_2}{\mathrm{d}\eta} + æ_3^{(k)} u_1 - \vartheta_3 = 0\,;\\
&\frac{\mathrm{d}u_3}{\mathrm{d}\eta} + \vartheta_2 = 0
\end{aligned} \tag{5.175}$$

or, in vector form,

$$\mathbf{Y}' + \mathbf{A}\mathbf{Y} = 0\,, \tag{5.176}$$

where

$$\mathbf{Y} = (Q_1, Q_2, Q_3, M_1, M_2, M_3, \vartheta_1, \vartheta_2, \vartheta_3, u_1, u_2, u_3)^{\mathrm{T}}.$$

The solution of (5.176) is as follows:

$$\mathbf{Y} = \mathrm{K}(\eta)\,\mathbf{C}\,, \qquad \mathrm{K}(0) = \mathrm{E}\,. \tag{5.177}$$

For dead loads $\mathbf{P}^{(1)}$ and $\mathbf{T}$, the components of the vector $\mathbf{Y}$ must satisfy the following boundary conditions:

$$\begin{aligned}
&(1) \quad \text{at} \quad \eta = 0\,, \quad y_7 = y_8 = y_9 = y_{10} = y_{11} = y_{12} = 0\,,\\
&\qquad\qquad\qquad \text{hence}, \quad C_7 = C_8 = \ldots = C_{12} = 0\,;\\
&(2) \quad \text{at} \quad \eta = 1\,, \quad y_1 = P_2^{(k)} y_9\,, \quad y_2 = P_1^{(k)} y_9\,,\\
&\qquad\qquad\qquad y_3 = -P_2^{(k)} Y_7 + P_1^{(1)} y_8\,, \quad y_4 = -T_3^{(k)} y_8\,,\\
&\qquad\qquad\qquad y_5 = T_3^{(k)} y_7\,, \quad y_6 = 0\,;
\end{aligned}$$

here $P_1^{(k)}$, $P_2^{(k)}$, and $T_3^{(k)}$ are the values of the force $\mathbf{P}^{(1)}$ and the moment $\mathbf{T}$ at the kth step of loading.

The boundary conditions at $\eta = 1$ yield the following system of six equations:

$$\begin{aligned}
&\sum_{j=1}^{6} (k_{1j} - P_2^{(k)} k_{9j})\, c_j = 0\,;\\
&\sum_{j=1}^{6} (k_{2j} - P_1^{(k)} k_{9j})\, \mathbf{c}_j = 0\,;\\
&\sum_{j=1}^{6} (k_{3j} - P_2^{(k)} k_{7j}) + P_1^{(k)} k_{8j} = 0\,;\\
&\sum_{j=1}^{6} (k_{4j} + T_3^{(k)} k_{8j})\, c_j = 0\,;\\
&\sum_{j=1}^{6} (k_{5j} - T_3^{(k)} k_{7j}) = 0\,;\\
&\sum_{j=1}^{6} k_{6j} c_j = 0\,;
\end{aligned} \tag{5.178}$$

here

$$\begin{aligned}
P_1^{(k)} &= -P_{x_1}^{(k)} \cos(\vartheta_3^{(k)} + \vartheta_{30}) - P_{x_2}^{(k)} \sin(\vartheta_3^{(k)} + \vartheta_{30})\,;\\
P_2^{(k)} &= P_{x_1}^{(k)} \sin(\vartheta_3^{(k)} + \vartheta_{30}) - P_{x_2}^{(k)} \cos(\vartheta_3^{(k)} + \vartheta_{30})\,.
\end{aligned}$$

If the determinant D of the system of equations (5.178) at the $(k-1)$th and kth steps has different signs, then we conclude that the critical loads are subject to the inequalities

$$P_j^{(k-1)} \leq P_{j*} \leq P_j^{(k)}; \qquad T_3^{(k-1)} \leq T_{3*} \leq T_3^{(k)}.$$

Applying the method of iterations, we get the values $(P_{j*})_1$ and $(T_{3*})_1$ for which the determinant D equals zero.

To illustrate numerical determination of the critical values P_* and T_*, let us consider a rod whose axial line is a logarithmic spiral. Assume that $\mathbf{P} = -P_{x_1}\mathbf{i}_1$.

To perform numerical analysis, we must know the functions $æ_{30}(\eta)$, $\vartheta_{30}(\eta)$, and $\mathbf{e}_{10}(\eta)$. The determination of the functions is given in Appendix 5. Let the nondimensional stiffnesses be $A_{33} = 1$, $A_{22} = 0.25$, and $A_{11} = 0.275$ and the nondimensional length of the rod be equal to unity.

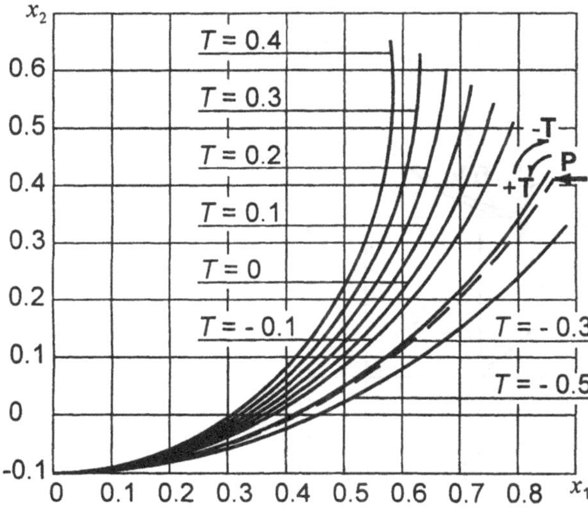

Fig. 5.22.

The rod axis in the critical state obtained for different values of T_3 are shown in Fig. 5.22 as solid lines. The axial line in the stress-free configuration is indicated as a dashed line. The critical force P_* and the absolute values of the displacement vectors of the free end (the point K) obtained for different values of T_3 are presented in Table 5.1.

Table 5.1.

T_3	-0.5	-0.3	-0.1	0	0.1	0.2	0.3	0.4	0.5
P_*	1.054	0.846	0.740	0.711	0.696	0.694	0.706	0.734	0.786
$\lvert\mathbf{u}_k\rvert$	0.167	0.001	0.1	0.150	0.202	0.256	0.311	0.368	0.405

Thus, for $T = 0.4$, the displacement of the point K makes about 40% of the rod length. For large displacements, we must use the nonlinear equilibrium equations to describe the shape of the rod. The critical value of the dead force $\mathbf{P}$ depends just slightly on the type of behavior of the moment $\mathbf{T}$ (i.e. it makes no difference whether the moment is dead or follower). If the force $\mathbf{P}$ is a follower one, then the loss of stability of the rod never takes place (it does not matter whether the moment $\mathbf{T}$ is dead or follower). Anyway, when a rod is subjected to a follower force, a so-called *loss of dynamical stability* can take place. This phenomenon is discussed in books devoted to dynamics of rods (e.g. Svetlitsky (1987)).

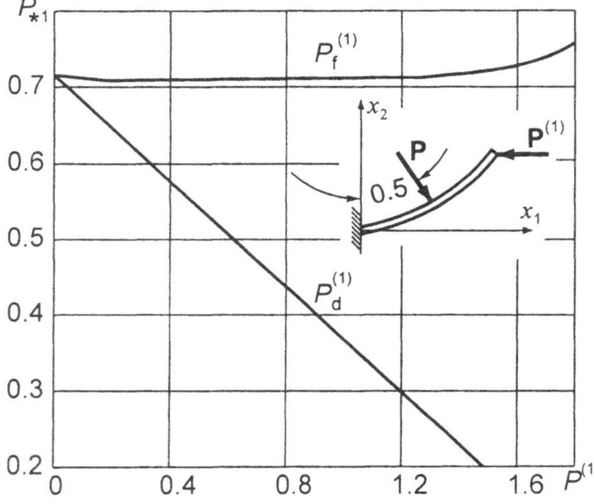

Fig. 5.23.

Suppose that in addition to the force $\mathbf{P}$, a force $\mathbf{P}^{(1)}$ is applied to the cross section $\eta = 0.5$. In this case, the results of equilibrium analysis are shown in Fig. 5.23. Terms containing $\mathbf{P}_*^{(1)}$ appear in the first two equations of (5.173). As a result, we have

$$\frac{\mathrm{d}Q_{1*}}{\mathrm{d}\eta} - æ_{3*}Q_{2*} + P_{1*}^{(1)}\,\delta(\eta - 0.5) = 0\,;$$

$$\frac{\mathrm{d}Q_{2*}}{\mathrm{d}\eta} + æ_{3*}Q_{1*} + P_{2*}^{(1)}\,\delta(\eta - 0.5) = 0\,,$$

where $P_{1*}^{(1)}$ and $P_{2*}^{(1)}$ are the projections of the vector $\mathbf{P}^{(1)}$ on the attached axes. These projections depend on the behavior of $\mathbf{P}^{(1)}$ during deformation. For example, if $\mathbf{P}^{(1)}$ is a follower load, then the projections $P_1^{(1)}$ and $P_2^{(1)}$ are functions of the absolute value of $\mathbf{P}^{(1)}$ alone. If the force $\mathbf{P}^{(1)} = P_{x_1}\mathbf{i}_1 + P_{x_2}\mathbf{i}_2$ is dead, then its projections on the attached axes are as follows:

$$P_1^{(1)} = P_{x_1}\cos(\vartheta_3 + \vartheta_{30}) + P_{x_2}\sin(\vartheta_3 + \vartheta_{30})\,;$$
$$P_2^{(1)} = -P_{x_1}\sin(\vartheta_3 + \vartheta_{30}) + P_{x_2}\cos(\vartheta_3 + \vartheta_{30})\,.$$

The terms

$$\Delta P_1^{(1)} = P_2^{(1)}\vartheta_3^{(k)}\,\delta\,;$$
$$\Delta P_2^{(1)} = -P_1^{(1)}\vartheta_3^{(k)}\,\delta\,;$$
$$\Delta P_3^{(1)} = (P_1^{(1)}\vartheta_2 - P_2^{(1)}\vartheta_1)\,\delta$$

appear in the left-hand sides of the first three equations of (5.175).

When solving the problem numerically, one should remember that when passing through the cross section $\eta = 0.5$, the component Q_{1*} is to be replaced by $P_1^{(1)}$, the components Q_{2*} and Q_1 are to be replaced by $P_2^{(1)}$, and the components Q_2 and Q_3 are to be replaced by ΔP_3^1.

For a dead force $\mathbf{P}$, let us consider the two cases:

(1) $\mathbf{P}^{(1)} = \mathbf{P}_{\mathrm{f}}^{(1)}$ is a follower force;

(2) $\mathbf{P}^{(1)} = \mathbf{P}_{\mathrm{d}}^{(1)}$ is a dead force.

The plots of the critical force P_{*1}, the first eigenvalue, versus $P_{\mathrm{f}}^{(1)}$ and $P_{\mathrm{d}}^{(b)}$ are given in Fig. 5.23. It is seen from the plots that the greater the dead force $P_{\mathrm{d}}^{(1)}$, the smaller the value P at which the loss of stability takes place. For values of the follower force $P_{\mathrm{f}}^{(1)}$ such that $0 \le P_{\mathrm{f}}^{(1)} \le 0.8$, we see that the critical force P_{*1} decreases insignificantly but then, as $P_{\mathrm{f}}^{(1)}$ continues to increase, the critical force P_{*1} also starts growing. If $\mathbf{P}_{\mathrm{f}}^{(1)}$ is a follower force, then the critical values of $\mathbf{P}$ are higher than those corresponding to the same values of the dead force $P_{\mathrm{f}}^{(1)}$. Hence, we may conclude that follower loads improve stability characteristics of rods. For example, for $P_{\mathrm{f}}^{(1)} = 1.2$, the critical value P_{*1} increases by 55% the critical value corresponding to the dead force $P_{\mathrm{d}}^{(1)} = 1.2$.

5.6 Problems

5.1. A rod of plane natural configuration (see Fig. 5.24*b*) is inserted into a conduit (Fig. 5.24*a*) such that $\Omega_{30}^{\mathrm{c}} = \text{const}$ and $\Omega_{30}^{\mathrm{r}} = \text{const}$. The bending stiffnesses of the rod are not equal, $A_{22} \neq A_{33}$. The rod is subjected to a twisting moment M_{10} applied at the cross section $\eta = 0$. Determine the twisting moment $M_{1\mathrm{K}}$ at the end cross section $\eta = 1$, the angle of relative rotation of the end cross sections, and the components of the distributed contact forces q_2 and q_3. Assume that the angles of rotation ϑ_0 of the rod cross sections are small.

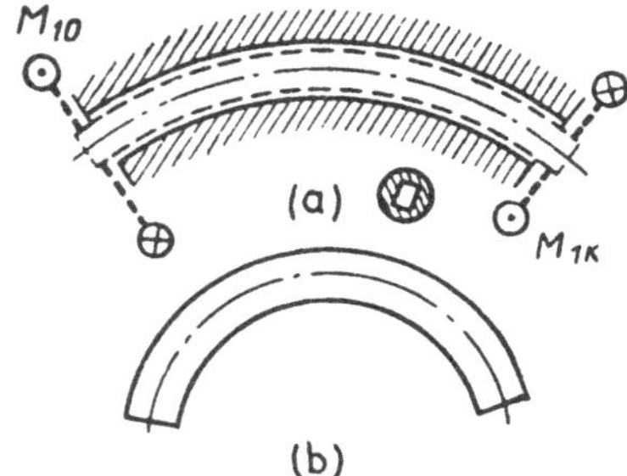

Fig. 5.24.

5.2. A rod, whose axial line is a plane curve ($\bar{\Omega}^{\mathrm{r}}_{30} = \mathrm{const}$ and $A_{22} \neq A_{33}$) is inserted into a conduit. The axial line of the conduit is a helical curve of small angle of helix. Determine the relative angle of rotation of the end cross sections, the moments $M_i(\eta)$, the forces $Q_i(\eta)$, and the distributed contact forces q_2 and q_3. Find the conditions under which the rod remains in equilibrium. Assume that the angles ϑ_{10} are small (this assumption is reasonable because of smallness of the angle of helix).

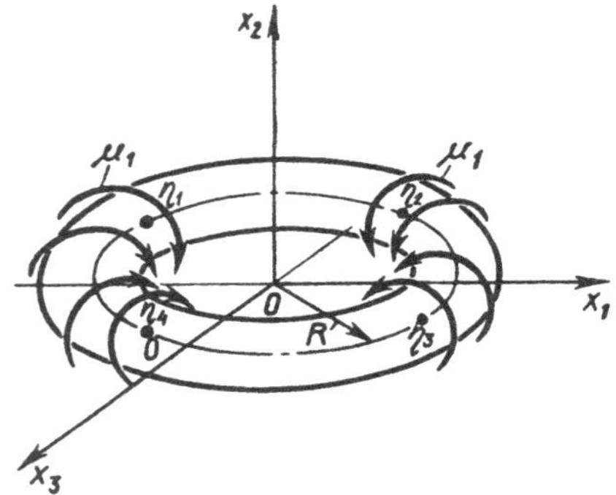

Fig. 5.25.

5.3. A ring is subjected to a distributed twisting moment μ_1 as indicated in Fig. 5.25. The moments are applied at the segments $(\eta_1\,,\,\eta_2) = (0.25\,,\,0.5)$ and $(\eta_3\,,\,\eta_4) = (0.75\,,\,1.0)$. Determine the stress-strain state of the ring.

6. Rods Interacting with Liquid or Air Flows

This chapter is devoted to the problems of interaction between rods and external or internal flows of liquid or air.

6.1 Introduction

The general linear and nonlinear equilibrium equations valid for all types of space-curved rods were derived in Chapters 1, 2, and 3. The only restriction was that the stress-strain state caused by the external loads was governed by the generalized Hooke's law. As it was mentioned above, this book as well as its continuation (Svetlitsky (1987)) deal with problems that are linear in physical sense.

In most applied problems, external loads ($\mathbf{q}$, $\mathbf{P}^{(j)}$, $\boldsymbol{\mu}$, and $\mathbf{T}^{(\nu)}$) are known a priori. For example, dead forces (see Figs. 4.14 and 4.19*a*), follower forces (see Figs. 4.18 and 4.26), electromagnetic attracting forces (see Fig. 4.16), etc.

Forces may be either dependent on rod deformations (more precisely, dependent on the displacements u_j and the angles of rotation of the attached axes ϑ_j as shown in Figs. 4.15 and 4.16) or independent of rod deformations (see Figs. 4.14 and 4.19). Forces that depend on u_j and ϑ_j are termed *feedback forces*. For most feedback forces, the relationship between the forces and moments on one hand and the displacements u_j and the angles ϑ_j on the other hand can be explicitly obtained. Hence, no difficulties arise in integration of the equilibrium equations and the equations of motion. The only possible difficulty is the choice of the most efficient numerical method. Serious difficulties indeed appear when we cannot determine the forces as functions of displacements of the axial points and angles of rotation. For example, this is the case in the problem of interaction between a drilling bit and a metal (see Fig. 0.20) and in the problem of interaction between a space-curved rod and a liquid or air flow (see Fig. 0.18).

Experimental data on interaction between rods and liquid or air flows, as a rule, can be obtained for a straight rod when the rod axis is perpendicular to the vector of the flow velocity. For other angles between the vectors $\mathbf{v}_0$ and $\mathbf{e}_1$ (the angles φ_a in Fig. 6.1), there is no comprehensive experimental data. Needless to say that it is a hard problem to obtain these expressions for curvilinear rods for an arbitrary angle φ_a. However, lacking the explicit

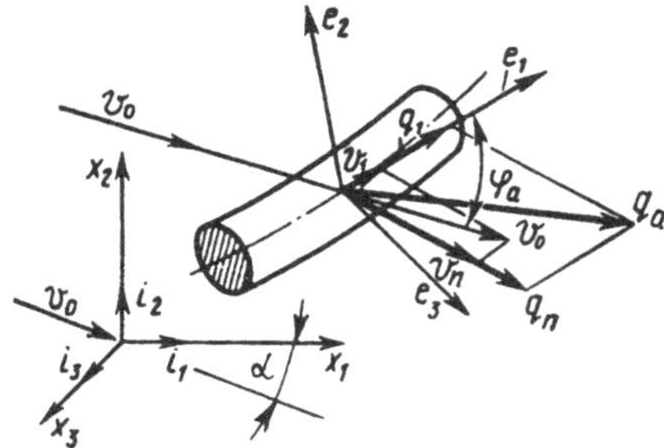

Fig. 6.1.

relations for interaction forces, we cannot determine the stress-strain state of a curvilinear rod and we cannot study its static and dynamic stability. Thus, a practically important problem of determination of approximate formulas for aerohydrodynamic forces arises. Of course, for particular cases $\varphi_a = 0$ and $\varphi_a = \pi/2$, the approximate formulas must be identical to those obtained from the experimental investigations for straight rods. In what follows, aerohydrodynamic forces and moments are called *aerodynamic loads*.

The prime objective of this chapter is the derivation of the components of aerodynamic loads in the Cartesian and in the attached coordinate systems (Sects. 6.4 and 6.6). It is assumed that the rod is placed into a uniform flow. Rods of circular cross section are discussed in Sect. 6.4. For these rods, the center of gravity of each cross section coincides with its stiffness center. Rods of noncircular cross section (when the center of gravity and the stiffness center do not coincide) are considered in Sect. 6.6. Basic concepts and notions of aerohydrodynamics used in this chapter and experimentally obtained formulas for aerodynamic loads are collected at the beginning of Sects. 6.2 and 6.3.

A rod element of circular cross section in an air flow is shown in Fig. 6.1. The vector of the flow velocity is denoted by $\mathbf{v}_0$. The element is subjected to a distributed aerodynamic force $\mathbf{q}_a = \mathbf{q}_n + \mathbf{q}_1$. A rod placed into a flow may take a shape that differs much from its original shape. This original shape corresponds to the equilibrium configuration of the rod when the medium is at rest. The aerodynamic loads depend on the shape of the rod axis. More precisely, the forces depend on the angle φ_a between the tangent to the rod axis (the vector $\mathbf{e}_1$) and the vector of the flow velocity $\mathbf{v}_0$ as shown in Fig. 6.1.

It is worth noting that the prime difficulty which appears in the interaction problems is that the aerodynamic forces depend on the shape of the rod and on the orientation of the axial line in the flow.

For large values of flow velocity, a rod of small bending stiffnesses may take an equilibrium configuration that differs dramatically from the natural configuration. Hence, nonlinear problems of statics of rods in a flow occur. Usually, it is assumed that the flow around the rod is steady and vortex-free. The latter is true only for a bounded range of the flow velocity values and for rods of comparatively smooth profile (e.g. for rods of circular cross section). For a rod of triangular or rectangular cross section, even a small rotation of the rod about its axial line may cause a considerable change of the aerodynamic forces applied to the rod, say, $\mathbf{q}_n$ and $\mathbf{q}_1$ as shown in Fig. 6.1.

It should be noted that for large values of the Reynolds number, the flow is unsteady, and, because of the Kármán forces, the rod may break into oscillation with respect to its equilibrium configuration. Strictly speaking, this problem is not the problem of statics. On the other hand, the problem can be treated by methods of statics because the shape of the rod axis in this equilibrium configuration can be determined from the equilibrium equations of the rod in which the aerodynamic forces $\mathbf{q}_n$ and $\mathbf{q}_1$ are replaced by their averaged, constant values.

6.2 Basic Concepts of Aerohydrodynamics

6.2.1 Eulerian and Lagrangian Representations

A motion of a liquid or a gas can be either steady or unsteady. Consider a fixed point in space. If pressure, density, and the vector of the flow velocity at this point are time-independent parameters, then we can say that a *steady* motion of the medium takes place. If some of these parameters change in time, then the motion is termed *unsteady*.

A motion of a liquid or a gas can be described by use of the Eulerian variables or the Lagrangian ones. The Lagrangian representation of motion implies that we trace an individual particle of medium and determine the parameters (velocity, density, pressure, etc.) of this particle. Within the Eulerian representation, we observe the parameters of medium at a fixed point and pay no attention to the behavior of individual particles.

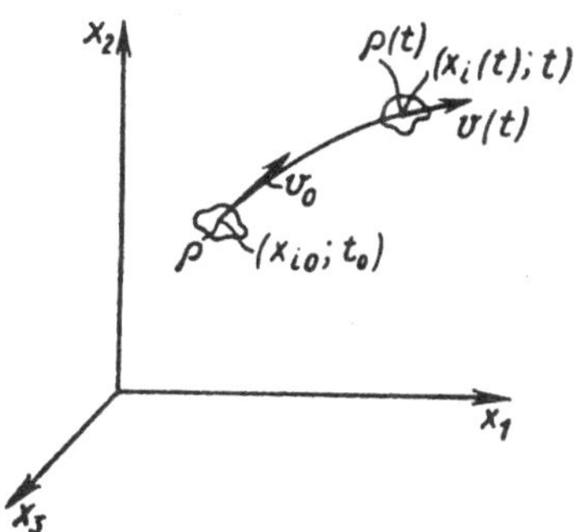

Fig. 6.2.

Lagrangian Variables Within the framework of the Lagrangian representation, we are to study the motion of individual particles of medium (see Fig. 6.2). This approach comprises the analysis of the following problems:

(1) The determination of the vector parameters (velocity $\mathbf{v}$ and acceleration $\dot{\mathbf{v}}$[1]) and the scalar parameter (density ρ) of a material particle as functions of time t.

(2) The derivation of the relationship between these parameters for different material particles.

First, we are to learn to distinguish different material particles. This can be done as follows: we ascribe to each material particle the coordinates of the point that this particle occupied at time t_0, e.g. the Cartesian coordinates x_{i0} as shown in Fig. 6.2. The Cartesian coordinates of the material particle at any time t are

$$x_i(t) = x_i(x_{10}\,,\, x_{20}\,,\, x_{30}\,,\, t)\,. \tag{6.1}$$

Similarly, the other characteristics of motion can be expressed through x_{i0} and t. Thus, x_{i0} and t are called *Lagrangian variables.*

The components of the velocity vector $\mathbf{v}$ and of the acceleration vector $\dot{\mathbf{v}}$, density and pressure are related to the Lagrangian variables as follows:

$$\dot{x}_i = \frac{\partial x_i(x_{j0}\,,\, t)}{\partial t}\,; \tag{6.2}$$

$$\ddot{x}_i = \frac{\partial^2 x_i(x_{j0}\,,\, t)}{\partial t^2}\,; \tag{6.3}$$

$$\rho = \rho\,(x_{j0}\,,\, t)\,; \tag{6.4}$$

$$p = p\,(x_{j0}\,,\, t)\,. \tag{6.5}$$

It should be noted that unlike the notion *point* accepted in mathematics, a *material particle* should be thought of as a nonzero volume of medium. This volume is small as compared to the total volume that the medium occupies. The volume of a material particle may change during motion of the medium while its mass remains the same. Hence, the density ρ of a particle may also change.

The Lagrangian representation is convenient when displacements of material particles are small, e.g. in the case of small oscillations of a medium.

Eulerian Variables Let us investigate the change of the parameters (velocity, pressure, etc.) of a medium at a fixed point. In the Lagrangian representation, an observer moves along with an individual material particle and measures its characteristics while in the Eulerian representation, an observer watches a fixed spatial place and measures characteristics of the medium at this point. Within the framework of the Eulerian approach, the following problems arise:

[1] The upper dot signifies the differentiation with respect to time t.

(1) The determination of vector and scalar characteristics at a fixed point as functions of time t;

(2) The derivation of relationships between these characteristics of neighboring particles.

The Eulerian variables are the current coordinates x_i of material particles and time t (see Fig. 6.2). We have

$$\mathbf{v} = \mathbf{v}\,(x_i(t)\,,\,t)\,; \qquad \rho = \rho\,(x_i(t)\,,\,t)\,; \qquad p = p\,(x_i(t)\,,\,t)\,. \tag{6.6}$$

The Eulerian approach enables us to determine the vector and scalar fields, i.e. the velocity $\mathbf{v}$, the acceleration $\dot{\mathbf{v}}$, density ρ, and pressure p.

Using the Eulerian variables, we can express the total derivative of the velocity $\mathbf{v}$ with respect to time t (so-called *material* or *substantial* derivative) as follows:

$$\frac{\mathrm{d}\mathbf{v}}{\mathrm{d}t} = \frac{\partial \mathbf{v}}{\partial t} + \sum_{i=1}^{3} \frac{\partial \mathbf{v}}{\partial x_i}\, x_i\,. \tag{6.7}$$

The same rule of differentiation applies in the case of any other vector field (not necessarily $\mathbf{v}$).

The physical meaning for the terms in the right-hand side of (6.7) is as follows: $\dfrac{\partial \mathbf{v}}{\partial t}$ is a partial derivative which indicates the rate of change of $\mathbf{v}$ at a fixed point; $\dfrac{\partial \mathbf{v}}{\partial x_i}$, $i = 1\,,2\,,3$, are partial derivatives which indicate the rate of change of $\mathbf{v}$ as the coordinates x_i vary while time t is held constant.

In terms of the Lagrangian variables, we have

$$\frac{\mathrm{d}\mathbf{v}}{\mathrm{d}t} = \frac{\partial \mathbf{v}}{\partial t}$$

because x_{j0} do not depend on t.

Information on the behavior of individual material particles is of little interest in most applied problems. Consequently, the Eulerian approach has some advantages over the Lagrangian one.

A set of scalar functions of coordinates x_1, x_2, and x_3 may be treated as one vector-function. This vector-function assigns a vector to a point with the coordinates $(x_1,\, x_2,\, x_3)$. Hence, we obtain a set of vectors that is called a *vector field.* To illustrate a vector field, one should draw the lines of the field. A line of a vector field (*field line*) is a curve that is tangent to this field (see Fig. 6.3). For any point A, there exists a unique field line passing through A. The field lines of a velocity field are called *streamlines.* If the velocity at a point A is equal to zero, then this point is termed a *critical point* of the velocity field. The field at such a point has a singularity. Thus, the name *critical.*

Streamlines provide insight into an instantaneous state of a liquid. It is especially advantageous in the problems concerned with bodies interacting

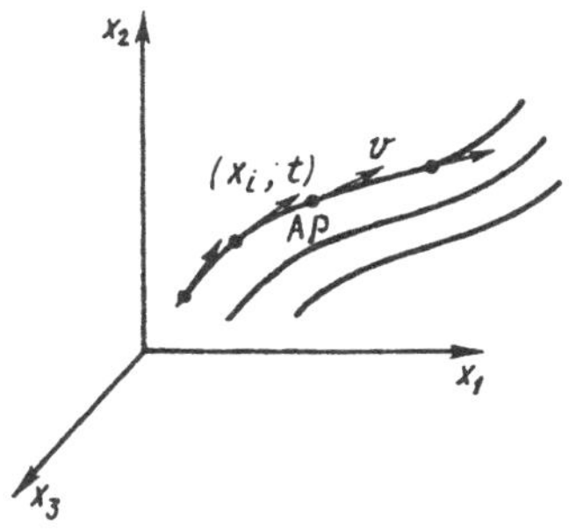

Fig. 6.3.

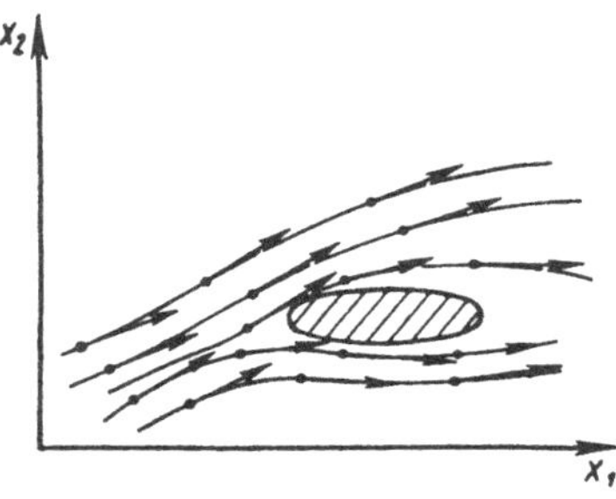

Fig. 6.4.

with flows (see Fig. 6.4). For a steady motion of a medium, the streamlines and the path lines coincide (see Fig. 6.3). A *stream tube* is the collection of streamlines that intersect a closed curve as illustrated in Fig. 6.5. Since the material particles move in the direction of the velocity field, they cannot cross the boundary of a stream tube. In the case of a steady motion, representing the flow as a set of stream tubes, we can understand many subtle points and reduce the analysis of the spatial flow to the analysis of a one-dimensional flow. Then, using the law of conservation of mass and the law of conservation of energy for each tube, we can examine the problem of interaction.

6.2.2 Basic Principles of Aerodynamics

Equation of Continuity Let us consider a steady motion of a fluid. According to the law of conservation of mass, the mass of liquid or gas carried through each cross section of a stream tube is the same for equal time intervals, that is,

$$\rho\, v(s)\, F(s) = \text{const}\,, \tag{6.8}$$

where ρ is the density of liquid, $v(s)$ is the magnitude of the average velocity in the tube, and $F(s)$ is the cross-sectional area of the tube. For incompressible fluids, we have $\rho = \text{const}$, hence,

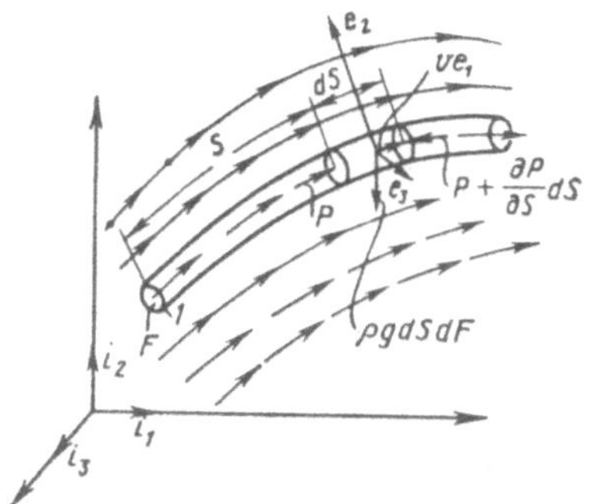

Fig. 6.5.

$$vF = \text{const}\,. \tag{6.9}$$

The relations (6.8) and (6.9) are valid if applied to the motion of ideal liquid in a pipeline of varying cross section $F(s)$, i.e. the condition (6.8) is not local but integral one because it contains the finite area F.

Let us derive the equation of continuity in terms of the Eulerian variables. Consider a liquid element of mass $\delta m = \rho\,\delta v$ and volume δv. During motion, the mass of the elementary volume remains unchanged, that is,

$$\rho\,\delta v = \rho_0\,\delta v_0 = \text{const}\,, \tag{6.10}$$

where ρ_0 and δv_0 are the volume and the density at time t_0.

Considering the total derivative of (6.10), we get

$$\frac{\mathrm{d}\rho}{\mathrm{d}t} + \rho\,\frac{\mathrm{d}\,(\delta v)}{\delta v\,\mathrm{d}t} = 0\,. \tag{6.11}$$

The term

$$\frac{\mathrm{d}\,(\delta v)}{\delta v\,\mathrm{d}t}$$

measures a rate of change of the volume at a fixed point. In some textbooks on hydrodynamics, this ratio is shown to be equal to the *divergence of the vector field* $\mathbf{v}$ at this point,

$$\frac{\mathrm{d}\,(\delta v)}{\delta v\,\mathrm{d}t} = \operatorname{div}\mathbf{v} = \sum_{i=1}^{3}\frac{\partial v_{x_i}}{\partial x_i}\,. \tag{6.12}$$

In terms of the Eulerian variables, we have

$$\frac{\mathrm{d}\rho}{\mathrm{d}t} = \frac{\partial\rho}{\partial t} + \sum_{i=1}^{3}\frac{\partial\rho}{\partial x_i}\,\dot{x}_i\,, \qquad \dot{x}_i = v_{x_i}\,; \tag{6.13}$$

in view of (6.13), we obtain the following equation of continuity:

$$\frac{\mathrm{d}\rho}{\mathrm{d}t} + \sum_{i=1}^{3} \frac{\partial (\rho v_{x_i})}{\partial x_i} = 0\,. \tag{6.14}$$

For an incompressible liquid ($\rho = \text{const}$), the relation (6.14) yields

$$\frac{\partial v_{x1}}{\partial x_1} + \frac{\partial v_{x2}}{\partial x_2} + \frac{\partial v_{x3}}{\partial x_3} = 0\,. \tag{6.15}$$

Bernoulli's Equation An elementary volume of liquid moving with a velocity $\mathbf{v}$ is shown in Fig. 6.5. Since the liquid is ideal, there is no friction forces and thus the volume is subjected to the gravity force and difference of pressure multiplied by the area of the tube $\mathrm{d}F$. We have

$$\rho\,\mathrm{d}s\,\mathrm{d}F\,\frac{\mathrm{d}\mathbf{v}}{\mathrm{d}t} = -\frac{\partial \mathbf{p}}{\partial s}\,\mathrm{d}s\,\mathrm{d}F + \rho\mathbf{g}\,\mathrm{d}s\,\mathrm{d}F$$

or, by use of the Eulerian variables,

$$\rho\left(\frac{\partial \mathbf{v}}{\partial t} + \frac{\partial \mathbf{v}}{\partial s}\frac{\mathrm{d}s}{\mathrm{d}t}\right) = -\frac{\partial \mathbf{p}}{\partial s} + \rho\mathbf{g}\,, \tag{6.16}$$

where

$$\mathbf{v} = v\mathbf{e}_1\,; \qquad \frac{\mathrm{d}s}{\mathrm{d}t} = v\,; \qquad \mathbf{p} = p\mathbf{e}_1\,; \qquad \mathbf{g} = -g\mathbf{i}_2\,;$$

here $\mathbf{e}_1$ is a unit vector tangent to the axial line of the stream tube.

Using local derivatives, we can rearrange (6.16) as follows:

$$\left[\frac{\tilde{\partial}\mathbf{v}}{\partial t} + \boldsymbol{\omega}\times\mathbf{v} + \frac{\tilde{\partial}\mathbf{v}}{\partial s}v + (\text{æ}\times\mathbf{v})\,v\right] = -\frac{\tilde{\partial}\mathbf{p}}{\partial s} - \text{æ}\times\mathbf{p} - \rho g\mathbf{i}_2\,. \tag{6.17}$$

For steady motions, the tube keeps its spatial position and the vector and scalar parameters of the flow (velocity $\mathbf{v}$, density ρ, and pressure p) do not depend on time t, i.e. $\partial\mathbf{v}/\partial t = 0$ and $\boldsymbol{\omega} = 0$. Thus, from (6.17) it follows that

$$\rho\,\frac{\partial v}{\partial s}\,v\mathbf{e}_1 + \rho v^2 \text{æ}_3\mathbf{e}_2 = -\frac{\partial p}{\partial s}\,\mathbf{e}_1 - \text{æ}_3 p\mathbf{e}_2 - \rho g\mathbf{i}_2\,. \tag{6.18}$$

Multiplying both sides of (6.18) by $\mathbf{e}_1$, we get

$$\frac{1}{2}\,\rho\,\frac{\partial}{\partial s}\,(v^2) + \frac{\partial p}{\partial s} + \rho g\,\frac{\partial x_2}{\partial s} = 0\,, \tag{6.19}$$

where

$$\frac{\partial x_2}{\partial s} = (\mathbf{i}_2\cdot\mathbf{e}_1)\,.$$

Integrating (6.19), we obtain Bernoulli's equation

$$\frac{\rho v^2}{2} + p + \rho g x_2 = \text{const}$$

or

$$\frac{\rho v_1^2}{2} + p_1 + \rho g x_2(0) = \frac{\rho v^2}{2} + p + \rho g x_2 \,, \tag{6.20}$$

where v_1, p_1, and $x_2(0)$ are the velocity, the pressure, and the coordinate x_2 at cross section *1* (see Fig. 6.5). From (6.20) it follows that in the case of a steady motion of an incompressible liquid, the sum of the kinetic energy, $\frac{\rho v_1^2}{2}$, and the potential energy, $p + \rho g x_2$, for a material volume remains unchanged.

For an unsteady motion, we have $\frac{\tilde{\partial} v}{\partial t} \neq 0$. Anyway, whenever the tube occupies the same spatial position during motion ($\omega = 0$), we have

$$\frac{\rho v^2}{2} + p + \rho g x_2 + \int_0^s \frac{\tilde{\partial} v}{\partial t}\, \mathrm{d}s = \text{const}\,. \tag{6.21}$$

This equation is valid for any tube provided the velocity is a function of time t alone (the velocity does not depend on the coordinate s). This is true for incompressible liquids when the cross-sectional area of the tube remains the same. In this case, using (6.21), we get

$$\frac{\rho v^2}{2} + p + \rho g x_2 + \frac{\partial v}{\partial t} = \text{const}\,. \tag{6.22}$$

Equation (6.22) can be used for analysis of nonstationary flows in pipelines.

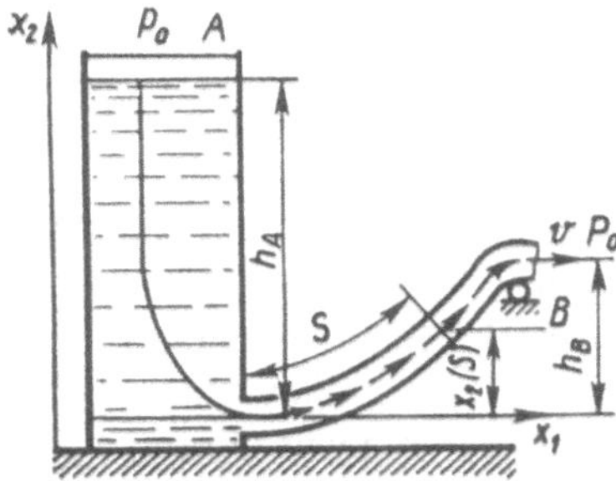

Fig. 6.6.

Consider some examples which illustrate an application of Bernoulli's equation. A vessel supplied with a pipeline is shown in Fig. 6.6. The liquid runs out of the vessel through the pipeline. Our purpose is to determine the outflow velocity v and the pressure as a function of the coordinate $x_2(s)$. All the stream tubes take their origin at the free surface A. The initial velocity of the liquid in each tube is equal to zero, and the original pressure is the atmospheric pressure p_0. One such a tube is illustrated in Fig. 6.6. The outflow velocity of liquid particles is denoted by v (the cross section B). The pressure at the output cross section B is equal to p_0.

For the cross section B, Bernoulli's equation gives

$$\frac{\rho v^2}{2} = \rho g\,(h_A - h_B)\,. \tag{6.23}$$

This equation can be solved for v. Since the liquid is incompressible and the pipeline is of constant cross section, we see that the velocity v is the same for any cross sections of the pipeline. For an arbitrary cross section, Bernoulli's equation takes the form

$$p_0 + \rho g h_A = \rho g x_2(s) + \frac{\rho v^2}{2} + p\,(s)\,. \tag{6.24}$$

Combining (6.23) and (6.24), we arrive at

$$p\,(s) = p_0 + \rho g\,[\,h_B - x_2(s)\,]\,.$$

Note that in this example, the flow is steady provided the level of liquid in the vessel does not change (h_A = const). If this condition is not satisfied, then the flow is unsteady and the functions $v(t)$ and $p(s\,,\,t)$ are to be determined from (6.22).

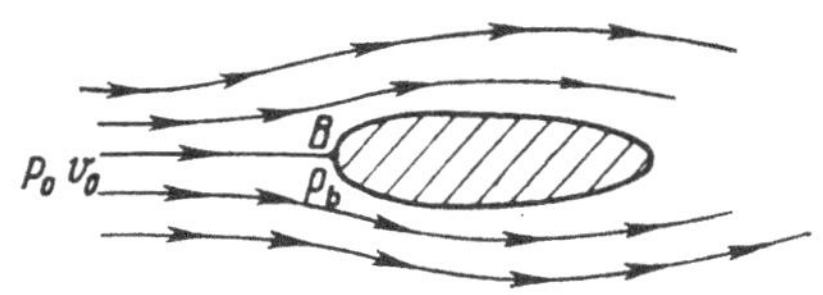

Fig. 6.7.

As the second example, let us consider a rod of noncircular cross section in a fluid flow (Fig. 6.7). There exists a streamline that ends at the point B. Since the point B is stationary, the velocity of the liquid particle that is carried to the point B must equal zero. Assuming that the flow tube ending at the point B is a horizontal one, we determine the increment of pressure at the point B as follows:

$$\Delta p = p_B - p_0 = \frac{\rho v_0^2}{2}\,;$$

here v_0 and p_0 are the velocity and the pressure of the flow.

Let δF be a small part of the rod surface containing the point B. The flow acts on the rod element with a force δX (so-called *drag force*) given by the relation

$$\delta X = \frac{\delta F \rho v_0^2}{2} .$$

Consequently, we can conclude that the aerodynamic force that acts on a rod is linear in density and quadratic in flow velocity. This fact is in excellent agreement with the experimental results.

In this example, Bernoulli's equation enables us to determine the pressure only at the stationary point B. To determine the pressure at the other points of the rod surface, one should use the general equations of aeroelasticity. Unfortunately, it seems to be impossible to derive the explicit formulas for aerodynamic forces. Consequently, experimental studies are of great importance.

6.3 Experimental Results

A cross section of a rod immersed into a plane homogeneous liquid or air flow is shown in Fig. 6.8. The cross section is subjected to a distributed pressure p. If the flow velocity equals zero, then the pressure is constant and the resultant force and the resultant moment acting on an element of unit length equal zero. For a nonzero velocity, the pressure is no longer constant, i.e. elements of equal length may experience different resultant loads. Hence, the cross section is subjected to a nonzero moment μ_1 and a force with projections q_{x_2} and q_{x_3} on the reference axes x_2Ox_3.

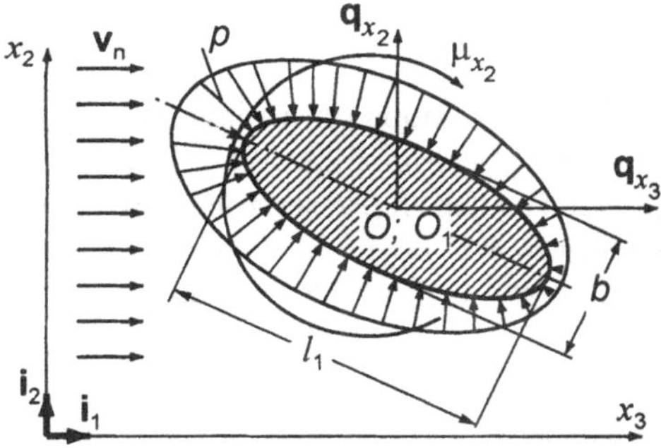

Fig. 6.8.

We assume that the aerodynamic forces q_{x_1} and q_{x_2} and the moment μ_1 are applied to the center of gravity of the cross section (the point O) as illustrated in Fig. 6.8. The aerodynamic forces q_{x_2} and q_{x_3} are termed a *lifting force* and a *drag force*, respectively. The distribution of pressure over the boundary is governed by the Reynolds number $\mathrm{Re} = vl/\nu$, where ν is a coefficient of kinematic viscosity of the liquid or air and l is a characteristic of the boundary size (e.g. the diameter of a circular cross section).

For a rod of an arbitrary cross section, the explicit formulas for the forces q_{x_i} and the moment μ_1 can hardly be obtained theoretically. However, using the theory of dimension and similarity, we are able to derive approximate formulas.

First, as the most simple case, let us consider incompressible ideal liquid. A motion of the liquid can be characterized by the following parameters: velocity v, density ρ, and characteristic length l. Assume that the force q_{x_3} is defined by the relation

$$q_{x_3} = \frac{1}{2} c_{n_1} v_n^a \rho^b l^c , \tag{6.25}$$

where c_{n_1} is a nondimensional coefficient, a, b, and c are unknown constants, and v_n is the projection of the vector $\mathbf{v}_0$ on a plane that is perpendicular to the rod axis. The left-hand side of (6.25) has the dimension of distributed force, hence, the right-hand side must be of the same dimension:

$$\mathrm{LMT}^{-2}\mathrm{L}^{-1} = c_{n_1} \frac{1}{2} (\mathrm{LT}^{-1})^a (\mathrm{L}^{-3}\mathrm{M})^b \mathrm{L}^c ; \tag{6.26}$$

here L, M, and T have the dimension of length, mass, and time, respectively.

The relation (6.26) implies the following equations in the unknowns a, b, and c:

$$0 = a - 3b + c ; \qquad b = 1 ; \qquad a = 2 .$$

As a result, we have $c = 1$ and

$$q_{x_3} = \frac{1}{2} c_{n_1} \rho v_n^2 l . \tag{6.27}$$

Consider the projections of the rod cross sections on a plane that is orthogonal to the velocity vector $\mathbf{v} = -v\mathbf{i}_1$ (see Fig. 6.8). We can take the square of the least area of these projections as the characteristic length l. For example, for a rod of circular cross section, we have $l = d$, where d is the diameter of a cross section. For the rod shown in Fig. 6.8, we can put $l = b$. Finally, we arrive at

$$q_{x_3} = c_{n_1} b \frac{\rho v_n^2}{2} , \tag{6.28}$$

where c_{n_1} is the nondimensional coefficient of head resistance.

Similar reasoning gives us the formulas for the distributed force q_{x_2} and the moment μ_1,

$$q_{x_2} = c_l \frac{\rho v_n^2 b}{2} ; \tag{6.29}$$

$$\mu_1 = c_m \frac{\rho v_n^2 b l_1}{2} , \tag{6.30}$$

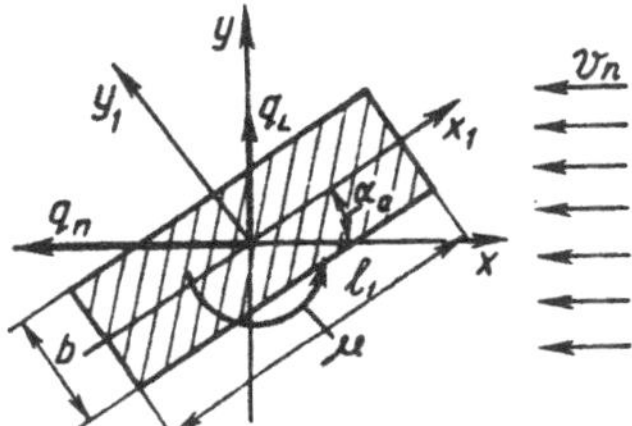

Fig. 6.9.

where l_1 is the characteristic size of the rod cross section.

For symmetrical cross sections, the aerodynamic coefficients c_{n_1}, c_l, and c_m in the right-hand sides of (6.28)–(6.30) are functions of the Reynolds number and the angle of attack α_a (see Figs. 6.8 and 6.9). These coefficients are to be determined experimentally.

To account for the fact that the moment depends on the sign of the angle α_a, we write

$$\mu_1 = c_m \frac{1}{2} \rho v_n^2 b l_1 \operatorname{sgn} \alpha_a . \tag{6.31}$$

In the case of a rod of circular cross section, the aerodynamic moment μ_1 is equal to zero. Moreover, if the Reynolds number varies within a certain interval, the aerodynamic coefficients c_n and c_l remain the same (Grafskii, Kazakevich (1983), Devnin (1975), and Kazakevich (1977)).

6.4 Aerodynamic Forces Acting on Rods of Circular Cross Section

Determine the projections of the aerodynamic forces when the flow velocity vector $\mathbf{v}_0$ is parallel to the plane x_1Ox_3 as illustrated in Fig. 6.10. We have

$$\mathbf{v}_0 = v_0 \cos\alpha \cdot \mathbf{i}_1 + v_0 \sin\alpha \cdot \mathbf{l}_3 ,$$

where α is the angle between the vectors $\mathbf{i}_1$ and $\mathbf{e}_v$ (Svetlitsky (1982)).

An element of a rod of circular cross section and the distributed aerodynamic forces acting on this element are shown in Fig. 6.10. Consider rods of circular cross sections (see Fig. 6.11*a,b*). Let the center of gravity of each cross section (the point O) coincide with its stiffness center (the point O_1). The case when these centers do not coincide is considered in Sect. 6.6.

Let φ_a be the angle between the tangent vector $\mathbf{e}_1$ and the vector $\mathbf{v}_0$. Assume that the magnitudes of the aerodynamic forces are as follows:

$$|\mathbf{q}_n| = \frac{1}{2} c_n \rho v_n^2 d , \qquad v_n = |\mathbf{v}_0| \sin\varphi_a ; \tag{6.32}$$

$$|\mathbf{q}_1| = \frac{1}{2} c_1 \rho v_1^2 d , \qquad v_1 = |\mathbf{v}_0| \cos\varphi_a ; \tag{6.33}$$

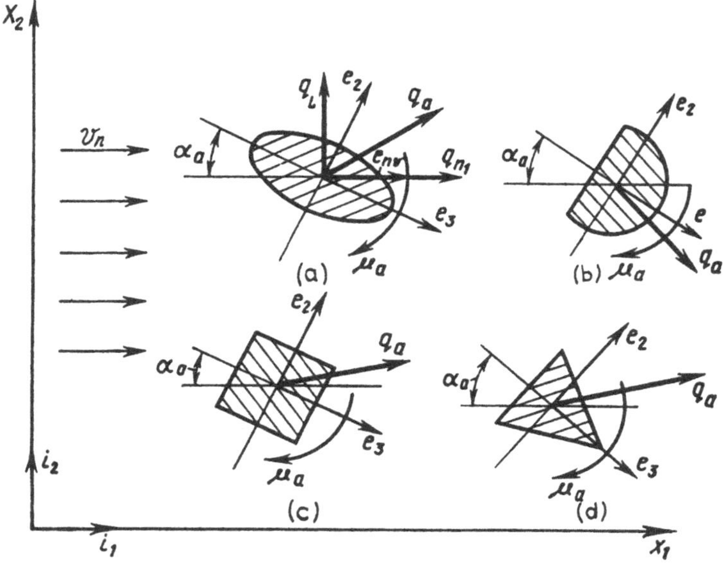

Fig. 6.10.

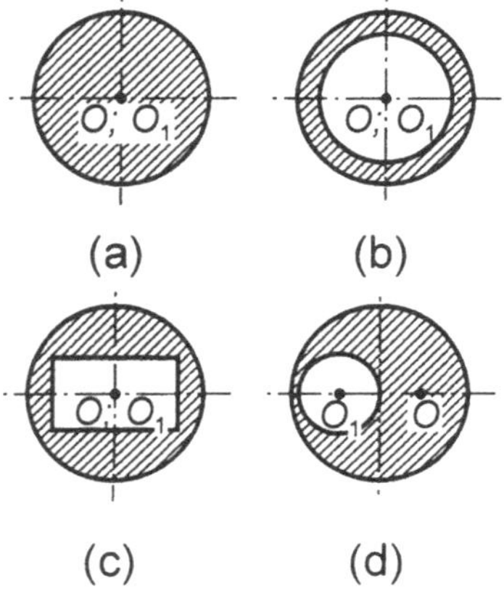

Fig. 6.11.

here c_n and c_l are aerodynamic coefficients, $\mathbf{v}_n$ is the normal component of the vector $\mathbf{v}_0$, and $\mathbf{v}_l$ is the tangent component of $\mathbf{v}_0$. The angle φ_a satisfies the relation

$$\cos\varphi_a = \frac{(\mathbf{v}_0 \cdot \mathbf{e}_1)}{|\mathbf{v}_0|} = x_1' \cos\alpha + x_3' \sin\alpha\,.$$

Determine now the projections of the drag force vector $\mathbf{q}_n$ on the immovable coordinate axes. The following equalities must hold:

$$(\mathbf{q}_n \cdot \mathbf{e}_1) = 0\,, \qquad \mathbf{e}_1 = \sum_{j=1}^{3} x_j' \mathbf{i}_j\,; \tag{6.34}$$

$$\mathbf{q}_n(\mathbf{e}_v \times \mathbf{e}_1) = 0\,; \tag{6.35}$$

the unit vector $\mathbf{e}_v$ is parallel to the flow velocity vector $\mathbf{v}_0$.

From (6.35) it follows that $\mathbf{e}_1$, $\mathbf{e}_v$, and $\mathbf{q}_n$ are coplanar vectors. The reason for this is that the vectors $\mathbf{q}_n$ and $\mathbf{v}_n$ equal in direction.

For a rod of circular cross section, the aerodynamic force $\mathbf{q}_n$ is orthogonal to the rod axis and directed along the vector $\mathbf{v}_n$, while the lifting force $\mathbf{q}_l$ vanishes. Equation (6.32) can be written as follows:

$$|\mathbf{q}_n|^2 = q_{nx_1}^2 + q_{nx_2}^2 + q_{nx_3}^2 = q_{n0}^2 \sin^4\varphi_a\,, \qquad q_{n0} = \frac{c_n}{2}\rho v_0^2 d\,; \tag{6.36}$$

here q_{nx_i} are the vector components in the immovable frame. In components, the relations (6.34) and (6.35) read

$$q_{nx_1}\frac{dx_1}{ds} + q_{nx_2}\frac{dx_2}{ds} + q_{nx_3}\frac{dx_3}{ds} = 0\,; \tag{6.37}$$

$$q_{nx_1}\frac{dx_2}{ds}\sin\alpha + q_{nx_2}\left(\frac{dx_3}{ds}\cos\alpha - \frac{dx_1}{ds}\sin\alpha\right) - q_{nx_3}\frac{dx_2}{ds}\cos\alpha = 0\,. \tag{6.38}$$

Equations (6.36)–(6.38) can be solved for the three components q_{nx_i} as follows:

$$q_{nx_1} = q_{n0}\sin\varphi_a\left(\cos\alpha - \frac{dx_1}{ds}\cos\varphi_a\right)\,; \tag{6.39}$$

$$q_{nx_2} = -q_{n0}\sin\varphi_a\cos\varphi_a\frac{dx_2}{ds}\,; \tag{6.40}$$

$$q_{nx_3} = q_{n0}\sin\varphi_a\left(\sin\alpha - \frac{dx_3}{ds}\cos\varphi_a\right)\,. \tag{6.41}$$

Let us obtain the components of $\mathbf{q}_n$ in the attached coordinate system. We have

$$\mathbf{q}_n = \sum_{j=1}^{3} q_{nx_j}\mathbf{i}_j\,.$$

Using the transformation matrix $L^{(1)}$, we can write

$$\mathbf{i}_j = \sum_{j=1}^{3} l_{kj}^{(1)} \mathbf{e}_k ,$$

therefore,

$$\mathbf{q}_n = \sum_{j=1}^{3} q_{nx_j} \left(\sum_{k=1}^{3} l_{kj}^{(1)} \mathbf{e}_k \right)$$

or

$$\mathbf{q}_n = \sum_{k=1}^{3} \left(\sum_{j=1}^{3} q_{nx_j} l_{kj}^{(1)} \right) \mathbf{e}_k = \sum_{k=1}^{3} q_{nk} \mathbf{e}_k ,$$

where q_{nk} are the components of the vector $\mathbf{q}_n$ in the attached basis,

$$q_{n\nu} = q_{nx_1} l_{\nu 1}^{(1)} + q_{nx_2} l_{\nu 2}^{(1)} + q_{nx3} l_{\nu 3}^{(1)} , \qquad \nu = 1 , 2 , 3 . \tag{6.42}$$

Recall that

$$l_{11}^{(1)} = x_1' ; \qquad l_{12}^{(1)} = x_2' ; \qquad l_{13}^{(1)} = x_3' ; \qquad \sum_{j=1}^{3} (x_j')^2 = 1 .$$

Let us show that $q_{n_1} = 0$. In the attached frame, the vector $\mathbf{q}_n$ is orthogonal to the base vector $\mathbf{e}_1$ of the attached coordinate system. Substituting the expressions for q_{nx_j} into (6.43) and setting $\nu = 1$, we arrive at

$$\begin{aligned} q_{n1} = q_{n0} \sin\varphi_a [& \cos\alpha x_1' - (x_1')^2 \cos\varphi_a \\ & - (x_2')^2 \cos\varphi_a + x_3' \sin\alpha - (x_3')^2 \cos\varphi_a] . \end{aligned}$$

Since

$$\cos\alpha x_1' + \sin\alpha x_3' = \cos\varphi_a ,$$

we get

$$q_{n1} = q_{n0} \sin\varphi_a \{ \cos\varphi_a - [(x_1')^2 + (x_2')^2 + (x_3')^2] \cos\varphi_a \} \equiv 0 .$$

The formulas for the components of the drag force $\mathbf{q}_n$ can be obtained without resort to the expressions for q_{nx_j}. In the attached axes, the relation (6.34) becomes an identity because the vector $\mathbf{e}_1$ is orthogonal to the vectors $\mathbf{e}_2$ and $\mathbf{e}_3$. In the Cartesian basis, the vector $\mathbf{e}_v$ can be written as follows:

$$\mathbf{e}_v = \cos\alpha \mathbf{i}_1 + \sin\alpha \mathbf{i}_3 ;$$

hence, in the attached basis, we get

$$\mathbf{e}_v = (l_{11}^{(1)} \cos\alpha + l_{13}^{(1)} \sin\alpha)\,\mathbf{e}_1 + (l_{21}^{(1)} \cos\alpha + l_{23}^{(1)} \sin\alpha)\,\mathbf{e}_2 + (l_{31}^{(1)} \cos\alpha + l_{33}^{(1)} \sin\alpha)\,\mathbf{e}_3 .$$

Therefore, (6.35) and (6.36) give two equations in the unknowns q_{n_2} and q_{n_3}:

$$q_{n2}(l_{31}^{(1)} \cos\alpha + l_{33}^{(1)} \sin\alpha) - q_{n3}(l_{21}^{(1)} \cos\alpha + l_{23}^{(1)} \sin\alpha) \equiv 0 ;$$
$$q_{n2}^2 + q_{n3}^2 = q_{n0}^2 \sin^4\varphi_a .$$

On rearrangement, we finally obtain

$$q_{n2} = q_{n0} \sin\varphi_a \left(l_{21}^{(1)} \cos\alpha + l_{23}^{(1)} \sin\alpha\right) ;$$
$$q_{n3} = q_{n0} \sin\varphi_a \left(l_{31}^{(1)} \cos\alpha + l_{33}^{(1)} \sin\alpha\right) . \tag{6.43}$$

Thus we can use the expressions for the components of the aerodynamic forces either in the form (6.42) (referred to the Cartesian basis) or in the form (6.43) (referred to the attached basis).

Determine now the projections of the tangent force vector $\mathbf{q}_1$ on the Cartesian axes. We have

$$q_{1x_1} = (\mathbf{q}_1 \cdot \mathbf{i}_j) = |\mathbf{q}_1|\,(\mathbf{e}_1 \cdot \mathbf{i}_j) , \qquad j = 1, 2, 3 . \tag{6.44}$$

In the attached coordinate system, we have

$$\mathbf{q}_1 = q_1 \mathbf{e}_1 , \qquad (\mathbf{q}_1)_2 = (\mathbf{q}_1)_3 = 0 , \tag{6.45}$$

where

$$q_1 = q_{10} \cos^2\varphi_a \mathrm{sgn}\,(\cos\varphi_a) , \qquad q_{10} = \frac{c_1}{2} \frac{\rho v_0^2 l^3 d}{A_{33}(0)} .$$

In view of (6.44), we arrive at

$$q_{1x_1} = q_{10} \cos^2\varphi_a \frac{dx_1}{ds} \,\mathrm{sgn}(\cos\varphi_a) ;$$
$$q_{1x_2} = q_{10} \cos^2\varphi_a \frac{dx_2}{ds} \,\mathrm{sgn}(\cos\varphi_a) ;$$
$$q_{1x_3} = q_{10} \cos^2\varphi_a \frac{dx_3}{ds} \,\mathrm{sgn}(\cos\varphi_a) , \tag{6.46}$$

where

$$q_{10} = \frac{c_1}{2} \rho v_0^2 d .$$

The factor $\mathrm{sgn}\,(\cos\varphi_a)$ is introduced to account for the direction of the vector $\mathbf{q}_1$, that is,

$$\text{if} \quad \operatorname{sgn}(\cos\varphi_a) > 0\,, \quad \text{then} \quad \mathbf{q}_1 = |\mathbf{q}_1|\mathbf{e}_1\,;$$
$$\text{if} \quad \operatorname{sgn}(\cos\varphi_a) < 0\,, \quad \text{then} \quad \mathbf{q}_1 = -|\mathbf{q}_1|\mathbf{e}_1\,.$$

Thus, the total aerodynamic force acting on a rod element of unit length has the following components in the Cartesian coordinate system:

$$q_{ax_1} = q_{n0}\sin\varphi_a(\cos\alpha - x_1'\cos\varphi_a) + q_{10}^{(1)}x_1'\cos^2\varphi_a\,; \tag{6.47}$$
$$q_{ax_2} = -\frac{1}{2}\,q_{n0}\sin 2\varphi_a x_2' + q_{10}^{(1)}x_2'\cos^2\varphi_a\,; \tag{6.48}$$
$$q_{ax_3} = q_{n0}\sin\varphi_a(\sin\alpha - x_3'\cos\varphi_a) + q_{10}^{(1)}x_3'\cos^2\varphi_a\,. \tag{6.49}$$

For further analysis, we introduce the nondimensional counterparts of the components of the aerodynamic forces (q_{nx_i}, q_{n2}, q_{n3}, q_{1x_i}, and q_1):

$$\tilde{q}_{n_j} = q_{n_j}\frac{l^3}{A_{33}(0)}\,; \qquad \tilde{q}_1 = q_1\frac{l^3}{A_{33}(0)}\,; \qquad \tilde{q}_{ax_i} = q_{ax_i}\frac{l^3}{A_{33}(0)}\,.$$

The nondimensional aerodynamic force $\mathbf{q}_a$ has the following components (here the tilde is dropped out):

$$q_{ax_1} = q_{10}^{(1)}x_1'\cos^2\varphi_a + q_{n0}\sin\varphi_a(\cos\alpha - x_1'\cos\varphi_a)\,; \tag{6.50}$$
$$q_{ax_2} = q_{10}^{(1)}x_2'\cos^2\varphi_a - \frac{1}{2}\,q_{n0}x_2'\sin 2\varphi_a\,; \tag{6.51}$$
$$q_{ax_3} = q_{10}^{(1)}x_3'\cos^2\varphi_a + q_{n0}\sin\varphi_a(\sin\alpha - x_3'\cos\varphi_a)\,; \tag{6.52}$$

here

$$q_{10}^{(1)} = \frac{c_1}{2}\,\frac{\rho v_0^2 l^3 d}{A_{33}(0)}\operatorname{sgn}(\cos\alpha_a)\,; \qquad q_{n0} = \frac{c_n}{2}\,\frac{\rho v_0^2 l^3 d}{A_{33}(0)}\,. \tag{6.53}$$

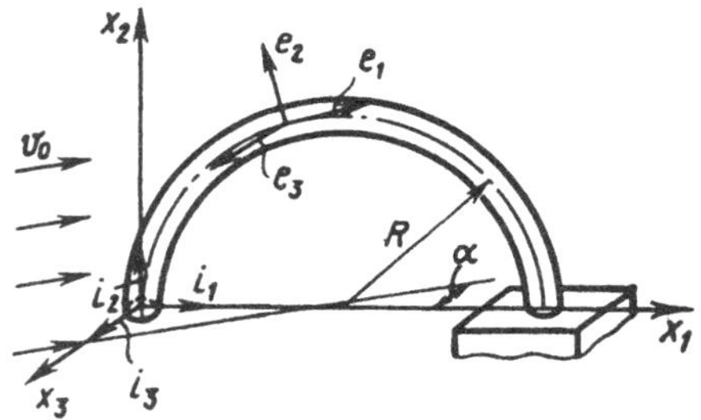

Fig. 6.12.

Consider an example. Let us determine the aerodynamic forces q_{1x_j}, q_{nx_j}, q_1, q_{n2}, and q_{n3} acting on a circular rod of circular cross section (see Fig. 6.12). The vector of flow velocity lies in the plane x_1Ox_3. The aerodynamic forces depend on the direction cosines x_j' of the vector $\mathbf{e}_1$ (see (6.39)–(6.41) and

(6.43)–(6.45)). In the case under consideration, the direction cosines depend on the shape of the rod axis. Let us assume that the displacements of the points of the axial line are small. Hence, we can determine x'_j using the method given in Appendix 5. The form of the axial line in the natural configuration is assumed to be known. In this example, we have

$$(x_1 - R_0)^2 + x_2^2 = R_0^2\,, \qquad R_0 = \frac{1}{\pi}\,, \tag{6.54}$$

where x_i and R_0 are nondimensional quantities. Differentiating (6.54) with respect to η, we get

$$(x_1 - R_0)\,x'_1 + x_2 x'_2 = 0\,. \tag{6.55}$$

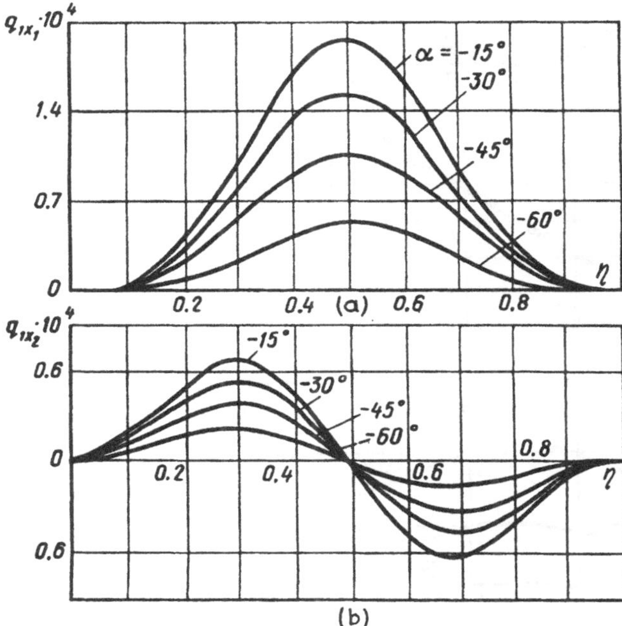

Fig. 6.13.

Since the rod axis is a plane curve, we have

$$x_1^{2'} + x_2^{2'} = 1\,. \tag{6.56}$$

Using (6.54) and eliminating x_2 from (6.55), we arrive at

$$(x_1 - R_0)\,x'_1 + \sqrt{R_0^2 - (x_1 - R_0)^2}\,x'_2 = 0\,. \tag{6.57}$$

Solving (6.57) for x_2' and thus eliminating x_2' from (6.56), we get

$$x_1' = \sqrt{\frac{R_0^2 - (x_1 - R_0)^2}{R_0^2}}\,. \tag{6.58}$$

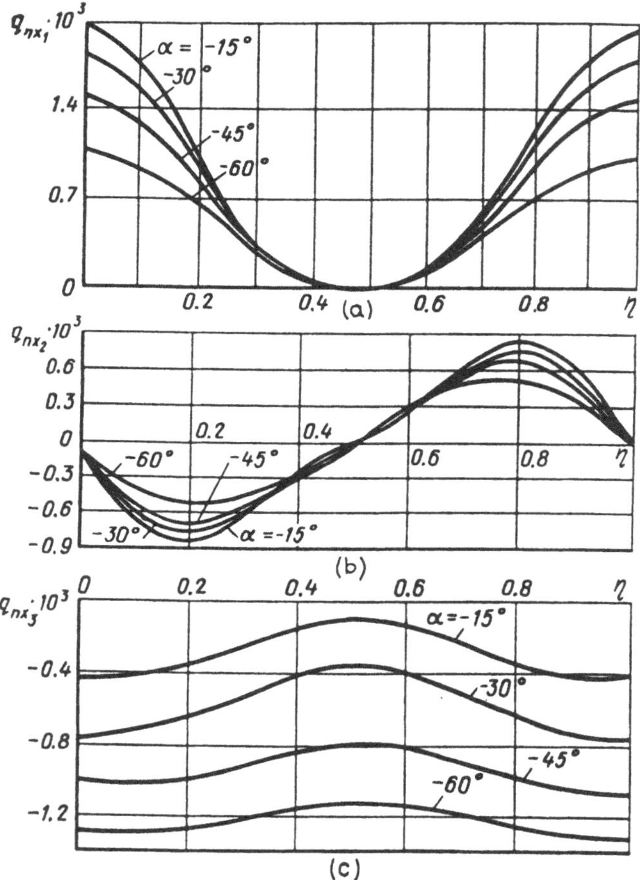

Fig. 6.14.

Integration of (6.58) gives us the relationships $x_1 = x_1(\eta)$ and $x_1' = x_1'(\eta)$. From (6.56) it follows that

$$x_2' = \sqrt{1 - x_1^{2'}} = \frac{(x_1 - R_0)}{R_0}\,.$$

The problem under consideration was solved for the following numerical values: $c_n = 1$, $c_1 = 0.1$, $l = 2.0\,\mathrm{m}$, $\rho = 1.0\,\mathrm{kg} \cdot \mathrm{m}^{-3}$, $d = 0.1\,\mathrm{m}$, $E = 2 \cdot 10^{11}\,\mathrm{N} \cdot \mathrm{m}^{-2}$, and $A_{33}(0) = 4 \cdot 10^{-2}\,\mathrm{N} \cdot \mathrm{m}^2$.

For some values of the angle α and $|\mathbf{v}_0| = 10^3\,\mathrm{cm} \cdot \mathrm{s}^{-1}$, the plots of q_{1x_j} versus η are shown in Fig. 6.13*a,b*. For the same angles, the plots of q_j versus η are shown in Fig. 6.14*a-c*. The relationships between the components of the aerodynamic force q_j in the attached axes and the coordinate η are illustrated in Fig. 6.15*a-c*. The plot of the absolute value of the total aerodynamic force acting on the rod versus the nondimensional coordinate η is shown in Fig. 6.16. The expressions obtained for the aerodynamic forces enable us to determine Q_j, M_j, ϑ_j, and u_j from the equilibrium equations of the rod.

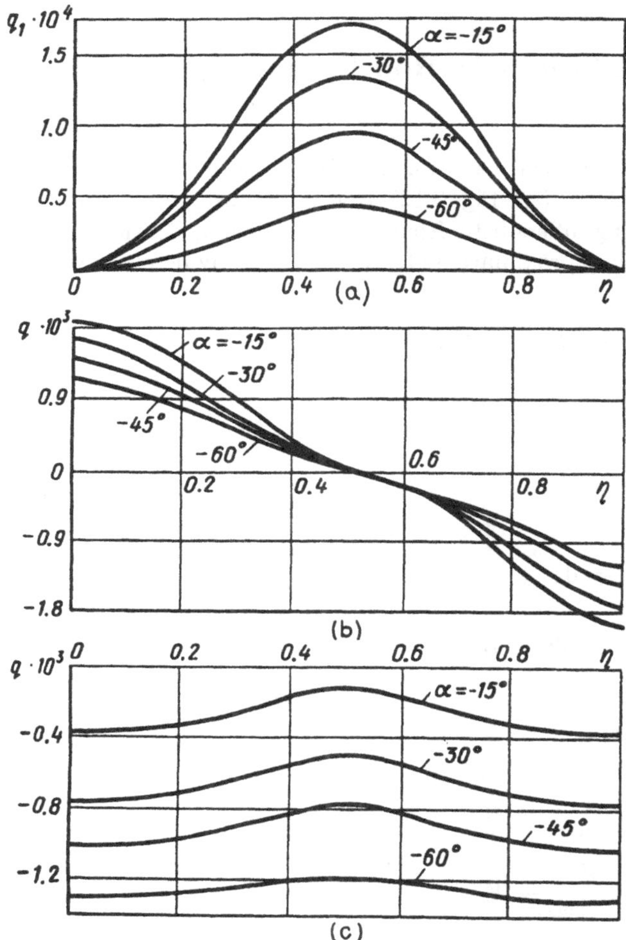

Fig. 6.15.

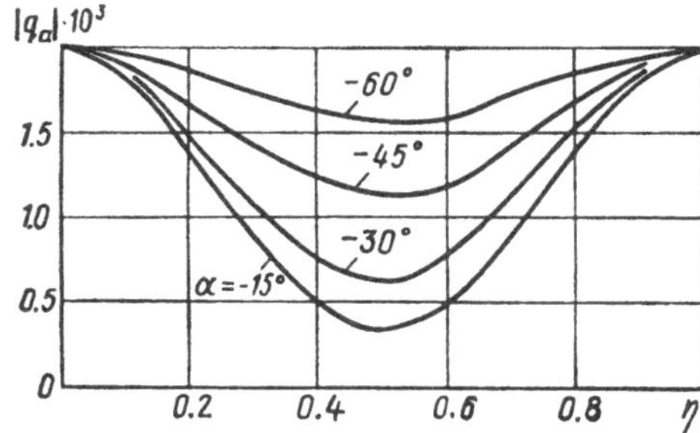

Fig. 6.16.

6.5 Stress-Strain State of a Rod Interacting with an Air Flow

First, let us consider a special case when deformation of a rod caused by aerodynamic forces is small (see Fig. 6.12), that is, we assume that the shape of the rod under loading differs little from its natural shape. In this case, the equations of the zeroth approximation (1.116)–(1.119) may be used. We have

$$
\begin{aligned}
&\frac{\mathrm{d}Q_1^{(0)}}{\mathrm{d}\eta} - æ_{30}Q_2^{(0)} = -q_1\,, \qquad æ_{30} = \pi\,;\\
&\frac{\mathrm{d}Q_2^{(0)}}{\mathrm{d}\eta} + æ_{30}Q_1^{(0)} = -q_2\,;\\
&\frac{\mathrm{d}Q_3^{(0)}}{\mathrm{d}\eta} = -q_3\,;\\
&\frac{\mathrm{d}M_1^{(0)}}{\mathrm{d}\eta} - æ_{30}M_2^{(0)} = 0\,;\\
&\frac{\mathrm{d}M_2^{(0)}}{\mathrm{d}\eta} + æ_{30}M_1^{(0)} - Q_3^{(0)} = 0\,;\\
&\frac{\mathrm{d}M_3^{(0)}}{\mathrm{d}\eta} + Q_2^{(0)} = 0\,; \qquad\qquad (6.59)\\
&\frac{\mathrm{d}\vartheta_1^{(0)}}{\mathrm{d}\eta} - æ_{30}\vartheta_2^{(0)} - \frac{M_1^{(0)}}{A_{11}} = 0\,;\\
&\frac{\mathrm{d}\vartheta_2^{(0)}}{\mathrm{d}\eta} + æ_{30}\vartheta_1^{(0)} - \frac{M_2^{(0)}}{A_{22}} = 0\,;\\
&\frac{\mathrm{d}\vartheta_3^{(0)}}{\mathrm{d}\eta} - \frac{M_3^{(0)}}{A_{33}} = 0\,;
\end{aligned}
$$

$$\frac{du_1^{(0)}}{d\eta} - æ_{30} u_2^{(0)} = 0\,;$$
$$\frac{du_2^{(0)}}{d\eta} + æ_{30} u_1^{(0)} - \vartheta_3^{(0)} = 0\,;$$
$$\frac{du_3^{(0)}}{d\eta} + \vartheta_2^{(0)} = 0\,. \tag{6.60}$$

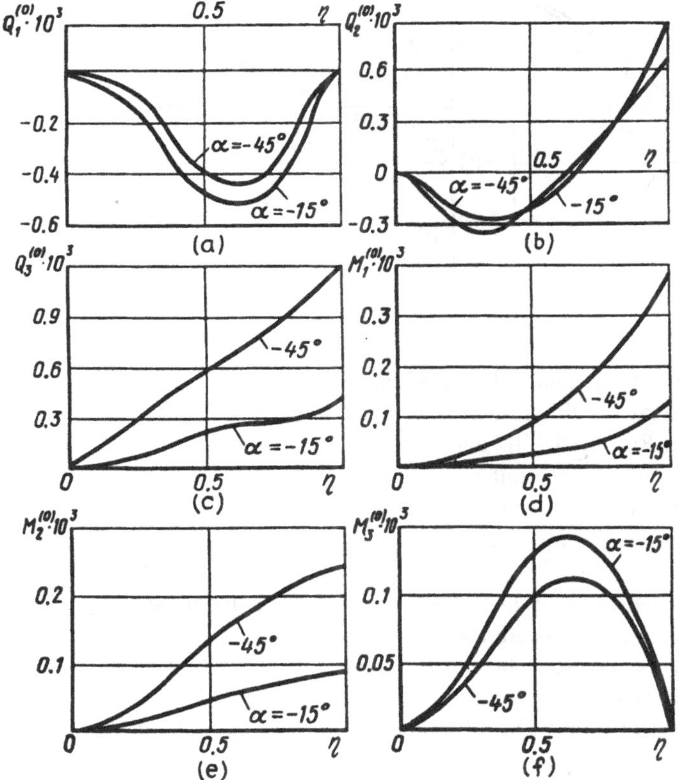

Fig. 6.17.

Equations (6.59) are to be integrated for the zero initial data. The plots of $Q_j^{(0)}$ and $M_j^{(0)}$ versus η for some values of the angle α are shown in Fig. 6.17*a-f*. Substituting a new variable $\eta_1 = 1 - \eta$ into (6.60), we integrate it for the zero initial data. The plots of $\vartheta_j^{(0)}$ and $u_j^{(0)}$ are illustrated in Fig. 6.18.

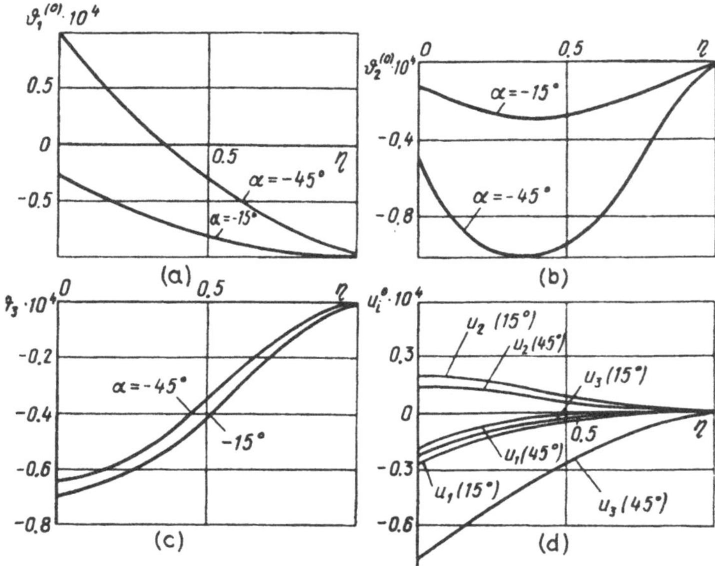

Fig. 6.18.

Consider now the most general case when the stress-strain state of a rod is governed by nonlinear equilibrium equations. A *rigid* hosepipe in a liquid or air flow is shown in Fig. 6.19. The aerodynamic forces acting on the hosepipe result in displacement of its axial points and rotation of the attached axes. Thus, the nonlinear equilibrium equations need be used.

Hosepipes designed for liquid transfer are widely used in engineering, e.g. various fueling systems (see Figs. 0.13 and 0.18), systems for lifting concretions from the sea bed (nodule pipes), systems of oil transfer (see Fig. 0.14). Traditionally, a hosepipe is considered as an absolutely flexible rod. Of course, this is just approximately so. The matter is that the bending and torsional stiffnesses of a real hosepipe are actually small but still different from zero. That is why the notion *rigid* hosepipe has appeared. Thus, from the mathematical point of view, there is no difference between a rod and a *rigid* hosepipe.

If the liquid is at rest, then the axial line of the hosepipe is a curve in the plane x_1Ox_2. The fluid flow results in the distributed aerodynamic forces acting on the hosepipe. These forces depend on the magnitude of the vector of flow velocity as well as on its direction. Assume that the flow is steady and vortex-free. This assumption is introduced for simplicity but it allows us to determine the stress-strain state of the hosepipe for large deflection of the rod axis.

The axial line of the loaded hosepipe is shown in Fig. 6.19 as a dot-and-dash line.

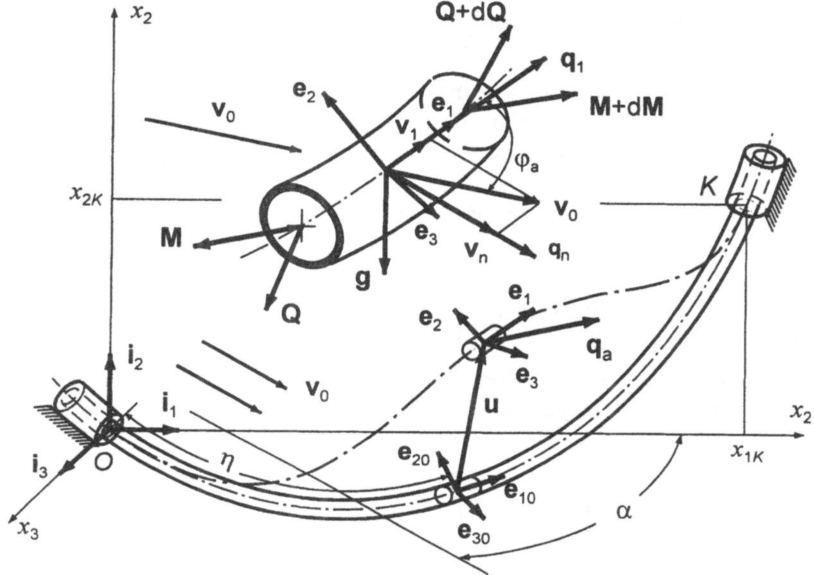

Fig. 6.19.

The nonlinear equations governing the stress-strain state of the hosepipe are as follows (see Sect. 1.3):

$$\frac{d\mathbf{Q}}{d\eta} + \text{æ} \times \mathbf{Q} + \mathbf{q}_1 + \mathbf{q}_n + \mathbf{g} = 0;$$

$$\frac{d\mathbf{M}}{d\eta} + \text{æ} \times \mathbf{M} + \mathbf{e}_1 \times \mathbf{Q} = 0;$$

$$\frac{d\boldsymbol{\vartheta}}{d\eta} + \mathrm{L}_1^{-1}\mathrm{L}_2\text{æ}_0^{(1)} - \mathrm{L}_1^{-1}\mathrm{A}^{-1}\mathbf{M} = 0;$$

$$\frac{d\mathbf{u}}{d\eta} + \text{æ} \times \mathbf{u} + (l_{11}^{(1)} - 1)\,\mathbf{e}_1 + l_{21}^{(1)}\mathbf{e}_2 + l_{31}^{(1)}\mathbf{e}_3 = 0;$$

$$\mathbf{M} = \mathrm{A}\,(\text{æ} - \text{æ}_0^{(1)})$$

or

$$\mathbf{Z}' + \mathbf{f}_1(\mathbf{Q}\,, \mathbf{M}\,, \boldsymbol{\vartheta}\,, \mathbf{u}) = \mathbf{f}_2(\mathbf{q}_1\,, \mathbf{q}_n\,, \mathbf{g})\,; \tag{6.61}$$

here $\mathbf{q}_1$ and $\mathbf{q}_n$ are the distributed aerodynamic forces and

$$\mathbf{g} = -\sum_{i=1}^{3} g\,(\mathbf{l}_2 \cdot \mathbf{e}_i)\,\mathbf{e}_i$$

is the distributed gravity force acting on the hosepipe.

As already proved, the components of the vectors $\mathbf{q}_1$ and $\mathbf{q}_n$ in the attached basis satisfy the relations (6.42), (6.43), and (6.45). The variables $l_{j1}^{(1)}$ in (6.61) are the elements of the matrix $\mathrm{L}^{(1)}$ (see (A.57)). Equations (6.61) are solved numerically by the method of step-by-step loading (see Sect. 2.3). Whatever numerical method is used, we must first define the initial approximation to the sought variables.

As an initial approximation of the state vector,

$$\mathbf{Z}^{(0)} = (\mathbf{Q}^{(0)}\,,\, \mathbf{M}^{(0)}\,,\, \boldsymbol{\vartheta}^{(0)}\,,\, \mathbf{u}^{(0)})^{\mathrm{T}}\,,$$

let us take the one that corresponds to an absolutely flexible rod subjected only to the gravity force ($\mathbf{v}_0 = 0$). As a result, we obtain the zeroth approximations $Q_1^{(0)}$ and $x_j^{(0)\prime}$ (recall that $\dot{x}_j^{(0)\prime}$, $j = 1\,,2\,,3$, appear in the expressions for the aerodynamic forces).

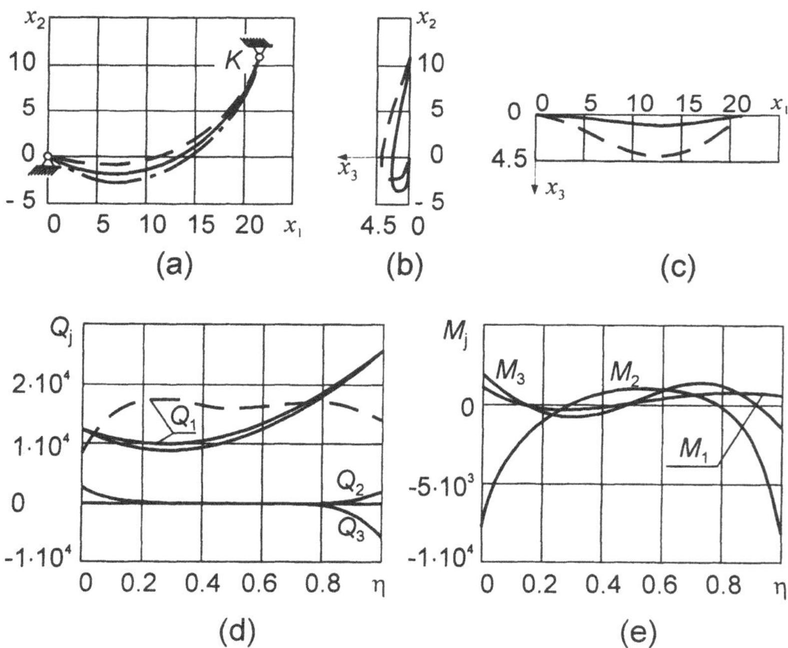

Fig. 6.20.

Now we can solve the nonlinear equations by the method of iterations. If $x_j^{(0)\prime}$, $j = 1\,,2\,,3$, are found, then, specifying the components of the vector of flow velocity $\mathbf{v}_0^{(j)}$, $j = 1\,,2\,,\ldots\,,n$, we can evaluate the corresponding aerodynamic forces $\mathbf{q}_1^{(0)}$ and $\mathbf{q}_n^{(0)}$. Thus, having found $v_0^{(j)}$, we obtain the first approximation of the state vector $\mathbf{Z}^{(1)}$ as follows:

$$\mathbf{Z}^{(1)'} + \mathbf{f}_1(\mathbf{Q}^{(1)}, \mathbf{M}^{(1)}, \boldsymbol{\vartheta}^{(1)}, \mathbf{u}^{(1)}) = \mathbf{f}_2(\mathbf{q}_1^{(0)}, \mathbf{q}_n^{(0)}, \mathbf{g}^{(0)}) . \tag{6.62}$$

Once $\vartheta_j^{(1)}$ and $u_j^{(1)}$ are found, we determine $x_j^{(1)}$, $\mathbf{q}_1^{(1)}$, $\mathbf{q}_n^{(1)}$, and $\mathbf{g}^{(1)}$ and, therefore, we obtain the second approximation

$$\mathbf{Z}^{(2)'} + \mathbf{f}_1(\mathbf{Q}^{(2)}, \mathbf{M}^{(2)}, \boldsymbol{\vartheta}^{(2)}, \mathbf{u}^{(2)}) = \mathbf{f}_2(\mathbf{q}_1^{(1)}, \mathbf{q}_n^{(1)}, \mathbf{g}^{(1)}) ,$$

and so on.

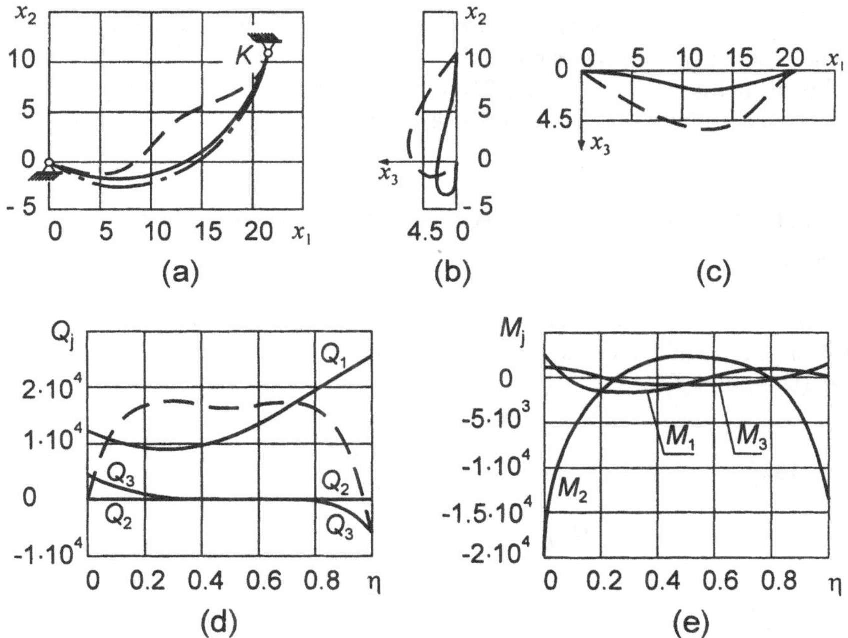

Fig. 6.21.

This example illustrates application of the explicit relations for the aerodynamic forces obtained in Sect. 6.4 for the case of arbitrary angles between the vectors $\mathbf{v}_0$ and $\mathbf{e}_1$. Nowadays there exist software packages for solution of two-point boundary-value problems. These packages may be useful in engineering design provided the explicit expressions for the aerodynamic forces are given.

The above problem was solved for the following data: $l = 30\,\text{m}$, $x_{1k} = 21\,\text{m}$, $x_{2k} = 12\,\text{m}$, $A_{11} = 7.36 \cdot 10^4\,\text{N} \cdot \text{m}^2$, and $A_{22} = A_{33} = 2.2 \cdot 10^5\,\text{N} \cdot \text{m}^2$. The flow velocity v_0 was taken $30\,\text{m} \cdot \text{s}^{-1}$ (air flow) and $3\,\text{m} \cdot \text{s}^{-1}$ (liquid flow). The solution was evaluated for the angles $\alpha = 45°$ and $\alpha = 90°$.

The numerical results are illustrated in Figs. 6.20 and 6.21. The shape of the rod axis for $v_0 = 0$ is shown in Figs. 6.20*a* and 6.21*a* as a dash-and-dot line. The projections of the axial line on the coordinate planes for $\alpha = 45°$ and $v_0 = 30\,\mathrm{m \cdot s^{-1}}$ (air flow) are illustrated in Fig. 6.20*a-c* as a solid line and for $v_0 = 3\,\mathrm{m \cdot s^{-1}}$ (liquid flow) as a dashed line.

It is seen from the plots that the hosepipe in the air flow does not deflect much out of the vertical plane (the projections of the maximum deflection does not exceed 5% of the hosepipe length). In the case of the liquid flow (for $v_0 = 3\,\mathrm{m \cdot s^{-1}}$), this deflection makes about 15%. The plots for the case of $\alpha = 90°$ are shown in Fig. 6.21*a-c*. In this case, the out-of-plane deflections are greater than those for $\alpha = 45°$. For the hosepipe in the liquid flow, the projections of the maximum deflections make 22.5% of the hosepipe length.

The plots of Q_j and M_j versus η for $\alpha = 45°$ and $\alpha = 90°$ ($v_0 = 30\,\mathrm{m \cdot s^{-1}}$) are shown in Figs. 6.20*d,e* and 6.21*d,e*. For the hosepipe in the liquid flow, the relationship between the axial force Q_1 and η are shown as dashed lines.

6.6 Aerodynamic Forces Acting on Rods of Noncircular Cross Section

An element of a rod of noncircular cross section is shown in Fig. 6.22. The center of gravity of the cross section (the point O) coincides with its stiffness center (the point O_1).

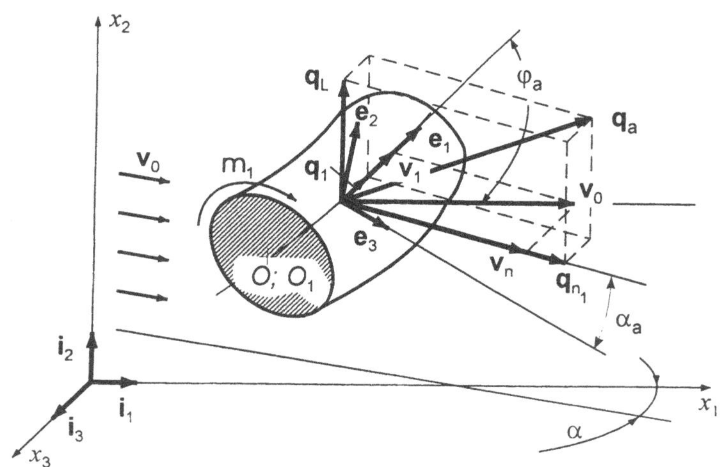

Fig. 6.22.

In general, these two points are different as indicated in Fig. 6.23.

A rod of circular cross section (Fig. 6.1) is subjected to the drag force $\mathbf{q}_{n_1}$ and the axial force $\mathbf{q}_1$. A rod of noncircular cross section (Fig. 6.22) is

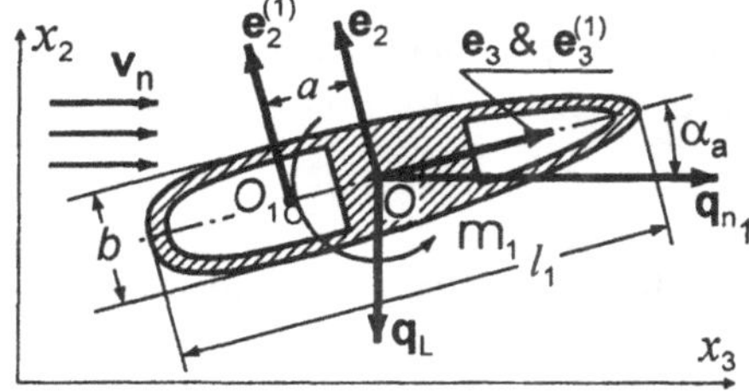

Fig. 6.23.

additionally subjected to the lifting force $\mathbf{q}_l$, which is orthogonal to the vector $\mathbf{q}_{n_1}$, and the aerodynamic moment μ_1 (see Sect. 6.3). According to the sign convention, the cross section in Fig. 6.23 is rotated counterclockwise by an angle $\alpha_a > 0$. Hence, the lifting force is directed downwards.

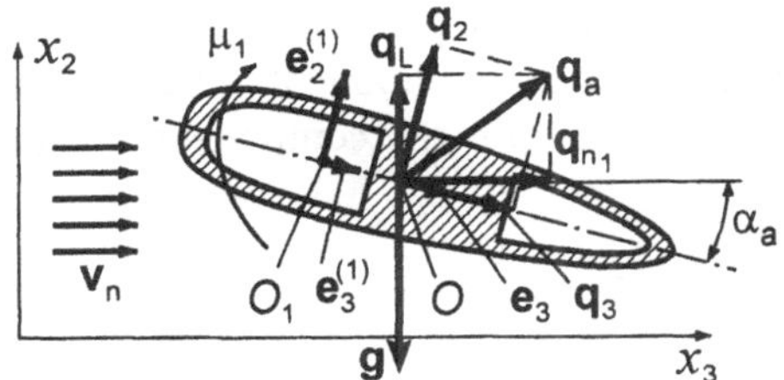

Fig. 6.24.

The aerodynamic forces acting on a rod are the drag force $\mathbf{q}_{n_1}$ and the lifting force $\mathbf{q}_l$. The vectors of these forces lie in the plane perpendicular to the rod axis. These forses depend on the normal component of velocity $\mathbf{v}_n$ (see Figs. 6.23 and 6.24). The vectors $\mathbf{q}_{n_1}$ and $\mathbf{v}_n$ are codirectional vectors. The vector $\mathbf{q}_l$ is perpendicular to the vector $\mathbf{q}_{n_1}$. Since the direction of the force $\mathbf{q}_l$ depends on the sign of the angle of attack α_a, we write $\mathbf{q}_l = |\mathbf{q}_l| \mathbf{e}_l \operatorname{sgn} \alpha_a$; here $\mathbf{e}_l$ is a unit vector collinear to the vector $\mathbf{q}_l$.

Experimental investigations show that the aerodynamic coefficients c_{n_1} and c_l depend on the angle of attack α_a. The nondimensional absolute values of the aerodynamic forces are as follows:

$$|\mathbf{q}_{n_1}| = \frac{c_{n_1}}{2} \frac{\rho v_n^2 b l^3}{A_{33}(0)}; \qquad |\mathbf{q}_l| = \frac{c_l}{2} \frac{\rho v_n^2 b l^3}{A_{33}(0)};$$

in view of the formula $v_n = v_0 \sin \varphi_a$, we have

$$g_{n_1} = |\mathbf{q}_{n_1}| = c_{n_1} g_{n_1}^{(1)} \sin^2 \varphi_a;$$
$$q_l = |\mathbf{q}_l| = c_l q_l^{(1)} \sin^2 \varphi_a, \tag{6.63}$$

where

$$\mathbf{q}_{n_1}^{(1)} = q_l^{(1)} = \frac{1}{2} \frac{\rho v_0^2 b l^3}{A_{33}(0)}.$$

When determining the forces $\mathbf{q}_{n_1}$ and $\mathbf{q}_l$ experimentally, we assume that these forces are applied to the center of gravity (the point O) of a cross section.

6.6.1 Components of $\mathbf{q}_{n_1}$ and $\mathbf{q}_l$ in the Cartesian Coordinate System

The drag force $\mathbf{q}_{n_1}$ satisfies the following relations:

$$\begin{aligned} &\mathbf{q}_{n_1} \cdot \mathbf{e}_1 = 0\,; \\ &\mathbf{q}_{n_1} \cdot (\mathbf{e}_v \times \mathbf{e}_1) = 0\,; \\ &|\mathbf{q}_{n_1}| = c_{n_1}(\alpha_a)\,\frac{\rho v_0^2 b l^3 \sin^2\varphi_a}{2A_{33}(0)}\,. \end{aligned} \tag{6.64}$$

These relations are identical to (6.34) and (6.35). Recall that we used (6.34) and (6.35) to derive the expressions for the components of $\mathbf{q}_n$ in the Cartesian basis. In view of (6.30)–(6.41), we obtain the components of $\mathbf{q}_{n_1}$ in this basis as follows:

$$\begin{aligned} q_{n_1 x_1} &= c_{n_1} q_{n_1}^{(1)} \sin\varphi_a(\cos\alpha - x_1' \cos\varphi_a)\,; \\ q_{n_1 x_2} &= -c_{n_1} q_{n_1}^{(1)} \sin\varphi_a x_1' \cos\varphi_a\,; \\ q_{n_1 x_3} &= c_{n_1} q_{n_1}^{(1)} \sin\varphi_a(\sin\alpha - x_3' \cos\varphi_a)\,. \end{aligned} \tag{6.65}$$

These relations are identical to those for $\mathbf{q}_n$, except that the aerodynamic coefficient c_n for rods of circular cross section does not depend on the angle of attack α_a, while the coefficient c_{n_1} depends on this angle. The lifting force $\mathbf{q}_l$ which acts on a rod of noncircular cross section is orthogonal to the drag force $\mathbf{q}_{n_1}$. Hence, the vector $\mathbf{q}_l$ must satisfy the following relations:

$$\mathbf{q}_l \cdot \mathbf{e}_1 = 0\,; \tag{6.66}$$

$$\mathbf{q}_l \cdot \mathbf{e}_n = 0\,; \tag{6.67}$$

$$|\mathbf{q}_l| = c_l q_l^{(1)} \sin^2\varphi_a\,, \qquad q_l^{(1)} = \frac{1}{2}\,\frac{\rho v_0^2 b l^3}{A_{33}(0)} \tag{6.68}$$

or

$$\begin{aligned} &q_{lx_1} x_1' + q_{lx_2} x_2' + q_{lx_3} x_3' = 0\,; \\ &q_{lx_1} e_{x_1} + q_{lx_2} e_{x_2} + q_{lx_3} e_{x_3} = 0\,; \\ &q_{lx_1}^2 + q_{lx_2}^2 + q_{lx_3}^2 = (c_l q_l \sin^2\varphi_a)^2\,. \end{aligned} \tag{6.69}$$

The unit vector $\mathbf{e}_n$ in (6.67) and the vector $\mathbf{v}_n$ are collinear. Let us now obtain the components e_{x_j} of the vector $\mathbf{e}_n$ in the Cartesian frame. We have

$$\mathbf{e}_n \cdot \mathbf{e}_1 = 0\,; \qquad \mathbf{e}_n \cdot (\mathbf{e}_n \times \mathbf{e}_1) = 0\,; \qquad |\mathbf{e}_n| = 1\,.$$

In a more expanded notation, we have the following equations in the three unknowns e_{x_j}:

$$e_{x_1}x_1' + e_{x_2}x_2' + e_{x_3}x_3' = 0\,;$$
$$-\sin\alpha x_2' e_{x_1} + (\sin\alpha x_1' - \cos\alpha x_3')\, e_{x_2} + \cos\alpha x_2' e_{x_3} = 0\,;$$
$$e_{x_1}^2 + e_{x_2}^2 + e_{x_3}^2 = 1\,. \tag{6.70}$$

Equations (6.70) are similar to (6.34)–(6.36), thus, on rearrangement, we get

$$e_{x_1} = (\cos\alpha - x_1'\cos\varphi_a)\,\frac{1}{\sin\varphi_a}\,;$$
$$e_{x_2} = -x_2'\cos\varphi_a\,\frac{1}{\sin\varphi_a}\,;$$
$$e_{x_3} = (\sin\alpha - x_3'\cos\varphi_a)\,\frac{1}{\sin\varphi_a}\,.$$

Solving (6.69) for q_{lx_j} and multiplying the result by $\operatorname{sgn}\alpha_a$, we arrive at

$$q_{lx_1} = -q_l^{(1)} c_l\,\Delta_1 \operatorname{sgn}\alpha_a\,;$$
$$q_{lx_2} = -q_l^{(1)} c_l\,\Delta_2 \operatorname{sgn}\alpha_a\,;$$
$$q_{lx_3} = -q_l^{(1)} c_l\,\Delta_3 \operatorname{sgn}\alpha_a\,, \tag{6.71}$$

where

$$\Delta_1 = e_{x_2}x_3' - e_{x_3}x_2'\,; \qquad \Delta_2 = e_{x_3}x_1' - e_{x_1}x_3'\,; \qquad \Delta_3 = e_{x_2}x_1' - e_{x_1}x_2'\,.$$

Substituting Δ_j and e_{x_j} into (6.71), we get

$$q_{lx_1} = q_l^{(1)} c_l \sin\varphi_\alpha \sin\alpha x_2' \operatorname{sgn}\alpha_a\,;$$
$$q_{lx_2} = q_l^{(1)} c_l \sin\varphi_\alpha(-\sin\alpha x_1' + \cos\alpha x_3') \operatorname{sgn}\alpha_a\,;$$
$$q_{lx_3} = -q_l^{(1)} c_l \sin\varphi_\alpha \cos\alpha x_2' \operatorname{sgn}\alpha_a\,. \tag{6.72}$$

The coefficient $c_l(\alpha_a)$ in (6.72) depends on the angle of attack and on the shape of a rod cross section. As noted above, the relationships between these coefficients and the angle of attack α_a can be obtained experimentally. The experimental data for the aerodynamic coefficients c_{n_1}, c_l, and c_m versus α_a are given in Kazakevich (1977). Knowing these relationships, we are able to solve the equilibrium equations numerically. If a rod deforms under the action of aerodynamic forces, then the angle of attack reads $\alpha_a = \alpha_{a0} + \alpha_{a1}$, where α_{a0} is the initial (given a priori) angle of attack and α_{a1} is an additional angle caused by deformation of the rod. The additional angle can be determined from the equilibrium equations. For small displacements of the axial points and small angles of rotation of the attached axes, the expression for the additional angle will be obtained later (see (6.89)).

The vectors $\mathbf{q}_{n_1}$ and $\mathbf{q}_\mathrm{l}$ must satisfy the orthogonality condition, that is, $\mathbf{q}_{n_1} \cdot \mathbf{q}_\mathrm{l} = 0$ or, in the expanded notation,

$$q_{n_1 x_1} q_{\mathrm{l} x_1} + q_{n_1 x_2} q_{\mathrm{l} x_2} + q_{n_1 x_3} q_{\mathrm{l} x_3} = 0\,.$$

From (6.65) and (6.72) we see that the orthogonality condition for the vectors $\mathbf{q}_{n_1}$ and $\mathbf{q}_\mathrm{l}$ holds.

To perform a numerical analysis of the equilibrium equations for rods of noncircular cross section, one must know the relationship between the aerodynamic coefficients c_{n_1} and c_l and the angle of attack α_a (see (6.65) and (6.72)). The sign of the angle α_a must also be known.

Let us obtain the explicit expressions for the angle α_a. For any location of the basis $\mathbf{e}_j$ relative to the vector $\mathbf{v}_0$, the angle α_a is the angle between the normal component of the velocity $\mathbf{v}_n$ and the base vector $\mathbf{e}_3$. Then,

$$\mathbf{v}_n = (\mathbf{v}_0 \cdot \mathbf{e}_2)\,\mathbf{e}_2 + (\mathbf{v}_0 \cdot \mathbf{e}_3)\,\mathbf{e}_3 = v_{n2}\mathbf{e}_2 + v_{n3}\mathbf{e}_3\,, \tag{6.73}$$

where

$$\mathbf{v}_0 = v_0(\cos\alpha \mathbf{i}_1 + \sin\alpha \mathbf{i}_3)\,.$$

Since

$$v_{n2} = |\mathbf{v}_n| \sin\alpha_a\,; \qquad v_{n3} = |\mathbf{v}_n| \cos\alpha_a\,,$$

we obtain

$$\sin\alpha_a = \frac{(\mathbf{v}_0 \cdot \mathbf{e}_2)}{|\mathbf{v}_n|}\,, \qquad \cos\alpha_a = \frac{(\mathbf{v}_0 \cdot \mathbf{e}_3)}{|\mathbf{v}_n|}\,.$$

where

$$|\mathbf{v}_n| = \sqrt{(\mathbf{v}_0\mathbf{e}_2)^2 + (\mathbf{v}_0\mathbf{e}_3)^2}\,.$$

The scalar products take the form

$$(\mathbf{v}_0 \cdot \mathbf{e}_2) = v_0(\cos\alpha l_{21}^{(1)} + \sin\alpha l_{23}^{(1)})\,;$$
$$(\mathbf{v}_0 \cdot \mathbf{e}_3) = v_0(\cos\alpha l_{31}^{(1)} + \sin\alpha l_{33}^{(1)})\,.$$

Therefore,

$$\sin\alpha_a = \frac{l_{21}^{(1)}\cos\alpha + l_{23}^{(1)}\sin\alpha}{\sqrt{\left(l_{21}^{(1)}\cos\alpha + l_{23}^{(1)}\sin\alpha\right)^2 + \left(l_{31}^{(1)}\cos\alpha + l_{23}^{(1)}\sin\alpha\right)^2}}\,;$$
$$\cos\alpha_a = \frac{l_{31}^{(1)}\cos\alpha + l_{33}^{(1)}\sin\alpha}{\sqrt{\left(l_{21}^{(1)}\cos\alpha + l_{23}^{(1)}\sin\alpha\right)^2 + \left(l_{31}^{(1)}\cos\alpha + l_{23}^{(1)}\sin\alpha\right)^2}}\,. \tag{6.74}$$

Using the properties of the matrix $\mathrm{L}^{(1)}$ (see (A.49)), we have

$$l_{11}^{(1)^2} + l_{21}^{(1)^2} + l_{31}^{(1)^2} = 1;$$
$$l_{13}^{(1)^2} + l_{23}^{(1)^2} + l_{33}^{(1)^2} = 1;$$
$$l_{11}^{(1)} l_{13}^{(1)} + l_{21}^{(1)} l_{23}^{(1)} + l_{31}^{(1)} l_{33}^{(1)} = 0.$$

The expression under the radical sign in (6.74) can be rearranged as follows:

$$\left(\cos\alpha l_{21}^{(1)} + \sin\alpha l_{23}^{(1)}\right)^2 + \left(\cos\alpha l_{31}^{(1)} + \sin\alpha l_{33}^{(1)}\right)^2 = 1 - \cos^2\varphi_a = \sin^2\varphi_a.$$

Finally, we get

$$\sin\alpha_a = \frac{l_{21}^{(1)}\cos\alpha + l_{23}^{(1)}\sin\alpha}{\sin\varphi_a}; \qquad \cos\alpha_a = \frac{l_{31}^{(1)}\cos\alpha + l_{33}^{(1)}\sin\alpha}{\sin\varphi_a}.$$

6.6.2 Components of $\mathbf{q}_{n_1}$ and $\mathbf{q}_l$ in the Attached Coordinate System

Recall that

$$\mathbf{q}_{n_1} = \sum_{j=1}^{3} q_{n_1 x_j}\mathbf{i}_j, \qquad \mathbf{q}_l = \sum_{j=1}^{3} q_{l x_j}\,\mathrm{sgn}\,(\alpha_a)\mathbf{i}_j;$$

after the transformation from the basis $\{\mathbf{i}_j\}$ to the basis $\{\mathbf{e}_j\}$, we get

$$\mathbf{q}_{n_1} = \sum_{K=1}^{3}\left(\sum_{j=1}^{3} q_{n_1 x_j} l_{K_j}^{(1)}\right)\mathbf{e}_K; \tag{6.75}$$

$$\mathbf{q}_l = \sum_{K=1}^{3}\left(\sum_{j=1}^{3} q_{l x_j}\,\mathrm{sgn}\,\alpha_a l_{K_j}^{(1)}\right)\mathbf{e}_K. \tag{6.76}$$

From (6.75) and (6.76) it follows that

$$q_{n_1 K} = \sum_{j=1}^{3} q_{n_1 x_j} l_{K_j}^{(1)}; \qquad q_{lK} = \sum_{j=1}^{3} q_{l x_j}\,\mathrm{sgn}\,\alpha_a l_{K_j}^{(1)}. \tag{6.77}$$

It can be shown that q_{n_1} and q_{l1} are equal to zero. The nonzero aerodynamic forces $\mathbf{q}_{n_1}$ and $\mathbf{q}_l$ in the attached frame are as follows:

$$q_{n_1 2} = -c_{n_1} q_{n_1}^{(1)}\sin\varphi_a(l_{21}^{(1)}\cos\alpha + l_{23}^{(1)}\sin\alpha);$$
$$q_{n_1 3} = c_{n_1} q_{n_1}^{(1)}\sin\varphi_a(l_{31}^{(1)}\cos\alpha + l_{33}^{(1)}\sin\alpha); \tag{6.78}$$
$$q_{l2} = -c_l q_l^{(1)}\sin\varphi_a(l_{31}^{(1)}\cos\alpha + l_{33}^{(1)}\sin\alpha)\,\mathrm{sgn}\,\alpha_a;$$
$$q_{l3} = -c_l q_l^{(1)}\sin\varphi_a(l_{21}^{(1)}\cos\alpha + l_{23}^{(1)}\sin\alpha)\,\mathrm{sgn}\,\alpha_a. \tag{6.79}$$

Let us consider a rod of noncircular symmetrical cross section with one axis of symmetry. In the attached frame, we have

$$\begin{aligned} q_1 &= q_{10}\cos^2\varphi_a \operatorname{sgn}(\cos\varphi_a);\\ q_2 &= q_{n_12} + q_{l2};\\ q_3 &= q_{n_13} + q_{l3};\\ \mu_1 &= \mu_{10}\sin^2\varphi_a \operatorname{sgn}\alpha_a, \qquad \mu_{10} = c_m\frac{\rho v_0^2 b l_1 l^2}{2A_{33}(0)}. \end{aligned} \tag{6.80}$$

Recall that the forces are assumed to be applied to the center of gravity of the cross section.

Up to now, the aerodynamic force have been considered as the sum

$$\mathbf{q}_a = \mathbf{q}_{n_1} + \mathbf{q}_l,$$

where $\mathbf{q}_{n_1}$ is the drag force and $\mathbf{q}_l$ is the lifting force. For a numerical analysis, it may be convenient to deal with the projections of the aerodynamic force on the principal axes of a cross section (it is especially so for axisymmetrical cross sections) (see Fig. 6.24). The aerodynamic coefficients c_2 and c_3 in the expression for the forces $\mathbf{q}_2$ and $\mathbf{q}_3$ are related to the coefficients c_{n_1} and c_l as follows:

$$\begin{aligned} c_2 &= c_l\cos\alpha_a + c_{n_1}\sin\alpha_a;\\ c_3 &= -c_l\sin\alpha_a + c_{n_1}\cos\alpha_a. \end{aligned} \tag{6.81}$$

Let us obtain the expressions for the aerodynamic forces and moments. The forces are assumed to be applied to the stiffness center O_1. As noted above, in experimental studies the aerodynamic forces and the moment are assumed to be reduced to the center of gravity (the point O in Fig. 6.23). The principal axes were used in the derivation of the equilibrium equations for rods of solid cross section, i.e. the forces were assumed to be applied to the center of gravity O. This point is also the stiffness center of the cross section (see Fig. 6.8).

A force that is directed along one of the principal axes results in a displacement of the point O in this direction. Likewise, a moment whose vector is directed along a principal axis of a cross section results in rotation of the cross section about this axis. For example, a force $\mathbf{P}$ that is collinear with the vector $\mathbf{e}_2$ as indicated in Fig. 6.24 results in a displacement of the center of gravity such that the displacement vector lies in the coordinate plane ($\mathbf{e}_2$, $\mathbf{e}_3$). Moreover, under the action of $\mathbf{P}$, the attached axes rotate relative to their original position. Now let a force $\mathbf{P}$ be directed along the base vector $\mathbf{e}_2^{(1)}$ referred to the center of stiffness O_1. In this case, the only effect produced by this force is a displacement along the vector $\mathbf{e}_2^{(1)}$. Suppose that a moment μ_1 is perpendicular to the plane of drawing and its vector passes through the point O. As a result, the cross section rotates remaining in the plane of drawing while the point O is at rest. Recall the relationship between the vector $\mathbf{M}$ and the curvatures (the vector æ) obtained on the assumption that each component of the vector $\mathbf{M}$ acts independently from the other components:

$$M_j = A_{jj}(æ_j - æ_{j0})\,;$$

here A_{jj} are nondimensional stiffness parameters related to the axes passing through the center of stiffness.

Now let the load be applied to the center of gravity; we have

$$\begin{aligned}
&\mathbf{q}_{n_1} = q_{n_1 2}\mathbf{e}_2^{(1)} + q_{n_1 3}\mathbf{e}_3^{(1)}\,;\\
&\mathbf{q}_{\mathrm{l}} = q_{\mathrm{l}2}\mathbf{e}_2^{(1)} - q_{\mathrm{l}3}\mathbf{e}_3^{(1)}\,;\\
&\mathbf{g} = -g_0\left(l_{12}^{(1)}\mathbf{e}_1^{(1)} + l_{22}^{(1)}\mathbf{e}_2^{(1)} + l_{32}^{(1)}\mathbf{e}_3^{(1)}\right),
\end{aligned} \tag{6.82}$$

where

$$g_0 = \frac{m_0 g l^3}{A_{33}(0)}$$

is the magnitude of the nondimensional gravity force. The twisting moment, which performs rotation about the axis $\mathbf{e}_1^{(1)}$, (the vector $\mathbf{e}_1^{(1)}$ is perpendicular to the plane of drawing, see Fig. 6.24) can be written as follows:

$$\boldsymbol{\mu}_{10_1} = \boldsymbol{\mu}_1 + \mathbf{a}\times(\mathbf{q}_{n_1} + \mathbf{q}_{\mathrm{l}} + \mathbf{g}) = \boldsymbol{\mu}_1 + \boldsymbol{\mu}_{1q_{n_1}} + \boldsymbol{\mu}_{q_{\mathrm{l}}} + \boldsymbol{\mu}_g\,; \tag{6.83}$$

here

$$\begin{aligned}
&\boldsymbol{\mu}_1 = \mu_{10}\sin^2\varphi_a\,\mathrm{sgn}\,\alpha_a\mathbf{e}_1^{(1)}\,;\\
&\boldsymbol{\mu}_{1q_{n_1}} = -aq_{n_1 2}\mathbf{e}_1^{(1)}\,;\\
&\boldsymbol{\mu}_{1q_{\mathrm{l}}} = -aq_{\mathrm{l}2}\,\mathrm{sgn}\,\alpha_a\mathbf{e}_1^{(1)}\,;\\
&\boldsymbol{\mu}_g = -ag_0 l_{22}^{(1)}\mathbf{e}_1^{(1)} + ag_0 l_{12}^{(1)}\mathbf{e}_2^{(1)}\,.
\end{aligned}$$

The explicit formulas for aerodynamic forces and moments derived in this section are valid for arbitrary deflections of the rod axis provided the stress-strain state of the rod is governed by the generalized Hooke's law.

6.7 Increments of Aerodynamic Forces at Small Displacements of Axial Points

In the preceding sections, the expressions for aerodynamic forces and moments valid for any displacements of axial points were obtained. The aerodynamic forces depend on the direction cosines of the vector $\mathbf{e}_1$. We have

$$x_j' = x_{j0}' + u_{x_j}',$$

where x'_{j0} are the direction cosines, which are characteristics of the shape of the axial line in the natural configuration, and u_{x_j} are finite changes of direction cosines caused by the aerodynamic forces.

Consider small displacements of the axial points. Let us establish the relationship between increments of the projections of the aerodynamic forces and small displacements of the axial points. Assuming that these increments are also small, we get

$$\begin{aligned} q_{1x_j} &= q_{1x_j0} + \Delta q_{1x_j}\,; \\ q_{nx_j} &= q_{nx_j0} + \Delta q_{nx_j}\,; \\ q_{lx_j} &= q_{lx_j0} + \Delta q_{lx_j}\,, \end{aligned} \tag{6.84}$$

where Δq_{1x_j}, Δq_{nx_j}, and Δq_{lx_j} are small quantities.

For a rod of circular cross section, we have $c_1 = \text{const}$, $c_n = \text{const}$, $c_l = 0$, and $c_m = 0$; consequently, the relations (6.39)–(6.41) and (6.46) give

$$\begin{aligned} \Delta q_{1x_j} &= \sum_{K=1}^{3} \left(\frac{\partial q_{1x_j}}{\partial \cos\varphi_a} \frac{\partial \cos\varphi_a}{\partial x'_K} + \frac{\partial q_{1x_j}}{\partial x'_K} \right)\Bigg|_{x'_K = x_{K0}} u'_{xK}\,; \\ \Delta q_{nx_j} &= \sum_{K=1}^{3} \left(\frac{\partial q_{nx_j}}{\partial \sin\varphi_a} \frac{\partial \sin\varphi_a}{\partial x'_K} + \frac{\partial q_{nx_j}}{\partial x'_K} \right)\Bigg|_{x'_K = x_{K0}} u'_{xK}\,. \end{aligned}$$

Let the vector $\mathbf{v}_0$ be parallel to the plane x_1Ox_3. We have

$$\begin{aligned} \Delta q_{1x_1} &= q_{10}^{(1)} \cos\varphi_{a0}(2x_{10}\cos\alpha + \cos\varphi_{a0}) \cdot u'_{x_1} \\ &\quad + q_{10}^{(1)} 2\cos\varphi_{a0} \sin\alpha x'_{10} u'_{x3}\,; \\ \Delta q_{1x_2} &= 2q_{10}^{(1)} \cos\varphi_{a0}\cos\alpha x'_{20} u'_x + q_{10}^{(1)} \cos^2\varphi_{a0} u'_{x2} \\ &\quad + 2q_{10}^{(1)} \cos\varphi_{a0} \sin\alpha x'_{20} u'_{x3}\,; \\ \Delta q_{1x3} &= 2q_{10}^{(1)} \cos\varphi_{a0}\cos\alpha x'_{30} u'_{x1} \\ &\quad + q_{10}^{(1)} \cos\varphi_{a0}(\cos\varphi_{a0} + 2\sin\alpha x'_{30})\, u'_{x3}\,; \end{aligned} \tag{6.85}$$

$$\begin{aligned} \Delta q_{nx_1} &= q_{n0} \left(\frac{x'_{10}\cos\alpha\cos 2\varphi_{a0} - \cos\varphi_{a0}\cos^2\alpha}{\sin\varphi_{a0}} - \frac{\sin 2\varphi_{a0}}{2} \right) u'_{x1} \\ &\quad + q_{n0}\sin\alpha \left(\frac{x'_{10}\cos 2\varphi_{a0} - \cos\varphi_{a0}\cos\alpha}{\sin\varphi_{a0}} \right) u'_{x3}\,; \\ \Delta q_{nx_2} &= q_{n0} x'_{20} \cos\alpha \frac{\cos 2\varphi_{a0}}{\sin\varphi_{a0}} u'_{x1} - q_{n0}\sin\varphi_{a0}\cos\varphi_{a0} u'_{x2} \\ &\quad + q_{n0} x'_{20} \sin\alpha \frac{\cos 2\varphi_{a0}}{\sin\varphi_{a0}} u'_{x3}\,; \\ \Delta q_{nx_3} &= q_{n0} \frac{\cos\alpha}{\sin\varphi_{a0}} (x'_{30}\cos 2\varphi_{a0} - \sin\varphi_{a0}\cos\varphi_{a0})\, u'_{x1} \\ &\quad + q_{n0} \left[\frac{\sin\alpha(x'_{30}\cos 2\varphi_{a0} - \sin\alpha\cos\varphi_{a0})}{\sin\varphi_{a0}} - \frac{\sin 2\varphi_{a0}}{2} \right] u'_{x3}\,, \end{aligned} \tag{6.86}$$

where

$$\cos \varphi_{a0} = x'_{10} \cos \alpha + x'_{30} \sin \alpha \, .$$

Equations (6.85) and (6.86) can be represented in vector form as follows:

$$\Delta \mathbf{q}_{1x} = \mathrm{A}^{(1)} \mathbf{u}'_x \, ; \qquad \Delta \mathbf{q}_{nx} = \mathrm{A}^{(2)} \mathbf{u}'_x \, ; \tag{6.87}$$

here the elements of the matrices $\mathrm{A}^{(1)}$ and $\mathrm{A}^{(2)}$ can be easily determined from (6.85) and (6.86).

6.7.1 Rods of Noncircular Cross Section

In the case of rods of noncircular symmetrical cross section, the coefficients c_{n_1}, c_l, and c_m depend on the angle of attack α_a. If the angle of attack varies little, we get

$$c_{n_1} = c_{n_1 0} + \frac{\partial c_{n_1}}{\partial \alpha_a} \Delta \alpha_a \, ;$$
$$c_l = c_{l0} + \frac{\partial c_l}{\partial \alpha_a} \Delta \alpha_a \, ;$$
$$c_m = c_{m0} + \frac{\partial c_m}{\partial \alpha_a} \Delta \alpha_a \, .$$

Let us obtain the relationship between $\Delta \alpha_a$ and small angles of rotation of the basis $\{\mathbf{e}_j\}$ relative to the basis $\{\mathbf{e}_{j0}\}$. The elements $l_{21}^{(1)}, l_{23}^{(1)}, l_{31}^{(1)}$, and $l_{33}^{(1)}$ of the matrix $\mathrm{L}^{(1)}$ in (6.79) are defined by (A.57). Recall that $\mathrm{L}^{(1)} = \mathrm{L}\mathrm{L}^{(0)}$; here $\mathrm{L}^{(0)}$ is the matrix of transformation from the basis $\{\mathbf{i}_j\}$ to the basis $\{\mathbf{e}_{j0}\}$ (see (A.55)), L is the matrix of transformation from the basis $\{\mathbf{e}_{j0}\}$ to the basis $\{\mathbf{e}_j\}$, and the basis $\{\mathbf{e}_{j0}\}$ is associated with the natural configuration of the rod. The elements $l_{ij}^{(0)}$ of the matrix $\mathrm{L}^{(0)}$ are assumed to be known. For small angles of rotation of the attached axes, the matrix L is as follows (see (A.46)):

$$\mathrm{L} = \begin{bmatrix} 1 & \vartheta_3 & -\vartheta_2 \\ -\vartheta_3 & 1 & \vartheta_1 \\ \vartheta_2 & -\vartheta_1 & 1 \end{bmatrix} .$$

Hence, we have

$$l_{21}^{(1)} = l_{21}^{(0)} + l_{31}^{(0)} \vartheta_1 - l_{11}^{(0)} \vartheta_3 \, ;$$
$$l_{23}^{(1)} = l_{23}^{(0)} + l_{33}^{(0)} \vartheta_1 - l_{13}^{(0)} \vartheta_3 \, ;$$
$$l_{31}^{(1)} = l_{31}^{(0)} - l_{21}^{(0)} \vartheta_1 + l_{11}^{(0)} \vartheta_2 \, ;$$
$$l_{33}^{(1)} = l_{33}^{(0)} - l_{23}^{(0)} \vartheta_1 + l_{12}^{(0)} \vartheta_2 \, . \tag{6.88}$$

Expanding $\sin \alpha_a$ defined by (6.79) into a series and using (6.88), we get (only linear terms are retained)

$$\sin\alpha_a = \sin\alpha_{a0} + c_1\vartheta_1 + c_2\vartheta_2 + c_3\vartheta_3 ,$$

where

$$c_1 = \frac{a_1}{D} - \frac{a_0(a_1-b_1)}{D^2}; \quad c_2 = -\frac{a_0 b_2}{D^2}; \quad c_3 = \frac{a_0 a_2}{D}, \quad D = \sqrt{a_0^2+b_0^2};$$
$$a_0 = l_{21}^{(0)}\cos\alpha + l_{23}^{(0)}\sin\alpha; \qquad b_0 = l_{31}^{(0)}\cos\alpha + l_{33}^{(0)}\sin\alpha;$$
$$a_1 = l_{31}^{(0)}\cos\alpha + l_{33}^{(0)}\sin\alpha; \qquad b_1 = l_{21}^{(0)}\cos\alpha + l_{23}^{(0)}\sin\alpha;$$
$$a_2 = l_{11}^{(0)}\cos\alpha + l_{13}^{(0)}\sin\alpha; \qquad b_2 = l_{11}^{(0)}\cos\alpha + l_{12}^{(0)}\sin\alpha .$$

Since $\sin\alpha_a = \sin\alpha_{a0} + \cos\alpha_{a0}\,\Delta\alpha_a$, we get

$$\Delta\alpha_a = \frac{c_1\vartheta_1 + c_2\vartheta_2 + c_3\vartheta_3}{\cos\alpha_{a0}}, \qquad \cos\alpha_{a0} = \frac{b_0}{D}. \tag{6.89}$$

The increments of the projections of the aerodynamic forces $q_{n_j x_j}$ and q_{1x_j} are related to the increments of the angle of attack $\Delta\alpha_a$ as follows:

$$\Delta q_{n_1 x_j} = \frac{\partial q_{n_1 x_j}}{\partial \alpha_a}\,\Delta\alpha_a + \sum_{k=1}^{3} \frac{\partial q_{n_1 x_j}}{\partial x'_k}\, u'_{x_k};$$
$$\Delta q_{1 x_j} = \frac{\partial q_{1 x_j}}{\partial \alpha_a}\,\Delta\alpha_a + \sum_{k=1}^{3} \frac{\partial q_{1 x_j}}{\partial x'_k}\, u'_{x_k}.$$

On rearrangement, we have

$$\begin{aligned}
\Delta q_{n_1 x_1} &= q_{n_1 00}\,\frac{\partial c_{n_1}}{\partial \alpha_0}\,(c_1\vartheta_1 + c_2\vartheta_2 + c_3\vartheta_3)\,\frac{\sin\varphi_{a0}}{\cos\varphi_{a0}}\,(\cos\alpha - x'_{10}\cos\varphi_{a0})\\
&+ q_{n_1 0}\left[\left(\frac{x'_{10}\cos\alpha\cos 2\varphi_{a0} - \cos\varphi_{a0}\cos^2\alpha}{\sin\varphi_{a0}} - \frac{\sin 2\varphi_{a0}}{2}\right) u'_{x_1}\right.\\
&\left. + \sin\alpha\,\frac{x'_{10}\cos 2\varphi_{a0} - \cos\varphi_{a0}\cos\alpha}{\sin\varphi_{a0}}\, u'_{x_3}\right];\\
\Delta q_{n_1 x_2} &= -q_{n_1 00}\,\frac{\partial c_{n_1}}{\partial \alpha_a}\,\frac{c_1\vartheta_1 + c_2\vartheta_2 + c_3\vartheta_3}{\cos\varphi_{a0}}\,\sin\varphi_{a0}\,\cos\varphi_{a0}\,x'_{20}\\
&+ q_{n_1 0}\left(x'_{20}\cos\alpha\,\frac{\cos 2\varphi_{a0}}{\sin\varphi_{a0}}\,u'_{x_1} - \sin\varphi_{a0}\cos\varphi_{a0} u'_{x_2}\right.\\
&\left. + x'_{20}\sin\alpha\,\frac{\cos 2\varphi_{a0}}{\sin\varphi_{a0}}\,u'_{x_3}\right);\\
\Delta q_{n_1 x_3} &= q_{n_1 00}\,\frac{\partial c_{n_1}}{\partial \alpha_a}\,\frac{c_1\vartheta_1 + c_2\vartheta_2 + c_3\vartheta_3}{\cos\varphi_{a0}}\,\sin\varphi_{a0}\,(\sin\alpha - x'_{30}\cos\varphi_{a0})\\
&+ q_{n_1 0}\left[\frac{\cos\alpha}{\sin\varphi_{a0}}\,(x'_{30}\cos 2\varphi_{a0} - \sin\varphi_{a0}\cos\varphi_{a0})\,u'_{x_1}\right.\\
&\left. + \left(\frac{\sin\alpha(x'_{30}\cos 2\varphi_{a0} - \sin\alpha\cos\varphi_{a0})}{\sin\varphi_{a0}} - \frac{\sin 2\varphi_{a0}}{2}\right) u'_{x_3}\right];
\end{aligned}$$

$$(6.90)$$

$$\begin{aligned}
\Delta q_{lx_1} &= q_l^0 \frac{\partial c_l}{\partial \alpha_a} \frac{c_1\vartheta_1 + c_2\vartheta_2 + c_3\vartheta_3}{\cos\varphi_{a0}} \sin\varphi_{a0} \sin\alpha x'_{20} \\
&+ q_l^0 c_{l0}^0 \sin\varphi_{a0} \sin\alpha u'_{x_2} \\
&- q_l^0 c_{l0}^0 \sin\alpha (x'_{10}\cos^2\alpha + \sin\alpha\cos\alpha x'_{30})\, u'_{x_1} \\
&- q_l^0 c_{l0}^0 \sin\alpha x'_{20} (x'_{30}\sin^2\alpha + \sin\alpha\cos\alpha x'_{10})\, u'_{x_3}\,; \\
\Delta q_{lx_2} &= q_l^0 \frac{\partial c_l}{\partial \alpha_a} \frac{c_1\vartheta_1 + c_2\vartheta_2 + c_3\vartheta_3}{\cos\varphi_{a0}} \sin\varphi_{a0} (\sin\alpha x'_{10} - \cos\alpha x'_{30}) \\
&+ q_l^0 c_{l0}^0 [\, (-\sin\alpha x'_{10} A_1 + \cos\alpha x'_{30} A_1 + \sin\alpha\sin\varphi_{a0})\, u'_{x_1} \\
&+ (-\sin\alpha x'_{10} A_2 + \cos\alpha x'_{30} A_2 - \cos\alpha\sin\varphi'_{a0})\, u'_{x_3}\,]\,; \\
\Delta q_{lx_3} &= q_l^0 \frac{\partial c_l}{\partial \alpha_a} \frac{c_1\vartheta_1 + c_2\vartheta_2 + c_3\vartheta_3}{\cos\varphi_{a0}} \sin\varphi_{a0} \cos\alpha x'_{20} \\
&+ q_l^0 c_{l0}^0 (\sin\varphi_{a0}\cos\alpha u'_{x_2} - A_1 x'_{20}\cos\alpha u'_{x_1} - A_2 x'_{20}\cos\alpha u'_{x_3})\,, \qquad (6.91)
\end{aligned}$$

where

$$q_{n_1 00} = \frac{1}{2}\rho d^{(1)} v_0^2\,; \qquad q_{n_1 0} = q_{n_1 00} c_{n_1 0}\,; \qquad c_{l0}^0 = c_{l0}\,\mathrm{sgn}\,\alpha_{a0}\,;$$
$$A_1 = x'_{10}\cos^2\alpha + \sin\alpha\cos\alpha x'_{30}\,; \qquad A_2 = x'_{30}\sin^2\alpha + \sin\alpha\cos\alpha x'_{10}\,.$$

In vector notation, we have

$$\begin{aligned}
\Delta \mathbf{q}_{n_1} &= \mathrm{A}^{(3)}\boldsymbol{\vartheta} + \mathrm{A}^{(2)}\mathbf{u}'_x\,; \\
\Delta \mathbf{q}_l &= \mathrm{A}^{(4)}\boldsymbol{\vartheta} + \mathrm{A}^{(5)}\mathbf{u}'_x\,.
\end{aligned} \qquad (6.92)$$

The increment of the moment $\Delta\mu_1$ can be expressed as follows:

$$\Delta\mu_1 = \left(\mu_{10}\frac{\partial c_m}{\partial \alpha_a}\frac{1}{c_{m_0}}\Delta\alpha_a \sin^2\varphi_a\right)\mathbf{e}_1\,; \qquad (6.93)$$

here

$$\mu_{10} = c_{m_0}\frac{\rho v_0^2 b l_1 l^2}{2A_{33}(0)}\,\mathrm{sgn}\,\alpha_a\,.$$

For small u_j and ϑ_j, the stress-strain state of a rod interacting with an air or liquid flow may be determined by use of the expressions for increments of the aerodynamic forces (6.90) and (6.92) and for an increment of the moment (6.93).

6.8 Rods Containing Internal Liquid Flows

There exists a great number of publications devoted to the problem of statics of rods containing an internal liquid flow. The extensive list of reference on the subject is available in the monograph Svetlitsky (1982).

In most publications, only rods of circular cross section (pipelines) are treated (see Fig. 6.25*a*). However, real constructions include hollow rods of noncircular cross section (see Fig. 6.25*b*-*d*) when the cross-sectional bending stiffnesses are different. For example, a tool for deep-hole drilling has a cross section shown in Fig. 6.25*c*. Such a shape of the cross section improves chip removal and provides efficient cooling of the tool during drilling.

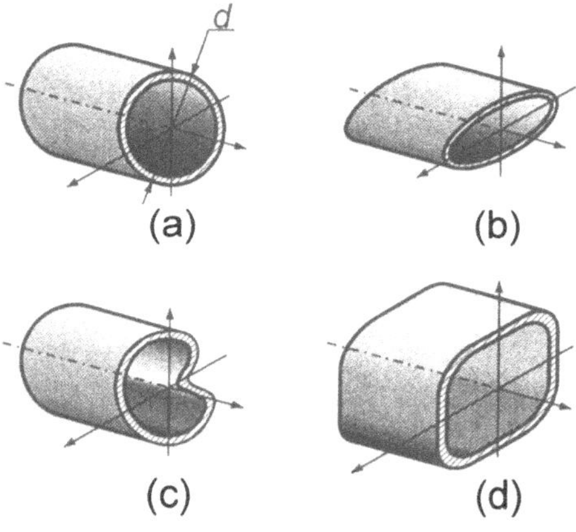

Fig. 6.25.

A liquid flow may result in additional forces applied to the rod. In the case of straight rods, these forces may lead to loss of static stability.

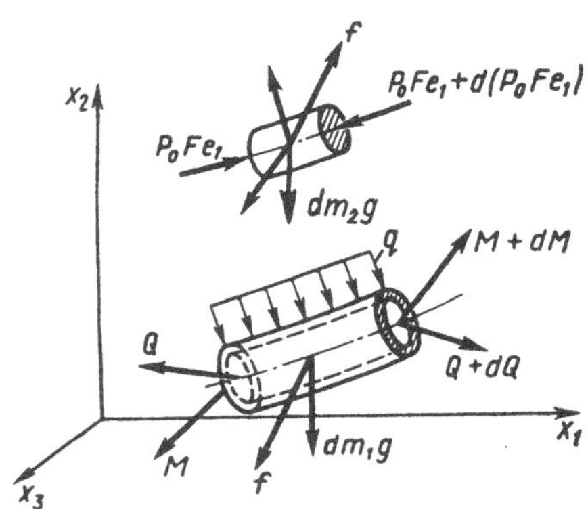

Fig. 6.26.

An element of a rod and an elementary volume of the liquid enclosed into this element as well as the forces applied to the rod and the liquid are shown in Fig. 6.26*a*,*b*. The elementary volume of the liquid moves with velocity $\mathbf{w}_0 = w_0 \mathbf{e}_1$, where w_0 is the velocity of liquid particles averaged over the cross section. According to the d'Alembert principle, we have

$$-\frac{\partial (P_0 \mathbf{e}_1)}{\partial s} - \mathbf{f} + m_2 g - m_2 \frac{\mathrm{d}\mathbf{w}_0}{\mathrm{d}t} = 0\,, \qquad P_0 = p_0 F\,, \tag{6.94}$$

where F is the cross-sectional area of the rod element, p_0 is the liquid pressure, m_2 is the mass of liquid enclosed into a rod element of unit length, and $\mathbf{f}$ is the distributed force of interaction between the liquid and the walls of the rod. In the case of ideal liquid, the vector $\mathbf{f}$ lies in the plane defined by the vectors $\mathbf{e}_2$ and $\mathbf{e}_3$, i.e. $\mathbf{f}$ is orthogonal to the vector $\mathbf{e}_1$. (This fact is valid both for statics and dynamics.) The flow velocity $\mathbf{w}$ is a function of t and s. Thus, using the Eulerian variables, we get the total derivative

$$\frac{\mathrm{d}\mathbf{w}_0}{\mathrm{d}t} = \frac{\partial \mathbf{w}_0}{\partial t} + \frac{\partial \mathbf{w}_0}{\partial s} w_0\,. \tag{6.95}$$

If the rod is in equilibrium and the absolute value of the velocity $\mathbf{w}_0$ remains unchanged (the motion of the liquid is steady), then $\partial \mathbf{w}_0 / \partial t = 0$. For $F = \text{const}$, from (6.109) it follows that

$$\frac{\mathrm{d}\mathbf{w}_0}{\mathrm{d}t} = w_0^2 \frac{\partial \mathbf{e}_1)}{\partial s} = w_0^2 æ_{30} \mathbf{e}_2 - w_0^2 æ_{20} \mathbf{e}_3\,. \tag{6.96}$$

Taking into consideration the force of gravity acting both on the rod and the liquid, we can represent the equilibrium equation as follows:

$$\frac{\mathrm{d}\mathbf{Q}}{\mathrm{d}s} + \mathbf{f} + m_1 g + \mathbf{q} + \sum_{i=1}^{n} \mathbf{P}^{(i)} \delta(s - s_i) = 0\,. \tag{6.97}$$

Eliminating the vector $\mathbf{f}$ from (6.94) and (6.97), we arrive at

$$\frac{\mathrm{d}\mathbf{Q}}{\mathrm{d}s} - \frac{\mathrm{d}}{\mathrm{d}s} [\,(P_0 + m_2 w_0^2)\,\mathbf{e}_1\,] + \gamma + \mathbf{q} + \sum_{i=1}^{n} \mathbf{P}^{(i)} \delta(s - s_i) = 0 \tag{6.98}$$

or

$$\frac{\mathrm{d}\mathbf{Q}^{(1)}}{\mathrm{d}s} + \gamma + \mathbf{q} + \sum_{i=1}^{n} \mathbf{P}^{(i)} \delta(s - s_i) = 0\,, \qquad \gamma = (m_1 + m_2)\,\mathbf{g}\,, \tag{6.99}$$

where

$$\mathbf{Q}^{(1)} = (Q_1 - P_0 - m_2 w_0^2)\,\mathbf{e}_1 + Q_2 \mathbf{e}_2 + Q_3 \mathbf{e}_3\,.$$

Summing the moments acting on the element, we get

$$\frac{\mathrm{d}\mathbf{M}}{\mathrm{d}s} + (\mathbf{e}_1 \times \mathbf{Q}) + \boldsymbol{\mu} + \sum_{\nu=1}^{\rho} \mathbf{T}^{(\nu)}\, \delta\,(s - s_\nu) = 0\,; \tag{6.100}$$

this equation is identical in form to (1.9). Since $\mathbf{e}_1 \times \mathbf{Q} = \mathbf{e}_1 \times \mathbf{Q}^{(1)}$, we can rewrite (6.100) as follows:

$$\frac{\mathrm{d}\mathbf{M}}{\mathrm{d}s} + (\mathbf{e}_1 \times \mathbf{Q}^{(1)}) + \boldsymbol{\mu} + \sum_{\nu=1}^{\rho} \mathbf{T}^{(\nu)}\, \delta\,(s - s_\nu) = 0\,; \tag{6.101}$$

here $\boldsymbol{\mu}$ is a distributed moment applied to the rod, say, the distributed aerodynamic moment.

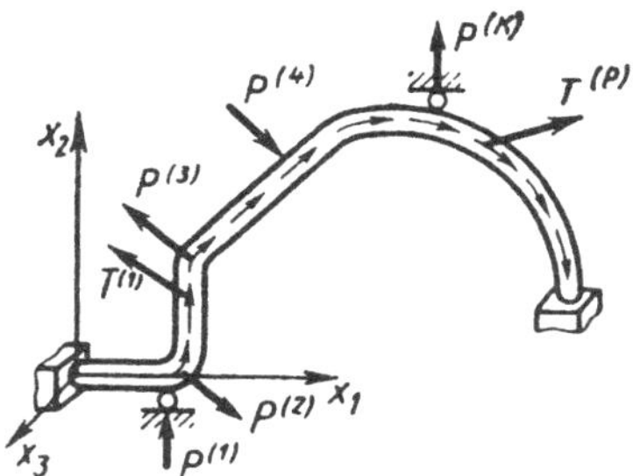

Fig. 6.27.

Concentrated forces $\mathbf{P}^{(i)}$ applied to a rod may be divided into two groups: forces produced by the liquid flow and other external forces. This is a key feature of problems of statics and dynamics of a rod that contains a liquid flow. One such a rod is shown in Fig. 6.27.

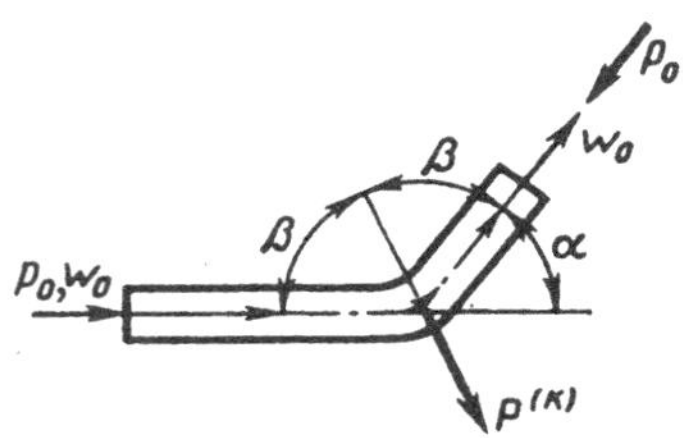

Fig. 6.28.

At the points where two tubes are joined together such that one tube makes an angle 2β with the other, concentrated forces $\mathbf{P}^{(2)}$ and $\mathbf{P}^{(3)}$ appear (see Fig. 6.28). Using the principle of balance of moment of momentum for fluids, we get the following expression for the absolute value of the concentrated force (see Fig. 6.28):

$$|\mathbf{P}^{(c)}| = 2P_0 \cos\beta + 2m_2 w_0^2 \cos\beta = 2\,(P_0 + m_2 w_0^2)\sin\frac{\alpha}{2}\,;$$

here P_0 is the pressure in the liquid.

The direction of the force $\mathbf{P}^{(c)}$ is shown in Fig. 6.28.

The displacement vector $\mathbf{u}$ and the vectors æ and $\boldsymbol{\vartheta}$ satisfy the relations (1.61) and (1.60), respectively. To obtain nondimensional equilibrium equations, we should put (see Sect. 1.1)

$$\eta = \frac{s}{l}\,; \qquad w_0 = \tilde{w}_0 l p_0\,; \qquad \mathbf{Q} = \frac{\tilde{\mathbf{Q}} A_{33}(0)}{l^2}\,; \qquad æ = \tilde{æ} l\,;$$

$$\mathbf{M} = \frac{\tilde{\mathbf{M}} A_{33}(0)}{l}\,; \qquad \mathbf{q} = \frac{\tilde{\mathbf{q}} A_{33}(0)}{l^3}\,; \qquad \boldsymbol{\mu} = \frac{\tilde{\boldsymbol{\mu}} A_{33}(0)}{l^2}\,;$$

$$\boldsymbol{\gamma} = \frac{\tilde{\boldsymbol{\gamma}} A_{33}(0)}{l^3}\,; \qquad A_{ii} = \frac{\tilde{A}_{ii}}{A_{33}(0)}\,; \qquad P_0 = \frac{\tilde{P}_0 A_{33}(0)}{l^2}\,;$$

$$\mathbf{P}^{(i)} = \frac{\tilde{\mathbf{P}}^{(i)} A_{33}(0)}{l^2}\,; \qquad \mathbf{T}^{(\nu)} = \frac{\tilde{\mathbf{T}}^{(\nu)} A_{33}(0)}{l}\,; \qquad p_0 = \sqrt{\frac{A_{33}(0)}{(m_1+m_2) l^4}}\,.$$

On rearrangement, the following nondimensional equilibrium equations for a rod of constant cross section interacting with an internal flow can be obtained (as before, a tilde is dropped out in the notation of nondimensional quantities):

$$\frac{d\mathbf{Q}^{(1)}}{d\eta} + \boldsymbol{\gamma} + \mathbf{q} + \sum_{i=1}^{n} \mathbf{P}^{(i)}\,\delta(\eta - \eta_i) = 0\,; \tag{6.102}$$

$$\frac{d\mathbf{M}}{d\eta} + (\mathbf{e}_1 \times \mathbf{Q}^{(1)}) + \boldsymbol{\mu} + \sum_{\nu=1}^{\rho} \mathbf{T}^{(\nu)}\,\delta(\eta - \eta_\nu) = 0\,; \tag{6.103}$$

$$\mathbf{M} = \mathrm{A}\,(æ - æ_0^{(1)})\,; \tag{6.104}$$

$$\mathrm{L}_1 \frac{d\boldsymbol{\vartheta}}{d\eta} + \mathrm{L}æ_0^{(1)} - æ = 0\,; \tag{6.105}$$

$$\frac{d\mathbf{u}}{d\eta} + (l_{11} - 1)\,\mathbf{e}_1 + l_{21}\mathbf{e}_2 + l_{31}\mathbf{e}_3 = 0\,; \tag{6.106}$$

here

$$\mathbf{Q}^{(1)} = [\,Q_1 - (P_0 + n_1 w_0^2)\,]\,\mathbf{e}_1 + Q_2\mathbf{e}_2 + Q_3\mathbf{e}_3\,, \qquad n_1 = \frac{m_2}{m_1+m_2}\,. \tag{6.107}$$

Equations (6.102)–(6.106) differ from (1.57)–(1.61) in that (6.102)–(6.107) contain the vector $\mathbf{Q}^{(1)}$ and the concentrated forces $\mathbf{P}^{(i)}$. The first component of the vector $\mathbf{Q}^{(1)}$ is a function of the flow velocity $\mathbf{w}_0$ and the pressure p_0 (see (6.107)). Owing to similarity of these systems of equations, we see that the problem under consideration can be solved by the numerical methods discussed in Chap. 2; however, the boundary conditions must be satisfied by the component $Q_1^{(1)} = Q_1 - (P_0 + n_1 w_0^2)$ instead of the component Q_1.

In the attached coordinate system, the equilibrium equations take the form

$$\frac{d\mathbf{Q}^{(1)}}{d\eta} + æ \times \mathbf{Q}^{(1)} + \boldsymbol{\gamma} + \mathbf{q} + \sum_{i=1}^{n} \mathbf{P}^{(i)}\, \delta(\eta - \eta_i) = 0\,; \tag{6.108}$$

$$\frac{d\mathbf{M}}{d\eta} + æ \times \mathbf{M} + \mathbf{e}_1 \times \mathbf{Q}^{(1)} + \boldsymbol{\mu} + \sum_{\nu=1}^{\rho} \mathbf{T}^{(\nu)}\, \delta(\eta - \eta_\nu) = 0\,; \tag{6.109}$$

$$\mathrm{L}_1 \frac{d\boldsymbol{\vartheta}}{d\eta} + \mathrm{L}æ_0^{(1)} - æ = 0\,; \tag{6.110}$$

$$\frac{d\mathbf{u}}{d\eta} + æ \times \mathbf{u} + (l_{11} - 1)\,\mathbf{e}_1 + l_{21}\mathbf{e}_2 + l_{31}\mathbf{e}_3 = 0\,; \tag{6.111}$$

$$\mathbf{M} = \mathrm{A}\,(æ - æ_0^{(1)})\,. \tag{6.112}$$

The expanded form of (6.108) is as follows:

$$\frac{d\mathbf{Q}}{d\eta} + æ \times \mathbf{Q} - æ \times [\,(P_0 + n_1 w_0^2)\,\mathbf{e}_1\,] + \boldsymbol{\gamma} + \mathbf{q} + \sum_{i=1}^{n} \mathbf{P}^{(i)}\, \delta(\eta - \eta_i) = 0\,; \tag{6.113}$$

here

$$æ \times [\,(P_0 + n_1 w_0^2)\,\mathbf{e}_1\,] = æ_3\,(P_0 + n_1 w_0^2)\,\mathbf{e}_2 - æ_2\,(P_0 + n_1 w_0^2)\,\mathbf{e}_3\,.$$

In view of the boundary conditions, from (6.108)–(6.112) it follows that

$$Q = Q_1^{(1)} + P_0 + n_1 w_0^2\,. \tag{6.114}$$

Let us consider some examples to illustrate how an internal liquid flow affects the stress-strain state and the static stability of a rod.

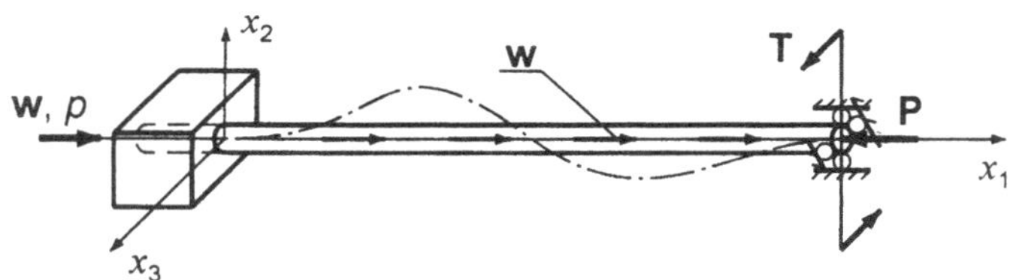

Fig. 6.29.

Consider a compressed-twisted straight rod whose axial line is not perfectly straight due to imperfections (Fig. 6.29). When the liquid is at rest, the shape of the axial line of the rod is characterized by the small curvatures

$\Delta æ_{20}$ and $\Delta æ_{30}$. Taking into account the initial twist $æ_{10}$, we can represent the curvature vector as follows:

$$æ = (æ_{10} + \Delta æ_1)\,\mathbf{e}_1 + (\Delta æ_2 + \Delta æ_{20})\,\mathbf{e}_2 + (\Delta æ_3 + \Delta æ_{30})\,\mathbf{e}_3\,. \qquad (6.115)$$

Assuming that $\Delta æ_j$ and $\Delta æ_{j0}$ are small and using (6.108)–(6.112), we obtain the linear equilibrium equations

$$\begin{aligned}
&\frac{\mathrm{d}Q_2}{\mathrm{d}\eta} - æ_{10}Q_3 + \frac{Q_1}{A_{33}}\,M_3 = \Delta æ_{30}Q_1^{(1)}\,;\\
&\frac{\mathrm{d}Q_3}{\mathrm{d}\eta} + æ_{10}Q_2 - \frac{Q_1}{A_{22}}\,M_3 = -\Delta æ_{20}Q_1^{(1)}\,;\\
&\frac{\mathrm{d}M_2}{\mathrm{d}\eta} + \left(\frac{M_{10}}{A_{33}} - æ_{10}\right)M_3 - Q_3 = -\Delta æ_{30}M_{10}\,;\\
&\frac{\mathrm{d}M_3}{\mathrm{d}\eta} + \left(æ_{10} - \frac{M_{10}}{A_{22}}\right)M_2 + Q_2 = \Delta æ_{20}M_{10}\,;\\
&\frac{\mathrm{d}\vartheta_2}{\mathrm{d}\eta} - \frac{M_2}{A_{22}} - æ_{10}\vartheta_3 = 0\,;\\
&\frac{\mathrm{d}\vartheta_3}{\mathrm{d}\eta} - \frac{M_3}{A_{33}} + æ_{10}\vartheta_2 = 0\,;\\
&\frac{\mathrm{d}u_2}{\mathrm{d}\eta} - \vartheta_3 - æ_{10}u_3 = 0\,;\\
&\frac{\mathrm{d}u_3}{\mathrm{d}\eta} + \vartheta_2 + æ_{10}u_2 = 0\,, \qquad (6.116)
\end{aligned}$$

where

$$M_{10} = T\,, \qquad æ_{10} = \frac{M_{10}}{A_{11}}\,;$$

these equations are valid in the general case when the bending stiffnesses of the rod are different, that is, $A_{22} \neq A_{33}$.

In vector form, the system of equations (6.116) reads

$$\mathbf{Z}' + \mathbf{A}\mathbf{Z} = \Delta\mathbf{b}\,, \qquad (6.117)$$

where

$$\begin{aligned}
&\mathbf{Z} = (Q_2\,, Q_3\,, M_2\,, M_3\,, \vartheta_2\,, \vartheta_3\,, u_2\,, u_3)^{\mathrm{T}}\,;\\
&\Delta\mathbf{b} = (\Delta æ_{30}Q_1^{(1)}\,, -\Delta æ_{20}Q_1^{(1)}\,, -\Delta æ_{30}M_{10}\,, \Delta æ_{20}M_{10}\,, 0\,, 0\,, 0\,, 0)^{\mathrm{T}}\,;
\end{aligned}$$

$$\mathrm{A} = \begin{bmatrix} 0 & -æ_{10} & 0 & \dfrac{Q_1^{(1)}}{A_{33}} & 0 & 0 & 0 & 0 \\ æ_{10} & 0 & -\dfrac{Q_1^{(1)}}{A_{22}} & 0 & 0 & 0 & 0 & 0 \\ 0 & -1 & 0 & \dfrac{M_{10}}{A_{33}} - æ_{10} & 0 & 0 & 0 & 0 \\ 1 & 0 & æ_{10} - \dfrac{M_{10}}{A_{22}} & 0 & 0 & 0 & 0 & 0 \\ 0 & 0 & -\dfrac{1}{A_{22}} & 0 & 0 & -æ_{10} & 0 & 0 \\ 0 & 0 & 0 & -\dfrac{1}{A_{22}} & æ_{10} & 0 & 0 & 0 \\ 0 & 0 & 0 & 0 & 0 & -1 & 0 & -æ_{10} \\ 0 & 0 & 0 & 0 & 1 & 0 & æ_{10} & 0 \end{bmatrix};$$

$$Q_1^{(1)} = Q_{10} - (P_0 + n_1 w_0^2), \qquad Q_{10} = -P.$$

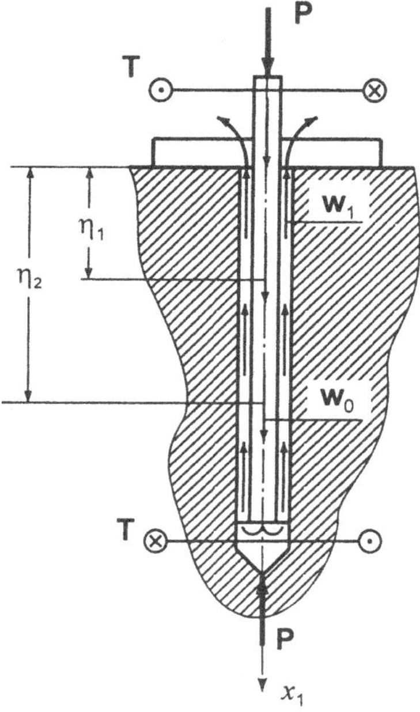

Fig. 6.30.

To account for the influence of the gravity force acting both on the rod and the liquid, one should treat Q_{10} and P_0 as functions of η. Equations (6.117) allow us to analyze the relationship between the stress-strain state of the rod and the small initial curvatures $\Delta æ_{20}$ and $\Delta æ_{30}$. As the liquid runs through curved segments of the rod ($\Delta æ_{20} \neq 0$ and $\Delta æ_{30} \neq 0$), centrifugal forces occur. These forces result in an additional deformation of the rod.

An oil derrick is shown in Fig. 6.30. A flow of a special liquid brings the turbodrill into rotation and lifts the rock. The cross-sectional area of the clearance between the turbodrill and the conduit is large as compared to the cross-sectional area of the hole in the turbodrill. Hence, for incompressible liquids, the velocity of the liquid in the turbodrill w_0 exceeds considerably the velocity of the liquid in the clearance w_1. Thus, the flow in the clearance may be neglected. Consequently, if the turbodrill has no additional supports, then the rod shown in Fig. 6.29 may be treated as a simplified model of such a turbodrill. Using (6.117), we can determine the limiting values of w_0 and P_0 (for specified values of $\Delta æ_{20}$ and $\Delta æ_{30}$) for which the magnitude of the displacement vector $|\mathbf{u}|$ satisfies the inequality

$$\max_{\eta} \sqrt{u_2^2 + u_3^2} \leq \Delta ,$$

where Δ is the radial clearance between the drilling bit and the conduit.

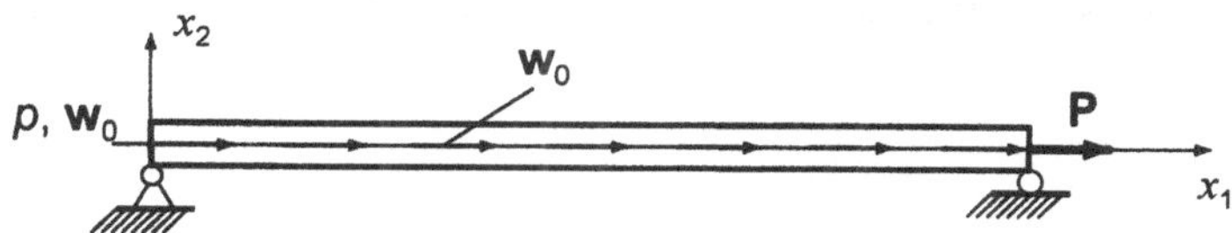

Fig. 6.31.

Let us examine the influence of liquid flows on static stability of compressed-twisted rods. First, consider a simple in-plane problem of stability of a straight rod which interacts with a liquid flow (see Fig. 6.31). Assume that the rod is subjected to a tensile force P. Setting $\Delta æ_{20} = \Delta æ_{30} = 0$ in (6.116), on rearrangement, we arrive at

$$u_2^{\mathrm{IV}} - Q_1^{(1)} u_2'' = 0 \tag{6.118}$$

Setting $Q_{10} = P$, we have

$$u_2^{\mathrm{IV}} + (-P + P_0 + n_1 w_0^2)\, u_2'' = 0 . \tag{6.119}$$

Solving this equation, we get

$$u_2 = c_1 \cos k\eta + c_2 \sin k\eta + c_3\eta + c_4 , \tag{6.120}$$

where

$$k = \sqrt{-P + P_0 + n_1 w_0^2}\,.$$

In the case of simply supported ends, we have $c_1 = c_3 = c_4 = 0$ and $c_2 k^2 \sin k = 0$. Since $k \neq 0$ and $c_2 \neq 0$, it follows that $\sin k = 0$. Thus $k = \pi n$ and

$$(\pi n)^2 = -P + (P_0 + n_1 w_0^2)\,. \tag{6.121}$$

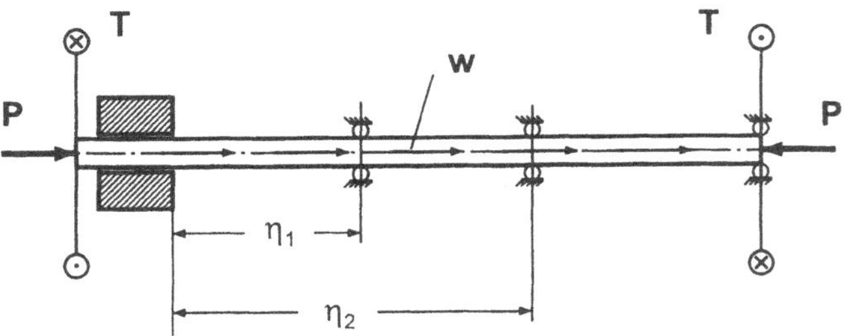

Fig. 6.32.

The critical values P, P_0, and $n_1 w_0^2$ correspond to $n = 1$. Hence, the critical tensile force is

$$P_{*1} = P_0 + n_1 w_0^2 - \pi^2\,. \tag{6.122}$$

Thus we see that a tensile force may result in loss of stability of the rod containing an internal liquid flow.

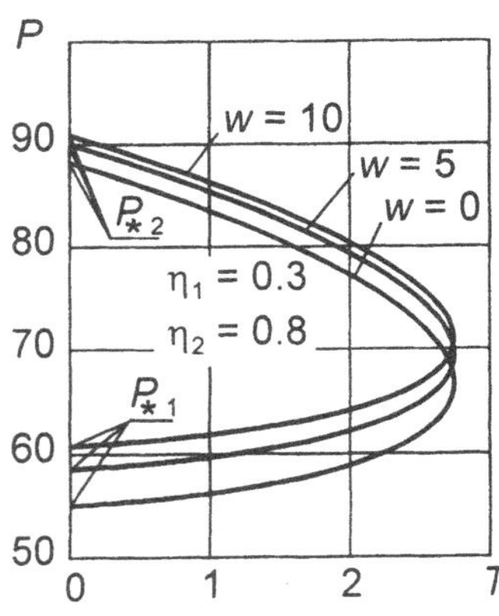

Fig. 6.33.

As another example, consider a compressed-twisted straight rod containing an internal liquid flow (see Fig. 6.32). To account for reaction forces, which appear at the intermediate supports at loss of stability, and determine the critical loads P and T, one can use (4.245). The term Q_{1*} of the matrix A in (4.245) should be replaced by $Q_{1*}^{(1)}$ such that

$$Q_{1*}^{(1)} = Q_{10} - (P_0 + n_1 w_0^2)\,.$$

Now the critical loads can be determined by the method discussed in Sect. 4.6. The only difference is the boundary condition at $\eta = 1$. In this example (Fig. 6.32), the rod is simply supported at $\eta = \eta_1$, $\eta = \eta_2$, and $\eta = 1$ such that only axial displacements of these points are allowed. Hence, the follower moment must satisfy the following boundary-condition equations:

$$Z_3(1) = 0\,, \qquad Z_4(1) = 0\,, \qquad Z_7(1) = 0\,, \qquad Z_8(1) = 0\,.$$

Solving the problem numerically, we obtain the plots of the first two eigenvalues P_{*1} and P_{*2} versus T: (1) for three values of w_0 and for $\eta = 0.3$ and $\eta = 0.8$ (see Fig. 6.33); (2) for three values of η and for $w = 5$ (see Fig. 6.34). These results were obtained with regard to the initial twist $æ_{1*} \neq 0$.

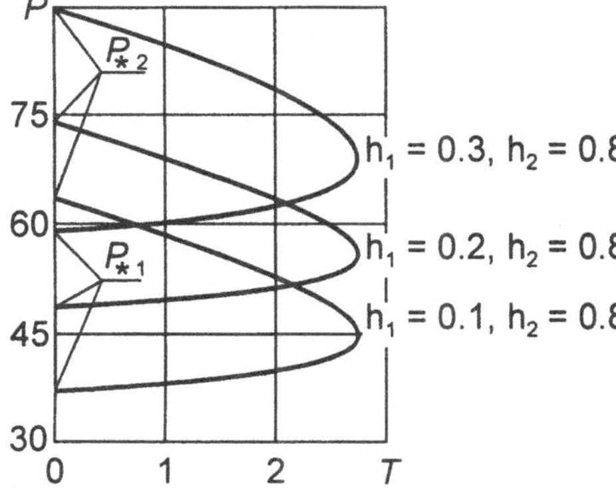

Fig. 6.34.

It can be seen from the plots that the critical force P_{*1} decreases as the flow velocity increases. For $A_{11} = 0.577$, $A_{22} = 0.664$, and $A_{33} = 1$, the critical force P_{*1} at $w_0 = 10$ is less than the critical force at $w_0 = 0$ by 9%. Consequently, the critical values of the force essentially depend on the flow velocity.

A. Appendices

A.1 Elements of Vector Algebra

A.1.1 Vector Bases; Coordinates of Vectors

A set of linearly independent vectors $\{\mathbf{e}_i\}$ forms a basis. In the three-dimensional space, any basis consists of three linearly independent vectors. An arbitrary vector $\mathbf{C}$ can be represented as follows:

$$\mathbf{C} = \sum_{i=1}^{3} c_i \mathbf{e}_i = c_i \mathbf{e}_i \,; \tag{A.1}$$

here c_i are the coordinates of the vector $\mathbf{C}$ in the basis $\{\mathbf{e}_i\}$.

We recognize three types of vectors:

(1) vectors whose one end is attached to a fixed point in space;
(2) vectors which can be moved along the line of their action;
(3) vectors whose ends are not restricted to any set of points in space.

Let us consider these types of vectors.

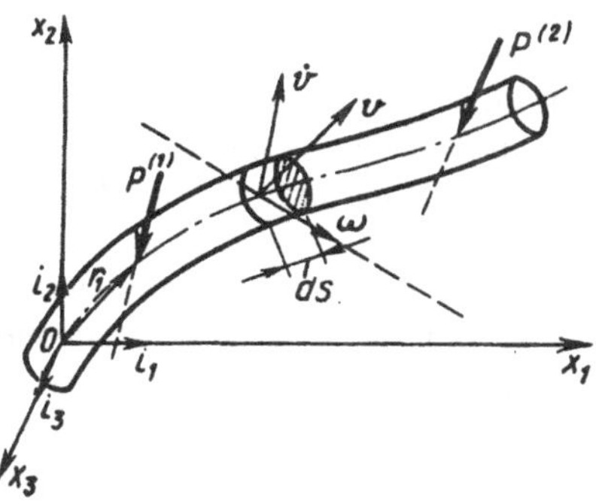

Fig. A.1.

Vectors whose one end is attached to a fixed point in space Such vectors are termed *bound vectors* or *polar vectors*. For example, concentrated forces $\mathbf{P}^{(i)}$ applied to specified points of an elastic body (Fig. A.1) are bound vectors. In structural mechanics, the vector of concentrated force can be neither moved parallel to itself nor displaced along its line of action. The velocity vector $\mathbf{v}$ and the acceleration vector $\dot{\mathbf{v}}$ of a rod element (Fig. A.1) are bound vectors.

Vectors which can be moved along the line of their action Such vectors are called *line vectors* or *sliding vectors* (or *pseudovectors*). For example, concentrated forces applied to a rigid body can be displaced along their lines of action. The instantaneous angular velocity $\boldsymbol{\omega}$ of a body (no matter rigid or deformable) is a sliding vector. Note that for deformable bodies, the concept *angular velocity* is well defined only for an infinitesimally small element of the body (see Fig. A.1).

Vectors whose ends are not restricted to any set of points in space Such vectors are called *free vectors*. For example, a moment vector $\mathbf{b}$ of a force couple $\mathbf{a}$ (Fig. A.2) is a free vector.

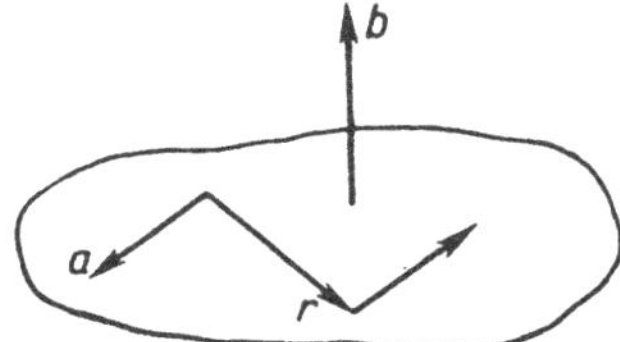

Fig. A.2.

Note that in (A.1) the following summation convention is used: *when an index appears twice in a term, summation with respect to that index over the range of the index must be performed.* A repeated index is called a *dummy index*. The use of other letters in place of dummy indices does not affect the value of the expression. Not repeated indices are called *free indices*.

Mutually orthogonal base vectors of unit magnitude are called *orthonormal base vectors*. In problems of mechanics of rods, two coordinate systems are usually introduced: the *immovable Cartesian coordinate system* whose axes are denoted by x_1, x_2, x_3 (see Fig. A.3) and the *moveable orthogonal coordinate system* x'_1, x'_2, x'_3 rigidly attached to the rod axis. These coordinate frames are assumed to be right-handed. A coordinate system whose base vectors are $\mathbf{e}_1$, $\mathbf{e}_2$, and $\mathbf{e}_3$ is said to be right-handed when a right-handed screw advancing in the $\mathbf{e}_3$-direction rotates $\mathbf{e}_1$ onto $\mathbf{e}_2$ by rotating through an angle less than 180°. The base vectors of the immovable frame are denoted by $\mathbf{i}_j$ and those of the attached frame are denoted by $\mathbf{e}_j$.

Two configurations of a rod are illustrated in Fig. A.3: the initial (unloaded) configuration *1* and a deformed configuration *2* (this deformation is

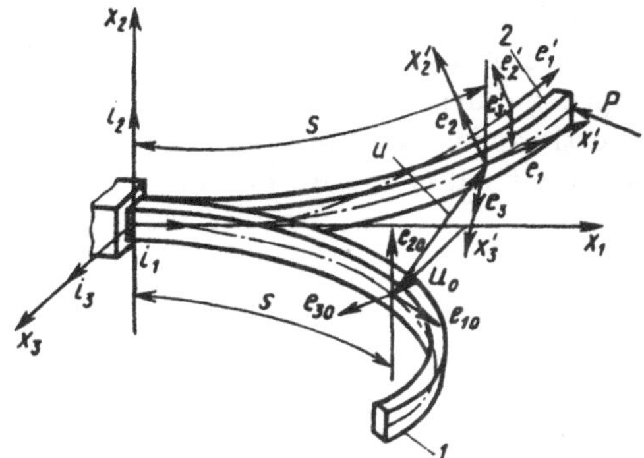

Fig. A.3.

caused by a force $\mathbf{P}$). In the attached coordinate system, the displacement vector $\mathbf{u}$ (see Fig. A.3) can be expressed as follows:

$$\mathbf{u} = \sum_{i=1}^{3} u_i \mathbf{e}_i \, . \tag{A.2}$$

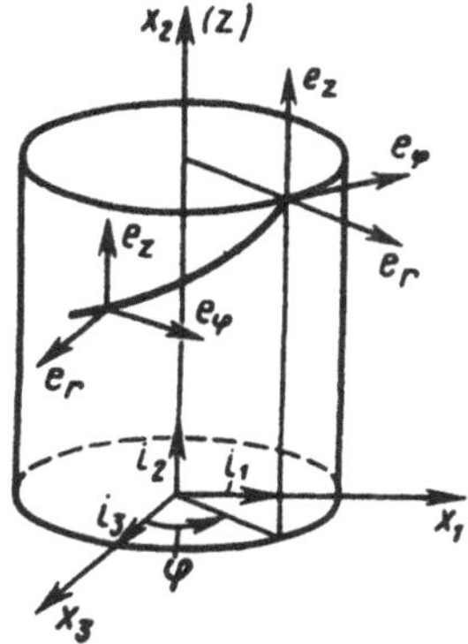

Fig. A.4.

Along with *rectilinear Cartesian coordinates*, so-called *curvilinear orthogonal coordinates* are widely used, these being *cylindrical coordinates, spherical coordinates*, etc. For example, cylindrical coordinates are best suited for derivation of the basic equations in the problem of motion of a flexible rod over a cylindrical surface. Cylindrical coordinates and its base vectors $\{\mathbf{e}_i\} = (\mathbf{e}_z \, , \, \mathbf{e}_r \, , \, \mathbf{e}_\varphi)$ are illustrated in Fig. A.4. In greater detail, curvilinear coordinates are discussed in Sect. A.2.8.

As pointed out above, the general equilibrium equations and the equations of motion represented in vector form are valid in any coordinate system. Anyway, for numerical study, the equations need be represented in coordinate form. Unlike the vector form, the coordinate form depends on the selected coordinate frame. The appropriate choice of a coordinate system may simplify the analysis of a problem. The prime difference between the orthogonal rectilinear coordinates $\{\mathbf{i}_j\}$ and the orthogonal curvilinear coordinates $\{\mathbf{e}_j\}$ is that the directions of the vectors $\mathbf{i}_j$ are fixed in space while the directions of $\mathbf{e}_j$ may vary.

A.1.2 Scalar Product

The *scalar product* $(\mathbf{a} \cdot \mathbf{b})$ of two vectors $\mathbf{a}$ and $\mathbf{b}$ is defined as the product of their magnitudes times the cosine of the angle between them,

$$(\mathbf{a} \cdot \mathbf{b}) = |\mathbf{a}|\,|\mathbf{b}| \cos(\widehat{\mathbf{a}\,,\,\mathbf{b}})\,. \tag{A.3}$$

From this definition,

$$\cos(\widehat{\mathbf{a}\,,\,\mathbf{b}}) = \frac{(\mathbf{a} \cdot \mathbf{b})}{|\mathbf{a}|\,|\mathbf{b}|}\,. \tag{A.4}$$

Thus the inequality

$$\frac{|\mathbf{a} \cdot \mathbf{b}|}{|\mathbf{a}|\,|\mathbf{b}|} \leq 1 \tag{A.5}$$

or

$$|\mathbf{a} \cdot \mathbf{b}| \leq |\mathbf{a}|\,|\mathbf{b}|\,. \tag{A.6}$$

From the definition it follows that the scalar product is *commutative*,

$$(\mathbf{a} \cdot \mathbf{b}) = (\mathbf{b} \cdot \mathbf{a})\,. \tag{A.7}$$

The necessary and sufficient condition for two nonzero vectors to be mutually orthogonal can be expressed as follows:

$$(\mathbf{a} \cdot \mathbf{b}) = 0\,. \tag{A.8}$$

According to (A.8), orthogonal base vectors of unit length, $|\mathbf{e}_j| = 1$, $j = 1\,,2\,,3$, satisfy the relation

$$(\mathbf{e}_i \cdot \mathbf{e}_j) = \delta_{ij} = \begin{cases} 1\,, & i = j\,; \\ 0\,, & i \neq j\,, \end{cases} \tag{A.9}$$

where δ_{ij} is a *Kronecker delta function.* Such a basis is termed an *orthonormal* basis.

If a_i and b_i are the rectangular components of the vectors $\mathbf{a}$ and $\mathbf{b}$, respectively, then, in view of (A.9), we get

$$(\mathbf{a} \cdot \mathbf{b}) = (a_i\mathbf{e}_i)\,(b_j\mathbf{e}_j) = (a_ib_j)\,(\mathbf{e}_i \cdot \mathbf{e}_j) = \sum_{i=1}^{3} a_ib_i = a_ib_i\,. \tag{A.10}$$

Consider two vectors $\mathbf{c}$ and $\mathbf{d}$ such that

$$\mathbf{c} = \mathrm{A}\mathbf{a}; \qquad \mathbf{d} = \mathrm{B}\mathbf{b}, \tag{A.11}$$

where A and B are square matrices. Consider the scalar product

$$m_1 = (\mathbf{c} \cdot \mathbf{d}) = (\mathrm{A}\mathbf{a} \cdot \mathrm{B}\mathbf{b}). \tag{A.12}$$

This relation may be rewritten as follows:

$$m_1 = (\mathrm{B}^{\mathrm{T}}\mathrm{A}\mathbf{a} \cdot \mathbf{b}); \qquad m_1 = (\mathbf{a} \cdot \mathrm{A}^{\mathrm{T}}\mathrm{B}\mathbf{b}); \tag{A.13}$$

here A^{T} and B^{T} are the transposes of the matrices A and B, respectively. The relations (A.13) are useful when dealing with transformations of coordinates. For $\mathbf{a} = \mathbf{b}$, the right-hand side of (A.12) is termed a *quadratic form.*

A.1.3 Vector Product

A vector product of two vectors $\mathbf{a}$ and $\mathbf{b}$ is defined as a vector $\mathbf{c} = \mathbf{a} \times \mathbf{b}$ such that $\mathbf{c}$ is perpendicular to both $\mathbf{a}$ and $\mathbf{b}$; the vectors $\mathbf{a}$, $\mathbf{b}$, and $\mathbf{c}$ form a right-handed system (see Fig. A.5); the magnitude of the vector $\mathbf{c}$ is equal to the area of the parallelogram whose sides are the vectors $\mathbf{a}$ and $\mathbf{b}$,

$$|\mathbf{c}| = |\mathbf{a}|\,|\mathbf{b}|\sin(\widehat{\mathbf{a}, \mathbf{b}}). \tag{A.14}$$

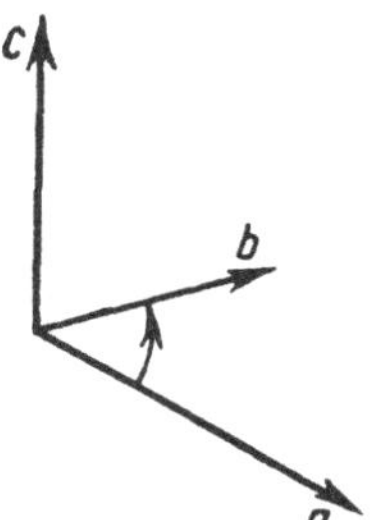

Fig. A.5.

From the definition it follows that whenever the vectors $\mathbf{a}$ and $\mathbf{b}$ are bound ones, the vector $\mathbf{c}$ is a sliding vector; the vector $\mathbf{c}$ is a bound vector whenever at least one of the vectors $\mathbf{a}$ and $\mathbf{b}$ is a sliding one. For example, the velocity vector $\mathbf{v}$ of a point of a rigid body is a bound vector given by the relation

$$\mathbf{v} = \boldsymbol{\omega} \times \mathbf{r}, \tag{A.15}$$

where $\mathbf{r}$ is the radius vector of the point (a bound vector) and $\boldsymbol{\omega}$ is the vector of the angular velocity of the body (a sliding vector).

If a vector $\mathbf{a}$ is parallel to a vector $\mathbf{b}$, then

$$\mathbf{a} \times \mathbf{b} = 0\,. \tag{A.16}$$

Vector multiplication is an anticommutative operation,

$$\mathbf{a} \times \mathbf{b} = -\mathbf{b} \times \mathbf{a}\,. \tag{A.17}$$

The vector product can be written as a determinant,

$$\mathbf{c} = \mathbf{a} \times \mathbf{b} = \begin{vmatrix} \mathbf{e}_1 & \mathbf{e}_2 & \mathbf{e}_3 \\ \mathbf{a}_1 & \mathbf{a}_2 & \mathbf{a}_3 \\ \mathbf{b}_1 & \mathbf{b}_2 & \mathbf{b}_3 \end{vmatrix} = c_i \mathbf{e}_i\,. \tag{A.18}$$

From this relation it follows that

$$\begin{aligned} c_1 &= a_2 b_3 - a_3 b_2\,; \\ c_2 &= a_3 b_1 - a_1 b_3\,; \\ c_3 &= a_1 b_2 - a_2 b_1\,. \end{aligned} \tag{A.19}$$

The orthonormal base vectors $\{\mathbf{e}_i\}$ satisfy the relations

$$\mathbf{e}_i \times \mathbf{e}_j = \varepsilon_{ijk} \mathbf{e}_k\,, \qquad i\,,j\,,k = 1\,,2\,,3\,, \tag{A.20}$$

where ε_{ijk} is a *Levi-Civita symbol* or a *permutation symbol.* The permutation symbols ε_{ijk} are defined as follows: $\varepsilon_{ijk} = 0$ if at least two indices are equal; $\varepsilon_{ijk} = 1$ if the sequence of numbers i, j, and k is the sequence 123 or an even permutation of the sequence (i.e. 312 and 231); and $\varepsilon_{ijk} = -1$ if the sequence of i, j, and k is an odd permutation of the sequence 123 (namely, 132, 321, and 213). Using the permutation symbols, we can rewrite (A.18) as

$$c_i = \varepsilon_{ijk} a_j b_k\,. \tag{A.21}$$

As an example, let us determine the moment $\mathbf{T}$ of the force $\mathbf{P}$ relative to the point O as indicated in Fig. A.1. Let $\mathbf{r}_1$ be the radius vector of the point of application of the force $\mathbf{P}$. Then,

$$\mathbf{T} = \mathbf{r}_1 \times \mathbf{P}^{(1)}\,. \tag{A.22}$$

Vector products often appear in the basic equations of the theory of rods. When deformation of a rod is negligibly small or when only static stability of a rod need be studied, then, as a rule, one factor in the vector product (say, the vector $\mathbf{a}$) is known a priori. In this case, it is convenient to rewrite (A.18) in the form

$$\mathbf{c} = \mathrm{A}\mathbf{b}\,, \tag{A.23}$$

where

$$\mathrm{A} = \begin{bmatrix} 0 & -a_3 & a_2 \\ a_3 & 0 & -a_1 \\ -a_2 & a_1 & 0 \end{bmatrix}. \tag{A.24}$$

A.1.4 Scalar Triple Product

The *scalar triple product* of three vectors **a**, **b**, and **c** is a number d defined as

$$d = \mathbf{c} \cdot (\mathbf{a} \times \mathbf{b}) \tag{A.25}$$

or, using a determinant,

$$d = \mathbf{c} \cdot (\mathbf{a} \times \mathbf{b}) = \begin{vmatrix} c_1 & c_2 & c_3 \\ a_1 & a_2 & a_3 \\ b_1 & b_2 & b_3 \end{vmatrix} . \tag{A.26}$$

If the three vectors **a**, **b**, and **c** lie in one plane (such vectors are called *coplanar*), then

$$\mathbf{c} \cdot (\mathbf{a} \times \mathbf{b}) = 0 . \tag{A.27}$$

This equality is also valid whenever at least two of the vectors **a**, **b**, and **c** are equal. For example, if $\mathbf{c} = \mathbf{a}$, then

$$\mathbf{a} \cdot (\mathbf{a} \times \mathbf{b}) = 0 . \tag{A.28}$$

The permutation symbols are related to the orthonormal base vectors as follows:

$$\varepsilon_{ijk} = \mathbf{e}_i \cdot (\mathbf{e}_j \times \mathbf{e}_k) . \tag{A.29}$$

A.1.5 Vector Triple Product

The vector triple product of the vectors **a**, **b**, and **c** is a vector **d** defined as

$$\mathbf{d} = \mathbf{a} \times (\mathbf{b} \times \mathbf{c}) \tag{A.30}$$

or

$$\mathbf{d} = \mathbf{a} \times (\mathbf{b} \times \mathbf{c}) = (\mathbf{a} \cdot \mathbf{c}) \cdot \mathbf{b} - (\mathbf{a} \cdot \mathbf{b}) \cdot \mathbf{c} . \tag{A.31}$$

Consider two vectors **e** and **f** obtained by a cyclic rearrangement of the vectors **a**, **b**, and **c** in (A.31),

$$\mathbf{e} = \mathbf{b} \times (\mathbf{c} \times \mathbf{a}) = (\mathbf{b} \cdot \mathbf{a}) \cdot \mathbf{c} - (\mathbf{b} \cdot \mathbf{c}) \cdot \mathbf{a} ; \tag{A.32}$$

$$\mathbf{f} = \mathbf{c} \times (\mathbf{a} \times \mathbf{b}) = (\mathbf{c} \cdot \mathbf{b}) \cdot \mathbf{a} - (\mathbf{c} \cdot \mathbf{a}) \cdot \mathbf{b} . \tag{A.33}$$

Summing the vectors **d**, **e**, and **f**, we obtain the Jacobi identity,

$$\mathbf{a} \times (\mathbf{b} \times \mathbf{c}) + \mathbf{b} \times (\mathbf{c} \times \mathbf{a}) + \mathbf{c} \times (\mathbf{a} \times \mathbf{b}) = 0 . \tag{A.34}$$

A.1.6 Transformation of Base Vectors

During deformation, the spatial location of a rod (described, for example, by the displacement vector $\mathbf{u}$) and the orientation of the basis $\{\mathbf{e}_i\}$ attached to the rod axis are defined with reference to the coordinate system x_i, $i = 1, 2, 3$ (Fig. A.3). However, the structural analysis may be simplified if we use a coordinate frame attributed to an element of the structure, say, the frame $\{\mathbf{e}_{i0}\}$ in Fig. A.3. A frequently used tool is transformation of coordinates. Thus, dealing with a set of coordinate systems, we need to know the relationship between the base vectors of these systems.

Let us obtain formulas of transition from one rectangular frame to another. Let $\{\mathbf{e}_{i0}\}$ and $\{\mathbf{e}_i\}$, $i = 1, 2, 3$, be the base vectors of two coordinate systems attached to the same cross section of a rod before and after deformation, respectively (see Fig. A.3). The vectors $\mathbf{e}_i$ can be expressed in terms of the vectors $\mathbf{e}_{i0}$ as follows:

$$\begin{aligned}\mathbf{e}_1 &= l_{11}\mathbf{e}_{10} + l_{12}\mathbf{e}_{20} + l_{13}\mathbf{e}_{30}\,;\\ \mathbf{e}_2 &= l_{21}\mathbf{e}_{10} + l_{22}\mathbf{e}_{20} + l_{23}\mathbf{e}_{30}\,;\\ \mathbf{e}_3 &= l_{31}\mathbf{e}_{10} + l_{32}\mathbf{e}_{20} + l_{33}\mathbf{e}_{30}\end{aligned}$$

or

$$\mathbf{e}_i = \sum_{j=1}^{3} l_{ij}\mathbf{e}_{j0} = l_{ij}\mathbf{e}_{j0}\,, \qquad i = 1, 2, 3; \tag{A.35}$$

here l_{ij} are the projections of the vectors $\mathbf{e}_i$ on the straight lines defined by the vectors $\mathbf{e}_{j0}$. The coefficients l_{ij} in (A.35) can be organized into a matrix

$$\mathrm{L} = [\,l_{ij}\,] = \begin{bmatrix} l_{11} & l_{12} & l_{13}\\ l_{21} & l_{22} & l_{23}\\ l_{31} & l_{32} & l_{33}\end{bmatrix}; \tag{A.36}$$

this matrix *transforms* the basis $\{\mathbf{e}_{j0}\}$ into the basis $\{\mathbf{e}_j\}$. The relation (A.35) can be rewritten as

$$\mathbf{e}_i = \mathrm{L}^{\mathrm{T}}\mathbf{e}_{i0}\,, \tag{A.37}$$

where L^{T} is the transpose of the matrix L.

For the inverse transformation, we have

$$\mathbf{e}_{i0} = \mathrm{L}\mathbf{e}_i\,.$$

Consider the structure of a transformation matrix in greater detail. Let the matrix

$$\mathrm{L} = [\,l_{ij}\,] = \begin{bmatrix} l_{11} & l_{12} & l_{13}\\ l_{21} & l_{22} & l_{23}\\ l_{31} & l_{32} & l_{33}\end{bmatrix}$$

be a matrix of transformation from a basis $\{\mathbf{a}_i\}$ to a basis $\{\mathbf{b}_j\}$. In expanded notation, the vectors of the second, new basis can be represented in terms of the vectors of the first, old basis as follows:

$$\begin{aligned}
\mathbf{b}_1 &= l_{11}\mathbf{a}_1 + l_{21}\mathbf{a}_2 + l_{31}\mathbf{a}_3 ;\\
\mathbf{b}_2 &= l_{12}\mathbf{a}_1 + l_{22}\mathbf{a}_2 + l_{32}\mathbf{a}_3 ;\\
\mathbf{b}_3 &= l_{13}\mathbf{a}_1 + l_{23}\mathbf{a}_2 + l_{33}\mathbf{a}_3 .
\end{aligned}$$

Thus, in order to represent the new basis vector in terms of the old basis, we must represent it as linear combination of the old basis vectors such that the coefficients of this linear combination are equal to the elements of the respective column in the transformation matrix. In other words, the columns of the transformation matrix are composed of the elements of the new basis vectors represented in terms of the old basis.

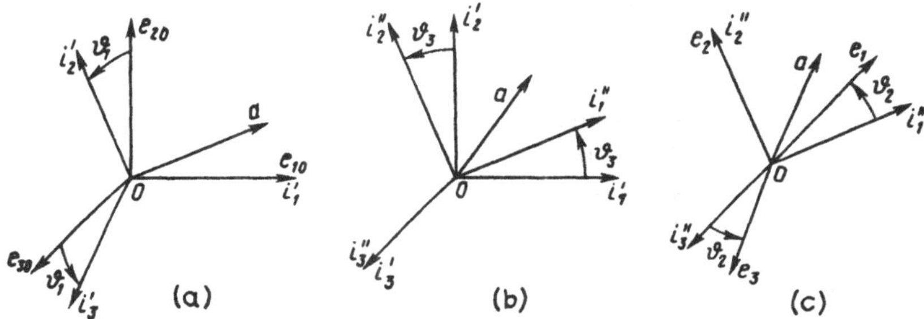

Fig. A.6.

As an example, let us evaluate the transformation matrix when the basis $\{\mathbf{e}_{i0}\}$ is first translated and then arbitrarily rotated (Fig. A.6). Evidently, the translation changes neither magnitudes of the vectors nor their directions, consequently, only rotations need be taken into consideration. An arbitrary rotation of a coordinate frame is equivalent to a composition of three independent rotations. Suppose that our basis is rotated by a positive angle ϑ_1 about the vector $\mathbf{e}_{10}$ (see Fig. A.6). We have

$$\begin{aligned}
\mathbf{i}'_1 &= \mathbf{e}_{10} ;\\
\mathbf{i}'_2 &= \cos\vartheta_1 \cdot \mathbf{e}_{20} + \sin\vartheta_1 \cdot \mathbf{e}_{30} ;\\
\mathbf{i}'_3 &= -\sin\vartheta_1 \cdot \mathbf{e}_{20} + \cos\vartheta_1 \cdot \mathbf{e}_{30} .
\end{aligned} \tag{A.38}$$

The corresponding transformation matrix reads

$$\mathrm{L}_{\vartheta_1} = \begin{bmatrix} 1 & 0 & 0 \\ 0 & \cos\vartheta_1 & \sin\vartheta_1 \\ 0 & -\sin\vartheta_1 & \cos\vartheta_1 \end{bmatrix} . \tag{A.39}$$

The elements of the matrix L_{ϑ_1} are the direction cosines of the vectors $\mathbf{i}_j$, $j = 1, 2, 3$, with respect to the vectors $\mathbf{e}_j$, $j = 1, 2, 3$. This statement is valid for any transformation matrix.

The rotation by an angle ϑ_3 about the vector $\mathbf{i}'_3$ (see Fig. A.6*b*) can be described as

$$\begin{aligned}
\mathbf{i}''_1 &= \cos\vartheta_3 \cdot \mathbf{i}'_1 + \sin\vartheta_3 \cdot \mathbf{i}'_2 ;\\
\mathbf{i}''_2 &= -\sin\vartheta_3 \cdot \mathbf{i}'_1 + \cos\vartheta_3 \cdot \mathbf{i}'_2 ;\\
\mathbf{i}''_3 &= \mathbf{i}'_3 .
\end{aligned} \tag{A.40}$$

In this case, the transformation matrix is as follows:

$$L_{\vartheta_3} = \begin{bmatrix} \cos\vartheta_3 & \sin\vartheta_3 & 0 \\ -\sin\vartheta_3 & \cos\vartheta_3 & 0 \\ 0 & 0 & 1 \end{bmatrix} . \tag{A.41}$$

The last-named rotation by a positive angle ϑ_2 about the vector $\mathbf{i}''_2 = \mathbf{e}_2$ rotates the vectors $\mathbf{i}''_j$ onto the vectors $\mathbf{e}_j$ (Fig. A.6*c*). The transformation matrix for this rotation is

$$L_{\vartheta_2} = \begin{bmatrix} \cos\vartheta_2 & 0 & -\sin\vartheta_2 \\ 0 & 1 & 0 \\ \sin\vartheta_2 & 0 & \cos\vartheta_2 \end{bmatrix} . \tag{A.42}$$

Thus, the matrix L of the transformation from the basis $\{\mathbf{e}_{i0}\}$ to the basis $\{\mathbf{e}_i\}$ is as follows:

$$L = [\, l_{ij} \,] = L_{\vartheta_2} L_{\vartheta_3} L_{\vartheta_1} \tag{A.43}$$

or

$$L = \begin{bmatrix} \cos\vartheta_2\cos\vartheta_3 & \begin{matrix}\cos\vartheta_2\sin\vartheta_3\cos\vartheta_1 \\ +\sin\vartheta_2\sin\vartheta_1\end{matrix} & \begin{matrix}\cos\vartheta_2\sin\vartheta_3\sin\vartheta_1 \\ -\sin\vartheta_2\cos\vartheta_1\end{matrix} \\ -\sin\vartheta_3 & \cos\vartheta_1\cos\vartheta_3 & \cos\vartheta_3\sin\vartheta_1 \\ \sin\vartheta_2\cos\vartheta_3 & \begin{matrix}\sin\vartheta_2\sin\vartheta_3\cos\vartheta_1 \\ -\cos\vartheta_2\sin\vartheta_1\end{matrix} & \begin{matrix}-\sin\vartheta_2\sin\vartheta_3\sin\vartheta_1 \\ +\cos\vartheta_2\cos\vartheta_1\end{matrix} \end{bmatrix} . \tag{A.44}$$

If we perform the rotations in the sequence $\vartheta_2 \to \vartheta_3 \to \vartheta_1$ (Fig. A.7), then the transformation matrix reads

$$L = \begin{bmatrix} \begin{matrix}\cos\vartheta_3\cos\vartheta_2 \\ -\sin\vartheta_1\sin\vartheta_3\sin\vartheta_2\end{matrix} & \begin{matrix}\sin\vartheta_3\cos\vartheta_2 \\ +\sin\vartheta_1\cos\vartheta_3\sin\vartheta_2\end{matrix} & -\cos\vartheta_1\sin\vartheta_2 \\ -\cos\vartheta_1\sin\vartheta_3 & \cos\vartheta_1\cos\vartheta_3 & \sin\vartheta_1 \\ \begin{matrix}\cos\vartheta_3\sin\vartheta_2 \\ +\sin\vartheta_1\sin\vartheta_3\cos\vartheta_2\end{matrix} & \begin{matrix}\sin\vartheta_3\sin\vartheta_2 \\ -\sin\vartheta_1\cos\vartheta_3\cos\vartheta_2\end{matrix} & \cos\vartheta_1\cos\vartheta_2 \end{bmatrix} . \tag{A.45}$$

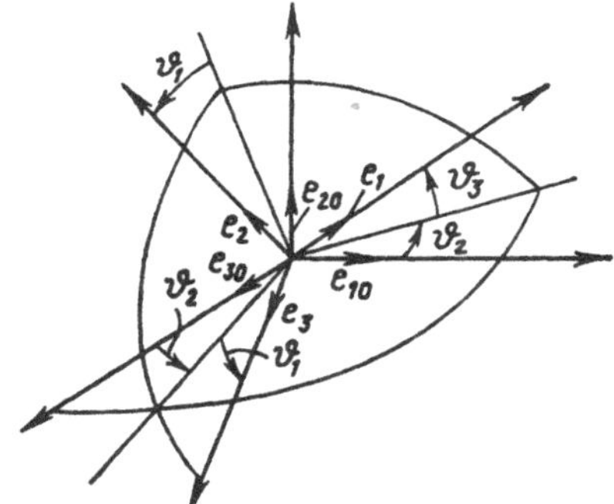

Fig. A.7.

In spite of similarity in notation, the angles in (A.44) and (A.45) differ essentially.

For small rotations, the matrices (A.44) and (A.45) take the form

$$\Delta \mathrm{L} = \begin{bmatrix} 1 & \vartheta_3 & -\vartheta_2 \\ -\vartheta_3 & 1 & \vartheta_1 \\ \vartheta_2 & -\vartheta_1 & 1 \end{bmatrix} \tag{A.46}$$

or

$$\Delta \mathrm{L} = \mathrm{E} + \Delta \mathrm{L}_1 \,, \tag{A.47}$$

where

$$\Delta \mathrm{L}_1 = \begin{bmatrix} 0 & \vartheta_3 & -\vartheta_2 \\ -\vartheta_3 & 0 & \vartheta_1 \\ \vartheta_2 & -\vartheta_1 & 0 \end{bmatrix} .$$

Let us prove that the matrices L defined by (A.44) and (A.45) satisfy the relation

$$\mathrm{L}^{-1} = \mathrm{L}^{\mathrm{T}} \,, \tag{A.48}$$

where L^{-1} is the inverse of the matrix L.

Eliminating $\mathbf{e}_{i0}$ from (A.37), we arrive at $\mathbf{e}_i = \mathrm{LL}^{\mathrm{T}}\mathbf{e}_i = \mathrm{E}\mathbf{e}_i$, therefore, $\mathrm{LL}^{\mathrm{T}} = \mathrm{E}$ and, finally, $\mathrm{L}^{-1} = \mathrm{L}^{\mathrm{T}}$. Since $\mathrm{LL}^{-1} = \mathrm{LL}^{\mathrm{T}} = \mathrm{E}$, we see that the elements l_{ij} of the matrices (A.44) and (A.45) satisfy the following six equations:

$$\begin{aligned} &l_{11}^2 + l_{12}^2 + l_{13}^2 = 1 \,; \\ &l_{21}^2 + l_{22}^2 + l_{23}^2 = 1 \,; \\ &l_{31}^2 + l_{32}^2 + l_{33}^2 = 1 \,; \\ &l_{11}l_{21} + l_{12}l_{22} + l_{13}l_{23} = 0 \,; \\ &l_{21}l_{31} + l_{22}l_{32} + l_{23}l_{33} = 0 \,; \\ &l_{31}l_{11} + l_{32}l_{12} + l_{33}l_{13} = 0 \end{aligned}$$

or

$$l_{ij} l_{kj} = \delta_{ik} \,. \tag{A.49}$$

The equality

$$\mathrm{L}^{\mathrm{T}} \mathrm{L} = \mathrm{E} \tag{A.50}$$

implies six additional relations

$$l_{kj} l_{ij} = \delta_{ki} \,. \tag{A.51}$$

The transformation matrix establishes the relation between the coordinates of a vector in these bases,

$$a_i = a_{k0} l_{ik} \,, \tag{A.52}$$

where a_{k0} are the projections of the vector $\mathbf{a}$ in the frame $\{\mathbf{e}_{i0}\}$ and a_i are the projections of this vector in the frame $\{\mathbf{e}_i\}$. In vector notation, we have

$$\mathbf{a} = \mathrm{L}\mathbf{a}_0 \,. \tag{A.53}$$

Then, we can express the projections of the vector $\mathbf{a}$ in the basis $\{\mathbf{e}_{i0}\}$ in terms of its projections in the basis $\{\mathbf{e}_i\}$ as follows:

$$\mathbf{a}_0 = \mathrm{L}^{\mathrm{T}} \mathbf{a} \,. \tag{A.54}$$

We see that the matrix L relates the base vectors of the bases $\{\mathbf{e}_{i0}\}$ and $\{\mathbf{e}_i\}$. These bases are attributed to an arbitrary cross section of the rod in the unloaded (initial) and a loaded (deformed) configurations, respectively (Fig. A.3).

Let us derive the matrix of transformation L^0 from the basis $\{\mathbf{i}_j\}$ to the basis $\{\mathbf{e}_{j0}\}$. The basis $\{\mathbf{i}_j\}$ coincides with the basis $\{\mathbf{e}_{j00}\}$ associated with the initially straight axial line of the rod. The basis $\{\mathbf{e}_{j0}\}$ is attributed to the axial line of the rod in a deformed configuration. Arguing as above, we introduce the angles of rotation ϑ_j^0 and obtain the matrix L^0. This expression is identical in form to the expression for the matrix L, that is,

$$\mathrm{L}^0 = \begin{bmatrix} \cos\vartheta_2^0 \cos\vartheta_3^0 & \begin{array}{c} \cos\vartheta_2^0 \sin\vartheta_3^0 \cos\vartheta_1^0 \\ + \sin\vartheta_2^0 \sin\vartheta_1^0 \end{array} & \begin{array}{c} \cos\vartheta_2^0 \sin\vartheta_3^0 \sin\vartheta_1^0 \\ - \sin\vartheta_2^0 \cos\vartheta_1^0 \end{array} \\ -\sin\vartheta_3^0 & \cos\vartheta_1^0 \cos\vartheta_3^0 & \cos\vartheta_3^0 \sin\vartheta_1^0 \\ \sin\vartheta_2^0 \cos\vartheta_3^0 & \begin{array}{c} \sin\vartheta_2^0 \sin\vartheta_3^0 \cos\vartheta_1^0 \\ - \cos\vartheta_2^0 \sin\vartheta_1^0 \end{array} & \begin{array}{c} \sin\vartheta_2^0 \sin\vartheta_3^0 \sin\vartheta_1^0 \\ + \cos\vartheta_2^0 \cos\vartheta_1^0 \end{array} \end{bmatrix} . \tag{A.55}$$

Let us now obtain the matrix of transformation from the basis $\{\mathbf{i}_j\}$ to the basis $\{\mathbf{e}_j\}$ (Fig. A.8). Recall that $\{\mathbf{i}_j\} \equiv \{\mathbf{e}_{j00}\}$. We have

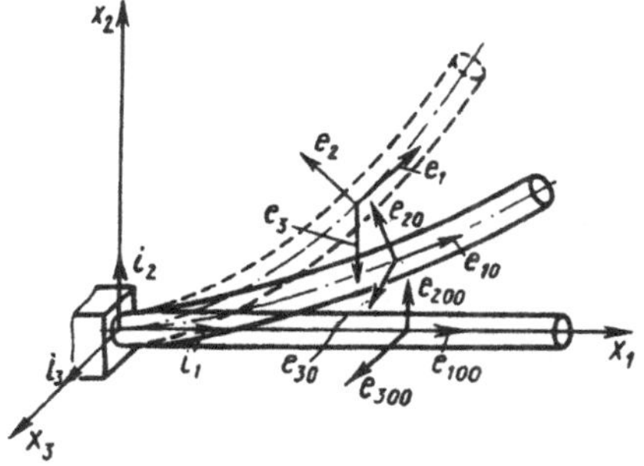

Fig. A.8.

$$\mathbf{e}_{j0} = l_{j1}^0 \mathbf{i}_1 + l_{j2}^0 \mathbf{i}_2 + l_{j3}^0 \mathbf{i}_3 ;$$
$$\mathbf{e}_k = l_{k1}\mathbf{e}_{10} + l_{k2}\mathbf{e}_{20} + l_{k3}\mathbf{e}_{30} . \tag{A.56}$$

Eliminating $\mathbf{e}_{j0}$ from (A.56), we arrive at

$$\mathbf{e}_k = \sum_{j=1}^{3}\sum_{\nu=1}^{3} l_{kj} l_{j\nu}^0 \mathbf{i}_\nu = \sum_{\nu=1}^{3} l_{k\nu}^{(1)} \mathbf{i}_\nu ,$$

where $l_{k\nu}^{(1)}$ are the elements of the matrix

$$\mathrm{L}^{(1)} = \mathrm{L}\mathrm{L}^0 . \tag{A.57}$$

Recall that the elements of the matrix L^0 are known a priori. These elements are the characteristics of the shape of the rod axis in the natural configuration. The matrix L is the matrix of transformation from the basis $\{\mathbf{e}_{j0}\}$ to the basis $\{\mathbf{e}_j\}$. If the axial line of a rod in the natural configuration is a straight line, then $\mathrm{L}^0 = \mathrm{E}$.

Using (A.57), we obtain the following relationship between the base vectors of the bases $\{\mathbf{e}_j\}$ and $\{\mathbf{i}_j\}$:

$$\mathbf{e}_j = (\mathrm{L}^{(1)})^{\mathrm{T}} \mathbf{i}_j = (\mathrm{L}^0)^{\mathrm{T}} \mathrm{L}^{\mathrm{T}} \mathbf{i}_j ; \tag{A.58}$$

this relationship is identical in form to (A.37).

Let the projections of the vector $\mathbf{a}$ in the basis $\{\mathbf{e}_j\}$ be organized into a column $\mathbf{a}_e$. Similarly, let us introduce a column $\mathbf{a}_x$ composed from the projections of the vector $\mathbf{a}$ in the basis $\{\mathbf{i}_j\}$. Using (A.57), we can write

$$\mathbf{a}_e = \mathrm{L}^{(1)} \mathbf{a}_x \tag{A.59}$$

or, in coordinate form,

$$a_j = \sum_{j=1}^{3} a_{xj} l_{ij}^{(1)} .$$

The inverse transformation reads

$$\mathbf{a}_x = (\mathrm{L}^{(1)})^{\mathrm{T}} \mathbf{a}_e \,. \tag{A.60}$$

The relationships (A.59) and (A.60) simplify rearrangement of equilibrium equations under coordinate transformations.

A.2 Basics of Differential Geometry

A.2.1 The Derivative of a Radius Vector

Consider two points A and B on a plane curve. Let the coordinates of the points be s and $s + \Delta s$, respectively (Fig. A.9). This curve is the hodograph of a radius vector. The corresponding increment of the vector $\mathbf{r}$ is

$$\Delta \mathbf{r} = \mathbf{r}_1 - \mathbf{r} = \mathbf{AB} \,.$$

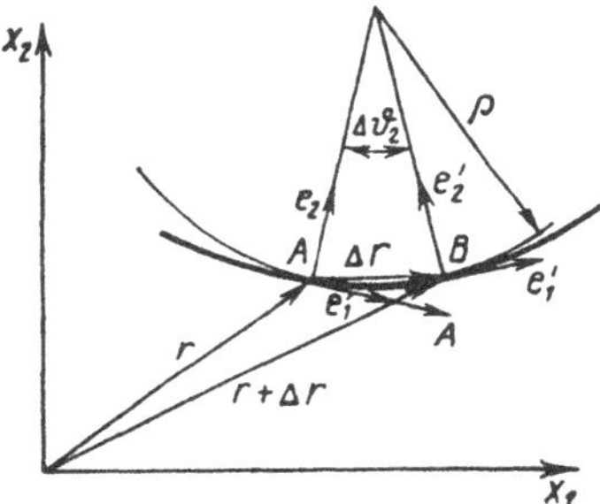

Fig. A.9.

As Δs tends to zero, the point B goes towards the point A and thus the vector $\mathbf{AB}$ rotates about the point A. In the limit, the vector $\mathbf{AB}$ is directed along the vector $\mathbf{A}$ that is tangent to the curve at the point A, that is,

$$\lim_{\Delta s \to 0} \frac{\Delta \mathbf{r}}{\Delta s} = \frac{\mathrm{d}\mathbf{r}}{\mathrm{d}s} = \mathbf{A} = |\mathbf{A}|\, \mathbf{e}_1 \,.$$

We see that the derivative of the vector function $\mathbf{r}(s)$ with respect to the scalar argument s is a vector that is tangent to the curve.

Let us prove that $\mathbf{A}$ is a unit vector. For small values of Δs, the corresponding element of the curve can be replaced by an arc of the osculating circle of radius ρ as shown in Fig. A.9. Consequently, we have

$$|\mathbf{A}| = \lim_{\Delta s \to 0} \left| \frac{\Delta \mathbf{r}}{\Delta s} \right| = \lim_{\Delta \vartheta_2 \to 0} \left| \frac{\rho \sin \Delta \vartheta_2}{\rho \, \Delta \vartheta_2} \right| = 1$$

and, finally,

$$\frac{\mathrm{d}\mathbf{r}}{\mathrm{d}s} = \mathbf{e}_1 . \tag{A.61}$$

Let us prove that the differentiation of the vector $\mathbf{e}_1$ yields a vector which is orthogonal to the vector $\mathbf{e}_1$. Differentiating the identity $(\mathbf{e}_1 \cdot \mathbf{e}_1) = 1$, we get

$$2\left(\mathbf{e}_1 \cdot \frac{\mathrm{d}\mathbf{e}_1}{\mathrm{d}s}\right) = 0 ; \tag{A.62}$$

hence, our statement is proved.

Since the vector $\mathrm{d}\mathbf{e}_1/\mathrm{d}s$ is orthogonal to the vector $\mathbf{e}_1$, we see that the vector $\mathrm{d}\mathbf{e}_1/\mathrm{d}s$ is directed along the vector $\mathbf{e}_2$. We have

$$\frac{\mathrm{d}\mathbf{e}_1}{\mathrm{d}s} = \lim_{\Delta s \to 0} \frac{\Delta \mathbf{e}_1}{\Delta s} = \lim_{\Delta s \to 0} \left| \frac{\Delta \mathbf{e}_1}{\Delta s} \right| \mathbf{e}_2 . \tag{A.63}$$

In view of the relations $\Delta \mathbf{e}_1 = \Delta\vartheta_2 \, |\mathbf{e}_1|$ and $\Delta s = \Delta\vartheta_2 \, \rho$, we have

$$\frac{\mathrm{d}^2\mathbf{r}}{\mathrm{d}s^2} = \frac{\mathrm{d}\mathbf{e}_1}{\mathrm{d}s} = \frac{1}{\rho}\,\mathbf{e}_2 . \tag{A.64}$$

A.2.2 Spatial Curves

The definition of the tangent to a spatial curve is similar to that definition for a plane curve. Unlike a plane curve, for a spatial curve and a fixed point A on it, there exist infinitely many vectors that are perpendicular to the curve at that point; these vectors form a plane α that is perpendicular to the vector $\mathbf{e}_1$ at the point A (Fig. A.10).

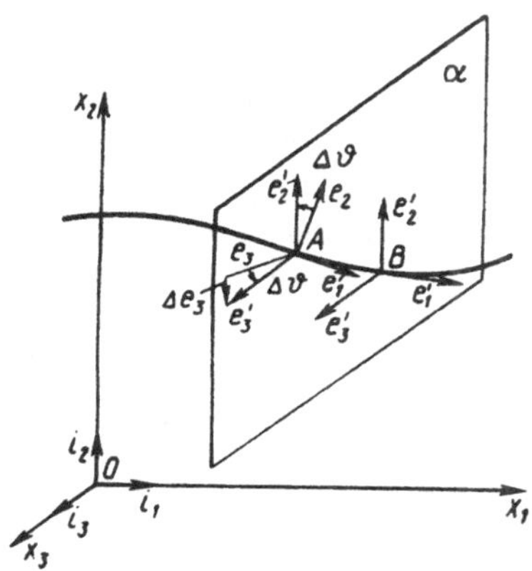

Fig. A.10.

Take a point B in the vicinity of the point A (Fig. A.11). The small element of the curve AB can be regarded as a plane curve lying in a plane β such that the tangent **AM** and the point B lie in this plane. When the point B approaches the point A, the plane β rotates and in the limit occupies a certain location in space. Among other planes passing through the point A, this limit plane yields the best local approximation of the curve. That is why the limit plane is called the *osculating plane* of the curve at the point A. The *principal normal* to the curve at the point A is the straight line **AN** which is perpendicular to the curve and lies in the osculating plane. At each point of the curve there exists a unique principal normal. Let the base vector $\mathbf{e}_2$ be directed along the principal normal. The vector $\mathbf{e}_3$ is so defined that the vectors $\mathbf{e}_i$ form a right-handed coordinate frame. The segment **AL** perpendicular to the principal normal is termed a *binormal*. The vector $\mathbf{e}_3$ is directed along the binormal. The axes directed along the vectors $\mathbf{e}_1$, $\mathbf{e}_2$, and $\mathbf{e}_3$ are called *natural axes*.

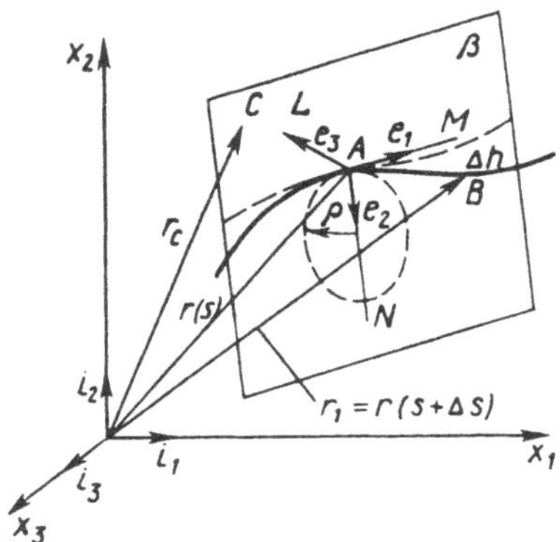

Fig. A.11.

The curve segment AB can be approximated by an arc of the osculating circle. The center of this circle lies on the principal normal. Let us prove that the second derivative of the radius vector with respect to s is a vector lying in the osculating plane. The osculating plane β containing the point A is shown in Fig. A.11. Consider a point B that lies on the curve near the point A. The distance from the point B to the plane β can be written as

$$\Delta h = \mathbf{e}_3 \cdot \mathbf{AB} = \mathbf{e}_3 \cdot (\mathbf{r}_1 - \mathbf{r}) \,, \tag{A.65}$$

where the vector $\mathbf{e}_3$ is perpendicular to the plane β. The Taylor expansion of the vector $\mathbf{r}_1$ is as follows:

$$\Delta h = \mathbf{e}_3 \cdot \left(\mathbf{r}' \, \Delta s + \mathbf{r}'' \, \frac{\Delta s^2}{2} + \mathbf{r}''' \, \frac{\Delta s^3}{6} + \dots \right) \tag{A.66}$$

or

$$\Delta h = (\mathbf{e}_3 \cdot \mathbf{r}') \, \Delta s + (\mathbf{e}_3 \cdot \mathbf{r}'') \, \frac{\Delta s^2}{2} + (\mathbf{e}_3 \cdot \mathbf{r}''') \, \frac{\Delta s^3}{6} + \dots \,. \tag{A.67}$$

Since the vector $\mathbf{r}'$ is orthogonal to the vector $\mathbf{e}_3$ and $\mathbf{r}' = \mathbf{e}_1$, we have

$$\Delta h = (\mathbf{e}_3 \cdot \mathbf{r}'')\,\frac{\Delta s^2}{2} + (\mathbf{e}_3 \cdot \mathbf{r}''')\,\frac{\Delta s^3}{6} + \dots . \tag{A.68}$$

The vector $\mathbf{r}$ is orthogonal to the vector $\mathbf{e}_1$, hence it is clear that Δh has a minimum when $(\mathbf{e}_3 \cdot \mathbf{r}'') = 0$ (in the vicinity of the point of contact, the curve adjoins tightly to the plane β). Thus, we see that the vector $\mathbf{r}''$ is orthogonal to the vector $\mathbf{e}_1$, that is, the vector $\mathbf{r}''$ lies in the osculating plane. The projection of the curve segment AB on the osculating plane is a plane curve. Considering the osculating circle of this curve and using the results obtained above, we get

$$\frac{\mathrm{d}^2\mathbf{r}''}{\mathrm{d}s^2} = \frac{\mathrm{d}\mathbf{e}_1}{\mathrm{d}s} = \frac{1}{\rho}\,\mathbf{e}_2 . \tag{A.69}$$

It should be noted that ρ in (A.69) signifies the radius of curvature of a spatial curve. The radius of curvature is the radius of the osculating circle. The curvature of a curve vanishes (the radius of curvature becomes equal to infinity) at *points of inflection*. The radius of curvature is an *invariant* because it remains unchanged under coordinate transformations.

A.2.3 Derivatives of the Base Vectors

Consider derivatives of the unit base vectors $\mathbf{e}_i$ with respect to the coordinate s. Differentiating a vector with respect to a scalar variable, we obtain a vector. Let us represent this new vector as a linear combination of the base vectors $\mathbf{e}_i$. As above, assume that the vector $\mathbf{e}_1$ is tangent to a curve (say, an axial line of a rod (Fig. A.10)). We have

$$\frac{\mathrm{d}\mathbf{e}_i}{\mathrm{d}s} = \sum_{j=1}^{3} æ_{ij}\mathbf{e}_j = æ_{ij}\mathbf{e}_j , \tag{A.70}$$

where $æ_{ij}$ are the elements of the matrix $[æ_{ij}]$. Let us prove that the matrix $[æ_{ij}]$ is skew-symmetrical, that is, $æ_{ij} = -æ_{ji}$. Multiplying both sides of (A.70) by $\mathbf{e}_k$, we get

$$\frac{\mathrm{d}\mathbf{e}_i}{\mathrm{d}s}\,\mathbf{e}_k = æ_{ik} . \tag{A.71}$$

Differentiating (A.71) with respect to s and using the identity $(\mathbf{e}_i \cdot \mathbf{e}_k) = \delta_{jk}$, we have

$$\frac{\mathrm{d}\mathbf{e}_i}{\mathrm{d}s}\,\mathbf{e}_k + \frac{\mathrm{d}\mathbf{e}_k}{\mathrm{d}s}\,\mathbf{e}_i = 0 . \tag{A.72}$$

Combining (A.71) and (A.72), we arrive at

$$æ_{ik} = -æ_{ki} . \tag{A.72}$$

From this relation it is clear that

$$æ_{ii} = 0\,, \qquad i = 1\,,2\,,3\,. \tag{A.74}$$

Thus, the elements of the matrix $[æ_{ij}]$ are expressed in terms of three independent parameters as follows:

$$[\,æ_{ij}\,] = \begin{bmatrix} 0 & -æ_3 & æ_2 \\ æ_3 & 0 & -æ_1 \\ -æ_2 & æ_1 & 0 \end{bmatrix}. \tag{A.75}$$

The elements of the matrix $[æ_{ij}]$ are geometrical characteristics of the curve. The geometrical meaning of the quantities $æ_1$, $æ_2$, and $æ_3$ is discussed in Sect. A.2.4. For the basis $\{\mathbf{e}_{i0}\}$, we get the similar relations,

$$\frac{\mathrm{d}\mathbf{e}_{i0}}{\mathrm{d}s} = æ_{ij0}\,\mathbf{e}_{\rho 0}\,. \tag{A.76}$$

The matrices $[æ_{ij}]$ in (A.71) and (A.76) can be replaced by the vectors $\boldsymbol{æ}$ and $\boldsymbol{æ}_0$, respectively, that is,

$$\frac{\mathrm{d}\mathbf{e}_i}{\mathrm{d}s} = \boldsymbol{æ} \times \mathbf{e}_i = \varepsilon_{kji} æ_j \mathbf{e}_k\,, \qquad \boldsymbol{æ} = æ_j \mathbf{e}_j\,; \tag{A.77}$$

$$\frac{\mathrm{d}\mathbf{e}_{\nu 0}}{\mathrm{d}s} = \boldsymbol{æ}_0 \times \mathbf{e}_0 = \varepsilon_{n\rho\nu} æ_{\rho 0} \mathbf{e}_{n0}\,, \tag{A.78}$$

where

$$\boldsymbol{æ} = \sum_{j=1}^{3} æ_j \mathbf{e}_j\,. \tag{A.79}$$

In expanded form, the system of equations (A.77) is as follows:

$$\begin{aligned} \frac{\mathrm{d}\mathbf{e}_1}{\mathrm{d}s} &= æ_3 \mathbf{e}_2 - æ_2 \mathbf{e}_3\,; \\ \frac{\mathrm{d}\mathbf{e}_2}{\mathrm{d}s} &= -æ_1 \mathbf{e}_3 + æ_3 \mathbf{e}_1\,; \\ \frac{\mathrm{d}\mathbf{e}_3}{\mathrm{d}s} &= æ_2 \mathbf{e}_1 - æ_1 \mathbf{e}_2\,. \end{aligned} \tag{A.80}$$

A.2.4 Geometrical Meaning of the Components of the Vector æ

No restrictions were imposed on the vectors $\mathbf{e}_2$ and $\mathbf{e}_3$ when we derived the relations (A.80). Therefore, these relations are the most general ones. Now let the vectors $\mathbf{e}_i$ in (A.80) be directed along the natural axes. In this case, we have (notation $\boldsymbol{\Omega}$ is used instead of æ)

$$\frac{\mathrm{d}\mathbf{e}_1}{\mathrm{d}s} = \frac{1}{\rho}\,\mathbf{e}_2 = \Omega_3 \mathbf{e}_2 - \Omega_2 \mathbf{e}_3\,. \tag{A.81}$$

Consequently, $\Omega_2 = 0$ and $\Omega_3 = 1/\rho$. We see that the component Ω_3 of the vector $\boldsymbol{\Omega}$ is the *curvature of the spatial curve*. Finally, considering the derivative of the unit vectors of the natural frame, we arrive at

$$\frac{\mathrm{d}\mathbf{e}_1}{\mathrm{d}s} = \Omega_3 \mathbf{e}_2 \, ; \tag{A.82}$$

$$\frac{\mathrm{d}\mathbf{e}_2}{\mathrm{d}s} = -\Omega_3 \mathbf{e}_1 + \Omega_1 \mathbf{e}_3 \, ; \tag{A.83}$$

$$\frac{\mathrm{d}\mathbf{e}_3}{\mathrm{d}s} = -\Omega_1 \mathbf{e}_2 \, . \tag{A.84}$$

The relations (A.82)–(A.84) are known as the *Serret-Frenet formulas*.

Let us clarify the geometrical meaning of the component Ω_1. The vector $\mathrm{d}\mathbf{e}_3/\mathrm{d}s$ is orthogonal to the vector $\mathbf{e}_3$ and lies in the plane α. This plane is perpendicular to the osculating plane β (Fig. A.10). We have

$$\frac{\mathrm{d}\mathbf{e}_3}{\mathrm{d}s} = \lim_{\Delta s \to 0} \left(\frac{\Delta \mathbf{e}_3}{\Delta s} \right) \mathbf{e}_2 = - \lim_{\Delta s \to 0} \left| \frac{\Delta \vartheta_{10}}{\Delta s} \right| \mathbf{e}_2 = - \frac{\Delta \vartheta_{10}}{\Delta s} \mathbf{e}_2 \, , \tag{A.85}$$

therefore,

$$\Omega_1 = \frac{\mathrm{d}\vartheta_1}{\mathrm{d}s} \, . \tag{A.86}$$

The component Ω_1 is the measure of deviation of the curve from the osculating plane. That is why Ω_1 is called the *second curvature* or the *twist* of a spatial curve.

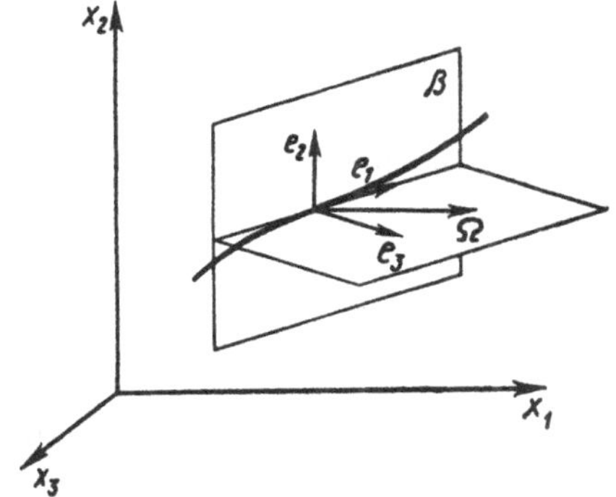

Fig. A.12.

When the point A moves along the curve, the natural frame with origin at A changes its orientation. The rotation of the natural frame is characterized by the *Darboux vector*. Unlike the vector æ considered in Sect. A.2.3, the Darboux vector has only two nonzero components Ω_1 and Ω_3 ($\Omega_2 = 0$). The

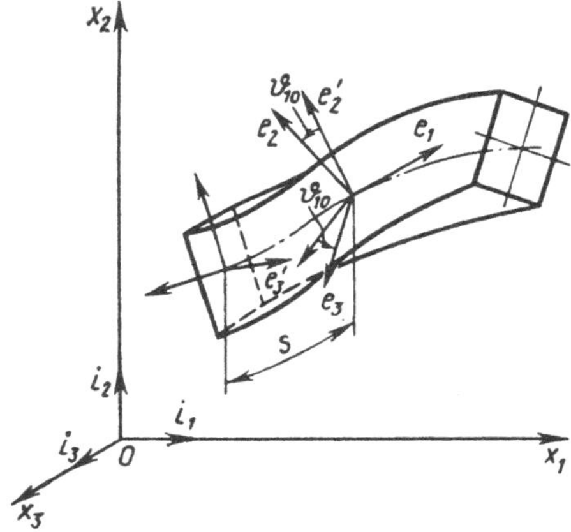

Fig. A.13.

vector $\boldsymbol{\Omega}$ lies in the plane which is perpendicular to the osculating plane β (see Fig. A.12).

The vector æ in Sect. A.2.3 defines rotation of an arbitrary coordinate frame as its origin moves along the curve. The vectors $\mathbf{e}_2$ and $\mathbf{e}_3$ are not necessarily directed along the normal and the binormal of the curve, respectively. Suppose that these vectors are directed along the principal axes of a cross section of a rod (Fig. A.13). The principal axes $\mathbf{e}_2$ and $\mathbf{e}_3$ are rotated by an angle ϑ_{10} relative to the natural axes. Determine the projections of the vector æ in the coordinate frame associated with the principal axes of the cross section. The matrix of transformation from the basis $\{\mathbf{e}_i'\}$ to the basis $\{\mathbf{e}_i\}$ is as follows:

$$\mathrm{L}^{(1)} = \begin{bmatrix} 1 & 0 & 0 \\ 0 & \cos\vartheta_{10} & \sin\vartheta_{10} \\ 0 & -\sin\vartheta_{10} & \cos\vartheta_{10} \end{bmatrix} . \tag{A.87}$$

Then, we can rewrite (A.79) in terms of projections in the basis $\{\mathbf{e}_i\}$,

$$æ = æ_1\mathbf{e}_1 + \Omega_3 \sin\vartheta_{10}\mathbf{e}_2 + \Omega_3\cos\vartheta_{10}\mathbf{e}_3 \, , \tag{A.88}$$

where $\Omega_3 \sin\vartheta_{10} = æ_2$ and $\Omega_3\cos\vartheta_{10} = æ_3$.

The quantity $æ_1$ in (A.88) is a sum of two quantities,

$$æ_1 = \Omega_1 + \frac{\mathrm{d}\vartheta_{10}}{\mathrm{d}s} \, , \tag{A.89}$$

where Ω_1 is the twist of the rod (caused by the moment $\mathbf{M}$) and $\mathrm{d}\vartheta_{10}/\mathrm{d}s$ is the natural twisting of the rod.

For initially-twisted rods (e.g. drilling bits), the angle ϑ_{10} does not necessarily result from a deformation of the rod.

Let us now establish the relationship between the projections Ω_3 (curvature) and Ω_1 (twist) in terms of the Cartesian coordinates of the point A. Since

$$\frac{\mathrm{d}^2\mathbf{r}}{\mathrm{d}s^2} = \Omega_3\mathbf{e}_2' , \qquad \mathbf{r} = x_1\mathbf{i}_1 + x_2\mathbf{i}_2 + x_3\mathbf{i}_3 , \tag{A.90}$$

we have

$$\Omega_3 = \left|\frac{\mathrm{d}^2\mathbf{r}}{\mathrm{d}s^2}\right| = \sqrt{x_1''^2 + x_2''^2 + x_3''^2} . \tag{A.91}$$

In the case of a plane curve ($x_3 = 0$), we have

$$\Omega_3 = \sqrt{x_1''^2 + x_2''^2} \tag{A.92}$$

or, assuming that x_2 is a function of x_1,

$$\Omega_3 = \frac{\dfrac{\mathrm{d}^2x_2}{\mathrm{d}x_1^2}}{\left[1 + \left(\dfrac{\mathrm{d}^2x_2}{\mathrm{d}x_1}\right)^2\right]^{3/2}} . \tag{A.93}$$

Now let us obtain another formula for the curvature Ω_3 of a plane curve. Since

$$\frac{\mathrm{d}x_1}{\mathrm{d}s} = \cos\varphi ; \qquad \frac{\mathrm{d}x_2}{\mathrm{d}s} = \sin\varphi , \tag{A.94}$$

we see that the differentiation of (A.94) with respect to s yields

$$x_1'' = -\frac{1}{\rho}\sin\varphi ; \qquad x_2'' = \frac{1}{\rho}\cos\varphi . \tag{A.95}$$

In view of (A.94) and (A.95), we have

$$x_1'x_2'' = \frac{1}{\rho}\cos^2\varphi ; \qquad x_2'x_1'' = -\frac{1}{\rho}\sin^2\varphi . \tag{A.96}$$

Finally, we arrive at

$$\Omega_3 = \frac{1}{\rho} = x_1'x_2'' - x_2'x_1'' . \tag{A.97}$$

Considering that

$$\mathbf{e}_2 = \frac{1}{\Omega_3}\mathbf{r}'' ; \qquad \mathbf{e}_3 = (\mathbf{e}_1 \times \mathbf{e}_2) = \frac{1}{\Omega_3}(\mathbf{r}' \times \mathbf{r}'') \tag{A.98}$$

and using (A.84), we get

$$-(\mathbf{e}_2 \cdot \mathbf{e}_3') = \Omega_1 . \tag{A.99}$$

Combining (A.98) and (A.99), we arrive at

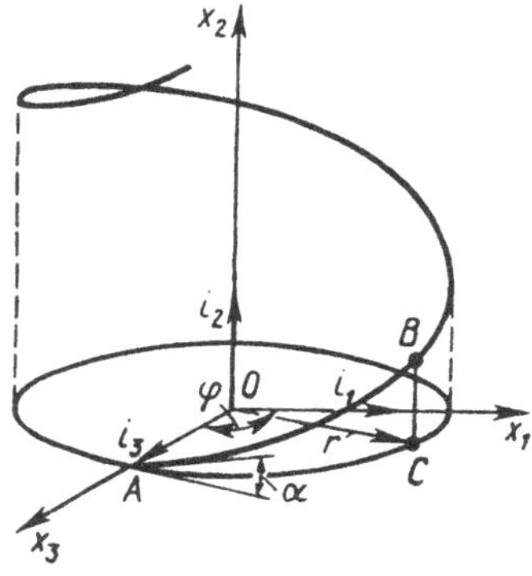

Fig. A.14.

$$\Omega_1 = \frac{\mathbf{r}' \cdot (\mathbf{r}'' \times \mathbf{r}''')}{\mathbf{r}''^2} = \frac{1}{\sum_{j=1}^{3} x_j''^2} \begin{vmatrix} x_1' & x_2' & x_3' \\ x_1'' & x_2'' & x_3'' \\ x_1''' & x_2''' & x_3''' \end{vmatrix} . \tag{A.100}$$

As an example, let us evaluate the curvature and the twist for a helical curve (e.g. the axial line of a cylindrical spring (Fig. A.14)). The coordinates of the point B are

$$x_1 = r \sin\varphi ; \qquad x_2 = r\varphi \tan\alpha ; \qquad x_3 = r\cos\varphi . \tag{A.101}$$

The coordinate x_2 is related to the length of the segment $AB = s$ as follows:

$$s = \varphi\sqrt{r^2 + r^2\tan^2\alpha} = \varphi k , \qquad k = r\sqrt{1+\tan^2\alpha} ; \tag{A.102}$$

here α is the angle of helix. We have

$$x_1 = r\sin\frac{s}{k} ; \qquad x_2 = r\,\frac{s}{k}\tan\alpha ; \qquad x_3 = r\cos\frac{s}{k} . \tag{A.103}$$

Substituting the derivatives of x_i with respect to s into (A.97) and (A.100), we get

$$\Omega_3 = \frac{r}{r^2 + r^2\tan^2\alpha} = \frac{1}{r}\cos^2\alpha = \text{const} ; \tag{A.104}$$

$$\Omega_1 = \frac{r\tan\alpha}{r^2 + r^2\tan^2\alpha} = \frac{1}{r}\,\frac{\sin 2\alpha}{2} = \text{const} . \tag{A.105}$$

A.2.5 Relationship Between $æ_i$ and ϑ_j

As mentioned above, the base vectors of the natural coordinate frame change their directions when the origin moves along the curve. As it was proved in Sect. A.1.8, a new location of the natural basis relative to its original location is well defined if the three angles of rotation ϑ_1, ϑ_2, and ϑ_3 are specified. Consequently, the components of the vector æ, which characterizes the rotation of the natural frame, can be expressed in terms of these angles. Let us derive the relationship between $æ_i$ and ϑ_j using (A.70), (A.76), and the relations

$$\mathbf{e}_i = l_{i\nu}\mathbf{e}_{\nu 0}\,; \qquad \mathbf{e}_{\nu 0} = l_{k\nu}\mathbf{e}_k\,. \tag{A.106}$$

Differentiating $\mathbf{e}_i$ with respect to s and using (A.70), we get

$$\frac{\mathrm{d}\mathbf{e}_i}{\mathrm{d}s} = \frac{\mathrm{d}l_{i\nu}}{\mathrm{d}s}\mathbf{e}_{\nu 0} + l_{i\nu}\frac{\mathrm{d}\mathbf{e}_{\nu 0}}{\mathrm{d}s} = \varepsilon_{kji}æ_j\mathbf{e}_k\,. \tag{A.107}$$

Eliminating $\mathbf{e}_{\nu 0}$ and $\mathrm{d}\mathbf{e}_{\nu 0}/\mathrm{d}s$ from (A.107) by use of (A.106) and (A.107), we arrive at

$$\frac{\mathrm{d}\mathbf{e}_i}{\mathrm{d}s} = \frac{\mathrm{d}l_{i\nu}}{\mathrm{d}s}\,l_{m\nu}\mathbf{e}_m + l_{i\nu}\varepsilon_{n\rho\nu}æ_{\rho 0}\mathbf{e}_{n0} = \varepsilon_{kji}æ_j\mathbf{e}_k\,. \tag{A.108}$$

Finally, elimination of $\mathbf{e}_{n0} = l_{kn}\mathbf{e}_k$ yields

$$\frac{\mathrm{d}\mathbf{e}_i}{\mathrm{d}s} = \frac{\mathrm{d}l_{i\nu}}{\mathrm{d}s}\,l_{k\nu}\mathbf{e}_k + l_{i\nu}\varepsilon_{n\rho\nu}æ_{\rho 0}l_{kn}\mathbf{e}_k = \varepsilon_{kji}æ_j\mathbf{e}_k\,. \tag{A.109}$$

The repeated indices can be replaced by any other letters, e.g. in the first term in the left-hand side of the equality, the index ν may be replaced by n. The result is

$$\left(\frac{\mathrm{d}l_{in}}{\mathrm{d}s}\,l_{kn} + l_{i\nu}\varepsilon_{n\rho\nu}æ_{\rho 0}l_{kn}\right)\mathbf{e}_k = \varepsilon_{kji}æ_j\mathbf{e}_k\,, \tag{A.110}$$

hence,

$$\varepsilon_{kji}æ_j = \left(\frac{\mathrm{d}l_{in}}{\mathrm{d}s} + \varepsilon_{n\rho\nu}l_{i\nu}æ_{\rho 0}\right)l_{kn}\,. \tag{A.111}$$

Let us obtain $æ_1$. Setting $k = 3$ and $i = 2$, we get

$$\begin{aligned} æ_1 = \left(\frac{\mathrm{d}l_{2n}}{\mathrm{d}s} + \varepsilon_{n\rho\nu}l_{2\nu}æ_{\rho 0}\right)l_{3n} &= \left(\frac{\mathrm{d}l_{21}}{\mathrm{d}s} + \varepsilon_{1\rho\nu}l_{2\nu}æ_{\rho 0}\right)l_{31} \\ &+ \left(\frac{\mathrm{d}l_{22}}{\mathrm{d}s} + \varepsilon_{2\rho\nu}l_{2\nu}æ_{\rho 0}\right)l_{32} + \left(\frac{\mathrm{d}l_{23}}{\mathrm{d}s} + \varepsilon_{3\rho\nu}l_{2\nu}æ_{\rho 0}\right)l_{33}\,. \end{aligned} \tag{A.112}$$

The second terms enclosed in the parenthesis are as follows:

$$\begin{aligned} \varepsilon_{1\rho\nu}l_{2\nu}æ_{\rho 0} &= l_{23}æ_{20} - l_{22}æ_{30}\,; \\ \varepsilon_{2\rho\nu}l_{2\nu}æ_{\rho 0} &= l_{21}æ_{30} - l_{23}æ_{10}\,; \\ \varepsilon_{3\rho\nu}l_{2\nu}æ_{\rho 0} &= l_{22}æ_{10} - l_{21}æ_{20}\,. \end{aligned} \tag{A.113}$$

As a result, we have

$$æ_1 = \left(\frac{\mathrm{d}l_{21}}{\mathrm{d}s} + l_{23}æ_{20} - l_{22}æ_{30}\right) l_{31} + \left(\frac{\mathrm{d}l_{22}}{\mathrm{d}s} + l_{21}æ_{30} - l_{23}æ_{10}\right) l_{32} + \left(\frac{\mathrm{d}l_{32}}{\mathrm{d}s} + l_{22}æ_{10} - l_{21}æ_{20}\right) \tag{A.114}$$

or

$$æ_1 = \frac{\mathrm{d}l_{21}}{\mathrm{d}s} l_{31} + \frac{\mathrm{d}l_{22}}{\mathrm{d}s} l_{32} + \frac{\mathrm{d}l_{23}}{\mathrm{d}s} l_{33} + (l_{22}l_{33} - l_{23}l_{32})\, æ_{10} + (l_{23}l_{31} - l_{21}l_{33})\, æ_{20} + (l_{21}l_{32} - l_{22}l_{31})\, æ_{30}\,. \tag{A.115}$$

The components $æ_2$ and $æ_3$ can be obtained similarly:

$$æ_2 = \frac{\mathrm{d}l_{31}}{\mathrm{d}s} l_{11} + \frac{\mathrm{d}l_{32}}{\mathrm{d}s} l_{12} + \frac{\mathrm{d}l_{33}}{\mathrm{d}s} l_{13} + (l_{32}l_{13} - l_{33}l_{12})\, æ_{10} + (l_{33}l_{11} - l_{31}l_{13})\, æ_{20} + (l_{31}l_{12} - l_{32}l_{11})\, æ_{30}\,; \tag{A.116}$$

$$æ_3 = \frac{\mathrm{d}l_{11}}{\mathrm{d}s} l_{21} + \frac{\mathrm{d}l_{12}}{\mathrm{d}s} l_{22} + \frac{\mathrm{d}l_{13}}{\mathrm{d}s} l_{23} + (l_{12}l_{23} - l_{13}l_{22})\, æ_{10} + (l_{13}l_{21} - l_{11}l_{23})\, æ_{20} + (l_{11}l_{22} - l_{12}l_{21})\, æ_{30}\,. \tag{A.117}$$

Representing l_{ij} in (A.115)–(A.117) in terms of the angles ϑ_1, ϑ_2, and ϑ_3, we arrive at

$$æ_1 = \left(\frac{\mathrm{d}\vartheta_1}{\mathrm{d}s} + æ_{10}\right)\cos\vartheta_2\cos\vartheta_3 - \frac{\mathrm{d}\vartheta_3}{\mathrm{d}s}\sin\vartheta_2 + (\sin\vartheta_2\sin\vartheta_1 + \cos\vartheta_2\sin\vartheta_3\cos\vartheta_1)\, æ_{20} + (\cos\vartheta_2\sin\vartheta_3\sin\vartheta_1 - \sin\vartheta_2\cos\vartheta_1)\, æ_{30}\,; \tag{A.118}$$

$$æ_2 = \frac{\mathrm{d}\vartheta_2}{\mathrm{d}s} - \left(\frac{\mathrm{d}\vartheta_1}{\mathrm{d}s} + æ_{10}\right)\sin\vartheta_3 + \cos\vartheta_3\cos\vartheta_1 æ_{20} + \cos\vartheta_3\sin\vartheta_1 æ_{30}\,; \tag{A.119}$$

$$æ_3 = \frac{\mathrm{d}\vartheta_3}{\mathrm{d}s}\cos\vartheta_2 + \left(\frac{\mathrm{d}\vartheta_1}{\mathrm{d}s} + æ_{10}\right)\sin\vartheta_2\cos\vartheta_3 + (\sin\vartheta_2\sin\vartheta_3\cos\vartheta_1 - \cos\vartheta_2\sin\vartheta_1)\, æ_{20} + (\cos\vartheta_2\cos\vartheta_1 + \sin\vartheta_2\sin\vartheta_3\sin\vartheta_1)\, æ_{30}\,. \tag{A.120}$$

The angles ϑ_1, ϑ_2, and ϑ_3 in (A.118)–(A.120) are measured from the reference frame $\{\mathbf{e}_{i0}\}$ attributed to the natural state of the rod. The initial state of the rod is assumed to be identical to its natural state. To simplify rearrangement, let us write the relations (A.118)–(A.120) as a vector equation

$$\boldsymbol{æ} = \mathrm{L}_1 \frac{\mathrm{d}\tilde{\boldsymbol{\vartheta}}}{\mathrm{d}s} + \mathrm{L}æ_0^{(1)}\,, \qquad æ_0^{(1)} = æ_{i0}\mathbf{e}_i\,, \tag{A.121}$$

where

$$\boldsymbol{\vartheta} = \begin{bmatrix} \vartheta_1 \\ \vartheta_2 \\ \vartheta_3 \end{bmatrix}; \qquad \frac{\tilde{\mathrm{d}}\boldsymbol{\vartheta}}{\mathrm{d}s} = \begin{bmatrix} \vartheta_1' \\ \vartheta_2' \\ \vartheta_3' \end{bmatrix};$$

$$\mathrm{L}_1 = \begin{bmatrix} \cos\vartheta_2\cos\vartheta_3 & 0 & -\sin\vartheta_2 \\ -\sin\vartheta_3 & 1 & 0 \\ \sin\vartheta_2\cos\vartheta_3 & 0 & \cos\vartheta_2 \end{bmatrix}; \tag{A.122}$$

here $\frac{\tilde{\mathrm{d}}}{\mathrm{d}s}$ signifies the local derivative.

The vector $\text{æ}_0^{(1)}$ is not the same as the vector æ_0. The latter characterizes the shape of the curve in the initial configuration. The projections of the vector $\text{æ}_0^{(1)}$ in the basis $\{\mathbf{e}_i\}$ are equal to the projections of the vector æ_0 in the basis $\{\mathbf{e}_{i0}\}$. If the initial shape of the curve described by the vector æ_{i0} is known a priori, then the relations (A.121) help us understand how the shape of the rod axial line (described by the vector æ_0) changes.

Let us derive the expressions for the vector æ_0. We assume that the Cartesian components of the base vectors $\mathbf{e}_{i0}$ at any point of the curve are given (Fig. A.14), that is, the matrix L^0 (see (A.55)) is known.

The vector $\mathbf{e}_{i0}$ can be represented as follows:

$$\mathbf{e}_{i0} = \sum_{j=1}^{3} l_{i\rho}^0 \mathbf{i}_\rho .$$

Since the vectors $\mathbf{i}_\rho$ do not depend on s, we see that the derivatives of $\mathbf{e}_{i0}$ with respect to s take the form

$$\frac{\mathrm{d}\mathbf{e}_{i0}}{\mathrm{d}s} = \frac{\mathrm{d}l_{i\rho}^0}{\mathrm{d}s}\mathbf{i}_\rho = \varepsilon_{kji}\text{æ}_{j0}\mathbf{e}_{k0} . \tag{A.123}$$

Using the relation $\mathbf{i}_\rho = l_{k\rho}^{(0)}\mathbf{e}_{k0}$, we can rewrite (A.123) as follows:

$$\varepsilon_{kji}\text{æ}_{j0} = \frac{\mathrm{d}l_{i\rho}^0}{\mathrm{d}s} l_{k\rho}^0 . \tag{A.124}$$

In expanded notation, we have

$$\begin{aligned} \text{æ}_{10} &= \frac{\mathrm{d}l_{21}^0}{\mathrm{d}\eta} l_{31}^0 + \frac{\mathrm{d}l_{22}^0}{\mathrm{d}\eta} l_{32}^0 + \frac{\mathrm{d}l_{23}^0}{\mathrm{d}\eta} l_{33}^0 ; \\ \text{æ}_{20} &= \frac{\mathrm{d}l_{31}^0}{\mathrm{d}\eta} l_{11}^0 + \frac{\mathrm{d}l_{32}^0}{\mathrm{d}\eta} l_{12}^0 + \frac{\mathrm{d}l_{33}^0}{\mathrm{d}\eta} l_{13}^0 ; \\ \text{æ}_{30} &= \frac{\mathrm{d}l_{11}^0}{\mathrm{d}\eta} l_{21}^0 + \frac{\mathrm{d}l_{12}^0}{\mathrm{d}\eta} l_{22}^0 + \frac{\mathrm{d}l_{13}^0}{\mathrm{d}\eta} l_{23}^0 \end{aligned} \tag{A.125}$$

or

$$
\begin{aligned}
æ_{10} &= \frac{d\vartheta_1^0}{ds}\cos\vartheta_2^2\cos\vartheta_3^0 - \frac{d\vartheta_3^0}{ds}\sin\vartheta_2^0\,;\\
æ_{20} &= \frac{d\vartheta_2^0}{ds} - \frac{d\vartheta_1^0}{ds}\sin\vartheta_3^0\,;\\
æ_{30} &= \frac{d\vartheta_3^0}{ds}\cos\vartheta_2^0 + \frac{d\vartheta_1^0}{ds}\sin\vartheta_2^0\cos\vartheta_1^0\,. \qquad \text{(A.126)}
\end{aligned}
$$

In vector notation, equations (A.126) can be rewritten as follows:

$$
æ_0 = \mathrm{L}_1^0\,\frac{d\boldsymbol{\vartheta}^0}{ds}\,; \qquad \text{(A.127)}
$$

here

$$
\mathrm{L}_1^0 = \begin{bmatrix} \cos\vartheta_2^0\cos\vartheta_3^0 & 0 & -\sin\vartheta_2^0 \\ -\sin\vartheta_3^0 & 1 & 0 \\ \sin\vartheta_2^0\cos\vartheta_1^0 & 0 & \cos\vartheta_2^0 \end{bmatrix}.
$$

A.2.6 Derivatives of a Vector in the Attached Coordinate System

Differentiating a vector **a** with respect to s, we get

$$
\frac{d\mathbf{a}}{ds} = \sum_{i=1}^{3}\frac{da_i}{ds}\,\mathbf{e}_i + \sum_{i=1}^{3} a_i\frac{d\mathbf{e}_i}{ds}\,. \qquad \text{(A.128)}
$$

Using (A.70) in (A.128), we have

$$
\frac{d\mathbf{a}}{ds} = \frac{\tilde{d}\mathbf{a}}{ds} + æ\times\mathbf{a}\,, \qquad \text{(A.129)}
$$

where $\frac{\tilde{d}}{ds}$ signifies the local derivative; this derivative describes the change of the components of the vector **a** in the movable reference frame. If the components of the vector æ are known, then it is more convenient to represent the vector product æ × **a** as follows:

$$
æ\times\mathbf{a} = \mathrm{A}_{æ}\mathbf{a}\,; \qquad \text{(A.130)}
$$

here

$$
\mathrm{A}_{æ} = \begin{bmatrix} 0 & -æ_3 & æ_2 \\ æ_3 & 0 & -æ_1 \\ -æ_2 & æ_1 & 0 \end{bmatrix}.
$$

If the movable frame is identical to the natural frame (i.e. the vector æ is a Darboux vector), then the matrix $\mathrm{A}_{æ}$ reads

$$
\mathrm{A}_{æ} = \begin{bmatrix} 0 & -\Omega_3 & 0 \\ \Omega_3 & 0 & -\Omega_1 \\ 0 & \Omega_1 & 0 \end{bmatrix}. \qquad \text{(A.131)}
$$

It should be noted that the matrices $A_{æ}$ in (A.130) and (A.131) are degenerate (their determinants are equal to zero).

Finally, the derivative of the vector $\mathbf{a}$ with respect to s takes the form

$$\frac{d\mathbf{a}}{ds} = \frac{\tilde{d}\mathbf{a}}{ds} + A_{æ}\mathbf{a}\,. \tag{A.132}$$

Recall that the derivative of the base vectors $\mathbf{e}_j$ (A.80) and the derivative of an arbitrary vector $\mathbf{a}$ (A.132) are represented in terms of local derivatives. These formulas are useful in representing equations in terms of projections in the attached coordinate system.

A.3 Increments of the Components of a Vector under Transformation of the Attached Coordinate System

When deriving equilibrium equations, we are to operate with vectors of external and internal loads and vectors that describe a shape of the rod axis. Thus, it is important to obtain the formulas for the changes of the components of the vectors under transformations of the attached coordinate system. Initial and final configurations of an element of a rod and the attached coordinate system are shown in Fig. A.15.

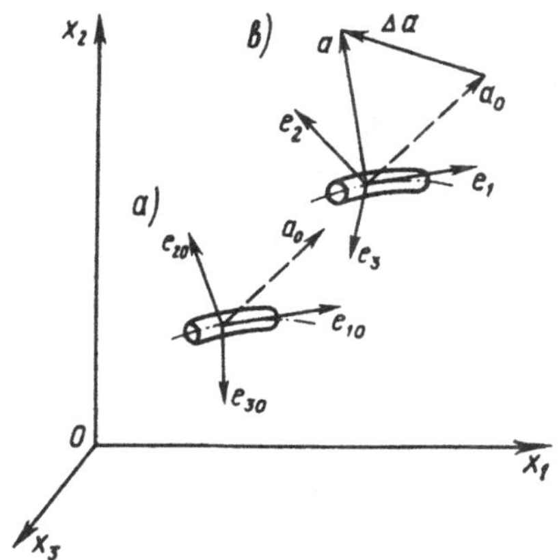

Fig. A.15.

When the basis $\{\mathbf{e}_{i0}\}$ is translated and rotated, the projections of a vector $\mathbf{a}_0$ in this basis change. We have

$$\Delta a_i = a_i - a_{i0}\,, \tag{A.133}$$

where a_{i0} are the projections of the vector $\mathbf{a}_0$ in the initial attached frame (Fig. A.15*a*) and a_i are the projections of the vector $\mathbf{a}$ in the new attached frame (Fig. A.15*б*). The increments Δa_i are assumed to be finite.

Consider a vector $\Delta\mathbf{a}$ whose components are Δa_i. We have

$$\Delta\mathbf{a} = \mathbf{a} - \mathbf{a}_0 . \tag{A.134}$$

The vectors $\mathbf{a}$ and $\mathbf{a}_0$ are given by their projections in the different frames,

$$\mathbf{a} = \sum_{i=1}^{3} a_i \mathbf{e}_i ; \qquad \mathbf{a}_0 = \sum_{i=1}^{3} a_{i0} \mathbf{e}_{i0} , \tag{A.135}$$

consequently, the relation (A.134) cannot be derived directly from (A.133). Let us introduce a vector $\mathbf{a}_0^{(1)}$ such that

$$\mathbf{a}_0^{(1)} = \sum_{i=1}^{3} \mathbf{a}_{i0} \mathbf{e}_i , \qquad \mathbf{a}_0^{(1)} \neq \mathbf{a}_0 , \tag{A.136}$$

where a_{i0} are the projections of the vector $\mathbf{a}_0$ in the basis $\{\mathbf{e}_{i0}\}$. Let

$$\Delta\mathbf{a} = \mathbf{a} - \mathbf{a}_0^{(1)} . \tag{A.137}$$

In problems of static stability of rods, we often need to determine increments of external loads. At loss of stability, such a load keeps its original magnitude but changes its direction with reference to the attached frame such that $|\mathbf{a}_0| = |\mathbf{a}|$ but $\mathbf{a}_0 \neq \mathbf{a}$. Suppose that the initial configuration is a critical configuration of the rod (see Fig. A.15) and the final configuration corresponds to the new, postbuckling equilibrium configuration. Our task is to determine the increments of the components of the vector $\mathbf{a}$ under the condition $|\mathbf{a}| = |\mathbf{a}_0|$. In this case, the change of the vector components is caused only by rotations of the vector $\mathbf{a}$ relative to the attached frame. The projections of the vector $\mathbf{a}$ in the frames $\{\mathbf{e}_i\}$ and $\{\mathbf{e}_{i0}\}$ are related as follows: (see Sect. A.1.7):

$$\mathbf{a} = \mathrm{L}\mathbf{a}_0 . \tag{A.138}$$

Subtracting $\mathbf{a}_0^{(1)}$ from both sides of (A.138), we obtain $\Delta\mathbf{a}$ referred to the basis $\{\mathbf{e}_j\}$,

$$\Delta\mathbf{a} = \mathrm{L}\mathbf{a}_0 - \mathbf{a}_0^{(1)} , \tag{A.139}$$

where $\mathbf{a}_0^{(1)} = \mathrm{E}\mathbf{a}_0$, therefore, using (A.44), we get

$$\Delta\mathbf{a} = (\mathrm{L} - \mathrm{E})\,\mathbf{a}_0 , \tag{A.140}$$

where

$$\mathrm{L} - \mathrm{E} = \begin{bmatrix} \cos\vartheta_2\cos\vartheta_3 - 1 & \begin{matrix}\cos\vartheta_2\sin\vartheta_3\cos\vartheta_1 \\ +\sin\vartheta_2\sin\vartheta_1\end{matrix} & \begin{matrix}\cos\vartheta_2\sin\vartheta_3\sin\vartheta_1 \\ -\sin\vartheta_2\cos\vartheta_1\end{matrix} \\ -\sin\vartheta_3 & \cos\vartheta_1\cos\vartheta_3 - 1 & \cos\vartheta_3\sin\vartheta_1 \\ \sin\vartheta_2\cos\vartheta_3 & \begin{matrix}\sin\vartheta_2\sin\vartheta_3\cos\vartheta_1 \\ -\cos\vartheta_2\sin\vartheta_1\end{matrix} & \begin{matrix}\sin\vartheta_2\sin\vartheta_3\sin\vartheta_1 \\ +\cos\vartheta_2\cos\vartheta_1 - 1\end{matrix} \end{bmatrix} .$$

The relation (A.140) enables us to determine increments of the components of the vector $\mathbf{a}_0$ for any angles of rotation ϑ_j of the attached frame $\{\mathbf{e}_j\}$. Suppose that these angles of rotation are small; then, considering (A.46), we get

$$\Delta\mathbf{a} = \begin{bmatrix} 0 & \vartheta_3 & -\vartheta_2 \\ -\vartheta_3 & 0 & \vartheta_1 \\ \vartheta_2 & -\vartheta_1 & 0 \end{bmatrix} \mathbf{a}_0 \tag{A.141}$$

or

$$\Delta\mathbf{a} = (\vartheta_3 a_{20} - \vartheta_2 a_{30})\,\mathbf{e}_1 + (-\vartheta_3 a_{10} + \vartheta_1 a_{30})\,\mathbf{e}_2 + (\vartheta_2 a_{10} - \vartheta_1 a_{20})\,\mathbf{e}_3 \,. \tag{A.142}$$

If the components a_{j0} are known while the angles ϑ_j are unknown, it is convenient to write

$$\Delta\mathbf{a} = \mathrm{C}\boldsymbol{\vartheta}\,, \tag{A.143}$$

where

$$\mathrm{C} = \begin{bmatrix} 0 & -a_{30} & a_{20} \\ a_{30} & 0 & -a_{10} \\ -a_{20} & a_{10} & 0 \end{bmatrix}; \qquad \boldsymbol{\vartheta} = \begin{bmatrix} \vartheta_1 \\ \vartheta_2 \\ \vartheta_3 \end{bmatrix}.$$

A.4 Distributions

A.4.1 The δ-function

Consider a function $\Phi(z)$ that has a maximum at $z = 0$ and rapidly decreases with z (Fig. A.16). Let

$$\int_{-\infty}^{\infty} \Phi(z)\,\mathrm{d}z = 1\,. \tag{A.144}$$

There is a lot of functions which satisfy the formulated conditions, e.g.

$$\Phi_1(z) = \frac{1}{\pi}\,\frac{1}{1+z^2}\,; \qquad \Phi_2(z) = \frac{1}{\sqrt{\pi}}\,e^{-z^2}\,. \tag{A.145}$$

Let the value $\Phi_1(0)$ be m times increased while the plot of Φ_1 is m times contracted in the direction of the axis z, that is, instead of Φ_1 we consider a new function

$$\Phi_1(mz) = \frac{m}{\pi\,[\,1+(mz)^2\,]}\,. \tag{A.146}$$

For some values of $m > 1$, the plots of $\Phi_1(mz)$ are qualitatively illustrated in Fig. A.16. The functions $\Phi_1(mz)$ satisfy the condition (A.144) for any m.

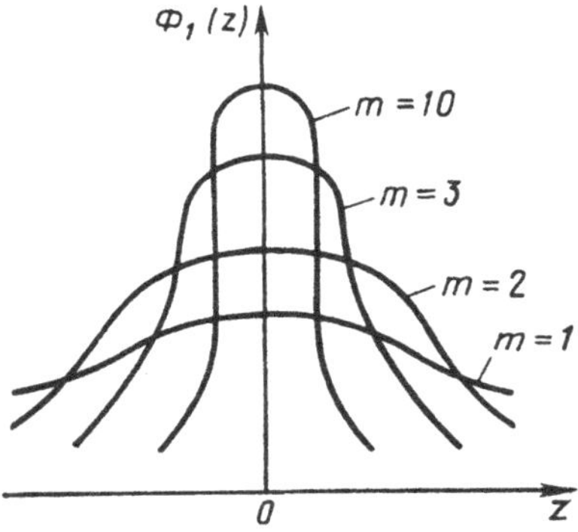

Fig. A.16.

As m tends to infinity, the formula (A.146) gives us a function $\delta(z)$ such that (see Fig. A.17)

$$\delta(z) = \lim_{m\to\infty} \Phi_1(mz) = \begin{cases} 0\,, & z < 0\,; \\ \infty\,, & z = 0\,; \\ 0\,, & z > 0 \end{cases} \tag{A.147}$$

and

$$\int_{-\infty}^{\infty} \delta(z)\,\mathrm{d}z = 1\,; \tag{A.148}$$

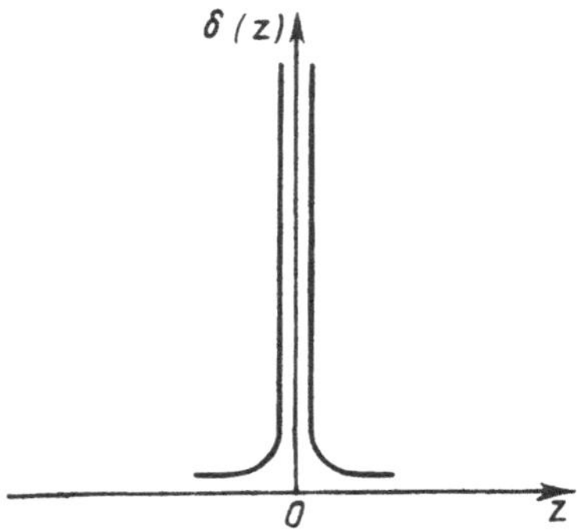

Fig. A.17.

thus defined function $\delta(z)$ is Dirac's delta. From (A.148) it is seen that the δ-function has dimensions of $1/[z]$, that is, $[\delta(z)] = 1/[z]$.

A.4.2 The Nondimensional δ-function

When nondimensional equilibrium equations are under consideration, we are to deal with a δ-function of the form $\delta(a\eta)$, where η is a nondimensional quantity. Let us prove that

$$\delta(a\eta) = \frac{1}{|a|}\,\delta(\eta)\,. \tag{A.149}$$

Integrating $\delta(\eta)$, we get

$$\int_{-\infty}^{\infty} \delta(\eta)\,\mathrm{d}\eta = \int_{-\infty}^{\infty} |a|\,\delta(a\eta)\,\mathrm{d}\eta\,.$$

For $a > 0$, we have

$$\int_{-\infty}^{\infty} \delta(a\eta)\,a\,\mathrm{d}\eta = \int_{-\infty}^{\infty} \delta(\eta_1)\,\mathrm{d}\eta_1 = 1\,, \qquad \eta_1 = a\eta\,,$$

and for $a < 0$,

$$-\int_{-\infty}^{\infty} \delta(a\eta)\,a\,\mathrm{d}\eta = -\int_{\infty}^{-\infty} \delta(\eta_1)\,\mathrm{d}\eta_1 = \int_{-\infty}^{\infty} \delta(\eta_1)\,\mathrm{d}\eta_1 = 1\,,$$

consequently, the function $|a|\,\delta(a\eta)$ has all the properties of the δ-function.

The derivatives of the δ-function can be introduced in a similar way:

$$\lim_{m\to\infty} \Phi_1'(m\eta) = \delta'(\eta)\,. \tag{A.150}$$

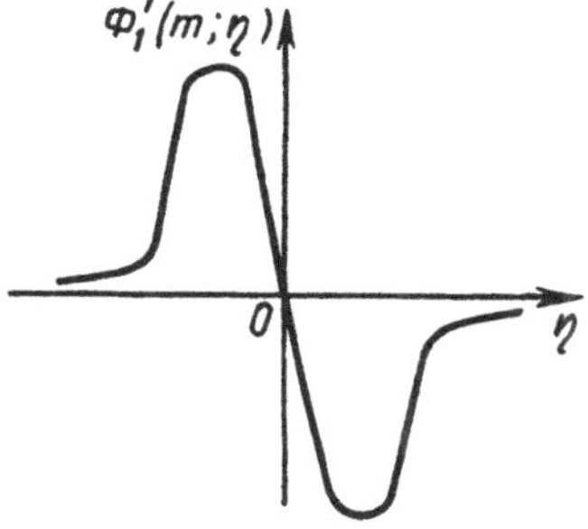

Fig. A.18.

The plot of $\Phi_1'(m\eta)$ for a finite m is qualitatively shown in Fig. A.18.

As $m \to \infty$, an nth derivative of $\Phi_1(mz)$ takes the form

$$\lim_{m\to\infty} \Phi_1^{(n)}(m\eta) = \delta^{(n)}(\eta)\,.$$

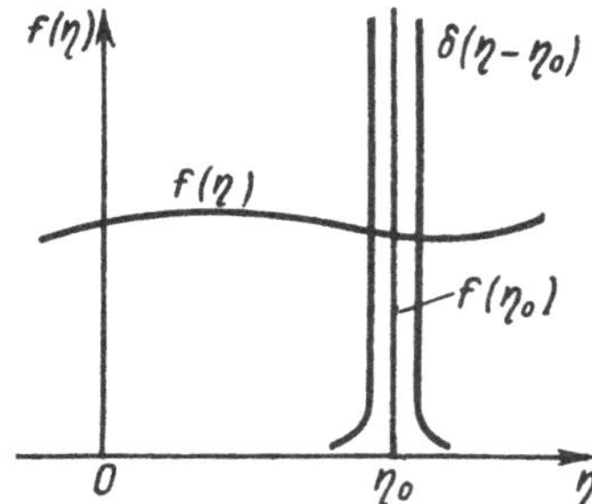

Fig. A.19.

Consider a function $\Phi_1[m(\eta - \eta_0)]$ (see Fig. A.19). As $m \to \infty$, we have

$$\int_{-\infty}^{\infty} \delta(\eta - \eta_0)\,\mathrm{d}\eta = 1\,. \tag{A.151}$$

The function $\Phi_1[m(\eta - \eta_0)]$ is termed a *shift function.*

A.4.3 The Heaviside Function

Consider a function that is related to the δ-function in the following way:

$$H(\eta - \eta_0) = \int_{-\infty}^{\eta} \delta(\zeta - \eta_0)\,\mathrm{d}\zeta\,. \tag{A.152}$$

The introduced function (the *Heaviside function*) satisfies the relation

$$H(\eta - \eta_0) = \begin{cases} 0\,, & \eta < \eta_0\,; \\ \frac{1}{2}\,, & \eta = \eta_0\,; \\ 1\,, & \eta > \eta_0\,. \end{cases} \tag{A.153}$$

From (A.152) it follows that

$$\frac{\mathrm{d}H(\eta - \eta_0)}{\mathrm{d}z} = \delta(\eta - \eta_0)\,. \tag{A.154}$$

Integration of the Heaviside function gives

$$\int_{-\infty}^{\eta} H(\zeta - \eta_0)\,\mathrm{d}\zeta = \int_{\eta_0}^{\eta} H(\zeta - \eta_0)\,\mathrm{d}\zeta = (\eta - \eta_0)\,H(\eta - \eta_0)\,. \tag{A.155}$$

A.4.4 Applications of the δ-function

Consider an integral

$$J = \int_{-\infty}^{\infty} f(\eta)\,\delta(\eta - \eta_0)\,\mathrm{d}\eta\,, \tag{A.156}$$

where $f(\eta)$ is a continuous function. The graphical representation of this integral is shown in Fig. A.19. We can see that the expression under the integral is different from zero only on the interval $(\eta_0 - \eta_1\,,\ \eta_0 + \eta_1)$, where η_1 is a small quantity. Within this interval, the function $f(\eta)$ varies little and approximately equals its value at the midspan $f(\eta_0)$, consequently, we can write

$$J = f(\eta_0) \int_{-\infty}^{\infty} \delta(\eta - \eta_0)\,\mathrm{d}\eta = f(\eta_0)\,. \tag{A.157}$$

Similarly, it can be shown that

$$f(\eta)\,\delta(\eta - \eta_0) = f(\eta_0)\,\delta(\eta - \eta_0)\,. \tag{A.158}$$

For an integral with a variable upper limit, we have

$$J = \int_{-\infty}^{\eta} f(\zeta)\,\delta(\zeta - \eta_0)\,\mathrm{d}\zeta = f(\eta_0)\,H(\eta - \eta_0)\,. \tag{A.159}$$

A.4.5 Integrals Containing Derivatives of the δ-function

Integrating by parts, we obtain

$$\int_{-\infty}^{\infty} f(\eta)\,\delta'(\eta - \eta_0)\,\mathrm{d}\eta = f(\eta)\,\delta(\eta - \eta_0)\,|_{-\infty}^{\infty} - \int_{-\infty}^{\infty} f'(\eta)\,\delta(\eta - \eta_0)\,\mathrm{d}\eta$$

or

$$\int_{-\infty}^{\infty} f(\eta)\,\delta'(\eta - \eta_0)\,\mathrm{d}\eta = -f(\eta_0)\,. \tag{A.160}$$

In the general case,

$$\int_{-\infty}^{\infty} f(\eta)\,\delta^{(n)}(\eta - \eta_0)\,\mathrm{d}\eta = (-1)^n f^{(n)}(\eta_0)\,. \tag{A.161}$$

In the case of an integral with a variable upper limit, we have

$$\int_{-\infty}^{\eta} f(\zeta)\,\delta'(\zeta - \eta_0)\,\mathrm{d}\zeta = f(\eta)\,\delta(\eta - \eta_0) - f'(\eta_0)\,H(\eta - \eta_0)\,. \tag{A.162}$$

A.5 Direction Cosines of the Unit Vector Tangent to a Rod Axis

A.5.1 Plane Curve

Dealing with problems of statics of curvilinear rods, one should know direction cosines $x'_j(s)$ of the unit vector $\mathbf{e}_1$, which is tangent to the rod axis, as functions of the axial coordinate s or the nondimensional coordinate $\eta = s/l$. Besides, one must know $æ_1(\eta)$ and $æ_3(\eta)$.

The equation of the rod axis can be written as follows:

$$f(x_1\,,\,x_2) = 0\,; \tag{A.163}$$

here x_1 and x_2 are Cartesian coordinates of the axial points. In what follows, the quantities x_i are assumed to be nondimensional. Differentiation of (A.163) with respect to η yields

$$\frac{\partial f}{\partial x_1}\,x'_1 + \frac{\partial f}{\partial x_2}\,x'_2 = 0\,. \tag{A.164}$$

The derivatives of x_j with respect to η satisfy the relation

$$x_1'^{\,2} + x_2'^{\,2} = 1\,. \tag{A.165}$$

Equations (A.164) and (A.165) can be solved for x'_1 and x'_2 as follows:

$$x'_1 = -\frac{\dfrac{\partial f}{\partial x_2}}{\sqrt{\left(\dfrac{\partial f}{\partial x_1}\right)^2 + \left(\dfrac{\partial f}{\partial x_2}\right)^2}} = F_1(x_1\,,\,x_2)\,; \tag{A.166}$$

$$x'_2 = -\frac{\dfrac{\partial f}{\partial x_1}}{\sqrt{\left(\dfrac{\partial f}{\partial x_1}\right)^2 + \left(\dfrac{\partial f}{\partial x_2}\right)^2}} = F_2(x_1\,,\,x_2)\,. \tag{A.167}$$

Solving for $x_j(\eta)$ and $x'_j(\eta)$ from the nonlinear differential equations (A.166)–(A.167), we get

$$æ_3(\eta) = \sqrt{x_1''^{2} + x_2''^{2}}\,. \tag{A.168}$$

Consider an example. A rod of circular form is shown in Fig. A.20. The equation of its axial line is as follows:

$$x_1^2 + (x_2 - R^0)^2 = R^{0^2}\,, \qquad R^0 = \frac{R}{l}\,. \tag{A.169}$$

Differentiating (A.169) with respect to η, we get

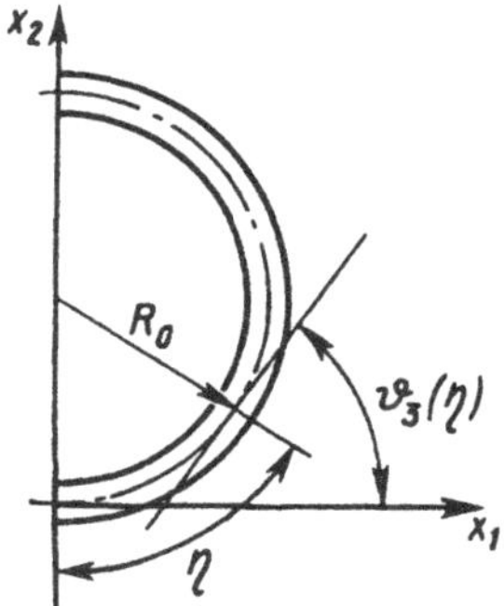

Fig. A.20.

$$x_1 x_1' + (x_2 - R^0)\, x_2' = 0\,. \tag{A.170}$$

Since ${x_1'}^2 + {x_2'}^2 = 1$, we have

$$x_1' = 1 - \frac{x_2}{R^0}\,; \qquad x_2' = 1 - \frac{x_1}{R^0}\,, \tag{A.171}$$

consequently,

$$x_1'' + \left(\frac{1}{R^0}\right)^2 x_1 = 0\,. \tag{A.172}$$

The solution of (A.172) is as follows:

$$x_1 = \frac{c_1 \cos\eta}{R^0} + \frac{c_2 \sin\eta}{R^0}\,. \tag{A.173}$$

Consequently,

$$x_2 = R^0 + \frac{c_1 \sin\eta}{R^0} - \frac{c_2 \cos\eta}{R^0}\,. \tag{A.174}$$

The arbitrary constants are to be determined from the zero conditions $x_1(0) = 0$ and $x_2(0) = 0$. We have $c_1 = 0$ and $c_2 = R^0$. Finally, we get

$$x_1(\eta) = \frac{R^0 \sin\eta}{R^0}\,; \qquad x_2(\eta) = R^0 - \frac{R^0 \cos\eta}{R^0}\,. \tag{A.175}$$

Differentiating (A.175) with respect to η, we arrive at

$$x_1' = \frac{\cos\eta}{R^0}\,; \qquad x_2' = \frac{\sin\eta}{R^0}\,. \tag{A.176}$$

The angle $\vartheta_3(\eta)$ (Fig. A.20) is related to η as follows:

$$\vartheta_3(\eta) = \frac{\eta}{R^0}\,. \tag{A.177}$$

A.5.2 Spatial Curve

The equations that define a spatial axial line may be written in the form

$$f_1(x_1\,,\,x_2\,,\,x_3) = 0\,; \tag{A.178}$$
$$f_2(x_1\,,\,x_2\,,\,x_3) = 0\,. \tag{A.179}$$

Differentiating (A.178) and (A.179) with respect to η, we get

$$\sum_{j=1}^{3} \frac{\partial f_1}{\partial x_j}\, x'_j = 0\,; \tag{A.180}$$
$$\sum_{j=1}^{3} \frac{\partial f_2}{\partial x_j}\, x'_j = 0\,. \tag{A.181}$$

The derivatives x'_j must satisfy the relation

$$\sum_{j=1}^{3} {x'_j}^2 = 1\,. \tag{A.182}$$

Solving for x'_j from (A.180)–(A.182), we can write

$$\begin{aligned} x'_1 &= F_1(x_1\,,\,x_2\,,\,x_3)\,;\\ x'_2 &= F_2(x_1\,,\,x_2\,,\,x_3)\,;\\ x'_3 &= F_3(x_1\,,\,x_2\,,\,x_3)\,. \end{aligned} \tag{A.183}$$

Solving the Cauchy problem for the system (A.183), we determine the functions $x_j(\eta)$ as well as their derivatives $x'_j(\eta)$, $x''_j(\eta)$, and $x'''_j(\eta)$. The curvature and the twist of a space-curved rod are as follows:

$$\Omega_3(\eta) = \sqrt{\sum_{j=1}^{3} {x''_j}^2}\,; \tag{A.184}$$
$$\Omega_1(\eta) = \frac{1}{\Omega_3^2} \begin{vmatrix} x'_1 & x'_2 & x'_3 \\ x''_1 & x''_2 & x''_3 \\ x'''_1 & x'''_2 & x'''_3 \end{vmatrix}\,. \tag{A.185}$$

A.6 Equations of the First and Higher Approximation

In terms of projections in the attached coordinate frame, the system of equations of the first approximation can be written as follows:

$$\frac{\mathrm{d}Q_1^{(1)}}{\mathrm{d}\eta} + Q_3^{(1)} æ_{20}^{(0)} - Q_2^{(1)} æ_{30}^{(1)} + Q_3^{(0)}\,\Delta æ_2^{(1)} - Q_2^{(0)}\,\Delta æ_3^{(1)}$$
$$+ k\left(q_{30}\vartheta_2^{(1)} - q_{20}\vartheta_3^{(1)}\right)$$
$$= Q_2^{(0)}\,\Delta æ_3^{(0)} - Q_3^{(0)}\,\Delta æ_2^{(0)} - \sum_{i=1}^{n} k\left(P_{30}^{(i)}\vartheta_2^{(1)} - P_{20}^{(i)}\vartheta_1^{(1)}\right)\delta(\eta - \eta_i)\,;$$

$$\frac{\mathrm{d}Q_2^{(1)}}{\mathrm{d}\eta} + Q_3^{(1)} æ_{30}^{(0)} - Q_3^{(1)} æ_{10}^{(0)} + Q_1^{(0)}\,\Delta æ_3^{(1)} - Q_3^{(0)}\,\Delta æ_1^{(1)}$$
$$+ k\left(q_{10}\vartheta_3^{(1)} - q_{30}\vartheta_1^{(1)}\right)$$
$$= Q_3^{(0)}\,\Delta æ_1^{(0)} - Q_1^{(0)}\,\Delta æ_3^{(0)} - \sum_{i=1}^{n} k\left(P_{10}^{(i)}\vartheta_3^{(1)} - P_{30}^{(i)}\vartheta_1^{(1)}\right)\delta(\eta - \eta_i)\,;$$

$$\frac{\mathrm{d}Q_3^{(1)}}{\mathrm{d}\eta} + Q_2^{(1)} æ_{10}^{(0)} - Q_1^{(1)} æ_{20}^{(0)} + Q_2^{(0)}\,\Delta æ_1^{(1)} - Q_1^{(0)}\,\Delta æ_2^{(1)}$$
$$+ k\left(q_{20}\vartheta_1^{(1)} - q_{10}\vartheta_2^{(1)}\right)$$
$$= Q_1^{(0)}\,\Delta æ_2^{(0)} - Q_2^{(0)}\,\Delta æ_1^{(0)} - \sum_{i=1}^{n} k\left(P_{20}^{(i)}\vartheta_1^{(1)} - P_{10}^{(i)}\vartheta_2^{(1)}\right)\delta(\eta - \eta_i)\,;$$
(A.186)

$$\frac{\mathrm{d}M_1^{(1)}}{\mathrm{d}\eta} + M_3^{(1)} æ_{20}^{(0)} - M_2^{(1)} æ_{30}^{(1)} + M_3^{(0)}\,\Delta æ_2^{(1)} - M_2^{(0)}\,\Delta æ_3^{(1)}$$
$$+ k\left(\mu_{30}\vartheta_2^{(1)} - \mu_{20}\vartheta_3^{(1)}\right)$$
$$= M_2^{(0)}\,\Delta æ_3^{(0)} - M_3^{(0)}\,\Delta æ_2^{(0)} - \sum_{\nu=1}^{n} k\left(T_{30}^{(\nu)}\vartheta_2^{(1)} - T_{20}^{(\nu)}\vartheta_3^{(1)}\right)\delta(\eta - \eta_\nu)\,;$$

$$\frac{\mathrm{d}M_2^{(1)}}{\mathrm{d}\eta} + M_1^{(1)} æ_{30}^{(0)} - M_3^{(1)} æ_{10}^{(0)} + M_1^{(0)}\,\Delta æ_3^{(1)} - M_3^{(0)}\,\Delta æ_1^{(1)} - Q_3^{(1)}$$
$$+ k\left(\mu_{10}\vartheta_3^{(1)} - \mu_{30}\vartheta_1^{(1)}\right)$$
$$= M_3^{(0)}\,\Delta æ_1^{(0)} - M_1^{(0)}\,\Delta æ_3^{(0)} - \sum_{\nu=1}^{n} k\left(T_{10}^{(\nu)}\vartheta_3^{(1)} - T_{30}^{(\nu)}\vartheta_1^{(1)}\right)\delta(\eta - \eta_\nu)\,;$$

$$\frac{\mathrm{d}M_3^{(1)}}{\mathrm{d}\eta} + M_2^{(1)} æ_{10}^{(0)} - M_1^{(1)} æ_{20}^{(0)} + M_2^{(0)}\,\Delta æ_1^{(1)} - M_1^{(0)}\,\Delta æ_2^{(1)} + Q_2^{(1)}$$
$$+ k\left(\mu_{20}\vartheta_1^{(1)} - \mu_{10}\vartheta_2^{(1)}\right)$$
$$= M_1^{(0)}\,\Delta æ_2^{(0)} - M_2^{(0)}\,\Delta æ_1^{(0)} - \sum_{\nu=1}^{n} k\left(T_{20}^{(\nu)}\vartheta_1^{(1)} - T_{10}^{(\nu)}\vartheta_2^{(1)}\right)\delta(\eta - \eta_\nu)\,;$$
(A.187)

$$M_1^{(1)} = A_{11}\,\Delta æ_1^{(1)}\,;$$

$$M_2^{(1)} = A_{22}\,\Delta æ_2^{(1)}\,;$$
$$M_3^{(1)} = A_{33}\,\Delta æ_3^{(1)}\,; \tag{A.188}$$

$$\frac{\mathrm{d}\vartheta_1^{(1)}}{\mathrm{d}\eta} + \vartheta_3^{(1)} æ_{20}^{(1)} - \vartheta_2^{(1)} æ_{30}^{(1)} - \frac{1}{A_{11}} M_1^{(1)} = 0\,;$$
$$\frac{\mathrm{d}\vartheta_2^{(1)}}{\mathrm{d}\eta} + \vartheta_1^{(1)} æ_{30}^{(1)} - \vartheta_3^{(1)} æ_{10}^{(1)} - \frac{1}{A_{22}} M_2^{(1)} = 0\,;$$
$$\frac{\mathrm{d}\vartheta_3^{(1)}}{\mathrm{d}\eta} + \vartheta_2^{(1)} æ_{10}^{(1)} - \vartheta_1^{(1)} æ_{20}^{(1)} - \frac{1}{A_{33}} M_3^{(1)} = 0\,; \tag{A.189}$$

$$\frac{\mathrm{d}u_1^{(1)}}{\mathrm{d}\eta} + u_3^{(1)} æ_{20}^{(1)} - u_2^{(1)} æ_{30}^{(1)} = 0\,;$$
$$\frac{\mathrm{d}u_2^{(1)}}{\mathrm{d}\eta} + u_1^{(1)} æ_{30}^{(1)} - u_3^{(1)} æ_{10}^{(1)} - \vartheta_3^{(1)} = 0\,;$$
$$\frac{\mathrm{d}u_3^{(1)}}{\mathrm{d}\eta} + u_2^{(1)} æ_{10}^{(1)} - u_1^{(1)} æ_{20}^{(1)} + \vartheta_2^{(1)} = 0\,. \tag{A.190}$$

If the applied forces are follower, then the factor k in (A.186) and (A.187) must be set to zero. If the forces are not follower (e.g. dead forces), one should put $k = 1$. Using tensor notation, we have

$$\frac{\mathrm{d}Q_j^{(1)}}{\mathrm{d}\eta} + \varepsilon_{jk\rho}\left(æ_{k0}^{(0)} Q_\rho^{(1)} + Q_k^{(0)}\,\Delta æ_\rho^{(1)} + k q_{k0}\vartheta_\rho^{(1)}\right)$$
$$= \varepsilon_{jki} Q_k^{(0)}\,\Delta æ_i^{(0)} - k\sum_{i=1}^{n} \varepsilon_{jk\rho} P_{k0}^{(i)} \vartheta_\rho^{(0)}\,\delta(\eta - \eta_i)\,;$$

$$\frac{\mathrm{d}M_j^{(1)}}{\mathrm{d}\eta} + \varepsilon_{jkm}\left(æ_{k0}^{(0)} M_m^{(1)} + M_k^{(0)}\,\Delta æ_m^{(1)} + k\mu_{k0}\vartheta_m^{(1)}\right) + \delta_{3j} Q_2^{(1)} - \delta_{2j} Q_3^{(1)}$$
$$= \varepsilon_{jki} M_k^{(0)}\,\Delta æ_i^{(0)} - k\sum_{k=1}^{\rho} \varepsilon_{jkm} T_{k0}^{(\nu)} \vartheta_m^{(1)}\,\delta(\eta - \eta_\nu)\,;$$

$$\Delta æ_j^{(1)} = \frac{M_j^{(1)}}{A_{jj}}\,;$$

$$\frac{\mathrm{d}\vartheta_j^{(1)}}{\mathrm{d}\eta} + \varepsilon_{jki} æ_{k0}^{(0)} \vartheta_i^{(1)} - \Delta æ_j^{(1)} = 0\,;$$

$$\frac{\mathrm{d}u_j^{(1)}}{\mathrm{d}\eta} + \varepsilon_{jki} æ_{k0}^{(0)} u_i^{(1)} - \delta_{2i}\vartheta_3^{(1)} + \delta_{3i}\vartheta_2^{(1)} = 0$$

Continuing this line of reasoning, we can obtain the equilibrium equations of the second approximation and so on. The equilibrium equations of an nth approximation are as follows:

$$\begin{aligned}
&\frac{\mathrm{d}\mathbf{Q}^{(n)}}{\mathrm{d}\eta} + \mathrm{A}_{æ}^{(n-1)}\mathbf{Q}^{(n)} + \mathrm{A}_{Q}^{(n-1)}\,\Delta\boldsymbol{æ}^{(n)} + \mathrm{C}^{(1)}\boldsymbol{\vartheta}^{(n)} \\
&= -\Delta\boldsymbol{æ}^{(n-1)} \times \boldsymbol{\vartheta}^{(n-1)} - \sum_{i=1}^{n} \mathrm{B}_i^{(1)}\boldsymbol{\vartheta}^{(n)}\,\delta(\eta-\eta_i)\,; \\
&\frac{\mathrm{d}\mathbf{M}^{(n)}}{\mathrm{d}\eta} + \mathrm{A}_{æ}^{(n-1)}\mathbf{M}^{(n)} + \mathrm{A}_{M}^{(n-1)}\,\Delta\boldsymbol{æ}^{(n)} + \mathrm{A}_1\mathbf{Q}^{(n)} + \mathrm{C}^{(3)}\boldsymbol{\vartheta}^{(n)} \\
&= -\Delta\boldsymbol{æ}^{(n-1)} \times \mathbf{M}^{(n-1)} - \sum_{\nu=1}^{\rho} \mathrm{B}_\nu^{(3)}\boldsymbol{\vartheta}^{(n)}\,\delta(\eta-\eta_\nu)\,; \\
&\mathbf{M}^{(n)} = \mathrm{A}\,\Delta\boldsymbol{æ}^{(n)}\,; \\
&\frac{\mathrm{d}\boldsymbol{\vartheta}^{(n)}}{\mathrm{d}\eta} + \mathrm{A}_{æ}^{(n-1)}\boldsymbol{\vartheta}^{(n)} - \mathrm{A}^{-1}\mathbf{M}^{(n)} = 0\,; \\
&\frac{\mathrm{d}\mathbf{u}^{(n)}}{\mathrm{d}\eta} + \mathrm{A}_{æ}^{(n-1)}\mathbf{u}^{(n)} + \mathrm{A}_1\boldsymbol{\vartheta}^{(n)} = 0\,;
\end{aligned} \tag{A.191}$$

here

$$\begin{aligned}
&\tilde{\mathbf{Q}}^{(n-1)} = \sum_{j=0}^{n-1} \mathbf{Q}^{(j)}\,; \\
&\tilde{\mathbf{M}}^{(n-1)} = \sum_{j=0}^{n-1} \mathbf{M}^{(j)}\,; \\
&\tilde{\boldsymbol{\vartheta}}^{(n-1)} = \sum_{j=0}^{n-1} \boldsymbol{\vartheta}^{(j)}\,; \\
&\widetilde{\Delta\boldsymbol{æ}}^{(n-1)} = \sum_{j=0}^{n-1} \Delta\boldsymbol{æ}^{(j)}\,; \\
&\mathrm{A}_{æ}^{(n-1)} = \begin{bmatrix} 0 & -æ_3^{(n-1)} & æ_2^{(n-1)} \\ æ_3^{(n-1)} & 0 & -æ_1^{(n-1)} \\ -æ_2^{(n-1)} & æ_1^{(n-1)} & 0 \end{bmatrix}; \\
&\mathrm{A}_{Q}^{(n-1)} = \begin{bmatrix} 0 & \tilde{Q}_3^{(n-1)} & -\tilde{Q}_2^{(n-1)} \\ -\tilde{Q}_3^{(n-1)} & 0 & \tilde{Q}_1^{(n-1)} \\ \tilde{Q}_2^{(n-1)} & -\tilde{Q}_1^{(n-1)} & 0 \end{bmatrix}; \\
&\mathrm{A}_{M}^{(n-1)} = \begin{bmatrix} 0 & \tilde{M}_3^{(n-1)} & -\tilde{M}_2^{(n-1)} \\ -\tilde{M}_3^{(n-1)} & 0 & \tilde{M}_1^{(n-1)} \\ \tilde{M}_2^{(n-1)} & -\tilde{M}_1^{(n-1)} & 0 \end{bmatrix}; \\
&æ_k^{(n-1)} = æ_k + \Delta æ_k^{(n-1)}\,;
\end{aligned}$$

$$\tilde{Q}_k^{(n-1)} = \sum_{j=0}^{n-1} Q_k^{(j)} \, ;$$

$$\tilde{M}_k^{(n-1)} = \sum_{j=0}^{n-1} M_k^{(j)} \, .$$

B. Solution of the Problems

B.1 To Chapter 1

1.1. For the rod axis to remain a *plane curve* during deformation, the applied loads must satisfy the following conditions: (1) the force vectors must lie in the plane containing the axial line; (2) the moment vectors must be perpendicular to this plane. In the case of a rod of circular or square cross section, these conditions are not only necessary but also sufficient ones. In general, for a rod of arbitrary cross section, these conditions are not sufficient and, consequently, an additional condition must be satisfied. For example, such a condition may restrict one of the principal axes of any cross section to lie in the plane of the axial line.

Equilibrium Equations Since the rod axis is a plane curve, we have $æ_1 = æ_2 = 0$, $æ_{10} = æ_{20} = 0$, $M_1 = M_2 = 0$, $Q_3 = 0$, and $\vartheta_1 = \vartheta_2 = 0$. Using (1.64)–(1.68), we obtain

$$\frac{\mathrm{d}Q_1}{\mathrm{d}\eta} - Q_2 æ_3 + q_1 + \sum_{i=1}^{n} P_1^{(i)}\,\delta(\eta - \eta_i) = 0\,;$$

$$\frac{\mathrm{d}Q_2}{\mathrm{d}\eta} + Q_1 æ_3 + q_2 + \sum_{i=1}^{n} P_2^{(i)}\,\delta(\eta - \eta_i) = 0\,;$$

$$\frac{\mathrm{d}M_3}{\mathrm{d}\eta} + Q_2 + \mu_3 + \sum_{\nu=1}^{\rho} T_3^{(\nu)}\,\delta(\eta - \eta_\nu) = 0\,;$$

$$M_3 = A_{33}(æ_3 - æ_{30})\,;$$

$$æ_3 = \frac{\mathrm{d}\vartheta_3}{\mathrm{d}\eta} + æ_{30}\,;$$

$$\frac{\mathrm{d}u_1}{\mathrm{d}\eta} - æ_3 u_2 - \cos\vartheta_3 + 1 = 0\,;$$

$$\frac{\mathrm{d}u_2}{\mathrm{d}\eta} + æ_3 u_1 - \sin\vartheta_3 = 0\,.$$

1.2. Since the displacements are small, we can use the general equilibrium equations of the zeroth approximation (1.116)–(1.119) written in terms of projections on the attached axes. Setting $æ_{10} = æ_{20} = 0$, $æ_{30} = 1/R$, $M_1 = M_2 = 0$, $Q_3 = 0$, $\vartheta_2 = \vartheta_3 = 0$, and $u_3 = 0$, we obtain (see Prob. 1.1)[1]

$$\frac{\mathrm{d}Q_1^{(0)}}{\mathrm{d}\eta} - \frac{Q_2^{(0)}}{R} + P_{10} + \Delta P_1^{(0)} = 0\,; \tag{1}$$

$$\frac{\mathrm{d}Q_2^{(0)}}{\mathrm{d}\eta} + \frac{Q_1^{(0)}}{R_0} + P_{20} + \Delta P_2^{(0)} = 0\,; \tag{2}$$

$$\frac{\mathrm{d}M_3^{(0)}}{\mathrm{d}\eta} + Q_2^{(0)} = 0\,; \tag{3}$$

$$\frac{\mathrm{d}\vartheta_3^{(0)}}{\mathrm{d}\eta} - \frac{M_3^{(0)}}{A_{33}} = 0\,; \tag{4}$$

$$\frac{\mathrm{d}u_1^{(0)}}{\mathrm{d}\eta} - \frac{u_2^{(0)}}{R_0} = 0\,; \tag{5}$$

$$\frac{\mathrm{d}u_2^{(0)}}{\mathrm{d}\eta} + \frac{u_1^{(0)}}{R_0} - \vartheta_3^{(0)} = 0\,. \tag{6}$$

Examine in greater detail the forces in (1) and (2). In general, the distributed follower load $\mathbf{q}$ has two nonzero components, $\mathbf{q} = q_1\mathbf{e}_1 + q_2\mathbf{e}_2$. The distributed load acts just on a part of the rod. In view of (1.24), we get

$$P_{10} + \Delta P_1^{(0)} = q_{10}\, H(\eta - \eta_K) + \left(P_{10}^{(1)} + \Delta P_1^{(1)}\right)\, \delta(\eta - \eta_K)\,; \tag{7}$$

$$P_{20} + \Delta P_2^{(0)} = q_{20}\, H(\eta - \eta_K) + \left(P_{20}^{(1)} + \Delta P_2^{(1)}\right)\, \delta(\eta - \eta_K)\,. \tag{8}$$

The force $\mathbf{P}^{(1)}$ is a *dead* one, hence, in the attached basis, its components vary during deformation. Recall that displacements of the points of the axial line and angles of rotation of the cross sections are small. Using (1.140), we get

$$P_i^{(1)} = \sum_{j=1}^{2} P_{x_j}^{(1)}\, (l_{ij_0}^{(1)} + \Delta l_{ij})\,, \tag{9}$$

where $l_{ij}^{(1)}$ are the elements of the matrix $\mathrm{L}^{(1)}$ (see (A.57)). In the case of in-plane deformations, this matrix reads

$$\mathrm{L}^{(1)} = \begin{bmatrix} \cos\vartheta_{30} - \vartheta_3^{(0)}\sin\vartheta_{30} & \sin\vartheta_{30} - \vartheta_3^{(0)}\cos\vartheta_{30} \\ -\sin\vartheta_{30} - \vartheta_3^{(0)}\cos\vartheta_{30} & \cos\vartheta_{30} - \vartheta_3^{(0)}\sin\vartheta_{30} \end{bmatrix}. \tag{10}$$

Consequently,

[1] In what follows, the formulas in the solutions of the problems of each chapter are enumerated independently.

$$P_1^{(1)} = P_{10}^{(1)} + \Delta P_1^{(0)}$$
$$= P_{x_1} \cos\vartheta_{30} + P_{x_2} \sin\vartheta_{30} + (P_{x_2} \cos\vartheta_{30} - P_{x_1} \sin\vartheta_{30})\,\vartheta_3^{(0)}\,; \qquad (11)$$
$$P_2^{(1)} = P_{10}^{(1)} + \Delta P_1^{(0)}$$
$$= -P_{x_1} \sin\vartheta_{30} + P_{x_2} \cos\vartheta_{30} - (P_{x_1} \cos\vartheta_{30} + P_{x_2} \sin\vartheta_{30})\,\vartheta_3^{(0)}\,, \qquad (12)$$

where $\vartheta_3^{(0)}$ is a small angle of rotation of the cross section K relative to the undeformed configuration and ϑ_{30} is an angle of rotation of the attached axes at $\eta = \eta_K$ as shown in Fig. 1.26. Substitution of (11) and (12) into (1) and (2) yields a system of linear first-order differential equations with variable coefficients in the six unknowns $Q_1^{(0)}$, $Q_2^{(0)}$, $M_3^{(0)}$, $u_1^{(0)}$, $\vartheta_3^{(0)}$, and $u_2^{(0)}$.

1.3. In this problem, the equilibrium equations are identical to those in Prob. 1.2 but the projections of the forces are different. The acceleration vector **a** does not lie in the plane of drawing, hence, in a deformed configuration, the rod axis is a spatial curve. For small angles of rotation of the attached axes, the matrix $\mathrm{L}^{(1)}$ takes the form (see (A.57))

$$\mathrm{L}^{(1)} = \begin{bmatrix} 1 & \vartheta_3 & -\vartheta_2 \\ -\vartheta_3 & 1 & \vartheta_1 \\ \vartheta_2 & -\vartheta_1 & 1 \end{bmatrix} \mathrm{L}^0\,,$$

where L^0 is the matrix (A.55).

The axial line is a plane curve, consequently, $\vartheta_{10} = \vartheta_{20} = 0$, $\vartheta_{30} \neq 0$, and

$$\mathrm{L}^{(0)} = \begin{bmatrix} \cos\vartheta_{30} & \sin\vartheta_{30} & 0 \\ -\sin\vartheta_{30} & \cos\vartheta_{30} & 0 \\ 0 & 0 & 1 \end{bmatrix}.$$

The matrix of transformation from the basis $\{\mathbf{i}_j\}$ to the basis $\{\mathbf{e}_j\}$ is as follows:

$$\mathrm{L}^{(1)} = \begin{bmatrix} \cos\vartheta_{30} - \vartheta_3 \sin\vartheta_{30} & \sin\vartheta_{30} + \vartheta_3 \cos\vartheta_{30} & -\vartheta_2 \\ -\vartheta_{30} \cos\vartheta_{30} - \sin\vartheta_{30} & \cos\vartheta_{30} - \vartheta_3 \sin\vartheta_{30} & 1 \\ \vartheta_2 \cos\vartheta_{30} + \vartheta_1 \sin\vartheta_{30} & \vartheta_2 \sin\vartheta_{30} - \vartheta_1 \cos\vartheta_{30} & 1 \end{bmatrix}.$$

In the attached coordinate system, we have

$$\mathbf{i}_1 = (\cos\vartheta_{30} - \vartheta_3 \sin\vartheta_{30})\,\mathbf{e}_1 - (\sin\vartheta_{30} + \vartheta_3 \cos\vartheta_{30})\,\mathbf{e}_2$$
$$+ (\vartheta_2 \cos\vartheta_{30} + \vartheta_1 \sin\vartheta_{30})\,\mathbf{e}_3\,;$$
$$\mathbf{i}_2 = (\sin\vartheta_{30} + \vartheta_3 \cos\vartheta_{30})\,\mathbf{e}_1 + (\cos\vartheta_{30} - \vartheta_3 \sin\vartheta_{30})\,\mathbf{e}_2$$
$$+ (\vartheta_2 \sin\vartheta_{30} - \vartheta_1 \cos\vartheta_{30})\,\mathbf{e}_3\,.$$

The distributed load reads

$$\mathbf{q} = m_0 a_{x_1} \mathbf{i}_1 + m_0 a_{x_2} \mathbf{i}_2 = q_{x_1} \mathbf{i}_1 + q_{x_2} \mathbf{i}_2\,,$$

where $\mathbf{a}_{x_j}$ are known quantities. In the attached frame, we have

$$q_1 = q_{x_1} \cos\vartheta_{30} + q_{x_2} \sin\vartheta_{30} + (-q_{x_1} \sin\vartheta_{30} + q_{x_2} \cos\vartheta_{30})\,\vartheta_3\,;$$
$$q_2 = -q_{x_1} \sin\vartheta_{30} + q_{x_2} \cos\vartheta_{30} - (q_{x_1} \cos\vartheta_{30} + q_{x_2} \sin\vartheta_{30})\,\vartheta_3\,;$$
$$q_2 = (q_{x_1} \sin\vartheta_{30} - q_{x_2} \cos\vartheta_{30})\,\vartheta_1 + (q_{x_1} \cos\vartheta_{30} + q_{x_2} \sin\vartheta_{30})\,\vartheta_2$$

or

$$q_i = q_{i0} + \Delta q_i\,.$$

The concentrated force **P** is

$$\mathbf{P} = m\mathbf{a} = P_{x_1}\mathbf{i}_1 + P_{x_2}\mathbf{i}_2 = \sum_{j=1}^{3} P_j \mathbf{e}_j\,,$$

where (for brevity, only P_1 is written out)

$$P_1 = P_{x_1} \cos\vartheta_{30} + P_{x_2} \sin\vartheta_{30} + (-P_{x_1} \sin\vartheta_{30} + P_{x_2} \cos\vartheta_{30})\vartheta_3\,.$$

Let us derive the equilibrium equations for the rod in the attached coordinate system. We have $\vartheta_{10} = \vartheta_{20} = 0$, $æ_{10} = æ_{20} = 0$, and $æ_{30} = 1/R_0$. Using (1.116)–(1.119), we obtain the system of equations of the zeroth approximation:

$$\frac{dQ_1^{(0)}}{d\eta} - \frac{Q_2^{(0)}}{R_0} + q_{10} + \Delta q_1 + (P_{10} + \Delta P_1)\,\delta(\eta - \eta_K) = 0\,;$$
$$\frac{dQ_2^{(0)}}{d\eta} + \frac{Q_1^{(0)}}{R_0} + q_{20} + \Delta q_2 + (P_{20} + \Delta P_2)\,\delta(\eta - \eta_K) = 0\,;$$
$$\frac{dQ_3^{(0)}}{d\eta} + \Delta q_3 + \Delta P_3\,\Delta(\eta - \eta_K) = 0\,;$$
$$\frac{dM_1^{(0)}}{d\eta} - \frac{M_2^{(0)}}{R_0} = 0\,;$$
$$\frac{dM_2^{(0)}}{d\eta} + \frac{M_1^{(0)}}{R_0} - Q_3^{(0)} = 0\,;$$
$$\frac{dM_3^{(0)}}{d\eta} + Q_2^{(0)} = 0\,;$$
$$\frac{d\vartheta_1^{(0)}}{d\eta} - \frac{\vartheta_2^{(0)}}{R_0} - \frac{M_1^{(0)}}{A_{11}} = 0\,;$$
$$\frac{d\vartheta_2^{(0)}}{d\eta} + \frac{\vartheta_1^{(0)}}{R_0} - \frac{M_2^{(0)}}{A_{22}} = 0\,;$$
$$\frac{d\vartheta_3^{(0)}}{d\eta} - \frac{M_3^{(0)}}{A_{33}} = 0\,;$$
$$\frac{du_1^{(0)}}{d\eta} - \frac{u_2^{(0)}}{R_0} = 0\,;$$

$$\frac{du_2^{(0)}}{d\eta} + \frac{u_1^{(0)}}{R_0} - \vartheta_3^{(0)} = 0\,;$$
$$\frac{du_3^{(0)}}{d\eta} + \vartheta_2^{(0)} = 0\,.$$

1.4. From Fig. 1.28 it is seen that

$$\mathbf{q} = m_0\omega^2\boldsymbol{\rho} = m_0\omega^2(x_1\mathbf{i}_1 - a\mathbf{i}_2 + \mathbf{u})\,, \qquad \mathbf{u} = -u_1\mathbf{e}_1 - u_2\mathbf{e}_2\,.$$

We have $\mathrm{L}^0 = \mathrm{E}$ because the bases $\{\mathbf{e}_{j0}\}$ and $\{\mathbf{i}_j\}$ coincide.

Consider the attached axes in the plane of drawing. For finite angles of rotation of these axes, the matrix L is of the form

$$\mathrm{L}^{(1)} = \begin{bmatrix} \cos\vartheta_3 & \sin\vartheta_3 \\ -\sin\vartheta_3 & \cos\vartheta_3 \end{bmatrix}.$$

Therefore,

$$\mathbf{i}_1 = \cos\vartheta_3\mathbf{e}_1 - \sin\vartheta_3\mathbf{e}_2\,; \qquad \mathbf{i}_2 = \sin\vartheta_3\mathbf{e}_1 + \cos\vartheta_3\mathbf{e}_2\,.$$

Finally, we have

$$\begin{aligned} q_1 &= m_0\omega^2(x_1\cos\vartheta_3 - a\sin\vartheta_3) - u_1\,; \\ q_2 &= -m_0\omega^2(x_1\sin\vartheta_3 + a\cos\vartheta_3) - u_2\,. \end{aligned}$$

1.5. According to the results obtained in Sect. 1.2, the vector $\mathbf{r}$ can be written as follows (see Fig. 1.29):

$$\mathbf{r} = |\mathbf{r}|\,\mathbf{e}_r = \mathbf{r}_0 + \mathbf{u} = -a\mathbf{i}_2 + x_{1K}\mathbf{i}_1 + \mathbf{u}\,, \qquad x_{1K} = \eta_K\,.$$

In terms of projections on the Cartesian axes, the vector $\mathbf{u}$ reads

$$\begin{aligned} \mathbf{u} &= u_1\mathbf{e}_1 + u_2\mathbf{e}_2 \\ &= (u_1\cos\vartheta_3 - u_2\sin\vartheta_3)\,\mathbf{i}_1 + (u_1\sin\vartheta_3 + u_2\cos\vartheta_3)\,\mathbf{i}_2 \\ &= u_{x_1}\mathbf{i}_1 + u_{x_2}\mathbf{i}_2\,. \end{aligned}$$

Then,

$$\mathbf{e}_r = \frac{1}{|\mathbf{r}|}\left[\,(x_{1K} + u_{x_1})\,\mathbf{i}_1 + (-a + u_{x_2})\,\mathbf{i}_2\,\right] = e_{x_1}\mathbf{i}_1 + e_{x_2}\mathbf{i}_2\,,$$

where $|\mathbf{r}| = \sqrt{(x_{1K} + u_{x_1})^2 + (-a + u_{x_2})^2}$.

If $\mathbf{P}^{(1)} = |\mathbf{P}_0|\mathbf{e}_r$, then, in the immovable axes, we get

$$\mathbf{P}^{(1)} = |\mathbf{P}_0|e_{x_1}\mathbf{i}_1 + |\mathbf{P}_0|e_{x_2}\mathbf{i}_2\,,$$

while in the attached basis, we get

$$\mathbf{P}^{(1)} = |\mathbf{P}_0|\,[\,(e_{x_1}\cos\vartheta_3 + e_{x_2}\sin\vartheta_3)\,\mathbf{e}_1 + (-e_{x_1}\sin\vartheta_3 + e_{x_2}\cos\vartheta_3)\,\mathbf{e}_2\,]\,.$$

Thus, the projection $\mathbf{P}^{(1)}$ depends nonlinearly on the components u_{x_j} and u_j of the displacement vector and on the angle ϑ_3.

1.6. Consider the twisting of the rod. From the symmetry of the problem it follows that the angle of rotation of the cross section ϑ_{10} does not depend on the coordinate η, i.e. $\vartheta_{10} = \text{const}$. The rod axis remains a plane curve, consequently, $æ_1 = æ_{10} = æ_{20} = 0$, $æ_2 = æ_{30} \sin\vartheta_{10}$, and $æ_3 = æ_{30} \cos\vartheta_{10}$. During the twisting, the internal moments remain the same, consequently, we get $M_1 = 0$, $M_2 = A_{33} æ_{30} \sin\vartheta_{10}$, and $M_3 = A_{33}(æ_{30}\cos\vartheta_{10} - æ_{30})$. From the second and the third equations of (1.64) it follows that

$$Q_2 = Q_3 = 0\,. \tag{13}$$

The first equation of (1.65) yields

$$M_3 æ_2 - M_2 æ_3 + \mu_1 = 0\,. \tag{14}$$

Substituting $æ_j$ and M_j into (14), we have

$$A_{33} æ_{30}^2 (\cos\vartheta_{10} - 1)\sin\vartheta_{10} - A_{33} æ_{30}^2 \cos\vartheta_{10} \sin\vartheta_{10} + \mu_1 = 0$$

or

$$-A_{33} æ_{30}^2 \sin\vartheta_{10} + \mu_1 = 0\,. \tag{15}$$

As a result, we obtain the following relationship between the angle ϑ_{10} and μ_1:

$$\sin\vartheta_{10} = \frac{\mu_1}{A_{33} æ_{30}^2}\,. \tag{16}$$

1.7. Let $A_{22} \neq A_{33}$. From the first equation of (1.65) it follows that

$$(A_{33} - A_{22})\, æ_{30}^2 \cos\vartheta_{10} \sin\vartheta_{10} - A_{33} æ_{30}^2 \sin\vartheta_{10} + \mu_1 = 0\,. \tag{17}$$

The angle ϑ_{10*} such that $\mu_1(\vartheta_{10*})$ is a relative maximum can be found from the equality

$$A_{33}\cos\vartheta_{10*} - (A_{33} - A_{22})\cos 2\vartheta_{10*} = 0\,.$$

It is seen from (17) that, for $\vartheta_{10} = \pi$, the moment μ_1 vanishes. In this case, the ring is *turned inside out*, and the equilibrium configuration is unstable. For $A_{22} = A_{33}$ it is a well-known result.

B.2 To Chapter 3

3.1. The loss of stability may result in an out-of-plane deformation of the spring. If the rod axis is a plane curve, then the equilibrium equations for the critical configuration can be obtained from the general vector equations (3.10)–(3.14). In terms of projections on the attached axes, the equilibrium equations governing the behavior of the spring in the critical configuration take the form

$$\begin{aligned}
&\frac{dQ_{1*}}{d\eta} - æ_{3*}Q_{2*} + q_{1*} = 0\,;\\
&\frac{dQ_{2*}}{d\eta} + æ_{3*}Q_{1*} + q_{2*} = 0\,;\\
&\frac{dM_{3*}}{d\eta} + Q_{2*} = 0\,;\\
&\frac{d\vartheta_{3*}}{d\eta} - \frac{M_{3*}}{A_{33}} = 0\,;\\
&\frac{du_{1*}}{d\eta} - æ_{3*}u_{2*} = 0\,;\\
&\frac{du_{2*}}{d\eta} + æ_{3*}u_{1*} - \vartheta_{3*} = 0\,.
\end{aligned} \tag{1}$$

Suppose that the form of the axial line in the critical configuration differs little from that in the natural configuration. In this case, $æ_{3*} = 1/\rho_0^0(\eta)$ and $\vartheta_{3*} = \vartheta_{30}(\eta)$; here $\rho_0^0(\eta)$ is the nondimensional radius of curvature of the axial line (ρ_0^0 and ϑ_{30} are known functions of η). Hence, the system of equations (1) becomes a linear one. The projections of the distributed load are as follows:

$$q_{1*} = (\mathbf{q}_0 \cdot \mathbf{e}_{10}) = q_0(\mathbf{i}_1 \cdot \mathbf{e}_{10})\,; \qquad q_{2*} = q_0(\mathbf{i}_1 \cdot \mathbf{e}_{20})\,, \qquad q_0 = m_0|\mathbf{a}|\,.$$

The matrix of transformation from the basis $\{\mathbf{i}_j\}$ to the basis $\{\mathbf{e}_{i0}\}$ reads

$$\mathrm{L}^0 = \begin{bmatrix} \cos\vartheta_{30} & \sin\vartheta_{30} & 0 \\ -\sin\vartheta_{30} & \cos\vartheta_{30} & 0 \\ 0 & 0 & 1 \end{bmatrix}, \tag{2}$$

therefore,

$$q_{1*} = q_0 \cos\vartheta_{30}\,; \qquad q_{2*} = q_0 \sin\vartheta_{30}\,.$$

For a rod of constant cross section, the equilibrium equations after loss of stability (a particular case of (3.33)–(3.36)) take the form

$$\frac{dQ_{01}}{d\eta} - Q_{2*}M_{03} - \frac{1}{\rho_0^0} Q_{02} + \Delta P_1 = 0\,;$$
$$\frac{dQ_{02}}{d\eta} + Q_1 M_{03} + \frac{1}{\rho_0^0} Q_{01} + \Delta P_2 = 0\,;$$
$$\frac{dQ_{03}}{d\eta} + \frac{Q_{2*}}{A_{11}} M_{01} - \frac{Q_{1*}}{A_{22}} M_{02} + \Delta P_3 = 0\,;$$
$$\frac{dM_{01}}{d\eta} - \frac{1}{\rho_0^0} M_{02} + \frac{M_{3*}}{A_{22}} M_{02} = 0\,;$$
$$\frac{dM_{02}}{d\eta} + \frac{1}{\rho_0^0} M_{01} - \frac{M_{3*}}{A_{11}} M_{01} - Q_{03} = 0\,;$$
$$\frac{dM_{03}}{d\eta} + Q_{02} = 0\,;$$
$$\frac{d\vartheta_1}{d\eta} - \frac{M_{01}}{A_{11}} - \frac{1}{\rho_0^0} \vartheta_2 = 0\,;$$
$$\frac{d\vartheta_2}{d\eta} - \frac{M_{02}}{A_{22}} + \frac{1}{\rho_0^0} \vartheta_1 = 0\,;$$
$$\frac{d\vartheta_3}{d\eta} - M_{03} = 0\,;$$
$$\frac{du_1}{d\eta} - \frac{1}{\rho_0^0} u_2 = 0\,;$$
$$\frac{du_2}{d\eta} + \frac{1}{\rho_0^0} u_1 - \vartheta_3 = 0\,;$$
$$\frac{du_3}{d\eta} + \vartheta_2 = 0\,. \tag{3}$$

The spiral is subjected to a *dead* load, that is, ΔP_i are different from zero. Because of the absence of the distributed moment $\boldsymbol{\mu}$, we see that $\Delta P_i = \Delta q_i$ and $\Delta T_j = 0$.

Let us calculate Δq_i. We have $\Delta \mathbf{q} = \mathbf{q} - \mathbf{q}_0^{(1)}$, where $\mathbf{q} = q_0 \mathbf{i}_1$ and $\mathbf{q}_0^{(1)} = \sum_{i=1}^{3} q_{0i\mathbf{e}_i}$.

Let us represent the vector $\mathbf{i}_1$ in the basis $\{\mathbf{e}_j\}$. This basis is associated with the equilibrium configuration of the rod after the loss of stability. The axial line after the loss of stability is shown in Fig. 3.17 as a dashed line.

The matrix (A.57) of transformation from the basis $\{\mathbf{i}_j\}$ to the basis $\{\mathbf{e}_j\}$ is as follows: $\mathrm{L}^{(1)} = \mathrm{L}\mathrm{L}^0$; here L^0 is the matrix of transformation from the basis $\{\mathbf{i}_j\}$ to the basis $\{\mathbf{e}_{j0}\}$ and L is the matrix of transformation from the basis $\{\mathbf{e}_{j0}\}$ to the basis $\{\mathbf{e}_j\}$. For small angles of rotations of the attached axes, we get

$$\mathrm{L} = \begin{bmatrix} 1 & \vartheta_3 & -\vartheta_2 \\ -\vartheta_3 & 1 & \vartheta_1 \\ \vartheta_2 & -\vartheta_1 & 1 \end{bmatrix}.$$

Since $L^{(1)} = LL^{(0)}$, we have

$$L^{(1)} = \begin{bmatrix} \cos\vartheta_{30} - \vartheta_3 \sin\vartheta_{30} & \sin\vartheta_{30} + \vartheta_3 \cos\vartheta_{30} & -\vartheta_2 \\ -\vartheta_{30}\cos\vartheta_{30} - \sin\vartheta_{30} & \cos\vartheta_{30} - \vartheta_3 \sin\vartheta_{30} & \vartheta_1 \\ \vartheta_2 \cos\vartheta_{30} + \vartheta_1 \sin\vartheta_{30} & \vartheta_2 \sin\vartheta_{30} - \vartheta_1 \cos\vartheta_{30} & 1 \end{bmatrix} . \quad (4)$$

In the basis $\{\mathbf{e}_j\}$, the vector $\mathbf{q}_0$ can be written as

$$\begin{aligned} \mathbf{q}_0 = q\,(\cos\vartheta_{30} - \vartheta_3 \sin\vartheta_{30})\,\mathbf{e}_1 + q_0\,(-\vartheta_3 \cos\vartheta_{30} - \sin\vartheta_{30})\,\mathbf{e}_2 \\ + q_0(\vartheta_2 \cos\vartheta_{30} + \vartheta_1 \sin\vartheta_{30})\,\mathbf{e}_3\,, \end{aligned} \quad (5)$$

therefore, we get

$$\begin{aligned} \Delta\mathbf{q} = -q_0\vartheta_3 \sin\vartheta_{30}\mathbf{e}_1 - q_0\vartheta_3 \cos\vartheta_{30}\mathbf{e}_2 \\ + q_0(\vartheta_2 \cos\vartheta_{30} + \vartheta_1 \sin\vartheta_{30})\,\mathbf{e}_3\,, \end{aligned} \quad (6)$$

Thus, the components ΔP_i (see (3)) are as follows:

$$\begin{aligned} \Delta P_1 &= \Delta q_1 = -q_0\vartheta_3 \sin\vartheta_{30}\,; \\ \Delta P_2 &= \Delta q_2 = -q_0\vartheta_3 \cos\vartheta_{30}\,; \\ \Delta P_3 &= \Delta q_3 = q_0\vartheta_1 \sin\vartheta_{30} + q_0\vartheta_2 \cos\vartheta_{30}\,. \end{aligned} \quad (7)$$

If the form of the rod axis in a deformed configuration differs little from that in the natural configuration, then the absolute value of the distributed critical load q_{0*} can be determined from (1) and (3). The critical load is an eigenvalue of the boundary-value problem for (3).

When the critical load is found, we should check the validity of the assumption of smallness of the displacements u_j and the angles of rotation ϑ_3. For this purpose, one should insert the critical load into (1) and solve it. If it is deducible from the solution that the values u_j and ϑ_3 are small, then the eigenvalue obtained is the sought critical load. Otherwise (the assumption fails), the nonlinear equations (1), where $æ_{3*}$ and $\vartheta_{30} = \vartheta_{3*}$ are unknown, are to be used.

3.2. The equations governing the rod equilibrium in the critical configuration are the same as (1) in Prob. 3.1. If the shapes of the rod axis in a deformed and natural configurations differ little, then we put $æ_{3*} = 1/\rho_0^0$ and $\vartheta_{3*} = \vartheta_{30}$ in (1). After the loss of stability, the equilibrium equations are identical to (3) of Prob. 3.1 (see Fig. 3.18) except that in the problem under consideration the expressions for the increments of external loads are not the same as in Prob. 3.1. Suppose that before the loss of stability the displacements of the points of the rod axis are negligibly small. Recall that the distributed load depends on the displacement of the axial line. In this case, we can represent the distributed load in terms of its projections on the Cartesian axes as follows (see Fig. 3.18):

$$\mathbf{q} = m_0\omega^2(\mathbf{r}_0 + \mathbf{u}) = m_0\omega^2(x_{10}\mathbf{i}_1 + x_{20}\mathbf{i}_2) + m_0\omega^2(u_1\mathbf{e}_1 + u_2\mathbf{e}_2 + u_3\mathbf{e}_3)\,; \quad (8)$$

here x_{10} and x_{20} are known functions of η; these functions are the coordinates of the points of the axial line in the natural configuration. The increment of the vector $\mathbf{q}$ reads

$$\Delta\mathbf{q} = \mathbf{q} - \mathbf{q}_0^{(1)} .$$

Let us obtain the components of the vector $\mathbf{q}_0$ in the attached frame. Using the matrix of transformation from the basis $\{\mathbf{i}_j\}$ to the basis $\{\mathbf{e}_{j0}\}$, we get

$$\begin{aligned} \mathbf{q}_0 &= m_0\omega^2(x_{10}\mathbf{i}_1 + x_{20}\mathbf{i}_2) \\ &= m_0\omega^2(x_{10}\cos\vartheta_{30} + x_{20}\sin\vartheta_{30})\,\mathbf{e}_{10} \\ &\quad + m_0\omega^2(-x_{10}\sin\vartheta_{30} + x_{20}\cos\vartheta_{30})\,\mathbf{e}_{20} , \end{aligned} \tag{9}$$

consequently,

$$\begin{aligned} q_{10} &= m_0\omega^2(x_{10}\cos\vartheta_{30} + x_{20}\sin\vartheta_{30}) ; \\ q_{20} &= m_0\omega^2(-x_{10}\sin\vartheta_{30} + x_{20}\cos\vartheta_{30}) . \end{aligned} \tag{10}$$

We have

$$\mathbf{q}_0^{(1)} = q_{10}\mathbf{e}_1 + q_{20}\mathbf{e}_2 . \tag{11}$$

The matrix of transformation from the basis $\{\mathbf{i}_j\}$ to the basis $\{\mathbf{e}_i\}$ is obtained in Prob. 3.1. Using (4), we can represent the vector $\mathbf{q}$ in the basis $\{\mathbf{e}_j\}$ as follows:

$$\begin{aligned} \mathbf{q} = {} & q_{10}\mathbf{e}_1 + q_{20}\mathbf{e}_2 + m_0\omega^2(x_{20}\vartheta_3\cos\vartheta_{30} - x_{10}\vartheta_3\sin\vartheta_{30})\,\mathbf{e}_1 \\ & + m_0\omega^2(-x_{20}\vartheta_3\sin\vartheta_{30} - x_{10}\vartheta_3\cos\vartheta_{30})\,\mathbf{e}_2 \\ & + m_0\omega^2[\,x_{10}(\vartheta_2\cos\vartheta_{30} - \vartheta_1\sin\vartheta_{30}) + x_{20}(\vartheta_2\sin\vartheta_{30} - \vartheta_1\cos\vartheta_{30})]\,\mathbf{e}_3 \\ & + \sum_{j=1}^{3} m_0\omega^2 u_j\mathbf{e}_j . \end{aligned} \tag{12}$$

The vector $\Delta\mathbf{q}$ satisfies the relationship

$$\Delta\mathbf{q} = \mathbf{q} - q_{10}\mathbf{e}_1 - q_{20}\mathbf{e}_2 = \sum_{j=1}^{3} \Delta q_j\mathbf{e}_j , \tag{13}$$

where

$$\begin{aligned} \Delta q_1 &= m_0\omega^2[\,u_1 + (x_{20}\cos\vartheta_{30} - x_{10}\sin\vartheta_{30})\,\vartheta_3\,] ; \\ \Delta q_2 &= m_1\omega^2[\,u_2 - (x_{10}\cos\vartheta_{30} + x_{20}\sin\vartheta_{30})\,\vartheta_3\,] ; \\ \Delta q_3 &= m_0\omega^2[\,u_3 - (x_{10}\sin\vartheta_{30} - x_{20}\cos\vartheta_{30})\,\vartheta_1 \\ &\quad + (x_{10}\cos\vartheta_{30} + x_{20}\sin\vartheta_{30})\,\vartheta_2\,] . \end{aligned} \tag{14}$$

Let us obtain Δq_i in the case when the shape of the axial line in a deformed configuration differs considerably from the shape of the axial line in the natural configuration. We have

$$\mathbf{q}_0 = m_0\omega^2(x_{10}\mathbf{i}_1 + x_{20}\mathbf{i}_2 + u_{1*}\mathbf{e}_{1*} + u_{2*}\mathbf{e}_{2*}), \tag{15}$$

where $\mathbf{e}_{i*}$ are the unit vectors of the attached coordinate system in the critical configuration and u_{i*} are displacements of the points of the rod axis in this configuration.

In the critical configuration, the strain-stress state of the rod can be determined from the nonlinear equilibrium equations (1) (see Prob. 3.1). Let us derive the projections of the load q_{i*} in (1). For this purpose, the vectors $\mathbf{i}_j$ are to be represented in terms of their projections in the basis $\{\mathbf{e}_{j0}\}$. The matrix of transformation from the basis $\{\mathbf{i}_j\}$ to the basis $\{\mathbf{e}_{j*}\}$ is as follows:

$$\mathrm{L}_*^{(1)} = \mathrm{L}_*\mathrm{L}^0; \tag{16}$$

here L^0 is the matrix of transformation from the basis $\{\mathbf{i}_j\}$ to the basis $\{\mathbf{e}_{j0}\}$ and L_* is the matrix of transformation from the basis $\{\mathbf{e}_{j0}\}$ to the basis $\{\mathbf{e}_{j*}\}$. The matrix L_* is of the form

$$\mathrm{L}_* = \begin{bmatrix} \cos\vartheta_{3*} & \sin\vartheta_{3*} & 0 \\ -\sin\vartheta_{3*} & \cos\vartheta_{3*} & 0 \\ 0 & 0 & 1 \end{bmatrix},$$

consequently,

$$\mathrm{L}_*^{(1)} = \begin{bmatrix} \cos(\vartheta_{30}+\vartheta_{3*}) & \sin(\vartheta_{30}+\vartheta_{3*}) & 0 \\ -\sin(\vartheta_{30}+\vartheta_{3*}) & \cos(\vartheta_{30}+\vartheta_{3*}) & 0 \\ 0 & 0 & 1 \end{bmatrix}. \tag{17}$$

In the basis $\{\mathbf{e}_{j*}\}$, the vector $\mathbf{q}_{0*}$ can be written as

$$\begin{aligned}\mathbf{q}_{0*} = m_0\omega^2\{\,&[\,x_{10}\cos(\vartheta_{30}+\vartheta_{3*}) + x_{20}\sin(\vartheta_{30}+\vartheta_{3*}) + u_{1*}\,]\,\mathbf{e}_{1*} \\ &+ [\,-x_{10}\sin(\vartheta_{30}+\vartheta_{3*}) + x_{20}\cos(\vartheta_{30}+\vartheta_{3*}) + u_{2*}\,]\,\mathbf{e}_{2*}\}.\end{aligned}$$

After the loss of stability, the increment of the vector $\mathbf{q}$ reads

$$\Delta\mathbf{q} = \mathbf{q} - \mathbf{q}_{0*}^{(1)}, \qquad \mathbf{q}_0^{(1)} = \sum_{i=1}^{2} q_{i0*}\mathbf{e}_{i*}, \tag{18}$$

where

$$\mathbf{q} = m_0\omega^2(\mathbf{r}_0 + \mathbf{u}_*) + m_0\omega^2\mathbf{u} = \mathbf{q}_{0*} + m_0\omega^2\mathbf{u}. \tag{19}$$

To calculate the components of the vector $\Delta\mathbf{q}$ in the basis $\{\mathbf{e}_j\}$, one should represent the vector $\mathbf{q}_0$ in this basis. For small rotations of the attached axes, we can use the matrix of transformation from the basis $\{\mathbf{e}_{j0}\}$ to the basis $\{\mathbf{e}_j\}$ (the matrix L in Prob. 3.1). As a result, we have

$$\begin{aligned}\mathbf{e}_{1*} &= \mathbf{e}_1 - \vartheta_3\mathbf{e}_2 + \vartheta_3\mathbf{e}_3\,;\\ \mathbf{e}_{2*} &= \vartheta_3\mathbf{e}_1 + \mathbf{e}_2 - \vartheta_1\mathbf{e}_3\,,\end{aligned} \tag{20}$$

therefore,

$$\begin{aligned}\mathbf{q}_{0*} = {} & q_{10*}\mathbf{e}_1 + q_{20*}\mathbf{e}_2 + (q_{20*}\vartheta_3 + m_0\omega^2 u_1)\,\mathbf{e}_1\\ & + (-q_{10*}\vartheta_3 + m_0\omega^2 u_2)\,\mathbf{e}_2 + (q_{10*}\vartheta_2 - q_{20*}\vartheta_1 + m_0\omega^2 u_3)\,\mathbf{e}_3\,.\end{aligned} \tag{21}$$

Finally, we arrive at

$$\begin{aligned}\Delta\mathbf{q} = {} & (m_0\omega^2 u_1 + q_{20*}\vartheta_3)\,\mathbf{e}_1 + (m_0\omega^2 u_2 - q_{10*}\vartheta_3)\,\mathbf{e}_2\\ & + (m_0\omega^2 u_3 - q_{20*}\vartheta_1 + q_{10*}\vartheta_2)\,\mathbf{e}_3\,,\end{aligned} \tag{22}$$

where

$$\begin{aligned}q_{10*} &= m_0\omega^2[\,u_{1*} + x_{10}\cos(\vartheta_{30} + \vartheta_{3*}) + x_{20}\sin(\vartheta_{30} + \vartheta_{3*})\,]\,;\\ q_{20*} &= m_0\omega^2[\,u_{2*} - x_{10}\sin(\vartheta_{30} + \vartheta_{3*}) + x_{20}\cos(\vartheta_{30} + \vartheta_{3*})\,]\,.\end{aligned} \tag{23}$$

3.3. In this problem, two situations may occur: (1) in-plane loss of stability or (2) out-of-plane loss of stability.

In the case of in-plane loss of stability, the critical loads can be determined from (3.48). These equations do not contain the moment M_{30}, hence, the critical load do not depend on the initial strain-stress state of the rod.

To determine the critical loads in the case of out-of-plane loss of stability, we shall use (3.49). Terms containing the moment M_{30} are to be added to (3.49). These terms appear in the general equations (3.33)–(3.36) (recall that (3.48) and (3.49) were derived on the basis of (3.33)–(3.36)). As a result, we have

$$\begin{aligned}&\frac{\mathrm{d}Q_{03}}{\mathrm{d}\eta} - \frac{Q_{1*}}{A_{22}}\,M_{02} = 0\,, \qquad Q_{1*} = -q_{2*}R^0\,;\\ &\frac{\mathrm{d}M_{01}}{\mathrm{d}\eta} - \frac{1}{R^0}\,M_{02} + \frac{M_{30}}{A_{22}}\,M_{02} = 0\,;\\ &\frac{\mathrm{d}M_{02}}{\mathrm{d}\eta} + \frac{1}{R^0}\,M_{01} - \frac{M_{30}}{A_{11}}\,M_{01} - Q_{03} = 0\,;\\ &\frac{\mathrm{d}\vartheta_1}{\mathrm{d}\eta} - \frac{M_{01}}{A_{11}} - \frac{1}{R^0}\,\vartheta_2 = 0\,;\\ &\frac{\mathrm{d}\vartheta_2}{\mathrm{d}\eta} - \frac{M_{02}}{A_{22}} + \frac{1}{R^0}\,\vartheta_1 = 0\,;\\ &\frac{\mathrm{d}u_3}{\mathrm{d}\eta} + \vartheta_2 = 0\,.\end{aligned} \tag{24}$$

From the first three equations of the system (24) we have

$$M''_{02} + \left[\frac{1}{(R^0)^2} - \frac{(A_{11} + A_{22})}{R^0 A_{11} A_{22}} M_{30} + \frac{M_{30}^2}{A_{11} A_{22}} + \frac{q_{2*} R^2}{A_{22}} \right] M_{02} = 0 \,. \quad (25)$$

The solution of (25) must be a periodic function, thus the relation for determining q_{2*},

$$\cos k = 1 \,, \quad (26)$$

where

$$k = \sqrt{\frac{1}{(R^0)^2} - \frac{(A_{11} + A_{22})}{R^0 A_{11} A_{22}} M_{30} + \frac{M_{30}^2}{A_{11} A_{22}} + \frac{q_{2*} R^2}{A_{22}}} \,.$$

We have $R^0 = 1/(2\pi)$ and $M_{30} = 1/R^0$. Setting $k = 2\pi n$, we finally get the sought critical values

$$q_{2*} = 4\pi^2 \left[(n^2 - 1) + \frac{A_{11} + A_{22} - 1}{A_{11} A_{22}} \right] .$$

3.5. The load applied to the rod is a *dead* one, hence, the load increments ΔP_i in (3.33) are different from zero while the moment increments ΔT_i vanish. The equations governing equilibrium of the rod after the loss of stability are the same as (3) in Prob. 3.1. The load increments are as follows:

$$\Delta P_j = \Delta q_j + \Delta P_{j0}\, \delta(\eta - \eta_i) \,.$$

The increments Δq_i were obtained in Prob. 3.1,

$$\begin{aligned}
\Delta q_1 &= -q_0 \sin \vartheta_{30} \vartheta_3 \,; \\
\Delta q_2 &= -q_0 \cos \vartheta_{30} \vartheta_3 \,; \\
\Delta q_3 &= q_0 \sin \vartheta_{30} \vartheta_1 + q_0 \cos \vartheta_{30} \vartheta_2 \,, \qquad q_0 = a_0 m_0 \,.
\end{aligned}$$

Now we are to obtain the formulas for the increments of the concentrated loads. These formulas are identical in form to the formulas for Δq_i, that is,

$$\begin{aligned}
\Delta P_1 &= -P_0 \sin \vartheta_{30}(\eta_1)\, \vartheta_3(\eta_1) \,; \\
\Delta P_2 &= -P_0 \cos \vartheta_{30}(\eta_1)\, \vartheta_3(\eta_1) \,; \\
\Delta P_3 &= P_0 \sin \vartheta_{30}(\eta_1)\, \vartheta_1(\eta_1) + P_0 \cos \vartheta_{30}(\eta_1)\, \vartheta_2(\eta_1) \,, \qquad P_0 = a_0 m \,.
\end{aligned}$$

The equilibrium equations for the rod are as follows (here we write out only the first three equations of (3) of Prob. 3.1):

$$\begin{aligned}
&\frac{\mathrm{d}Q_{01}}{\mathrm{d}\eta} - Q_{2*} M_{03} - \frac{1}{R^0} Q_{02} - q_0 \sin \vartheta_{30} \vartheta_3 = P_0 \sin \vartheta_{30} \vartheta_3\, \delta(\eta - \eta_1) \,; \\
&\frac{\mathrm{d}Q_{02}}{\mathrm{d}\eta} + Q_{1*} M_{03} + \frac{1}{R^0} Q_{01} - q_0 \cos \vartheta_{30} \vartheta_3 = P_0 \cos \vartheta_{30} \vartheta_3\, \delta(\eta - \eta_1) \,; \\
&\frac{\mathrm{d}Q_{03}}{\mathrm{d}\eta} + \frac{Q_{2*}}{A_{11}} M_{01} - \frac{Q_{1*}}{A_{22}} M_{02} + q_0 \sin \vartheta_{30} \vartheta_1 + q_0 \cos \vartheta_{30} \vartheta_2 \\
&= -P_0 \sin \vartheta_{30} \vartheta_1\, \delta(\eta - \eta_1) + P_0 \cos \vartheta_{30} \vartheta_2\, \delta(\eta - \eta_1) \,.
\end{aligned}$$

B.3 To Chapter 4

4.1. Let us derive the equilibrium equations. For small displacements of the points of the rod axis, we get $\mathbf{u} = u_2\mathbf{e}_2 = u_{x_2}\mathbf{i}_2$. Therefore, the system of equations (4.103)–(4.107), which are associated with the Cartesian frame, can be used. We have

$$Q_{x_1} + P_{x_1} + \Delta P_{x_1} = 0\,; \tag{1}$$
$$Q'_{x_2} + P_{x_2} + \Delta P_{x_2} = 0\,; \tag{2}$$
$$M'_{x_2} + Q_{x_2} - \vartheta_3 Q_{x_1} = 0\,; \tag{3}$$
$$\vartheta'_3 - M_{x_3} = 0\,, \qquad M_{x_3} = \Delta æ_3\,; \tag{4}$$
$$u'_{x_2} - \vartheta_3 = 0\,. \tag{5}$$

In what follows, all the quantities are assumed to be replaced by their nondimensional counterparts. The nondimensional coordinate x_1 is denoted by η.

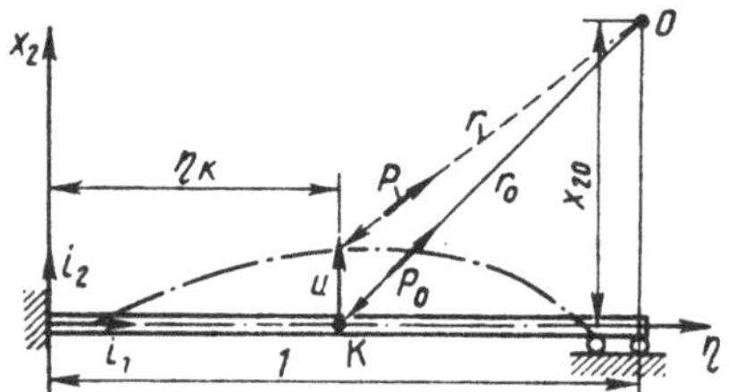

Fig. B.1.

Let us obtain P_{x_20} and ΔP_{x_j}. Consider a curvilinear rod illustrated in Fig. B.1. It is seen that

$$\mathbf{P} = |\mathbf{P}_0|\,\mathbf{e}_r\,, \tag{6}$$

where

$$\mathbf{e}_r = \frac{\mathbf{r}}{|\mathbf{r}_0 + \mathbf{u}|}\;(\mathbf{r} = \mathbf{r}_0 + \mathbf{u})\,. \tag{7}$$

Since

$$|\mathbf{r}_0 + \mathbf{u}| = \sqrt{r^2_{x_10} + (r_{x_20} + u_{x_2})^2}\,, \tag{8}$$

for small values of u_{x_2} we have

$$|\mathbf{r}_0 + \mathbf{u}| \approx r_0\left(1 + \frac{r_{x_10}}{r_0^2}\,u_{x_2}\right)\,, \qquad r_0 = \sqrt{r^2_{x_10} + r^2_{x_20}}\,, \tag{9}$$

consequently,

$$\mathbf{e}_r = (r_{x_10}\mathbf{i}_1 + r_{x_20}\mathbf{i}_2 + u_{x_2}\mathbf{i}_2)\,\frac{1}{r_0}\left(1 - \frac{r_{x_20}}{r_0^2}\,u_{x_2}\right). \tag{10}$$

Linearization of (10) with respect to u_{x_2} yields

$$\mathbf{e}_r = \left(\frac{r_{x_10}}{r_0} - \frac{r_{x_10}r_{x_20}}{r_0^3}\,u_{x_2}\right)\mathbf{i}_1 + \left(\frac{r_{x_20}}{r_0} + \frac{r_{x_10}^2}{r_0^3}\,u_{x_2}\right)\mathbf{i}_2\,, \tag{11}$$

where $r_{x_10} = 1 - \eta_i$ and $r_{x_20} = x_{20}$.

Using (11) in (6), we obtain

$$P_{x_10} = |\mathbf{P}_0|\,\frac{r_{x_10}}{r_0}\,;$$

$$P_{x_20} = |\mathbf{P}_0|\,\frac{r_{x_20}}{r_0}\,; \tag{12}$$

$$\Delta P_{x_1} = -|\mathbf{P}_0|\,\frac{r_{x_10}r_{x_20}}{r_0^3}\,u_{x_2} = -a_1u_{x_2}\,; \tag{13}$$

$$\Delta P_{x_2} = |\mathbf{P}_0|\,\frac{r_{x_10}^2}{r_0^3}\,u_{x_2} = a_2u_{x_2}\,; \tag{14}$$

By use of these relationships, (1) and (2) can be represented as follows:

$$Q'_{x_1} + P_{x_10}\,\delta(\eta-\eta_k) - a_1u_{x_2}\,\delta(\eta-\eta_k) = 0\,; \tag{15}$$

$$Q'_{x_2} + P_{x_20}\,\delta(\eta-\eta_k) + a_2u_{x_2}\,\delta(\eta-\eta_k) = 0\,. \tag{16}$$

Since Q_{x_1} in (3) is multiplied by a small quantity ϑ_3, we can omit the term containing u_{x_2} in (15). As a result, we get

$$Q_{x_1} = -P_{x_10}\,H(\eta-\eta_k) + c\,.$$

Since $Q_{x_1} = P_{x_10}$ for $\eta = 0$, we obtain

$$Q_{x_1} = P_{x_10} - P_{x_10}\,H(\eta-\eta_k)\,. \tag{17}$$

The system of equations (16) and (3)–(5) may be represented as one equation,

$$L = u_{x_2}^{\mathrm{IV}} - P_{x_10}\,[\,1 - H(\eta-\eta_k)\,]\,u''_{x_2} - a_2\,\delta(\eta-\eta_k)\,u_{x_2} - P_{x_20}\,\delta(\eta-\eta_k) = 0\,. \tag{18}$$

Recall that the function v_1 given by (4.187) satisfies the boundary conditions, therefore, the approximate solution of (18) is of the form

$$u_{\mathrm{t}x_2} = c_1v_1 = c_1\left(\eta^4 - \frac{5}{2}\eta^3 + \frac{3}{2}\eta^2\right) \tag{19}$$

and virtual displacements of the axial points are

$$\delta u = \delta b\,v_1\,. \tag{20}$$

According to the principle of virtual displacements, the constant c_1 can be determined from the equation

$$c_1 \left[\int_0^1 (v_1^{\mathrm{IV}} - P_{x_1 0}(1-H)\, v_1'')\, v_1 \, \mathrm{d}\eta - a_2 v_1^2(\eta_k) \right] - P_{x_2 0} v_1(\eta_k) = 0 \,. \tag{21}$$

4.2. Let us consider small vertical displacements of the rod. Note that only the projection of the force $\mathbf{P}$ on the axis x_2 performs nonzero work through these displacements. Consequently, the functional J_1 takes the form

$$J_1 = \frac{1}{2} \int_0^1 A_{33}\, u^{2'} \, \mathrm{d}\eta - P_{x_2} u\,(\eta_k)\,. \tag{22}$$

For small displacements of the axial points, we can write

$$P_{x_2} = P_{x_2 0} = |\mathbf{P}| \frac{x_{20}}{\sqrt{(1-\eta_k)^2 + x_{20}^2}}\,. \tag{23}$$

Let the function v_1 be of the form

$$v_1 = \eta^4 - \frac{5}{2}\eta^3 + \frac{3}{2}\eta^2\,. \tag{24}$$

Restrict our consideration to a one-term approximation. Substituting $u = a_1 v_1$ into (22), we get

$$a_1 \approx \frac{0.06\, P_{x_2 0}}{A_{33}}\,. \tag{25}$$

4.3. Let us use (1)–(5) from Prob. 4.1. Arguing as in Prob. 4.1 when we dealt with the projections of the concentrated force $\mathbf{P}$, we can obtain the distributed load in the Cartesian axes $x_1 = \eta$ and x_2. The only difference is that now the radius r_0 and its projections are functions of η,

$$r_0 = \sqrt{r_{x_1 0}^2(\eta) + r_{x_2 0}^2(\eta)}\,, \tag{26}$$

where $r_{x_1 0}(\eta) = \eta_0 - \eta$ and $r_{x_2 0}(\eta) = x_{20} = \text{const}$.

The components of the distributed load and the increments of these components are as follows:

$$q_{x_1 0} = q_0 \frac{r_{x_1 0}}{r_0}\,; \qquad q_{x_2 0} = q_0 \frac{r_{x_2 0}}{r_0}\,; \tag{27}$$

$$\Delta q_{x_1} = -q_0 \frac{r_{x_1 0} r_{x_2 0}}{r_0^3} u_{x_2} = -a_1(\eta)\, u_{x_2}\,; \tag{28}$$

$$\Delta q_{x_2} = q_0 \left(1 - \frac{r_{x_2 0}^2}{r_0^2}\right) \frac{u_{x_2}}{r_0} = a_2(\eta)\, u_{x_2}\,. \tag{29}$$

The equilibrium equations take the form

$$\frac{\mathrm{d}Q_{x_1}}{\mathrm{d}\eta} + q_0 \frac{r_{x_1 0}}{r_0} - a_1 u_{x_2} = 0\,; \tag{30}$$

$$\frac{\mathrm{d}Q_{x_2}}{\mathrm{d}\eta} + q_0 \frac{r_{x_2 0}}{r_0} + a_2 u_{x_2} = 0\,; \tag{31}$$

$$\frac{\mathrm{d}M_{x_3}}{\mathrm{d}\eta} + Q_{x_2} - \vartheta_3 Q_{x_1} = 0\,; \tag{32}$$

$$\frac{\mathrm{d}\vartheta_3}{\mathrm{d}\eta} - M_{x_3} = 0\,; \tag{33}$$

$$\frac{\mathrm{d}u_{x_2}}{\mathrm{d}\eta} - \vartheta_3 = 0\,. \tag{34}$$

Since Q_{x_1} in (32) is multiplied by a small quantity ϑ_3, we can omit the terms containing u_{x_2} in (30). Consequently, we get

$$Q_{x_1 0} = -q_0 \int \frac{r_{x_1 0}}{r_0}\,\mathrm{d}\eta + c_1 = q_0 \sqrt{(\eta_0 - \eta)^2 + x_{20}^2} + c_1\,.$$

Using the relation $Q_{x_1 0}(1) = 0$, we obtain

$$a_{x_1 0} = q_0 \left(\sqrt{(\eta_0 - \eta)^2 + x_{20}^2} - \sqrt{(\eta_0 - 1)^2 + x_{20}^2} \right)\,. \tag{35}$$

The system of equations (31)–(34) can be represented as one equation,

$$\mathrm{L} = u_{x_2}^{\mathrm{IV}} - Q_{x_1 0} u_{x_2}'' - a_2 u_{x_2} - q_0 \frac{r_{x_2 0}}{r_0} = 0\,. \tag{36}$$

Let $v_1(\eta)$ be of the form

$$v_1(\eta) = 6\eta^2 - 4\eta^3 + \eta^4\,. \tag{37}$$

This function satisfies all the boundary conditions of the problem. Setting $u_{\mathrm{t}x_2} = c_1 v_1(\eta)$ and $\delta u = \delta a\, v_1$ and using the principle of virtual displacements, we obtain the following equation in c_1:

$$\delta a \int_0^1 \mathrm{L}\,(u_{\mathrm{t}x_2})\, v_1\, \mathrm{d}\eta = 0 \tag{38}$$

or

$$c_1 \int_0^1 (v_1^{\mathrm{IV}} - Q_{x_1 0} v_1'' - a_2 v_1)\, v_1\, \mathrm{d}\eta - q_0 \int_0^1 \frac{r_{x_2 0}}{r_0} v_1\, \mathrm{d}\eta = 0\,. \tag{39}$$

4.4. Twice the work done by the external forces is as follows:

$$\int_0^1 (\mathbf{q} u_{x_2} \mathbf{i}_2)\, \mathrm{d}\eta = \int_0^1 q_{x_2 0} u_{x_2}\, \mathrm{d}\eta ; \tag{40}$$

here

$$q_{x_2 0} = q_0 \frac{x_{20}}{\sqrt{(\eta_0 - \eta)^2 + x_{20}^2}} .$$

The functional J_1 takes the form

$$J_1 = \frac{1}{2} \int_0^1 A_{33} (u''_{x_2})^2\, \mathrm{d}\eta - q_0 \int_0^1 \frac{x_{20} u_{x_2}\, \mathrm{d}\eta}{\sqrt{(\eta_0 - \eta)^2 + x_{20}^2}} . \tag{41}$$

Using a one-term approximation, we have

$$u_{\mathrm{t}x_2} = a_1 v_1 ,$$

where

$$v_1 = 6\eta^2 - 4\eta^3 + \eta^4 .$$

From the condition

$$\frac{\partial J_1}{\partial a_1} = 0$$

it follows that

$$a_1 = \frac{q_0 x_{20} \displaystyle\int_0^1 \frac{v_1\, \mathrm{d}\eta}{\sqrt{(\eta_0 - \eta)^2 + x_{20}^2}}}{\left(\displaystyle\int_0^1 A_{33} (v''_1)^2\, \mathrm{d}\eta \right)} . \tag{42}$$

4.5. Setting $P_{x_2 0} = R$, $P_{x_1 0} = \mu_{x_3 0} = T^{(1)}_{x_3 0} = 0$, and $A_{33} = 1$ in (4.138) (4.141) (R is the reaction force applied to the support), we get

$$\begin{aligned} &Q'_2 - k u_{x_2} = q_{x_2 0}\, H(\eta - 0.5) - R\,\delta(\eta - 0.5) ; \\ &M'_{x_3} + Q_{x_2} = 0 ; \\ &\vartheta'_3 - M_{x_3} = 0 ; \\ &u'_{x_2} - \vartheta_3 = 0 . \end{aligned} \tag{43}$$

For $b_2 = 0$, the system of equations (43) is identical to (4.151). Hence, using (4.161) and the fundamental matrix (4.148), we can represent the solution of (43) as follows:

$$\mathbf{Y} = \mathrm{K}(\eta)\, \mathbf{C} + \int_0^\eta \mathrm{K}(\eta - \zeta)\, \mathbf{b}\, \mathrm{d}\zeta ; \tag{44}$$

here

$$\mathbf{Y} = (Q_{x_2} M_{x_3} \vartheta_3 u_{x_2})^{\mathrm{T}} .$$

Recall that the components of the fundumental matrix (4.148) are expressed in terms of Krylov's functions.

From the boundary conditions at $\eta = 0$ it follows that $c_2 = c_4 = 0$. Using (4.161)–(4.168), we have

$$\mathbf{Y}_H = \int_0^\eta \mathrm{K}\,(\eta-\zeta)\,\mathbf{b}\,\mathrm{d}\zeta = \begin{bmatrix} \frac{q_{x_2 0}}{2\alpha_1^2}\,[\,K_2(\alpha_1\eta) - K_2(\alpha_1\cdot 0.5)\,]\,H(\eta-0.5) \\ -\,RK_1(\alpha_1\cdot 0.5)\,H(\eta-0.5) \\ \frac{q_{x_2 0}}{2\alpha_1^3}\,[\,K_3(\alpha_1\eta) - K_3(\alpha_1\cdot 0.5)\,]\,H \\ -\,\frac{R}{2\alpha_1}\,K_2(\alpha_1\cdot 0.5)\,H \\ \frac{q_{x_2 0}}{4\alpha_1^4}\,[\,K_4(\alpha_1\eta) - K_4(\alpha_1\cdot 0.5)\,]\,H \\ -\,\frac{R}{2\alpha_1^2}\,K_3(\alpha_1\cdot 0.5)\,H \\ \frac{q_{x_2 0}}{4\alpha_1^5}\,[\,K_1(\alpha_1\eta) - K_1(\alpha_1\cdot 0.5)\,]\,H \\ -\,\frac{R}{4\alpha_1^3}\,K_4(\alpha_1\cdot 0.5)\,H \end{bmatrix} . \tag{45}$$

In components, (44) reads

$$\begin{aligned} Y_1 &= Q_{x_2} = k_{11}c_1 + k_{13}c_3 + Y_{H1}\,; \\ Y_2 &= M_{x_3} = k_{21}c_1 + k_{23}c_3 + Y_{H2}\,; \\ Y_3 &= \vartheta_3 = k_{31}c_1 + k_{33}c_3 + Y_{H3}\,; \\ Y_4 &= u_{x_2} = k_{41}c_1 + k_{43}c_3 + Y_{H4}\,, \end{aligned} \tag{46}$$

where $k_{11} = K_1$, $k_{21} = \dfrac{K_2}{2\alpha_1}$, $k_{31} = \dfrac{K_3}{2\alpha_1^3}$, $k_{41} = \dfrac{K_4}{4\alpha_1^3}$, $k_{13} = -2\alpha_1^2 K_3$, $k_{23} = -\alpha_1 K_4$, $k_{33} = K_1$, and $k_{43} = \dfrac{K_2}{2\alpha_1}$.

The constants c_1 and c_3 and the reaction force at the support R are to be determined from the following conditions:

(1) at $\eta = 1$, $Q_{x_2} = 0$ and $M_{x_2=0}$;
(2) at $\eta = 0.5$, $u_{x_2} = 0$.

Using (46), we obtain the following system of inhomogeneous equations in the unknowns c_1, c_3, and R:

$$K_1(\alpha_1)\,c_1 - 2\alpha_1^2 K_3(\alpha_1)\,c_3 - K_1(\alpha_1 \cdot 0.5)\,R$$
$$= -\frac{q_{x20}}{2\alpha_1^2}\,[\,K_2(\alpha_1) - K_2(\alpha_1 \cdot 0.5)\,]\,;$$
$$\frac{K_2(\alpha_1)}{2\alpha_1}\,c_1 - \alpha_1 K_4(\alpha_1)\,c_3 - \frac{K_2(\alpha_1 \cdot 0.5)}{2\alpha_1}\,R$$
$$= -\frac{q_{x20}}{2\alpha_1^3}\,[\,K_3(\alpha_1) - K_3(\alpha_1 \cdot 0.5)\,]\,;$$
$$\frac{K_4(\alpha_1 \cdot 0.5)}{4\alpha_1^3}\,c_1 + \frac{K_2(\alpha_1 \cdot 0.5)}{2\alpha_1}\,c_3 = 0\,. \tag{47}$$

4.6. The equilibrium equations can be reduced to one equation as follows:

$$L = u_{x_2}^{\mathrm{IV}} + 4\alpha_1^4 u_{x_2}[\,H(\eta) - H(\eta - 0.5)\,] + T_{x_2}\,\delta'(\eta - 0.5) = 0\,.$$

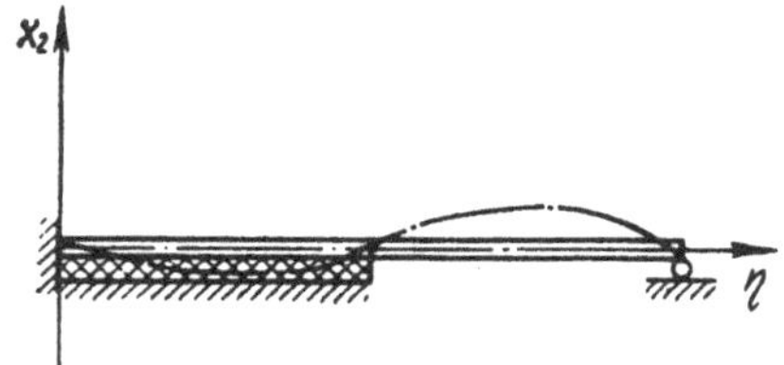

Fig. B.2.

After the deformation, the shape of the rod axis is illustrated (qualitatively) in Fig. 2. Let $u_{tx_2} = a_2 v_2(\eta)$, where

$$v_2(\eta) = \eta^5 - 2.2535\eta^4 + 1.1337\eta^3 + 0.1198\eta^2\,.$$

According to the principle of virtual displacements, we have

$$a_2 \int_0^1 \{v_2^{\mathrm{IV}} + 4\alpha_1^4 v_2[\,H(\eta) - H(\eta - 0.5)\,]\,\}\,v_2\,\mathrm{d}\eta + \int_0^1 T_{x_3} v_2\,\delta'\,\mathrm{d}\eta = 0$$

or

$$a_2 \int_0^1 \{v_2^{\mathrm{IV}} + 4\alpha_1^4 v_2[\,H(\eta) - H(\eta - 0.5)\,]\,\}\,v_2\,\mathrm{d}\eta - T_{x_3} v_2'(0.5) = 0\,.$$

Finally, we get

$$a_2 = \frac{T_{x_3} v_2'(0.5)}{\displaystyle\int_0^1 \{v_2^{\mathrm{IV}} + 4\alpha_1^4 v_2[\,H(\eta) - H(\eta - 0.5)\,]\,\}\,v_2\,\mathrm{d}\eta}\,.$$

4.7. The equilibrium equation of the rod is

$$L = u_{x_2}^{\mathrm{IV}} + 4\alpha_1^4 u_{x_2} + \gamma u_{x_2}^3 - q_{x_20}\, H(\eta - 0.5) = 0\,.$$

Let the approximate solution be

$$u_{\mathrm{t}x_2} = a_1 v_1(\eta)\,,$$

where

$$v_1(\eta) = 2\eta^4 - 5\eta^3 + 3\eta^2\,.$$

The coefficient a_1 is to be determined from the cubic equation

$$a_1^3 \int_0^1 \gamma v_1^4\, \mathrm{d}\eta + a_1 \int_0^1 (v_1^{\mathrm{IV}} + 4\alpha_1^4 v_1)\, v_1\, \mathrm{d}\eta - q_{x_20} \int_0^1 v_1\, \mathrm{d}\eta = 0\,.$$

B.4 To Chapter 5

5.1. For small angles ϑ_{10}, (5.154) takes the form

$$\vartheta_{10}'' + \frac{\Omega_{30}^{\mathrm{c}}}{A_{11}} [\,(A_{33} - A_{22})\, \Omega_{30}^{\mathrm{c}} - A_{33}\Omega_{30}^{\mathrm{r}}\,]\, \vartheta_{10} = 0\,. \tag{1}$$

The sign of the expression in the square brackets is dictated by the physical parameters of the rod. Hence, (1) can be rewritten as

$$\vartheta_{10}'' \pm k^2 \vartheta_{10} = 0\,. \tag{2}$$

The moments M_i are as follows:

$$\begin{aligned} M_1 &= A_{11}\vartheta_{10}'\,;\\ M_2 &= A_{22}\Omega_{30}^{\mathrm{c}} \sin\vartheta_{10}\,;\\ M_3 &= A_{33}(\Omega_{30}^{\mathrm{c}} \cos\vartheta_{10} - \Omega_{30}^{\mathrm{r}})\,. \end{aligned} \tag{3}$$

The moment M_3 can take one of the two following values:

$$M_3 = A_{33}(\Omega_{30}^{\mathrm{c}} \cos\vartheta_{10} - \Omega_{30}^{\mathrm{r}}) \tag{4}$$

or

$$M_3 = A_{33}(\Omega_{30}^{\mathrm{c}} \cos\vartheta_{10} + \Omega_{30}^{\mathrm{r}})\,. \tag{5}$$

For small angles ϑ_{10}, we have

$$M_1 = A_{11}\vartheta_{10}'\,; \qquad M_2 = A_{22}\Omega_{30}^{\mathrm{c}}\vartheta_{10}\,; \qquad M_3 = A_{33}(\Omega_{30}^{\mathrm{c}} - \Omega_{30}^{\mathrm{r}})\,.$$

Suppose that the expression in the square brackets in (1) is positive. The solution of (2) is

$$\vartheta_{10} = c_1 \cos k\eta + c_2 \sin k\eta \, . \tag{6}$$

Recall that our aim is to determine the angle of relative rotation of the end cross sections. We can put $\vartheta_{10}(0) = 0$ and thus $c_1 = 0$. From (3) and (6) it follows that

$$A_{11}\vartheta'_{10}(0) = M_{10} \, ; \qquad c_2 = \frac{M_{10}}{kA_{11}} \, . \tag{7}$$

Therefore,

$$\vartheta_{10}(\eta) = \frac{M_{10}}{kA_{11}} \sin k\eta \, . \tag{8}$$

The angle of rotation of the cross section $\eta = 1$ is the sought angle. We have

$$\vartheta_{10}(1) = \frac{M_{10}}{kA_{11}} \sin k \, . \tag{9}$$

The twisting moment M_{1k} is as follows:

$$M_{1k} = M_{10} \cos k \, . \tag{10}$$

The equality $M_{10} = M_{1k}$ holds for $k = 0$. Consequently,

$$k = \sqrt{\frac{\Omega^{\mathrm{c}}_{30}}{A_{11}} \left[(A_{33} - A_{22}) \, \Omega^{\mathrm{c}}_{30} - A_{33} \Omega^{\mathrm{r}}_{30} \right]} = 0 \, . \tag{11}$$

Let $\Omega^{\mathrm{c}}_{30} \neq 0$. There are two cases when the parameter k vanishes, namely,

$$\Omega^{\mathrm{r}}_{30} = 0 \, ; \qquad A_{22} = A_{33} \tag{12}$$

and

$$\frac{(A_{33} - A_{22})}{A_{33}} = \frac{\Omega^{\mathrm{r}}_{30}}{\Omega^{\mathrm{c}}_{30}} \, . \tag{13}$$

The first case (the rod is a straight one) is discussed in Sect. 5.3. If $A_{22} = A_{33}$ and $\Omega^{\mathrm{r}}_{30} = 0$, (5.154) is identical in form to (2). The twisting moment is the same at any cross section of the rod.

In the second case (the angles ϑ_{10} are small), only the moments at the end cross sections are of the same magnitude.

The load components q_2 and q_3 can be determined from (5.142), (5.143) and (5.139), (5.140). If the equality (4) holds, then, solving for Q_2 and Q_3 from (5.142) and (5.143), we get

$$Q_2 = (A_{33} - A_{22} + A_{11}) \, \Omega^{\mathrm{c}}_{30} \sin \vartheta_{10} \vartheta'_{10} \, ; \tag{14}$$

$$Q_3 = (A_{22} - A_{33} + A_{11}) \, \Omega^{\mathrm{c}}_{30} \cos \vartheta_{10} \vartheta'_{10} + A_{33} \Omega^{\mathrm{r}}_{30} \vartheta'_{10} \, . \tag{15}$$

The relations (11) and (12) are valid for arbitrary values of ϑ_{10}. For small angles, we have

$$Q_2 = (A_{33} - A_{22} + A_{11})\,\Omega^{\mathrm{c}}_{30}\vartheta_{10}\vartheta'_{10} \approx 0\,;$$
$$Q_3 = [\,(A_{22} - A_{33} + A_{11})\,\Omega^{\mathrm{c}}_{30} + A_{33}\Omega^{\mathrm{r}}_{30}\,]\,\vartheta'_{10}\,.$$

From (5.138) it follows that $Q'_1 = 0$ and $Q_1 = \text{const}$. The rod in the conduit may be in equilibrium when no forces are applied to its ends (i.e. $Q_1 \equiv 0$). Using (5.139) and (5.140), we arrive at

$$q_2 = -Q'_2 - Q_1 æ_3 + Q_3 æ_1 \approx 0\,;$$
$$q_3 = -Q'_3 - Q_2 æ_1 + Q_1 æ_2 = k_3 \frac{M_{10}k}{A_{11}} \sin k\eta\,,$$

where

$$K_3 = (A_{22} - A_{33} + A_{11})\,\Omega^{\mathrm{c}}_{30} = A_{33}\Omega^{\mathrm{r}}_{30}$$

Finally, the absolute value of the contact pressure is $|\mathbf{q}| = |q_3|$.

5.2. Let us use (5.152). For small values of the angles, we have

$$\vartheta''_{10} + \frac{\Omega^{\mathrm{c}}_{30}}{A_{11}}\,[\,(A_{33} - A_{22})\,\Omega^{\mathrm{c}}_{30} - A_{33}\Omega^{\mathrm{r}}_{30}\,]\,\vartheta_{10} = 0\,, \tag{16}$$

This equation is the same as (1) in Prob. 5.1.

For the moments M_i, we have

$$M_1 = A_{11}(\Omega^{\mathrm{c}}_{10} + \vartheta'_{10})\,;$$
$$M_2 = A_{22}\Omega^{\mathrm{c}}_{30}\vartheta_{10}\,;$$
$$M_3 = A_{33}(\Omega^{\mathrm{c}}_{30} - \Omega^{\mathrm{r}}_{30})\,.$$

Consider the following boundary conditions: at $\eta = 0$, $\vartheta_{10} = 0$ and $M_1(0) = M_{10}$. The solution of (16) is of the form (see Prob. 5.1)

$$\vartheta_{10} = \frac{M_{10}}{kA_{11}} \sin k\eta\,, \tag{17}$$

where

$$k = \sqrt{\frac{\Omega^{\mathrm{c}}_{30}}{A_{11}}\,[\,(A_{33} - A_{22})\,\Omega^{\mathrm{c}}_{30} - A_{33}\Omega^{\mathrm{r}}_{30}\,]}\,.$$

Solving for Q_2 and Q_3 from (5.142) and (5.143), we arrive at

$$Q_3 = K_3\vartheta'_{10} + K_4\,; \qquad Q_2 = K_5\vartheta_{10}\,, \tag{18}$$

where

$$K_3 = (A_{22} - A_{33} + A_{11})\,\Omega^{\mathrm{c}}_{30} + A_{33}\Omega^{\mathrm{r}}_{30}\,;$$
$$K_4 = A_{11}\Omega^{\mathrm{c}}_{30}\Omega^{\mathrm{r}}_{10} - A_{33}(\Omega^{\mathrm{c}}_{30} - \Omega^{\mathrm{r}}_{30})\,\Omega^{\mathrm{c}}_{10}\,;$$
$$K_5 = (A_{11} - A_{22})\,\Omega^{\mathrm{c}}_{30}\Omega^{\mathrm{c}}_{10}\,. \tag{19}$$

The component Q_1 can be determined from (5.138) as follows:

$$Q_1' = -[\,K_4\Omega_{30}^{\mathrm{c}} - K_5(\Omega_{30}^{\mathrm{c}} - \Omega_{30}^{\mathrm{r}})\,]\,\vartheta_{10}\,. \tag{20}$$

Integrating (20), we obtain

$$Q_1 = \frac{M_{10}}{k^2 A_{11}}\,K_6 \cos k\eta + Q_{10}\,, \tag{21}$$

where

$$K_6 = (A_{22} - A_{33})\,(\Omega_{30}^{\mathrm{c}})^2\Omega_{10}^{\mathrm{c}} + (A_{33} - A_{22} + A_{11})\,\Omega_{30}^{\mathrm{c}}\Omega_{10}^{\mathrm{c}}\Omega_{30}^{\mathrm{r}}\,. \tag{22}$$

If $Q_1(1) = 0$, then from (21) it follows that

$$Q_1 = \frac{K_6}{k^2}\,[\,\vartheta_{10}'(\eta) - \vartheta_{10}'(1)\,]\,. \tag{23}$$

It is seen from (23) that for the rod to be in equilibrium, an axial force must be applied at one of its ends. For example, if for some reason the axial force at $\eta = 1$ equals zero (the cross section B in Fig. 5.28), then a force

$$P = \frac{M_{10}}{k^2 A_{11}}\,K_6(1 - \cos k) \tag{24}$$

is to be applied to the cross section A. Solving for q_2 and q_3 from (5.139) and (5.140), we arrive at

$$\begin{aligned} q_2 &= \left[\,K_4 - K_5 + K_3\Omega_{10}^{\mathrm{c}} - \frac{K_6}{k^2}\,(\Omega_{30}^{\mathrm{c}} - \Omega_{30}^{\mathrm{r}})\,\frac{M_{10}}{A_{11}}\,\right]\cos k\eta \\ &\quad + \frac{K_6}{K_2}\,(\Omega_{30}^{\mathrm{c}} - \Omega_{30}^{\mathrm{r}})\,\frac{M_{10}}{A_{11}}\cos k\eta + K_4\Omega_{10}^{\mathrm{c}}\,; \\ q_3 &= \frac{M_{10}}{A_{11}}\left(K_3 k - \frac{K_5\Omega_{10}^{\mathrm{c}}}{k}\right)\sin k\eta\,. \end{aligned}$$

The magnitude of the distributed load is $|\mathbf{q}| = \sqrt{q_2^2 + q_3^2}$.

References

Alfutov, N.A. (2000): Stability of elastic structures. Springer-Verlag, Berlin, Heidelberg

Balabukh, L.I., Alfutov, N.A., Usyukin, V.I (1984): Structural mechanics of spacecraft. Vysshaya Shkola, Moscow. (Russian)

Bolotin, V.V. (1963): Nonconservative problems of the theory of elastic stability. Pergamon Press, London]

Devnin, S.I. (1975): Hydroelasticity of structures. Sudostroenie, Leningrad (Russian)

Feodosiev, V.I. (1996): Selected problems of the strength of materials. Nauka, Moscow (Russian)

Feodosiev, V.I. (1999): Strength of materials. BSTU Publishers, Moscow. (Russian)

Grafskii, I.Yu., Kazakevich, M.I. (1983): Aerodynamics of bodies. Naukova Dumka, Dnepropetrovsk (Russian)

Grigolyuk, E.I., Shalashilin, V.I. (1991): Problems of nonlinear deformation: The continuation method applied to nonlinear problems in solid mechanics. Kluwer Academic Publishers, Dordrecht, Boston, London

Kazakevich, M.I. (1977): Aerodynamic stability of pipelines. Nedra, Moscow (Russian)

Panovko, Ya.G., Gubanova, I.I. (1979): Stability and oscillations of elastic systems. Nauka, Moscow (Russian)

Ponomorev, S.D., Andreeva, L.E. (1980): Design of elastic elements. Mashinostroenie, Moscow (Russian)

Smirnov, A.F., Aleksandrov, A.V., Laschenkov, B.Ya., Shaposhnikov, N.N. (1981): Structural mechanics. Moscow (Russian)

Svetlitsky, V.A. (1982): Mechanics of pipelines. Mashinostroenie, Moscow (Russian)

Svetlitsky, V.A. (1987): Mechanics of rods. **2**: Dynamics. Mashinostroenie, Moscow (Russian)

Index

Foundations of Engineering Mechanics

Series Editors: Vladimir I. Babitsky, Loughborough University
Jens Wittenburg, Karlsruhe University

Palmov	Vibrations of Elasto-Plastic Bodies (1998, ISBN 3-540-63724-9)
Babitsky	Theory of Vibro-Impact Systems and Applications (1998, ISBN 3-540-63723-0)
Skrzypek/ Ganczarski	Modeling of Material Damage and Failure of Structures Theory and Applications (1999, ISBN 3-540-63725-7)
Kovaleva	Optimal Control of Mechanical Oscillations (1999, ISBN 3-540-65442-9)
Kolovsky	Nonlinear Dynamics of Active and Passive Systems of Vibration Protection (1999, ISBN 3-540-65661-8)
Guz	Fundamentals of the Three-Dimensional Theory of Stability of Deformable Bodies (1999, ISBN 3-540-63721-4)
Alfutov	Stability of Elastic Structures (2000, ISBN 3-540-65700-2)
Morozov/ Petrov	Dynamics of Fracture (2000, ISBN 3-540-64274-9)
Astashev/ Babitsky/ Kolovsky	Dynamics and Control of Machines (2000, ISBN 3-540-63722-2)

Zeitfracht Medien GmbH
Ferdinand-Jühlke-Straße 7
99095 Erfurt, Deutschland
produktsicherheit@kolibri360.de